TTL Combinational MSI Devices

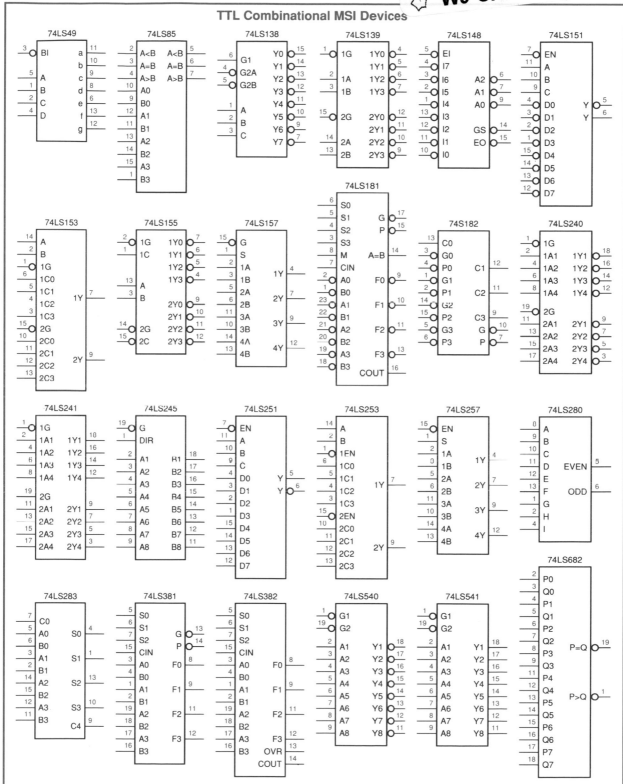

Digital Design
Principles and Practices

Prentice Hall Series in Computer Engineering
Edward J. McCluskey, Series Editor

Digital Design
Principles and Practices

JOHN F. WAKERLY

Stanford University

PRENTICE HALL, Englewood Cliffs, New Jersey 07632

Library of Congress Cataloging-in-Publication Data

Wakerly, John F.
   Digital design : principles and practices / John F. Wakerly.
      p.   cm. -- (Prentice Hall series in computer engineering)
   Includes bibliographical references.
   ISBN 0-13-212838-1
   1. Digital integrated circuits--Design and construction.   2. Logic
design.   I. Title.   II. Series.
   TK7874.W335   1990
   621.39'5--dc20   89-26498
                                                            CIP

Editorial/production supervision: Kathleen Schiaparelli
Cover design: Peggy Kenselaar
Cover painting: Casey Burley
Manufacturing buyer: Margaret Rizzi

PostScript is a trademark of Adobe Systems Incorporated.
Macintosh is a trademark of Apple Computer, Inc.
TRI-STATE is a trademark of National Semiconductor Corporation.
PAL is a trademark of Advanced Micro Devices, Inc.
ABEL is a trademark of Data I/O Corporation

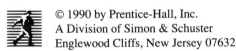 © 1990 by Prentice-Hall, Inc.
A Division of Simon & Schuster
Englewood Cliffs, New Jersey 07632

Printed in the United States of America

10 9 8 7 6 5 4 3 2

ISBN 0-13-212838-1

Prentice-Hall International (UK) Limited, *London*
Prentice-Hall of Australia Pty. Limited, *Sydney*
Prentice-Hall Canada Inc., *Toronto*
Prentice-Hall Hispanoamericana, S.A., *Mexico*
Prentice-Hall of India Private Limited, *New Delhi*
Prentice-Hall of Japan, Inc., *Tokyo*
Simon & Schuster Asia Pte. Ltd., *Singapore*
Editora Prentice-Hall do Brasil, Ltda, *Rio de Janiero*

To Susanne

The logical one

CONTENTS

PREFACE

This book is for everyone who wants to design and build real digital circuits. It is based on the idea that, in order to do this, you have to grasp the fundamentals, but at the same time you need to understand how things work in the real world. Hence, the "principles and practices" theme.

The material in this book is appropriate for introductory courses on digital logic design in electrical or computer engineering or computer science curricula. Computer science students who are unfamiliar with basic electronics concepts or who just aren't interested in the electrical behavior of digital devices may wish to skip Chapter 2; the rest of the book is written to be independent of this material as much as possible. On the other hand, *anyone* with a basic science background who wants to get up to speed on digital electronics can do so by reading Appendix A and Chapter 2.

introductory courses

electronics concepts

Although this book's level is introductory, it contains much more material than can be taught in a typical introductory course. Once I started writing, I found that I had many important things to say that wouldn't fit into a one-quarter course at Stanford or a 400-page book. Therefore, I have followed my usual practice of including *everything* that I think is at least moderately important, and leaving it up to the instructor or reader to decide what is most important in a particular environment. To help these decisions along, though, I've marked the headings of *optional sections* with an asterisk. In general, these sections can be skipped without any loss of continuity in the non-optional sections that follow.

optional sections

Undoubtedly, some people will use this book in advanced courses and in laboratory courses. Advanced students will want to skip the basics, of course, and get right into the fun stuff. Once you know the basics, the most important and fun stuff in this book is in Chapter 7, where you'll discover that your programming courses weren't a waste after all.

advanced courses
laboratory courses
fun stuff

working digital
designers

Another use of this book is as a self-study reference for a working digital designer, one of two kinds:

Novice If you're just getting started as a working digital designer, and you took a very "theoretical" logic design course in school, you should concentrate on Chapters 2, 4, 6, 7, and 8 to get prepared for the real world.

Old pro If you're experienced, you may not need all of the "practices" material in this book, but the principles in Chapters 1, 3, and 5 can help you organize your thinking, and the discussions there of what's important and what's not might relieve the guilt you feel for not having used a Karnaugh map in 10 years. The discussions of the IEEE standard in Chapters 4 and 6 will enable you to decipher those strange symbols that have appeared in the TI data books, and the design examples in Chapter 6 should give you additional insights into and appreciation for synchronous and state machine design methods. Finally, Chapter 7 may serve as your first organized introduction to programmable logic devices and PLD-based design.

marginal notes
marginal pun

All readers should make good use of the comprehensive index and of the *marginal notes* throughout the text that call attention to definitions and important topics. Maybe the highlighted topics in *this* section were more marginal than important, but I just wanted to show off.

CHAPTER DESCRIPTIONS

What follows is a list of short descriptions of the eight chapters in this book. This may remind you of the "For People Who Hate Reading Manuals" section in software guides. If you read this list, then I guess you don't have to read the rest of the book.

- Chapter 1 is an introduction to binary number systems and codes. Readers who are already familiar with binary number systems from a software course should still read Sections 1.10–1.13 to get an idea of how binary codes are used by hardware.

- Chapter 2 describes "everything you ever wanted to know about" digital circuit operation, with primary emphasis placed on the external electrical characteristics of logic devices. The starting point is a basic electronics background including voltage, current, and Ohm's law; readers unfamiliar with these concepts should consult Appendix A. This chapter may be omitted by readers who aren't interested in how to make real circuits work, or who have the luxury of employing someone else to do the dirty work.

- Chapter 3 teaches traditional combinational logic design principles, including switching algebra and combinational circuit analysis, synthesis, and minimization.

- Chapter 4 begins with a discussion of digital-system documentation standards, probably the most important practice for aspiring designers to start practicing. The rest of the chapter is devoted to describing commonly used combinational logic functions and corresponding TTL MSI devices and applications. A thorough but optional introduction to IEEE standard logic symbols is also included.

- Chapter 5 teaches traditional sequential logic design principles, starting with latches and flip-flops. The main emphasis in this chapter is on the analysis and design of clocked synchronous state machines. However, for the brave and daring, the chapter ends with an introduction to fundamental-mode circuits and the analysis and design of feedback sequential circuits.

- Chapter 6 is all about the practical design of sequential circuits. Like Chapter 4 before it, this chapter focuses on commonly used functions, and gives examples using TTL MSI devices and applications. Sections 6.6–6.8 discuss several practical approaches to state-machine design, and Sections 6.9 and 6.10 address the important problem of how to live synchronously in an asynchronous world.

- Chapter 7 is an introduction to today's most popular and successful digital design practice—structured design with programmable logic devices. Experienced digital designers and novice students alike may feel that "if it wasn't hard, they wouldn't call it hardware." This chapter shows that there *is* an easier way—you can design a logic function by writing a program!

- Chapter 8 discusses several miscellaneous real-world topics that are of interest to digital designers. When I started writing what I thought would be a 300-page book, I included this chapter in the outline to pad out the "core" material to a more impressive length. Well, the book is obviously long enough without it, but this material is useful just the same.

HOW THIS BOOK WAS PREPARED

The text of this book was prepared on a very cheap but reliable no-name PC clone. Unformatted text files were created using the very capable and speedy Epsilon text editor, an EMACS-like editor from Lugaru Software. They were then formatted by Don Knuth's TeX manuscript preparation program along with Leslie Lamport's LaTeX macro package and my own book-format customizations.

My PC ran the PCTEX implementation of TEX, written by Lance Carnes of Personal TEX Inc.

All of the figures in this book were created on an Apple Macintosh II using a very buggy drawing program from an almost defunct software company. The Macintosh and the PC were linked by using TOPS network software from Sun Microsystems. (My definition of multiprocessing: keeping the Mac and the PC busy at the same time. Some days, especially during the production cycle, I could use two of each.)

The figure files were actually stored on the PC's hard disk. In this way, the figures and text could be merged when PCTEX ran on the PC. Fast hard-disk to hard-disk backups of text and figures on the PC were conveniently made every night by the `PCBACKUP` program from Central Point Software, and were used more than once to recover from accidents, thank you.

PostScript page files were created from PCTEX's `DVI` output files by the ArborText `DVILASER/PS` output driver. The `psfigtex` macro package, written by Trevor Darrell, merged the page files with PostScript figure files created by the drawing program. Color separations were created from the page files by custom PostScript drivers written by the author, and the plates for the final pages were run on a 1270-dpi PostScript imagesetter.

ERRORS

Warning: This book may contain errors. The author assumes no liability for any damage—incidental, brain, or otherwise—caused by errors.

There, that should make the lawyers happy. Now, to make *you* happy, let me assure you that a great deal of care has gone into the preparation of this manuscript to make it as error free as possible. I am anxious to learn of the remaining errors so that they may be fixed in future printings, editions, and spin-offs. Therefore I will pay $5 to the first finder of each undiscovered error, be it technical, typographical, or otherwise. Please send your comments to me care of Computer Engineering Editor, Prentice-Hall, Inc., Englewood Cliffs, NJ 07632.

Any reader can request an up-to-date list of discovered errors by writing to me at the address above and enclosing a stamped self-addressed envelope. Only one ounce of first-class postage will be needed, I trust.

FOR INSTRUCTORS

A set of transparency masters including full-size black-and-white versions of all the figures and tables in this book is available from the publisher. Also available is an Instructor's Manual containing answers to exercises, sample exams, and a set of three-hour introductory laboratory assignments. Finally, you can get a Macintosh diskette containing the text files (in Microsoft Word) for the exams and lab assignments, and PC-compatible diskette containing the sources of all the Pascal and ABEL programs in Chapter 7 of the book.

All of Motorola's data books are available at a much reduced price for classroom use through their outstanding University Support Program. For more information, contact the program manager, Fritz Wilson, at Motorola Semiconductor Products Sector, Mail Drop 56-242, P.O. Box 52073, Phoenix, AZ 85072.

ACKNOWLEDGEMENTS

Many people helped make this book possible. My sponsoring editor at Prentice Hall, Tim Bozik, deserves the greatest thanks. Although he was 3000 miles away, I felt like he was constantly at my side; no, not nagging, but answering questions, making worthwhile suggestions, and promoting my project both inside and outside P-H.

For the ideas on the "principles" side of this book, I owe great thanks to my teacher, research advisor, and friend Ed McCluskey. On the "practices" side, I must acknowledge the influence of several outstanding digital designers over the years (in chronological order): Ed Davidson, Jim McClure, Mark Hahn, Courtenay Heater, Curt Widdoes, Prem Jain, Jeff Gilbert, Glenn Trewitt, Ted Tracy, Art Laursen, Rick Casey, and Dave Raaum. Special appreciation goes to the most practical digital, analog, mechanical, and network designer I know, Sam Wood. There were also some especially lousy designers who had even *more* influence than these, but they shall remain anonymous.

Over 300 students who took Stanford's introductory digital logic design lecture and laboratory course from 1987 through 1989 used drafts of this book; their patience and help are kindly acknowledged. Comments from the students and several anonymous reviewers were very helpful. In addition to suggesting or otherwise motivating many improvements, the students spotted dozens of typographical and technical errors whose fixes are incorporated in the typeset book. Several teaching asssistants, including Paul Ziemba, Edward Plumer, David Filo, and Eric Berman contributed directly to the development of the course and indirectly to the book. Kalon Holdbrook and Jon Newman undertook special-studies projects that contributed to the development of the material in Section 3.4.

Recently I have been giving students an incentive of $4 for each new error they find. I offered them the option of putting their names in the preface instead of giving them money, but I had only one taker. I've had similar experience with my other books, of course. In fact, I was so surprised to have even one taker that I lost the scrap of paper with her name on it, so now she's got nothing!

The most substantial direct contribution to this book (other than my own, of course) was made by Bruce Fleischer. During the few months of relative freedom Bruce had available between the completion of his Electrical Engineering Ph.D. dissertation at Stanford and the commencement of his real job at IBM Research, he agreed to work for slave wages preparing over 100 illustrations for Chapters 2, 4, and 6. More important, Bruce wrote and illustrated Appendix A, an introduction to basic electrical circuit concepts, that even I couldn't resist reading over and over for the pure enjoyment of it. Now, if I could only get him to write an undergraduate circuits book with me. . . .

About 100 more drawings were prepared by three other persons—Marian Langdon, Dennis Leung, and Matthew Kaufman—and the rest of the over 500 drawings in this book were prepared by the author. (Note that Dennis, not I, drew Figures 6–62 and 7–46!) Marian also assisted with the production work, and rode through the Quake of '89 with the computers and me.

A few other people helped out on various chores. For the book's second printing, Susan Gere entered all of the corrections, changes, and additions (except this one). Darlene Hadding helped me answer five years of book-related correspondence (all in the last two months!). For several years, Jim Klopp at CopiGraphics has done a nice job on short notice of printing and binding the draft manuscripts used by students and reviewers; I especially appreciated it when he moved his copy shop to a location only four blocks from my house!

The folks at Apple Computer, especially at Developer Services, were very helpful in providing me with Macintosh resources for this and my other writing projects. DataI/O Corporation supplied the ABEL software that was used to develop and check the PLD examples in Chapter 7. I also found ABEL to be handy for minimizing and checking logic equations in other parts the text.

Finally, I must thank my wife Kate for putting up with the late hours, frustration, crabbiness, preoccupation, eating and dieting binges, months of zero income, and euphoria (soon, I hope) that come with a writing project like this. We hope you enjoy the results.

John F. Wakerly
Mountain View, California

1

NUMBER
SYSTEMS
AND CODES

Digital systems are built from circuits that process binary digits—0s and 1s—yet very few real-life problems are based on binary numbers or any numbers at all. Therefore, a digital system designer must establish some correspondence between the binary digits processed by digital circuits and real-life numbers, events, and conditions. The purpose of this chapter is to show you how familiar numeric quantities can be represented and manipulated in a digital system, and how nonnumeric data, events, and conditions also can be represented.

The first nine sections describe binary number systems and show how addition, subtraction, multiplication, and division are performed in these systems. Sections 1.10–1.13 show how other things, such as decimal numbers, text characters, mechanical positions, and arbitrary conditions, can be encoded using strings of binary digits.

Section 1.14 introduces "n-cubes," which provide a way to visualize the relationship between different bit strings. The n-cubes are especially useful in the study of error-detecting codes in Section 1.15. We conclude the chapter with an introduction to codes for transmitting and storing data one bit at a time.

1.1   POSITIONAL NUMBER SYSTEMS

*positional number
system*

weight

The traditional number system that we learned in school and use every day in business is called a *positional number system.* In such a system, a number is represented by a string of digits where each digit position has an associated *weight.* The value of a number is a weighted sum of the digits, for example,

$$1734 = 1 \cdot 1000 + 7 \cdot 100 + 3 \cdot 10 + 4 \cdot 1$$

Each weight is a power of 10 corresponding to the digit's position. A decimal point allows negative as well as positive powers of 10 to be used:

$$5185.68 = 5 \cdot 1000 + 1 \cdot 100 + 8 \cdot 10 + 5 \cdot 1 + 6 \cdot 0.1 + 8 \cdot 0.01$$

In general, a number D of the form $d_1 d_0 . d_{-1} d_{-2}$ has the value

$$D = d_1 \cdot 10^1 + d_0 \cdot 10^0 + d_{-1} \cdot 10^{-1} + d_{-2} \cdot 10^{-2}$$

base

radix

Here, 10 is called the *base* or *radix* of the number system. In a general positional number system, the radix may be any integer r, and a digit in position i has weight r^i. The general form of a number in such a system is

$$d_{p-1} d_{p-2} \cdots d_1 d_0 . d_{-1} d_{-2} \cdots d_{-n}$$

radix point

where there are p digits to the left of the point and n digits to the right of the point, called the *radix point.* If the radix point is missing, it is assumed to be to the right of the rightmost digit. The value of the number is the sum of each digit multiplied by the corresponding power of the radix:

$$D = \sum_{i=-n}^{p-1} d_i \cdot r^i$$

Except for possible leading and trailing zeroes, the representation of a number in a positional number system is unique. (Obviously, 0185.6300 equals 185.63, and so on.) The leftmost digit in such a number is called the *high-order*

high-order digit

*most significant
digit*

low-order digit

*least significant
digit*

binary digit

bit

binary radix

or *most significant digit*; the rightmost is the *low-order* or *least significant digit.*

As we'll learn in Chapter 2, digital circuits have signals that are normally in one of only two conditions—low or high, charged or discharged, off or on. The signals in these circuits are interpreted to represent *binary digits* (or *bits*) that have one of two values, 0 and 1. Thus, the *binary radix* is normally used to represent numbers in a digital system. The general form of a binary number is

$$b_{p-1} b_{p-2} \cdots b_1 b_0 . b_{-1} b_{-2} \cdots b_{-n}$$

and its value is

$$B = \sum_{i=-n}^{p-1} b_i \cdot 2^i$$

In a binary number, the radix point is called the *binary point*. When dealing with binary and other nondecimal numbers, we use a subscript to indicate the radix of each number, unless the radix is clear from the context. Examples of binary numbers and their decimal equivalents are given below.

binary point

$$10011_2 \;=\; 1 \cdot 16 + 0 \cdot 8 + 0 \cdot 4 + 1 \cdot 2 + 1 \cdot 1 \;=\; 19_{10}$$

$$100010_2 \;=\; 1 \cdot 32 + 0 \cdot 16 + 0 \cdot 8 + 0 \cdot 4 + 1 \cdot 2 + 0 \cdot 1 \;=\; 34_{10}$$

$$101.001_2 \;=\; 1 \cdot 4 + 0 \cdot 2 + 1 \cdot 1 + 0 \cdot 0.5 + 0 \cdot 0.25 + 1 \cdot 0.125 \;=\; 5.125_{10}$$

The leftmost bit of a binary number is called the *high-order* or *most significant bit (MSB)*; the rightmost is the *low-order* or *least significant bit (LSB)*.

MSB
LSB

1.2   OCTAL AND HEXADECIMAL NUMBERS

Radix 10 is important because we use it in everyday business, and radix 2 is important because binary numbers can be processed directly by digital circuits. Numbers in other radices are not often processed directly, but may be important for documentation or other purposes. In particular, the radices 8 and 16 provide convenient shorthand representations for multibit numbers in a digital system.

The *octal number system* uses radix 8, while the *hexadecimal number system* uses radix 16. Table 1–1 shows the binary integers from 0 through 1111 and their octal, decimal, and hexadecimal equivalents. The octal system needs 8 digits, so it uses digits 0–7 of the decimal system. The hexadecimal system needs 16 digits, so it supplements decimal digits 0–9 with the letters *A–F*.

octal number system
hexadecimal number system

hexadecimal digits A–F

The octal and hexadecimal number systems are useful for representing multibit numbers because their radices are powers of 2. Since a string of three bits can take on eight different combinations, it follows that each 3-bit string can be uniquely represented by one octal digit, according to the third and fourth columns of Table 1–1. Likewise, a 4-bit string can be represented by one hexadecimal digit according to the fifth and sixth columns of the table.

Thus, it is very easy to convert a binary number to octal. Starting at the binary point and working left, we simply separate the bits into groups of three and replace each group with the corresponding octal digit:

binary to octal conversion

$$100011001110_2 \;=\; 100\ 011\ 001\ 110_2 \;=\; 4316_8$$

$$1110110111010100 1_2 \;=\; 011\ 101\ 101\ 110\ 101\ 001_2 \;=\; 355651_8$$

The procedure for binary to hexadecimal conversion is similar, except we use groups of four bits:

binary to hexadecimal conversion

$$100011001110_2 \;=\; 1000\ 1100\ 1110_2 \;=\; 8CE_{16}$$

$$1110110111010100 1_2 \;=\; 0001\ 1101\ 1011\ 1010\ 1001_2 \;=\; 1DBA9_{16}$$

Table 1–1
Binary, decimal, octal, and
hexadecimal numbers.

Binary	Decimal	Octal	3-Bit String	Hexadecimal	4-Bit String
0	0	0	000	0	0000
1	1	1	001	1	0001
10	2	2	010	2	0010
11	3	3	011	3	0011
100	4	4	100	4	0100
101	5	5	101	5	0101
110	6	6	110	6	0110
111	7	7	111	7	0111
1000	8	10	—	8	1000
1001	9	11	—	9	1001
1010	10	12	—	A	1010
1011	11	13	—	B	1011
1100	12	14	—	C	1100
1101	13	15	—	D	1101
1110	14	16	—	E	1110
1111	15	17	—	F	1111

In these examples we have freely added zeroes on the left to make the total number of bits a multiple of 3 or 4 as required.

If a binary number contains digits to the right of the binary point, we can convert them to octal or hexadecimal by starting at the binary point and working right. Both the left-hand and right-hand sides can be padded with zeroes to get multiples of three or four bits, as shown in the example below.

$$10.1011001011_2 \ = \ 010 \, . \, 101 \ 100 \ 101 \ 100_2 \ = \ 2.5454_8$$
$$= \ 0010 \, . \, 1011 \ 0010 \ 1100_2 \ = \ 2.B2C_{16}$$

octal or
hexadecimal to
binary conversion

Converting in the reverse direction, from octal or hexadecimal to binary, is very easy. We simply replace each octal or hexadecimal digit with the corresponding 3- or 4-bit string, as shown below:

$$1357_8 \ = \ 001 \ 011 \ 101 \ 111_2$$
$$2046.17_8 \ = \ 010 \ 000 \ 100 \ 110 \, . \, 001 \ 111_2$$
$$BEAD_{16} \ = \ 1011 \ 1110 \ 1010 \ 1101_2$$
$$9F.46C_{16} \ = \ 1001 \ 1111 \, . \, 0100 \ 0110 \ 1100_2$$

The octal number system was quite popular twenty years ago because of certain minicomputers that had their front-panel lights and switches arranged in groups of three. However, the octal number system is not used much today,

because of the preponderence of machines that process 8-bit *bytes*. It is difficult *byte*
to extract individual byte values in multibyte quantities in the octal representation;
for example, what are the octal values of the four 8-bit bytes in the 32-bit number
with octal representation 12345670123_8?

In the hexadecimal system, two digits neatly represent an 8-bit byte, and
$2n$ digits represent an n-byte word; each pair of digits constitutes exactly one
byte. For example, the 32-bit hexadecimal number $5678ABCD_{16}$ consists of
four bytes with values 56_{16}, 78_{16}, AB_{16}, and CD_{16}. In this context, a 4-bit
hexadecimal digit is sometimes called a *nibble*; a 32-bit (4-byte) number has   *nibble*
eight nibbles. Hexadecimal numbers are often used to describe a computer's
memory address space. For example, a computer with 16-bit addresses might
be described as having read/write memory installed at addresses $0–EFFF_{16}$, and
read-only memory at addresses $F000–FFFF_{16}$.

1.3   GENERAL POSITIONAL NUMBER SYSTEM CONVERSIONS

In general, conversion between two radices cannot be done by simple substitutions; arithmetic operations are required. In this section, we show how to convert
a number in any radix to radix 10 and vice versa, using radix-10 arithmetic.

In Section 1.1, we indicated that the value of a number in any radix is given   *radix-r to decimal*
by the formula   *conversion*

$$D = \sum_{i=-n}^{p-1} d_i \cdot r^i$$

where r is the radix of the number and there are p digits to the left of the
radix point and n to the right. Thus, the value of the number can be found by
converting each digit of the number to its radix-10 equivalent and expanding the
formula using radix-10 arithmetic. Some examples are given below:

$$1CE8_{16} = 1 \cdot 16^3 + 12 \cdot 16^2 + 14 \cdot 16^1 + 8 \cdot 16^0 = 7400_{10}$$

$$F1A3_{16} = 15 \cdot 16^3 + 1 \cdot 16^2 + 10 \cdot 16^1 + 3 \cdot 16^0 = 61859_{10}$$

$$436.5_8 = 4 \cdot 8^2 + 3 \cdot 8^1 + 6 \cdot 8^0 + 5 \cdot 8^{-1} = 286.625_{10}$$

$$132.3_4 = 1 \cdot 4^2 + 3 \cdot 4^1 + 2 \cdot 4^0 + 3 \cdot 4^{-1} = 30.75_{10}$$

A shortcut for converting whole numbers to radix 10 is obtained by rewriting the expansion formula as follows:

$$D = ((\cdots ((d_{p-1}) \cdot r + d_{p-2}) \cdot r + \cdots) \cdot r + d_1) \cdot r + d_0$$

That is, we start with a sum of 0; beginning with the leftmost digit, we multiply the sum by r and add the next digit to the sum, repeating until all digits have been processed. For example, we can write

$$\text{F1AC}_{16} \;=\; (((15) \cdot 16 + 1) \cdot 16 + 10) \cdot 16 + 12$$

decimal to radix-r
conversion

Although this formula is not too exciting in itself, it forms the basis for a very convenient method of converting a decimal number D to a radix r. Consider what happens if we divide the formula by r. Since the parenthesized part of the formula is evenly divisible by r, the quotient will be

$$Q \;=\; (\cdots((d_{p-1}) \cdot r + d_{p-2}) \cdot r + \cdots) \cdot r + d_1$$

and the remainder will be d_0. Thus, d_0 can be computed as the remainder of the long division of D by r. Furthermore, the quotient Q has the same form as the original formula. Therefore, successive divisions by r will yield successive digits of D from right to left, until all the digits of D have been derived. Examples are given below.

$$179 \div 2 = 89 \text{ remainder } 1 \quad (\text{LSB})$$
$$\div 2 = 44 \text{ remainder } 1$$
$$\div 2 = 22 \text{ remainder } 0$$
$$\div 2 = 11 \text{ remainder } 0$$
$$\div 2 = 5 \text{ remainder } 1$$
$$\div 2 = 2 \text{ remainder } 1$$
$$\div 2 = 1 \text{ remainder } 0$$
$$\div 2 = 0 \text{ remainder } 1 \quad (\text{MSB})$$
$$179_{10} = 10110011_2$$

$$467 \div 8 = 58 \text{ remainder } 3 \quad (\text{least significant digit})$$
$$\div 8 = 7 \text{ remainder } 2$$
$$\div 8 = 0 \text{ remainder } 7 \quad (\text{most significant digit})$$
$$467_{10} = 723_8$$

$$3417 \div 16 = 213 \text{ remainder } 9 \quad (\text{least significant digit})$$
$$\div 16 = 13 \text{ remainder } 5$$
$$\div 16 = 0 \text{ remainder } 13 \quad (\text{most significant digit})$$
$$3417_{10} = \text{D59}_{16}$$

Table 1–2 summarizes methods for converting among the most common radices.

Table 1–2 Conversion methods for common radices.

Conversion	Method	Example
Binary to		
Octal	Substitution	$10111011001_2 = 10\ 111\ 011\ 001_2 = 2731_8$
Hexadecimal	Substitution	$10111011001_2 = 101\ 1101\ 1001_2 = 5D9_{16}$
Decimal	Summation	$10111011001_2 = 1 \cdot 1024 + 0 \cdot 512 + 1 \cdot 256 + 1 \cdot 128 + 1 \cdot 64$
		$\qquad\qquad + 0 \cdot 32 + 1 \cdot 16 + 1 \cdot 8 + 0 \cdot 4 + 0 \cdot 2 + 1 \cdot 1 = 1497_{10}$
Octal to		
Binary	Substitution	$1234_8 = 001\ 010\ 011\ 100_2$
Hexadecimal	Substitution	$1234_8 = 001\ 010\ 011\ 100_2 = 0010\ 1001\ 1100_2 = 29C_{16}$
Decimal	Summation	$1234_8 = 1 \cdot 512 + 2 \cdot 64 + 3 \cdot 8 + 4 \cdot 1 = 668_{10}$
Hexadecimal to		
Binary	Substitution	$C0DE_{16} = 1100\ 0000\ 1101\ 1110_2$
Octal	Substitution	$C0DE_{16} = 1100\ 0000\ 1101\ 1110_2 = 1\ 100\ 000\ 011\ 011\ 110_2 = 140336_8$
Decimal	Summation	$C0DE_{16} = 12 \cdot 4096 + 0 \cdot 256 + 13 \cdot 16 + 14 \cdot 1 = 49374_{10}$
Decimal to		
Binary	Division	$108_{10} = 1101100_2$
		$108_{10} \div 2 = 54$ remainder 0 (LSB)
		$\div 2 = 27$ remainder 0
		$\div 2 = 13$ remainder 1
		$\div 2 = 6$ remainder 1
		$\div 2 = 3$ remainder 0
		$\div 2 = 1$ remainder 1
		$\div 2 = 0$ remainder 1 (MSB)
Octal	Division	$108_{10} = 154_8$
		$108_{10} \div 8 = 13$ remainder 4 (least significant digit)
		$\div 8 = 1$ remainder 5
		$\div 8 = 0$ remainder 1 (most significant digit)
Hexadecimal	Division	$108_{10} = 6C_{16}$
		$108_{10} \div 16 = 6$ remainder 12 (least significant digit)
		$\div 16 = 0$ remainder 6 (most significant digit)

1.4 ADDITION AND SUBTRACTION OF NONDECIMAL NUMBERS

Addition and subtraction of nondecimal numbers by hand uses the same technique that we learned in grammar school for decimal numbers; the only catch is that the addition and subtraction tables are different.

Table 1–3 is the addition and subtraction table for binary digits. To add *binary addition* two binary numbers X and Y, we add together the least significant bits with an initial carry (c_{in}) of 0, producing carry (c_{out}) and sum (s) bits according to the

Table 1–3
Binary addition and subtraction table.

c_{in} or b_{in}	x	y	c_{out}	s	b_{out}	d
0	0	0	0	0	0	0
0	0	1	0	1	1	1
0	1	0	0	1	0	1
0	1	1	1	0	0	0
1	0	0	0	1	1	1
1	0	1	1	0	1	0
1	1	0	1	0	0	0
1	1	1	1	1	1	1

table. We continue processing bits from right to left, including the carry out of each column in the next column's sum.

Two examples of decimal additions and the corresponding binary additions are shown in Figure 1–1, using a colored arrow to indicate a carry of 1. The same examples are repeated below along with two more, with the carries shown as a bit string C.

C		101111000		C		001011000
X	190	10111110		X	173	10101101
Y	+ 141	+ 10001101		Y	+ 44	+ 00101100
$X+Y$	331	101001011		$X+Y$	217	11011001

C		011111110		C		000000000
X	127	01111111		X	170	10101010
Y	+ 63	+ 00111111		Y	+ 85	+ 01010101
$X+Y$	190	10111110		$X+Y$	255	11111111

binary subtraction

Binary subtraction is performed similarly, using borrows (b_{in} and b_{out}) instead of carries between steps, and producing a difference bit d. Two examples of decimal subtractions and the corresponding binary subtractions are shown in Figure 1–2. As in decimal subtraction, the binary minuend values in the columns are modified when borrows occur, as shown by the colored arrows and bits. The

minuend
subtrahend

X	190	1 0 1 1 1 1 1 0
Y	+ 141	+ 1 0 0 0 1 1 0 1
$X+Y$	331	1 0 1 0 0 1 0 1 1

X	173	1 0 1 0 1 1 0 1
Y	+ 44	+ 0 0 1 0 1 1 0 0
$X+Y$	217	1 1 0 1 1 0 0 1

Figure 1–1 Examples of decimal and corresponding binary additions.

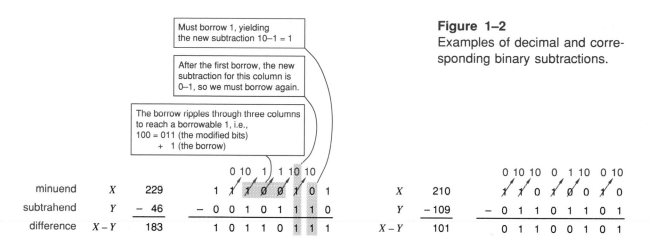

Figure 1–2

Examples of decimal and corresponding binary subtractions.

examples of the figure are repeated below along with two more, this time showing the borrows as a bit string B.

B		001111100		B		011011010
X	229	11100101		X	210	11010010
Y	− 46	− 00101110		Y	− 109	− 01101101
$X - Y$	183	10110111		$X - Y$	101	01100101

B		010101010		B		000000000
X	170	10101010		X	221	11011101
Y	− 85	− 01010101		Y	− 76	− 01001100
$X - Y$	85	01010101		$X - Y$	145	10010001

A very common use of subtraction in computers is to compare two numbers. For example, if the operation $X - Y$ produces a borrow out of the most significant bit position, then X is less than Y; otherwise X is greater than or equal to Y. The relationship between carries and borrows in adders and subtracters will be explored in Section 4.9.   *comparing numbers*

Addition and subtraction tables can be developed for octal and hexadecimal digits, or any other desired radix. However, few computer engineers bother to memorize these tables. If you must manipulate nondecimal numbers only infrequently, then it's easy enough to convert them to decimal, compute results, and convert back. On the other hand, if you must perform calculations in binary, octal, or hexadecimal frequently, then you should ask Santa for a programmer's "hex calculator" from Texas Instruments or Casio.

If the calculator's battery wears out, some mental shortcuts can be used to facilitate nondecimal arithmetic. In general, each column addition (or subtraction) can be done by converting the column digits to decimal, adding in decimal, and converting the result to corresponding sum and carry digits in the nondecimal radix. (A carry is produced whenever the column sum equals or exceeds the radix.) Since the addition is done in decimal, we rely on our knowledge of the decimal addition table; the only new thing that we need to learn is the conversion from decimal to nondecimal digits and vice versa. The sequence of steps for mentally adding two hexadecimal numbers is shown below.

hexadecimal addition

C		1 1 0 0			1	1	0	0	
X		1 9 B 9 $_{16}$			1	9	11	9	
Y		+ C 7 E 6 $_{16}$		+	12	7	14	6	
$X+Y$		E 1 9 F $_{16}$			14	17	25	15	
					14	16+1	16+9	15	
					E	1	9	F	

1.5   REPRESENTATION OF NEGATIVE NUMBERS

So far, we have dealt only with positive numbers, but there are many ways to represent negative numbers. In everyday business, we use the signed-magnitude system, discussed next. However, most computers use one of the complement number systems that we introduce later.

1.5.1   Signed-Magnitude Representation

signed-magnitude system

In the *signed-magnitude system*, a number consists of a magnitude and a symbol indicating whether the magnitude is positive or negative. Thus, we interpret decimal numbers +98, −57, +123.5, and −13 in the usual way, and we also assume that the sign is "+" if no sign symbol is written. There are two possible representations of zero, "+0" and "−0", but both have the same value.

sign bit

The signed-magnitude system is applied to binary numbers by using an extra bit position to represent the sign (the *sign bit*). Traditionally, the most significant bit (MSB) of a bit string is used as the sign bit (0 = plus, 1 = minus), and the lower-order bits contain the magnitude. Thus, we can write several 8-bit signed-magnitude integers and their decimal equivalents:

$$01010101_2 = +85_{10} \qquad 11010101_2 = -85_{10}$$
$$01111111_2 = +127_{10} \qquad 11111111_2 = -127_{10}$$
$$00000000_2 = +0_{10} \qquad 10000000_2 = -0_{10}$$

The signed-magnitude system contains an equal number of positive and negative integers. An n-bit signed-magnitude integer lies within the range $-(2^{n-1}-1)$ through $+(2^{n-1}-1)$, with two possible representations of zero.

Now suppose that we wanted to build a digital logic circuit that adds signed-magnitude numbers. The circuit must examine the signs of the addends to determine what to do with the magnitudes. If the signs are the same, it must add the magnitudes and give the result the same sign. If the signs are different, it must compare the magnitudes, subtract the smaller from the larger, and give the result the sign of the larger. All of these "ifs," "adds," "subtracts," and "compares" translate into a lot of logic-circuit complexity. Adders for complement number systems are much simpler, as we'll show next. Perhaps the one redeeming feature of a signed-magnitude system is that, once we know how to build a signed-magnitude adder, a signed-magnitude subtractor is almost trivial to build—it need only change the sign of the subtrahend and pass it along with the minuend to an adder.

signed-magnitude adder

signed-magnitude subtractor

1.5.2 Complement Number Systems

While the signed-magnitude system negates a number by changing its sign, a *complement number system* negates a number by taking its complement as defined by the system. Taking the complement is more difficult than changing the sign, but two numbers in a complement number system can be added or subtracted directly without the sign and magnitude checks required by the signed-magnitude system. We shall describe two complement number systems, called the "radix complement" and the "diminished radix-complement."

complement number system

In any complement number system, we normally deal with a fixed number of digits, say n. (However, we can increase the number of digits by "sign extension" as shown in Exercise 1.20, and decrease the number by truncating high-order digits as shown in Exercise 1.21.) We further assume that the radix is r, and that numbers have the form

$$D = d_{n-1}d_{n-2}\cdots d_1 d_0.$$

The radix point is on the right and so the number is an integer. If an operation produces a result that requires more than n digits, we throw away the extra high-order digit(s). If a number D is complemented twice, the result is D.

1.5.3 Radix-Complement Representation

In a *radix-complement system*, the complement of an n-digit number is obtained by subtracting it from r^n. In the decimal number system, the radix complement is called the *10's complement*. Some examples using 4-digit decimal numbers (and subtraction from 10,000) are shown in Table 1–4.

radix-complement system

10's complement

By definition, the radix complement of an n-digit number D is obtained by subtracting it from r^n. If D is between 1 and $r^n - 1$, this subtraction produces another number between 1 and $r^n - 1$. If D is 0, the result of the subtraction is

Table 1–4
Examples of 10's and 9s' complements.

Number	10's complement	9s' complement
1849	8151	8150
2067	7933	7932
100	9900	9899
7	9993	9992
8151	1849	1848
0	10000 (= 0)	9999

r^n, which has the form $100 \cdots 00$, where there are a total of $n + 1$ digits. We throw away the extra high-order digit and get the result 0. Thus there is only one representation of zero in a radix-complement system.

computing the radix complement

It seems from the definition that a subtraction operation is needed to compute the radix complement of D. However, this subtraction can be avoided by rewriting r^n as $(r^n - 1) + 1$ and $r^n - D$ as $((r^n - 1) - D) + 1$. The number $r^n - 1$ has the form $mm \cdots mm$, where $m = r - 1$ and there are n m's. For example, 10,000 equals 9,999 + 1. If we define the complement of a digit d as $r - 1 - d$, then $(r^n - 1) - D$ is obtained by complementing the digits of D. Therefore, the radix complement of a number D is obtained by complementing the individual digits of D and adding 1. For example, the 10's complement of

Table 1–5
Digit complements.

		Complement		
Digit	Binary	Octal	Decimal	Hexadecimal
0	1	7	9	F
1	0	6	8	E
2	–	5	7	D
3	–	4	6	C
4	–	3	5	B
5	–	2	4	A
6	–	1	3	9
7	–	0	2	8
8	–	–	1	7
9	–	–	0	6
A	–	–	–	5
B	–	–	–	4
C	–	–	–	3
D	–	–	–	2
E	–	–	–	1
F	–	–	–	0

1849 is $8150 + 1$, or 8151. You should confirm that this trick also works for the other 10's-complement examples above. Table 1–5 lists the digit complements for binary, octal, decimal, and hexadecimal numbers.

1.5.4 Two's-Complement Representation

For binary numbers, the radix complement is called the *two's complement*. The MSB of a number in this system serves as the sign bit; a number is negative if and only if its MSB is 1. The decimal equivalent for a two's-complement binary number is computed the same as for an unsigned number, except that the weight of the MSB is -2^{n-1} instead of $+2^{n-1}$. The range of representable numbers is $-(2^{n-1})$ through $+(2^{n-1} - 1)$. Some 8-bit examples are shown below.

two's complement

weight of MSB

$$17_{10} = 00010001_2$$
$$\Downarrow \quad \text{complement bits}$$
$$11101110$$
$$\underline{+1}$$
$$11101111_2 \ = \ -17_{10}$$

$$-99_{10} = 10011101_2$$
$$\Downarrow \quad \text{complement bits}$$
$$01100010$$
$$\underline{+1}$$
$$01100011_2 \ = \ 99_{10}$$

$$119_{10} = 01110111_2$$
$$\Downarrow \quad \text{complement bits}$$
$$10001000$$
$$\underline{+1}$$
$$10001001_2 \ = \ -119_{10}$$

$$-127_{10} - 10000001_2$$
$$\Downarrow \quad \text{complement bits}$$
$$01111110$$
$$\underline{+1}$$
$$01111111_2 \ = \ 127_{10}$$

$$0_{10} = 00000000_2$$
$$\Downarrow \quad \text{complement bits}$$
$$11111111$$
$$\underline{+1}$$
$$1\ 00000000_2 \ = \ 0_{10}$$

$$-128_{10} = 10000000_2$$
$$\Downarrow \quad \text{complement bits}$$
$$01111111$$
$$\underline{+1}$$
$$10000000_2 \ = \ -128_{10}$$

A carry out of the MSB position occurs in one case, as shown in color above. As in all two's-complement operations, this bit is ignored and only the low-order n bits of the result are used.

In the two's-complement number system, zero is considered positive because its sign bit is 0. Since two's complement has only one representation of zero, we end up with one extra negative number, -2^{n-1}, that doesn't have a positive counterpart.

extra negative number

We can convert an n-bit two's-complement number X into an m-bit one, but some care is needed. If $m > n$, we must append $m - n$ copies of X's sign bit to the left of X (see Exercise 1.20). That is, we pad a positive number with 0s and a negative one with 1s; this is called *sign extension*. If $m < n$, we discard

sign extension

X's $n - m$ leftmost bits; however, the result is valid only if all of the discarded bits are the same as the sign bit of the result (see Exercise 1.21).

Most computers and other digital systems use the two's-complement system to represent negative numbers. However, for completeness, we'll also describe the diminished radix-complement and ones'-complement systems.

* 1.5.5 Diminished Radix-Complement Representation

*diminished
radix-complement
system*

9s' complement

In a *diminished radix-complement system*, the complement of an n-digit number D is obtained by subtracting it from $b^n - 1$. This can be accomplished by complementing the individual digits of D, *without* adding 1 as in the radix-complement system. In decimal, this is called the *9s' complement*; some examples are given in the last column of Table 1–4.

* 1.5.6 Ones'-Complement Representation

ones' complement

The diminished radix-complement system for binary numbers is called the *ones' complement*. As in two's complement, the most significant bit is the sign, 0 if positive and 1 if negative. Thus there are two representations of zero, positive zero $(00 \cdots 00)$ and negative zero $(11 \cdots 11)$. Positive number representations are the same for both ones' and two's complements. However, negative number representations differ by 1. A weight of $-(2^{n-1} - 1)$, rather than -2^{n-1}, is given to the most significant bit when computing the decimal equivalent of a ones'-complement number. The range of representable numbers is $-(2^{n-1} - 1)$ through $+(2^{n-1} - 1)$. Some 8-bit numbers and their ones' complements are shown below.

$$17_{10} = 00010001_2 \qquad\qquad -99_{10} = 10011101_2$$
$$\Downarrow \qquad\qquad\qquad\qquad\qquad \Downarrow$$
$$11101110_2 = -17_{10} \qquad\qquad 01100010_2 = 99_{10}$$

$$119_{10} = 01110111_2 \qquad\qquad -127_{10} = 10000000_2$$
$$\Downarrow \qquad\qquad\qquad\qquad\qquad \Downarrow$$
$$10001000_2 = -119_{10} \qquad\qquad 01111111_2 = 127_{10}$$

$$0_{10} = 00000000_2 \text{ (positive zero)}$$
$$\Downarrow$$
$$11111111_2 = 0_{10} \text{ (negative zero)}$$

The main advantages of the ones'-complement system are its symmetry and the ease of complementation. However, an adder for ones'-complement

optional sections

[*]Throughout this book, *optional sections* are marked with an asterisk.

numbers is somewhat trickier than a two's-complement adder (see Exercise 5.42). Also, zero-detecting circuits in a ones'-complement system must check for both representations of zero, or must always convert $11 \cdots 11$ to $00 \cdots 00$.

* 1.5.7 Excess Representations

Yes, the number of different systems for representing negative numbers *is* excessive, but there's just one more for us to cover. In *excess-B representation*, an m-bit string whose unsigned integer value is M ($0 \le M < 2^m$) represents the signed integer $M - B$, where B is called the *bias* of the number system.

excess-B
representation
bias
excess-2^{m-1} system

For example, an *excess-2^{m-1} system* represents a number X in the range -2^{m-1} through $+2^{m-1} - 1$ by the m-bit binary representation of $X + 2^{m-1}$ (which is always nonnegative and less than 2^m). The range of this representation is exactly the same as that of m-bit two's-complement numbers. In fact, the representations of any number in the two systems are identical except for the sign bits, which are always opposite. (Note that this is true only when the bias is 2^{m-1}.)

The most common use of excess representations is in floating-point number systems (see References).

1.6 TWO'S-COMPLEMENT ADDITION AND SUBTRACTION

1.6.1 Addition Rules

A table of decimal numbers and their equivalents in different number systems, Table 1–6, reveals why the two's complement is preferred for arithmetic operations. If we start with 1000_2 (-8_{10}) and count up, we see that each successive two's-complement number all the way to 0111_2 ($+7_{10}$) can be obtained by adding 1 to the previous one, ignoring any carries beyond the fourth bit position. The same cannot be said of signed-magnitude and ones'-complement numbers. Because ordinary addition is just an extension of counting, two's-complement numbers can thus be added by ordinary binary addition, ignoring any carries beyond the MSB. The result will always be the correct sum as long as the range of the number system is not exceeded. Some examples of decimal addition and the corresponding 4-bit two's-complement additions confirm this:

two's-complement
addition

+3	0011		−2	1110
+ +4	+ 0100		+ −6	+ 1010
+7	0111		−8	1 1000
+6	0110		+4	0100
+ −3	+ 1101		+ −7	+ 1001
+3	1 0011		−3	1101

Table 1–6
Decimal and 4-bit numbers.

Decimal	Two's Complement	Ones' Complement	Signed Magnitude	Excess 2^{m-1}
−8	1000	—	—	0000
−7	1001	1000	1111	0001
−6	1010	1001	1110	0010
−5	1011	1010	1101	0011
−4	1100	1011	1100	0100
−3	1101	1100	1011	0101
−2	1110	1101	1010	0110
−1	1111	1110	1001	0111
0	0000	1111 or 0000	1000 or 0000	1000
1	0001	0001	0001	1001
2	0010	0010	0010	1010
3	0011	0011	0011	1011
4	0100	0100	0100	1100
5	0101	0101	0101	1101
6	0110	0110	0110	1110
7	0111	0111	0111	1111

1.6.2   A Graphical View

Another way to view the two's-complement system uses the 4-bit "counter" shown in Figure 1–3. Here we have shown the numbers in a circular or "modular" representation. The operation of this counter very closely mimics that of a real up/down counter circuit, which we'll study in Section 6.3. Starting with the arrow pointing to any number, we can add +n to that number by counting up n

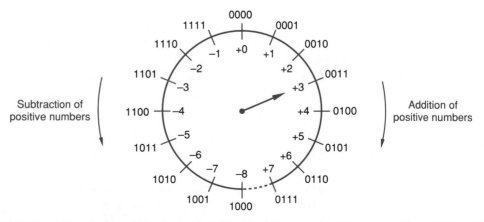

Figure 1–3   A modular counting representation of 4-bit two's-complement numbers.

times, that is, by moving the arrow n positions clockwise. It is also evident that we can subtract n from a number by counting down n times, that is, by moving the arrow n positions counterclockwise. Of course, these operations give correct results only if n is small enough that we don't cross the discontinuity between -8 and $+7$.

What is most interesting is that we can also subtract n (or add $-n$) by moving the arrow $16 - n$ positions clockwise. Notice that the quantity $16 - n$ is what we defined to be the 4-bit two's complement of n, that is, the two's-complement representation of $-n$. This graphically supports our earlier claim that a negative number in two's-complement representation may be added to another number simply by adding the 4-bit representations using ordinary binary addition. Adding a number in Figure 1–3 is equivalent to moving the arrow a corresponding number of positions clockwise.

1.6.3 Overflow

If an addition operation produces a result that exceeds the range of the number system, *overflow* is said to occur. In the modular counting representation of Figure 1–3, overflow occurs during addition of positive numbers when we count past $+7$. Addition of two numbers with different signs can never produce overflow, but addition of two numbers of like sign can, as shown by the following examples.

-3	1101	$+5$	0101
$+\ \ -6$	$+\ 1010$	$+\ \ +6$	$+\ 0110$
-9	$10111 = +7$	$+11$	$1011 = -5$

-8	1000	$+7$	0111
$+\ \ -8$	$+\ 1000$	$+\ \ +7$	$+\ 0111$
-16	$10000 = +0$	$+14$	$1110 = -2$

overflow (margin note)

Fortunately, there is a simple rule for detecting overflow in addition: an addition overflows if the signs of the addends are the same and the sign of the sum is different from the addends' sign. The overflow rule is sometimes stated in terms of carries generated during the addition operation: an addition overflows if the carry bits c_{in} into and c_{out} out of the sign position are different. Close examination of Table 1–3 shows that the two rules are equivalent—there are only two cases where $c_{in} \neq c_{out}$, and these are the only two cases where $x = y$ and the sum bit is different.

overflow rules (margin note)

1.6.4 Subtraction Rules

two's-complement
subtraction

Two's-complement numbers may be subtracted as if they were ordinary unsigned binary numbers, and appropriate rules for detecting overflow may be formulated. However, most subtraction circuits for two's-complement numbers do not perform subtraction directly. Rather, they negate the subtrahend by taking its two's complement, and then add it to the minuend using the normal rules for addition.

Negating the subtrahend and adding the minuend can be accomplished with only one addition operation as follows: perform a bit-by-bit complement of the subtrahend and add the complemented subtrahend to the minuend with an initial carry (c_{in}) of 1 instead of 0. Examples are given below.

$$
\begin{array}{rrr}
 & & 1 \longleftarrow c_{in} \\
+4 & 0100 & 0100 \\
-\ +3 & -\ 0011 & +\ 1100 \\
\hline
+1 & & 1\ 0001
\end{array}
\qquad
\begin{array}{rrr}
 & & 1 \longleftarrow c_{in} \\
+3 & 0011 & 0011 \\
-\ +4 & -\ 0100 & +\ 1011 \\
\hline
-1 & & 1111
\end{array}
$$

$$
\begin{array}{rrr}
 & & 1 \longleftarrow c_{in} \\
+3 & 0011 & 0011 \\
-\ -4 & -\ 1100 & +\ 0011 \\
\hline
+7 & & 0111
\end{array}
\qquad
\begin{array}{rrr}
 & & 1 \longleftarrow c_{in} \\
-3 & 1101 & 1101 \\
-\ -4 & -\ 1100 & +\ 0011 \\
\hline
+1 & & 1\ 0001
\end{array}
$$

Overflow in subtraction can be detected by examining the signs of the minuend and the *complemented* subtrahend, using the same rule as in addition. Or, using the technique in the preceding examples, the carries into and out of the sign position can be observed and overflow detected irrespective of the signs of inputs and output, again using the same rule as in addition.

An attempt to negate the "extra" negative number results in overflow according to the rules above, when we add 1 in the complementation process:

$$
\begin{array}{rl}
-(-8) = -1000 = & 0111 \\
+ & 0001 \\
\hline
& 1000 = -8
\end{array}
$$

However, this number can still be used in additions and subtractions as long as the final result does not exceed the number range:

$$
\begin{array}{rr}
+4 & 0100 \\
+\ -8 & +\ 1000 \\
\hline
-4 & 1100
\end{array}
\qquad
\begin{array}{rrr}
 & & 1 \longleftarrow c_{in} \\
-3 & 1101 & 1101 \\
-\ -8 & -\ 1000 & +\ 0111 \\
\hline
+5 & & 1\ 0101
\end{array}
$$

Figure 1–4   A modular counting representation of 4-bit unsigned numbers.

1.6.5   Two's-Complement and Unsigned Binary Numbers

Since two's-complement numbers are added and subtracted by the same basic binary addition and subtraction algorithms as unsigned numbers of the same length, a computer or other digital system can use the same adder circuit to handle numbers of both types. However, the results must be interpreted differently depending on whether the system is dealing with signed numbers (e.g., −8 through +7) or unsigned numbers (e.g., 0 through 15). *signed vs. unsigned numbers*

We introduced a graphical representation of the 4-bit two's-complement system in Figure 1–3. We can relabel this figure as shown in Figure 1–4 to obtain a representation of the 4-bit unsigned numbers. The binary combinations occupy the same positions on the wheel, and a number is still added by moving the arrow a corresponding number of positions clockwise, and subtracted by moving the arrow counterclockwise.

An addition operation exceeds the range of the 4-bit unsigned number system in Figure 1–4 if the arrow moves clockwise through the discontinuity between 0 and 15. In this case a *carry* out of the most significant bit position is said to occur. *carry*

Likewise a subtraction operation exceeds the range of the number system if the arrow moves counterclockwise through the discontinuity. In this case a *borrow* out of the most significant bit position is said to occur. *borrow*

From Figure 1–4 it is also evident that we may subtract an unsigned number *n* by counting *clockwise* $16 - n$ positions. This is equivalent to *adding* the 4-bit two's-complement of *n*. The subtraction produces a borrow if the corresponding addition of the two's complement *does not* produce a carry.

In summary, in unsigned addition the carry or borrow in the most significant bit position indicates an out-of-range result. In signed, two's-complement addition the overflow condition defined earlier indicates an out-of-range result. The carry from the most significant bit position is irrelevant in signed addition in the sense that overflow may or may not occur independently of whether or not a carry occurs.

* 1.7 ONES'-COMPLEMENT ADDITION AND SUBTRACTION

Another look at Table 1–6 helps to explain the rule for adding ones'-complement numbers. If we start at 1000_2 (-7_{10}) and count up, we obtain each successive ones'-complement number by adding 1 to the previous one, *except* at the transition from 1111_2 (negative 0) to 0001_2 ($+1_{10}$). To maintain the proper count, we must add 2 instead of 1 whenever we count past 1111_2. This suggests a technique for adding ones'-complement numbers: perform a standard binary addition, but add an extra 1 whenever we count past 1111_2.

ones'-complement addition

Counting past 1111_2 during an addition can be detected by observing the carry out of the sign position. Thus, the rule for adding ones'-complement numbers can be stated quite simply:

- Perform a standard binary addition; if there is a carry out of the sign position, add 1 to the result.

end-around carry

This rule is often called *end-around carry*. Examples of ones'-complement addition are given below; the last three include an end-around carry.

+3	0011		+4	0100		+5	0101
+ +4	+ 0100		+ −7	+ 1000		+ −5	+ 1010
+7	0111		−3	1100		−0	1111

−2	1101		+6	0110		−0	1111
+ −5	+ 1010		+ −3	+ 1100		+ −0	+ 1111
−7	1 0111		+3	1 0010		−0	1 1110
	+ 1			+ 1			+ 1
	1000			0011			1111

Following the two-step addition rule above, the addition of a number and its ones' complement produces negative 0. In fact, an addition operation using this rule can never produce positive 0 unless both addends are positive 0.

ones'-complement subtraction

As with two's-complement, the easiest way to do ones'-complement subtraction is to complement the subtrahend and add. Overflow rules for ones'-complement addition and subtraction are the same as for two's-complement.

Table 1–7 summarizes the rules that we presented in this and previous sections for negation, addition, and subtraction in binary number systems.

Table 1–7 Summary of addition and subtraction rules for binary numbers.

Number System	Addition rules	Negation rules	Subtraction rules
Unsigned	Add the numbers. Result is out-of-range if a carry out of the MSB occurs.	N/A	Subtract the subtrahend from the minuend. Result is out-of-range if a borrow out of the MSB occurs.
Signed magnitude	(same sign) Add the magnitudes; overflow occurs if a carry out of MSB occurs; result has the same sign. (opposite sign) Subtract the smaller magnitude from the larger; overflow is impossible; result has the sign of the larger.	Change the sign bit.	Change the sign bit of the subtrahend and proceed as in addition.
Two's complement	Add, ignoring any carry out of the MSB. Overflow occurs if the carries into and out of MSB are different.	Complement all bits of the subtrahend; add 1 to the result.	Complement all bits of the subtrahend and add to the minuend with an initial carry of 1.
Ones' complement	Add; if there is a carry out of the MSB, add 1 to the result. Overflow if carries into and out of MSB are different.	Complement all bits of the subtrahend.	Complement all bits of the subtrahend and proceed as in addition.

* 1.8 BINARY MULTIPLICATION

In grammar school we learned to multiply by adding a list of shifted multiplicands computed according to the digits of the multiplier. The same method can be used to obtain the product of two unsigned binary numbers:

shift-and-add multiplication

unsigned multiplication

```
       11              1011    multiplicand
     × 13            ×  1101    multiplier
     ────            ───────
       33              1011  ⎞
       11              0000  ⎟
     ────              1011  ⎬ shifted multiplicands
      143              1011  ⎠
                    ───────
                   10001111    product
```

Forming the shifted multiplicands is trivial in binary multiplication, since the only possible values of the multiplier digits are 0 and 1.

partial product

Instead of listing all the shifted multiplicands and then adding, in a digital system it is more convenient to add each shifted multiplicand as it is created to a *partial product*. Applying this technique to the previous example, four additions and partial products are used to multiply 4-bit numbers:

$$
\begin{array}{rl}
11 & \\
\times\ 13 & \\
\hline
\end{array}
\qquad
\begin{array}{rl}
1011 & \text{multiplicand} \\
\times\quad 1101 & \text{multiplier} \\
\hline
0000 & \text{partial product} \\
1011 & \text{shifted multiplicand} \\
\hline
01011 & \text{partial product} \\
0000{\downarrow} & \text{shifted multiplicand} \\
\hline
001011 & \text{partial product} \\
1011{\downarrow}{\downarrow} & \text{shifted multiplicand} \\
\hline
0110111 & \text{partial product} \\
1011{\downarrow}{\downarrow}{\downarrow} & \text{shifted multiplicand} \\
\hline
10001111 & \text{product} \\
\end{array}
$$

In general, when we multiply an n-bit number by an m-bit number, the resulting product requires at most $n + m$ bits to express. The shift-and-add algorithm requires m partial products and additions to obtain the result, but the first addition is trivial since the first partial product is zero. Although the first partial product has only n significant bits, after each addition step the partial product gains one more significant bit, since each addition may produce a carry. At the same time, each step yields one more partial product bit, starting with the rightmost and working towards the left, that does not change for the rest of the computation. The shift-and-add algorithm can be performed by a digital circuit that includes a shift register, an adder, and control logic, as shown in Section 6.9.2.

signed multiplication

Multiplication of signed numbers can be accomplished using unsigned multiplication and the usual grammar school rules: perform an unsigned multiplication of the magnitudes and make the product positive if the operands had the same sign, negative if they had different signs. This is very convenient in signed-magnitude systems, since the sign and magnitude are separate.

two's-complement multiplication

In the two's-complement system, obtaining the magnitude of a negative number and negating the unsigned product are nontrivial operations. This leads us to seek a more efficient way of performing two's-complement multiplication, described next.

Conceptually, unsigned multiplication is accomplished by a sequence of unsigned additions of the shifted multiplicands; at each step, the shift of the multiplicand corresponds to the weight of the multiplier bit. The bits in a two's-complement number have the same weights as in an unsigned number, except for the MSB, which has a negative weight (see Section 1.5.4). It follows that we can perform two's-complement multiplication by a sequence of two's-complement additions of shifted multiplicands, except for the last step, in which the shifted multiplicand corresponding to the MSB of the multiplier must be negated before it is added to the partial product. Our previous example is repeated below, this time interpreting the multiplier and multiplicand as two's-complement numbers:

```
       −5              1011     multiplicand
     × −3           ×  1101     multiplier
                      00000     partial product
                      11011     shifted multiplicand
                     111011     partial product
                     00000↓     shifted multiplicand
                    1111011     partial product
                    11011↓↓     shifted multiplicand
                   11100111     partial product
                   00101↓↓↓     shifted and negated multiplicand
                   00001111     product
```

Handling the MSBs is a little tricky, because we gain one significant bit at each step and we are working with signed numbers. Therefore, before adding each shifted multiplicand and k-bit partial product, we change them to $k+1$ significant bits by sign extension, as shown in color above. Each resulting sum has $k + 1$ bits; any carry out of the MSB of the $k + 1$-bit sum is ignored.

* 1.9 BINARY DIVISION

The simplest binary division algorithm is based on the shift-and-subtract method that we learned in grammar school. Table 1–8 gives examples of this method for unsigned decimal and binary numbers. In both cases, we mentally compare the reduced dividend with multiples of the divisor to determine which multiple of the shifted divisor to subtract. In the decimal case, we first pick 11 as the greatest multiple of 11 less than 21, and then pick 99 as the greatest multiple less than 107. In the binary case, the choice is somewhat simpler, since the only two choices are zero and the divisor itself. *shift-and-subtract division*

unsigned division

Division methods for binary numbers are somewhat complementary to binary multiplication methods. A typical division algorithm accepts an $n + m$-bit dividend and an n-bit divisor, and produces an m-bit quotient and an n-bit remainder. A division *overflows* if the divisor is zero or the quotient would take more than m bits to express. In most computer division circuits, $n = m$. *division overflow*

Table 1–8
Example of long division.

19	10011	quotient
11)217	1011)11011001	dividend
11	¯1011	shifted divisor
107	0101	reduced dividend
99	0000	shifted divisor
8	1010	reduced dividend
	0000	shifted divisor
	10100	reduced dividend
	¯1011	shifted divisor
	10011	reduced dividend
	¯1011	shifted divisor
	1000	remainder

signed division

Division of signed numbers can be accomplished using unsigned division and the usual grammar school rules: perform an unsigned division of the magnitudes and make the quotient positive if the operands had the same sign, negative if they had different signs. The remainder should be given the same sign as the dividend. As in multiplication, there are special techniques for performing division directly on two's-complement numbers; these techniques are often implemented in computer division circuits (see References).

1.10   BINARY CODES FOR DECIMAL NUMBERS

Even though binary numbers are the most appropriate for the internal computations of a digital system, most people still prefer to deal with decimal numbers. As a result, the external interfaces of a digital system may read or display decimal numbers, and some digital devices actually process decimal numbers directly.

The human need to represent decimal numbers doesn't change the basic nature of digital electronic circuits—they still process signals that take on one of only two states that we call 0 and 1. Therefore, a decimal number is represented in a digital system by a string of bits, where different combinations of bit values in the string represent different decimal numbers. For example, if we use a 4-bit string to represent a decimal number, we might assign bit combination 0000 to decimal digit 0, 0001 to 1, 0010 to 2, and so on.

code
code word

A set of n-bit strings in which different bit strings represent different numbers or other things is called a *code*. A particular combination of n bit-values is called a *code word*. As we'll see in the examples of decimal codes in this section, there may or may not be an arithmetic relationship between the bit values in a

Table 1–9
Decimal codes.

Decimal digit	BCD (8421)	2421	Excess-3	Biquinary	1-out-of-10
0	0000	0000	0011	0100001	1000000000
1	0001	0001	0100	0100010	0100000000
2	0010	0010	0101	0100100	0010000000
3	0011	0011	0110	0101000	0001000000
4	0100	0100	0111	0110000	0000100000
5	0101	1011	1000	1000001	0000010000
6	0110	1100	1001	1000010	0000001000
7	0111	1101	1010	1000100	0000000100
8	1000	1110	1011	1001000	0000000010
9	1001	1111	1100	1010000	0000000001
Unused code words					
	1010	0101	0000	0000000	0000000000
	1011	0110	0001	0000001	0000000011
	1100	0111	0010	0000010	0000000101
	1101	1000	1101	0000011	0000000110
	1110	1001	1110	0000101	0000000111
	1111	1010	1111	...	...

code word and the thing that it represents. Furthermore, a code that uses n-bit strings need not contain 2^n valid code words.

At least four bits are needed to represent the ten decimal digits. There are billions and billions of different ways to choose ten 4-bit code words,[1] but some of the more common decimal codes are listed in Table 1–9.

Perhaps the most "natural" decimal code is *binary-coded decimal (BCD)*, which encodes the digits 0 through 9 by their 4-bit unsigned binary representations, 0000 through 1001. The code words 1010 through 1111 are not used. Conversions between BCD and decimal representations are trivial, a direct substitution of four bits for each decimal digit. Some computer programs place two BCD digits in one 8-bit byte in *packed-BCD representation*; thus, one byte may represent the values from 0 to 99 as opposed to 0 to 255 for a normal unsigned 8-bit binary number. BCD numbers with any desired number of digits may be obtained by using one byte for each two digits.

binary-coded decimal (BCD)

packed-BCD representation

[1]The number of different ways to choose m items out of a set of n items is given by a *binomial coefficient*, denoted $\binom{n}{m}$, whose value is $\frac{n!}{m! \cdot (n-m)!}$. For a 4-bit decimal code, there are $\binom{16}{10}$ different ways to choose 10 out of 16 4-bit code words, and 10! ways to assign each different choice to the 10 digits. So there are $\frac{16!}{10! \cdot 6!} \cdot 10!$ or 29,059,430,400 different 4-bit decimal codes.

binomial coefficient

As with binary numbers, there are many possible representations of negative BCD numbers. Signed BCD numbers have one extra digit position for the sign. Both the signed-magnitude and 10's-complement representations are popular. In signed-magnitude BCD, the encoding of the sign bit string is arbitrary; in 10's-complement, 0000 indicates plus and 1001 indicates minus.

BCD addition

Addition of BCD digits is similar to adding 4-bit unsigned binary numbers, except that a correction must be made if a result exceeds 1001. The result is corrected by adding 6; examples are shown below:

```
      5      0101                          4      0100
   +  9   +  1001                       +  5   +  0101
  ─────   ──────                        ─────   ──────
     14      1110                          9      1001
          +  0110 — correction
          ──────
   10+4    1 0100
```

```
      5      0101                          9      1001
   + 11   +  1011                       +  9   +  1001
  ─────   ──────                        ─────   ──────
     16    1 0000                         18    1 0010
          +  0110 — correction                +  0110 — correction
          ──────                             ──────
   10+6    1 0110                       10+8    1 1000
```

Notice that the addition of two BCD digits produces a carry into the next digit position if either the initial binary addition or the correction factor addition produces a carry. Many computers perform packed-BCD arithmetic using special instructions that handle the carry correction automatically.

weighted code

Binary-coded decimal is a *weighted code*, because each decimal digit can be obtained from its code word by assigning a fixed weight to each code-word bit. The weights for the BCD bits are 8, 4, 2, and 1, and for this reason the code is

8421 code
2421 code
self-complementing code

sometimes called the *8421 code*. Another set of weights results in the *2421 code* shown in Table 1–9. This code has the advantage that it is *self-complementing*, that is, the code word for the 9s' complement of any digit may be obtained by complementing the individual bits of the digit's code word.

excess-3 code

Another self-complementing code shown in Table 1–9 is the *excess-3 code*. Although this code is not weighted, it has an arithmetic relationship with the BCD code—the code word for each decimal digit is the corresponding BCD code word plus 0011_2. Because the code words follow a standard binary counting sequence, standard binary counters can easily be made to count in excess-3 code, as we'll show in Section 6.3.

biquinary code

Decimal codes can have more than four bits; for example, the *biquinary code* in Table 1–9 uses seven. The first two bits in a code word indicate whether the number is in the range 0–4 or 5–9, and the last five bits indicate which of the five numbers in selected range is represented.

One potential advantage of using more than the minimum number of bits in a code is an error detecting property. In the biquinary code, if any one bit in a code word is accidentally changed to the opposite value, the resulting code word does not represent a decimal digit and can therefore be flagged as an error. Out of 128 possible 7-bit code words, only 10 are valid and recognized as decimal digits; the rest can be flagged as errors if they appear.

A *1-out-of-10 code* such as the one shown in the last column of Table 1–9   *1-out-of-10 code*
is the sparsest encoding for decimal digits, using 10 out of 1024 possible 10-bit code words.

1.11 GRAY CODE

In electromechanical applications of digital systems—such as machine tools, automotive braking systems, and copiers—it is sometimes necessary for an input sensor to produce a digital value that indicates a mechanical position. For example, Figure 1–5 is a conceptual sketch of an encoding disk and a set of contacts that produce one of eight 3-bit binary-coded values depending on the rotational position of the disk. The dark areas of the disk are connected to a signal source corresponding to logic 1, and the light areas are unconnected, which the contacts interpret as logic 0.

The encoder in Figure 1–5 has a problem when the disk is positioned at certain boundaries between the regions. For example, consider the boundary between the 001 and 010 regions of the disk; two of the encoded bits change here. What value will the encoder produce if the disk is positioned right on the theoretical boundary? Since we're on the border, both 001 and 010 are acceptable. However, because the mechanical assembly is not perfect, the two right-hand contacts may both touch a "1" region, giving an incorrect reading of 011. Likewise, a reading of 000 is possible. In general, this sort of problem can occur

Figure 1–5
A mechanical encoding disk
using a 3-bit binary code.

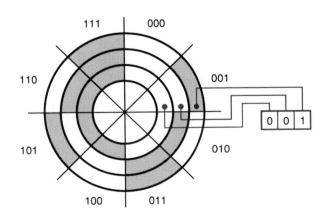

Table 1–10
A comparison of 3-bit binary code and Gray code.

Decimal number	Binary code	Gray code
0	000	000
1	001	001
2	010	011
3	011	010
4	100	110
5	101	111
6	110	101
7	111	100

at any boundary where more than one bit changes. The worst problems occur when all three bits are changing, as at the 000–111 and 011–100 boundaries.

Gray code

The encoding-disk problem can be solved by devising a digital code in which only one bit changes between each pair of successive code words. Such a code is called a *Gray code*; a 3-bit Gray code is listed in Table 1–10. We've redesigned the encoding disk using this code as shown in Figure 1–6. Only one bit of the new disk changes at each border, so borderline readings give us a value on one side or the other of the border.

reflected code

There are two convenient ways to construct a Gray code with any desired number of bits. The first method is based on the fact that Gray code is a *reflected code*; it can be defined (and constructed) inductively using the following rules:

(1) A 1-bit Gray code has two code words, 0 and 1.

(2) The first 2^n code words of an $n + 1$-bit Gray code equal the code words of an n-bit Gray code, written in order with a leading 0 appended.

(3) The last 2^n code words of an $n + 1$-bit Gray code equal the code words of an n-bit Gray code, written in reverse order with a leading 1 appended.

Figure 1–6
A mechanical encoding disk using a 3-bit Gray code.

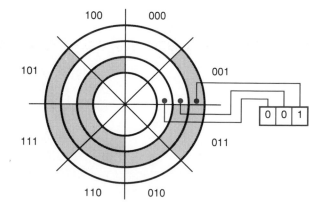

If we draw a line between rows 3 and 4 of Table 1–10, we can see that rules 2 and 3 are true for the 3-bit Gray code. Of course, to construct an n-bit Gray code for an arbitrary value of n with this method, we must also construct a Gray code of each length smaller than n.

The second method allows us to derive an n-bit Gray-code code word directly from the corresponding n-bit binary code word:

(1) The bits of an n-bit binary or Gray-code code word are numbered from right to left, from 0 to $n-1$.

(2) Bit i of a Gray-code code word is 0 if bits i and $i+1$ of the corresponding binary code word are the same, else bit i is 1. (When $i+1 = n$, bit n of the binary code word is considered to be 0.)

Again, inspection of Table 1–10 shows that this true for the 3-bit Gray code.

* 1.12   CHARACTER CODES

As we showed in the preceding section, a string of bits need not represent a number, and in fact most of the data processed by computers is nonnumeric. The most common type of nonnumeric data is *text*, strings of characters from some character set. Each character is represented in the computer by a bit string according to an established convention.

text

The most commonly used character code is *ASCII* (pronounced *ass'-key*), the American Standard Code for Information Interchange. ASCII represents each character by a 7-bit string, a total of 128 different characters shown in Table 1–11. The code contains the uppercase and lowercase alphabet, numerals, punctuation, and various nonprinting control characters. Thus, the text string "Yeccch!" is represented by a rather innocuous-looking list of seven 7-bit numbers:

ASCII

    1011001   1100101   1100011   1100011   1100011   1101000   0100001

1.13   CODES FOR ACTIONS, CONDITIONS, AND STATES

The codes that we've described so far are generally used to represent things that we would probably consider to be "data"—things like numbers, positions, and characters. Programmers know that dozens of different data types can be used in a single computer program.

In digital system design, we often encounter nondata applications where a string of bits must be used to control an action, to flag a condition, or to represent the current state of the hardware. Probably the most commonly used type of code for such an application is a simple binary code. If there are n different actions,

Table 1–11 American Standard Code for Information Interchange (ASCII), Standard No. X3.4–1968 of the American National Standards Institute.

$b_3b_2b_1b_0$	Row (hex)	000 0	001 1	010 2	011 3	100 4	101 5	110 6	111 7
						$b_6b_5b_4$ (column)			
0000	0	NUL	DLE	SP	0	@	P	`	p
0001	1	SOH	DC1	!	1	A	Q	a	q
0010	2	STX	DC2	"	2	B	R	b	r
0011	3	ETX	DC3	#	3	C	S	c	s
0100	4	EOT	DC4	$	4	D	T	d	t
0101	5	ENQ	NAK	%	5	E	U	e	u
0110	6	ACK	SYN	&	6	F	V	f	v
0111	7	BEL	ETB	'	7	G	W	g	w
1000	8	BS	CAN	(	8	H	X	h	x
1001	9	HT	EM	)	9	I	Y	i	y
1010	A	LF	SUB	*	:	J	Z	j	z
1011	B	VT	ESC	+	;	K	[	k	{
1100	C	FF	FS	,	<	L	\	l	|
1101	D	CR	GS	–	=	M	]	m	}
1110	E	SO	RS	.	>	N	^	n	~
1111	F	SI	US	/	?	O	_	o	DEL

Control codes

NUL	Null	DLE	Data link escape
SOH	Start of heading	DC1	Device control 1
STX	Start of text	DC2	Device control 2
ETX	End of text	DC3	Device control 3
EOT	End of transmission	DC4	Device control 4
ENQ	Enquiry	NAK	Negative acknowledge
ACK	Acknowledge	SYN	Synchronize
BEL	Bell	ETB	End transmitted block
BS	Backspace	CAN	Cancel
HT	Horizontal tab	EM	End of medium
LF	Line feed	SUB	Substitute
VT	Vertical tab	ESC	Escape
FF	Form feed	FS	File separator
CR	Carriage return	GS	Group separator
SO	Shift out	RS	Record separator
SI	Shift in	US	Unit separator
SP	Space	DEL	Delete or rubout

Table 1–12   States in a traffic-light controller.

State	N-S green	N-S yellow	N-S red	E-W green	E-W yellow	E-W red	Code word
			Lights				
N-S go	ON	off	off	off	off	ON	000
N-S wait	off	ON	off	off	off	ON	001
N-S delay	off	off	ON	off	off	ON	010
E-W go	off	off	ON	ON	off	off	100
E-W wait	off	off	ON	off	ON	off	101
E-W delay	off	off	ON	off	off	ON	110

conditions, or states, we can represent them with a binary code with $b = \lceil \log_2 n \rceil$ bits. (The brackets $\lceil \ \rceil$ denote the *ceiling function*—the smallest integer greater than or equal to the bracketed quantity. Thus, b is the smallest integer such that $2^b \geq n$.)

ceiling function

For example, consider a simple traffic-light controller. The signals at the intersection of a north-south (N-S) and an east-west (E-W) street might be in any of the six states listed in Table 1–12. These states can be encoded in three bits, as shown in the last column of the table. Only six of the eight possible 3-bit code words are used, and the assignment of the six chosen code words to states is arbitrary, so many other encodings are possible. An experienced digital designer chooses a particular encoding to minimize circuit cost or to optimize some other parameter (like design time   there's no need to try billions and billions of possible encodings).

Another application of a binary code is illustrated in Figure 1–7(a). Here, we have a system with n devices, each of which can perform a certain action. The characteristics of the devices are such that they may be enabled to operate only one at a time. The control unit produces a binary-coded "device select" word with $\lceil \log_2 n \rceil$ bits to indicate which device is enabled at any time. The "device select" code word is applied to each device, which compares it with its own "device ID" to determine whether it is enabled.

Although its code words have the minimum number of bits, a binary code isn't always the best choice for encoding actions, conditions, or states. Figure 1–7(b) shows how to control n devices with a *1-out-of-n code*, an n-bit code in which valid code words have one bit equal to 1 and the rest of the bits equal to 0. Each bit of the 1-out-of-n code word is connected directly to the enable input of a corresponding device. This simplifies the design of the devices, since they no longer have device IDs; they need only a single "enable" input bit.

1-out-of-n code

The code words of a 1-out-of-10 code were listed in Table 1–9. Sometimes an all-0s word may also be included in a 1-out-of-n code, to indicate that no

(a)

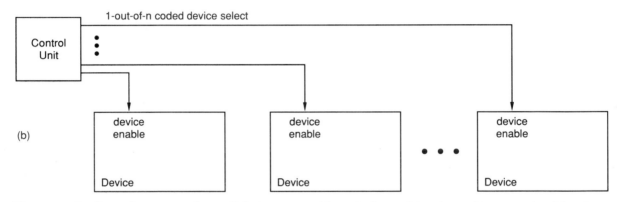

(b)

Figure 1–7 Control structure for a digital system with n devices: (a) using a binary code; (b) using a 1-out-of-n code.

inverted 1-out-of-n code

m-out-of-n code

device is selected. Another common code is an *inverted 1-out-of-n code*, in which valid code words have one 0 bit and the rest of the bits equal to 1.

An *m-out-of-n code* is a generalization of the 1-out-of-n code in which valid code words have m bits equal to 1 and the rest of the bits equal to 0. A valid m-out-of-n code word can be detected with an m-input AND gate, which produces a 1 output if all of its inputs are 1. This is fairly simple and inexpensive to do, yet for most values of m, an m-out-of-n code typically has far more valid code words than a 1-out-of-n code. The total number of code words is given by the binomial coefficient $\binom{n}{m}$, which has the value $\frac{n!}{m!\cdot(n-m)!}$. Thus, a 2-out-of-4 code has 6 valid code words, and a 3-out-of-10 code has 120.

In complex systems, a combination of coding techniques may be used in a single code. For example, consider a system similar to Figure 1–7(b), in which each of the n devices contains up to s subdevices. The control unit could produce a device select code word with a 1-out-of-n coded field to select a device, and a

Figure 1–8
n-cubes for *n* = 1, 2, 3, and 4.

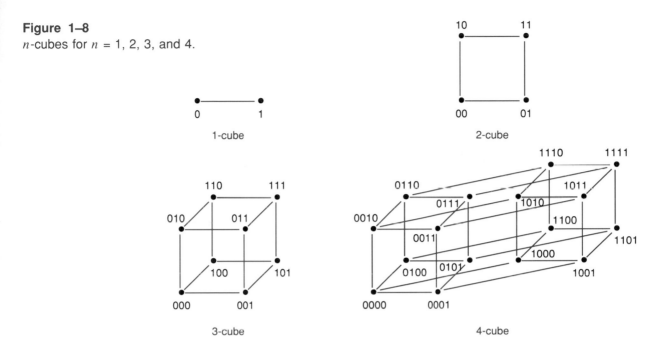

$\lceil \log_2 s \rceil$-bit binary coded field to select one of the *s* subdevices of the selected device. Or, if it were possible to enable *m* subdevices at once, an *m*-out-of-*s* code might be used to select subdevices.

* 1.14 *N*-CUBES AND DISTANCE

An *n*-bit string can be visualized geometrically, as a vertex of an object called an *n-cube*. Figure 1–8 shows *n*-cubes for *n* = 1, 2, 3, 4. An *n*-cube has 2^n vertices, each of which is labeled with an *n*-bit string. Edges are drawn so that each vertex is adjacent to *n* other vertices whose labels differ from the given vertex in only one bit. Beyond *n* = 4, *n*-cubes are really tough to draw. *n-cube*

    For reasonable values of *n*, *n*-cubes make it easy to visualize certain coding and logic minimization problems. For example, the problem of designing an *n*-bit Gray code is equivalent to finding a path along the edges of an *n*-cube, a path that visits each vertex exactly once. The paths for 3- and 4-bit Gray codes are shown in Figure 1–9.

    Cubes also provide a geometrical interpretation for the concept of *distance*, also called *Hamming distance*. The distance between two *n*-bit strings is the number of bit positions in which they differ. In terms of an *n*-cube, the distance is the minimum length of a path between the two corresponding vertices. Two *distance*

Hamming distance

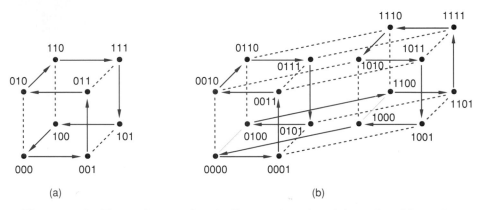

Figure 1–9 Traversing n-cubes in Gray-code order: (a) 3-cube; (b) 4-cube.

adjacent vertices have distance 1; vertices 001 and 100 in the 3-cube have distance 2. The concept of distance is crucial in the design and understanding of error-detecting codes, discussed in the next section.

m-subcube
An *m-subcube* of an n-cube is a set of 2^m vertices in which $n - m$ of the bits have the same value at each vertex, and the remaining m bits take on all 2^m combinations. For example, the vertices (000, 010, 100, 110) form a 2-subcube of the 3-cube. This subcube can also be denoted by a single string, xx0, where "x" denotes that a particular bit is a *don't-care*; any vertex whose bits match in the non-x positions belongs to this subcube. The concept of subcubes is particularly useful in visualizing algorithms that minimize the cost of combinational logic functions, as we'll show in Section 3.4.

don't-care

* 1.15 CODES FOR DETECTING AND CORRECTING ERRORS

error
failure
temporary failure
permanent failure
An *error* in a digital system is the corruption of data from its correct value to some other value. An error is caused by a physical *failure*. Failures can be either temporary or permanent. For example, a cosmic ray can cause a temporary failure of a memory circuit, changing the value of a bit stored in it. Letting a circuit get too hot or zapping it with static electricity can cause a permanent failure, so that it never works correctly again.

error model
independent error model
single error
multiple error
The effects of failures on data are predicted by *error models*. The simplest error model, which we consider here, is called the *independent error model*. In this model, a single physical failure is assumed to affect only a single bit of data; the corrupted data is said to contain a *single error*. Multiple failures may cause *multiple errors*—two or more bits in error—but multiple errors are normally assumed to be less likely than single errors.

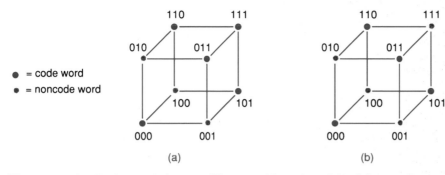

Figure 1–10   Code words in two different 3-bit codes: (a) minimum distance
= 1, does not detect all single errors; (b) minimum distance =
2, detects all single errors.

1.15.1   Error-Detecting Codes

Recall from our definitions in Section 1.10 that a code that uses n-bit strings need not contain 2^n valid code words; this is certainly the case for the codes that we now consider. An *error-detecting code* has the property that corrupting or garbling a code word will likely produce a bit string that is not a code word (a *noncode word*).

  A system that uses an error-detecting code generates, transmits, and stores only code words. Thus, errors in a bit string can be detected by a simple rule—if the bit string is a code word, it is assumed to be correct; if it is a noncode word, it contains an error.

  An n-bit code and its error-detecting properties under the independent error model are easily explained in terms of an n-cube. A code is simply a subset of the vertices of the n-cube. In order for the code to detect all single errors, no code-word vertex can be immediately adjacent to another code-word vertex.

  For example, Figure 1–10(a) shows a 3-bit code with five code words. Code word 111 is immediately adjacent to code words 110 and 011. Since a single failure could change 111 to 110 or 011, this code does not detect all single errors. If we make 111 a noncode word, we obtain a code that does have the single-error-detecting property, as shown in (b). No single error can change one code word into another.

  The ability of a code to detect single errors can be stated in terms of the concept of distance introduced in the preceding section:

- A code detects all single errors if the *minimum distance* between all possible pairs of code words is 2.

  In general, we need $n+1$ bits to construct a single-error-detecting code with 2^n code words. The first n bits of a code word, called *information bits*, may be

error-detecting code

noncode word

minimum distance

information bit

Table 1–13
Distance-2 codes with three information bits.

Information bits	Even-parity code	Odd-parity code
000	000 0	000 1
001	001 1	001 0
010	010 1	010 0
011	011 0	011 1
100	100 1	100 0
101	101 0	101 1
110	110 0	110 1
111	111 1	111 0

parity bit

even-parity code

odd-parity code
1-bit parity code

any of the 2^n n-bit strings. To obtain a minimum-distance-2 code, we add one more bit, called a *parity bit*, that is set to 0 if there are an even number of 1s among the information bits, and to 1 otherwise. This is illustrated in the first two columns of Table 1–13 for a code with three information bits. A valid $n + 1$-bit code word has an even number of 1s, and this code is called an *even-parity code*. We can also construct a code in which the total number of bits in a valid $n + 1$-bit code word is odd; this is called an *odd-parity code* and is shown in the third column of Table 1–13. These codes are also sometimes called *1-bit parity codes*, since they each use a single parity bit.

The 1-bit parity codes do not detect 2-bit errors, since changing two bits does not affect the parity. However, the codes can detect errors in any *odd* number of bits. For example, if three bits in a code word are changed, then the resulting word has the wrong parity and is a noncode word. This doesn't help us much, though. Under the independent error model, 3-bit errors are much less likely than 2-bit errors, which are not detectable. Thus, practically speaking, the 1-bit parity codes' error detection capability stops after 1-bit errors. Other codes, with minimum distance greater than 2, can be used to detect multiple errors.

1.15.2 Error-Correcting and Multiple-Error-Detecting Codes

By using more than one parity bit according to some well-chosen rules, we can create a code whose minimum distance is greater than 2. Before showing how this can be done, let's look at how such a code can be used to correct single errors or detect multiple errors.

Suppose that a code has a minimum distance of 3. Figure 1–11 shows a fragment of the n-cube for such a code. As shown, there are at least two noncode words between each pair of code words. Now suppose we transmit code words and assume that failures affect at most one bit of each received code word. Then a received noncode word with a 1-bit error will be closer to the originally transmitted code word than to any other code word. Therefore, when we receive

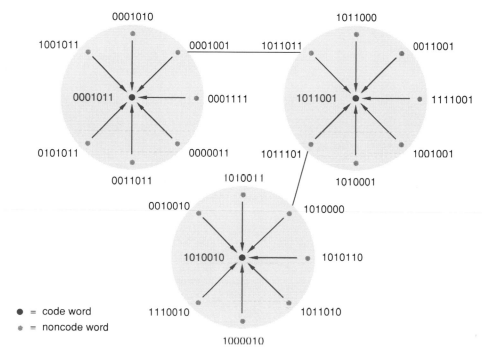

Figure 1–11 Some code words and noncode words in a 7-bit, distance-3 code.

a noncode word, we can *correct* the error by changing the received noncode word to the nearest code word, as indicated by the arrows in the figure. Deciding which code word was originally transmitted to produce a received word is called *decoding*, and the hardware that does this is an error-correcting *decoder*.[2]

error correction

decoding
decoder

A code that is used to correct errors is called an *error-correcting code*. In general, if a code has minimum distance $2c + 1$, it can be used to correct errors that affect up to c bits ($c = 1$ in the preceding example). If a code's minimum distance is $2c + d + 1$, it can be used to correct errors in up to c bits and to detect errors in up to d additional bits.

error-correcting code

For example, Figure 1–12(a) shows a fragment of the n-cube for a code with minimum distance 4 ($c = 1$, $d = 1$). Single-bit errors that produce noncode words 00101010 and 11010011 can be corrected. However, an error that produces 10100011 cannot be corrected, because no single-bit error can produce this noncode word, and either of two 2-bit errors could have produced it. So the code can detect a 2-bit error but it cannot correct it.

---

[2]The names make sense, since *decoding* and *decoder* are just distance-1 perturbations of *deciding* and *decider*.

Figure 1–12   Some code words and noncode words in an 8-bit, distance-4 code: (a) correcting 1-bit and detecting 2-bit errors; (b) incorrectly "correcting" a 3-bit error; (c) correcting no errors but detecting up to 3-bit errors.

When a noncode word is received, we don't know which code word was originally transmitted; we only know which code word is closest to what we've received. Thus, as shown in Figure 1–12(b), a 3-bit error may be "corrected" to the wrong value. The possibility of making this kind of mistake may be acceptable if 3-bit errors are very unlikely to occur. On the other hand, if we are concerned about 3-bit errors, we can change the decoding policy for the code. Instead of trying to correct errors, we just flag all noncode words as uncorrectable errors. Thus, as shown in (c), we can use the same distance-4 code detect up to 3-bit errors but correct no errors ($c = 0$, $d = 3$).

1.15.3 Hamming Codes

In 1950, R. W. Hamming described a general method for constructing codes with a minimum distance of 3, now called *Hamming codes*. For any value of i, his method yields a $2^i - 1$-bit code with i parity bits and $2^i - 1 - i$ information bits. Distance-3 codes with a smaller number of information bits are obtained by deleting information bits from a Hamming code with a larger number of bits.

Hamming code

The bit positions in a Hamming code word are numbered from 1 to $2^i - 1$. Any position whose number is a power of 2 is a parity bit, and the remaining positions are information bits. Each parity bit is grouped with a subset of the information bits, as specified by a *parity-check matrix*. As shown in Figure 1–13(b), each parity bit is grouped with the information positions whose numbers have a 1 in the same bit when expressed in binary. For example, parity bit 2 (010) is grouped with information bits 3 (011), 6 (110), and 7 (111). For a given combination of information-bit values, each parity bit is chosen so that the total number of 1s in its group is even.

parity-check matrix

Traditionally, the bit positions of a parity-check matrix and the resulting code words are rearranged so that all of the parity bits are on the right, as in Figure 1–13(b). The first two columns of Table 1–14 list the resulting code words.

We can prove that the minimum distance of a Hamming code is 3 by proving that at least a 3-bit change must be made to a code word to obtain another code word. That is, we'll prove that a 1-bit or 2-bit change in a code word yields a noncode word.

If we change one bit of a code word, in position j, then we change the parity of every group that contains position j. Since every information bit is contained in at least one group, at least one group has incorrect parity, and the result is a noncode word.

What happens if we change two bits, in positions j and k? Parity groups that contain both positions j and k will still have correct parity, since parity is unaffected when an even number of bits are changed. However, since j and k are different, their binary representations differ in at least one bit, corresponding

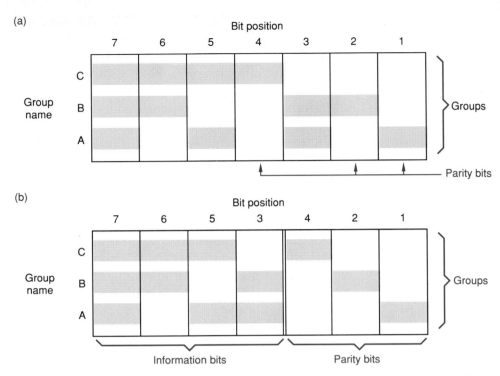

Figure 1–13   Parity-check matrices for 7-bit Hamming codes: (a) with bit positions in numerical order; (b) with parity bits and information bits separated.

to one of the parity groups. This group has only one bit changed, resulting in incorrect parity and a noncode word.

If you understand this proof, you should also see how the position numbering rules for constructing a Hamming code are a simple consequence of the proof. For the first part of the proof (1-bit errors), we required that the position numbers be nonzero. And for the second part (2-bit errors), we required that no two positions have the same number. Thus, with an i-bit position number, you can construct a Hamming code with up to $2^i - 1$ bit positions.

error-correcting
decoder

The proof also suggests how we can design an *error-correcting decoder* for a received Hamming code word. First, we check all of the parity groups; if all have even parity, then the received word is assumed to be correct. If one or more groups have odd parity, then a single error is assumed to have occurred. The

syndrome

pattern of groups that have odd parity (called the *syndrome*) must match one of the columns in the parity-check matrix; the corresponding bit position is assumed to contain the wrong value and is complemented. For example, using the code defined by Figure 1–13(b), suppose we receive the word 0101011. Groups A and B have odd parity, corresponding to position 6 of the parity-check matrix (the

Table 1–14
Code words in minimum-distance-3 and -4 Hamming codes with four information bits.

Minimum-distance-3 code		Minimum-distance-4 code	
Information bits	Parity bits	Information bits	Parity bits
0000	000	0000	0000
0001	011	0001	0111
0010	101	0010	1011
0011	110	0011	1100
0100	110	0100	1101
0101	101	0101	1010
0110	011	0110	0110
0111	000	0111	0001
1000	111	1000	1110
1001	100	1001	1001
1010	010	1010	0101
1011	001	1011	0010
1100	001	1100	0011
1101	010	1101	0100
1110	100	1110	1000
1111	111	1111	1111

syndrome is 110, or 6). By complementing the bit in position 6 of the received word, we determine that the correct word is 0001011.

A distance-3 Hamming code can easily be modified to increase its minimum distance to 4. We simply add one more parity bit, chosen so that the parity of all the bits, including the new one, is even. As in the 1-bit even-parity code, this bit ensures that all errors affecting an odd number of bits are detectable. In particular, any 3-bit error is detectable. We already showed that 1- and 2-bit errors are detected by the other parity bits, so the minimum distance of the modified code must be 4.

Distance-3 and distance-4 Hamming codes are commonly used to detect and correct errors in computer memory systems, especially in large mainframe computers where memory circuits account for the bulk of the system's failures. These codes are especially attractive for very wide memory words, since the required number of parity bits grows slowly with the width of the memory word, as shown in Table 1–15.

1.15.4 Two-Dimensional Codes

Another way to obtain a code with large minimum distance is to construct a *two-dimensional code*, as illustrated in Figure 1–14(a). The information bits are conceptually arranged in a two-dimensional array, and parity bits are provided to

two-dimensional code

Table 1–15 Word sizes of distance-3 and distance-4 Hamming codes.

Information bits	Minimum-distance-3 codes		Minimum-distance-4 codes	
	Parity bits	Total bits	Parity bits	Total bits
1	2	3	3	4
≤ 4	3	≤ 7	4	≤ 8
≤ 11	4	≤ 15	5	≤ 16
≤ 26	5	≤ 31	6	≤ 32
≤ 57	6	≤ 63	7	≤ 64
≤ 120	7	≤ 127	8	≤ 128

check both the rows and the columns. A code C_{row} with minimum distance d_{row} is used for the rows, and a possibly different code C_{col} with minimum distance d_{col} is used for the columns. That is, the row-parity bits are selected so that each row is a code word in C_{row} and the column-parity bits are selected so that each column is a code word in C_{col}. (The "corner" parity bits can be chosen according to either code.) The minimum distance of the two-dimensional code

Figure 1–14
Two-dimensional codes:
(a) general structure; (b) using
even parity for both the row and
column codes to obtain minimum
distance 4; (c) typical pattern of
an undetectable error.

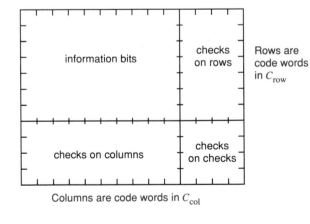

(a)

Columns are code words in C_{col}

(b)

(c)

is the product of d_{row} and d_{col}; in fact, two-dimensional codes are sometimes called *product codes*.

product code

As shown in Figure 1–14(b), the simplest two-dimensional code uses 1-bit even-parity codes for the rows and columns, and has a minimum distance of $2 \cdot 2$ or 4. You can easily prove that the minimum distance is 4 by convincing yourself that any pattern of one, two, or three bits in error causes incorrect parity of a row or a column or both. In order to obtain an undetectable error, at least four bits must be changed in a rectangular pattern as in (c).

To obtain an even larger minimum distance, a distance-3 or -4 Hamming code can be used for the row or column code or both. It is also possible to construct a code in three or more dimensions, with minimum distance equal to the product of the minimum distances in each dimension.

Two-dimensional codes are sometimes used to check computer memory systems, which are inherently two-dimensional structures—n words $\times$ b bits. Suppose we provide an even-parity check bit for each word in memory; this is a row code. We can use one extra memory word to create a column code. Each bit of the column check word stores the parity of the corresponding bit of all the words in memory. The bit-by-bit Exclusive OR instruction available in typical computer processors can be used n times to form the check word.

The error detecting and correcting procedures for the two-dimensional code described above are straightforward. As we read each byte from the memory, we check its row parity, and thereby detect any single-bit error. If an error is detected, we can't tell which bit is wrong from the row check alone. However, we can reconstruct the erroneous word by forming the bit-by-bit Exclusive OR of all of the words in memory, omitting the erroneous word, but including the column check word.

1.15.5 Checksum Codes

The parity checking operation that we've used in the previous subsections is essentially modulo-2 addition of bits—the sum modulo 2 of a group of bits is 0 if the number of 1s in the group is even, and 1 if it is odd. The technique of modular addition can be extended to other bases besides 2 to form check digits.

For example, a computer stores information as a set of 8-bit bytes. Each byte may be considered to have a decimal value from 0 to 255. Therefore, we can use modulo-256 addition to check the bytes. We form a single check byte, called a *checksum*, that is the sum modulo 256 of all the information bytes. The resulting *checksum code* can detect any single *byte* error, since such an error will cause a recomputed sum of bytes to disagree with the checksum.

checksum
checksum code

A checksum code can be combined with a 1-bit even-parity code to form a two-dimensional, distance-4, error-correcting code for a computer memory system. This code is similar to the code described at the end of the preceding subsection, except that the column check word is formed by adding all of the memory words, rather than forming their bit-by-bit Exclusive OR.

1.15.6   m-out-of-n Codes

The 1-out-of-n and m-out-of-n codes that we introduced in Section 1.13 have a minimum distance of 2, since changing only one bit changes the total number of 1s in a code word and therefore produces a noncode word.

unidirectional error

These codes have another useful error-detecting property—they detect unidirectional multiple errors. In a *unidirectional error*, all of the erroneous bits change in the same direction (0s change to 1s or vice versa). This property is very useful in systems where the predominant error mechanism tends to change all bits in the same direction.

1.16   CODES FOR SERIAL DATA TRANSMISSION AND STORAGE

1.16.1   Parallel and Serial Data

parallel data

Most computers and other digital systems transmit and store data in a *parallel* format. In parallel data transmission, a separate signal line is provided for each bit of a data word. In parallel data storage, all of the bits of a data word can be written or read simultaneously.

serial data

Parallel formats are not cost-effective for some applications. For example, parallel transmission of data bytes over the telephone network would require eight phone lines, and parallel storage of data bytes on a magnetic disk would require a disk drive with eight separate read/write heads. *Serial* formats allow data to be transmitted or stored one bit at a time, and reduce system cost in many applications.

bit rate

Figure 1–15 illustrates some of the basic ideas in serial data transmission. A repetitive clock signal, named CLOCK in the figure, defines the rate at which bits are transmitted, one bit per clock cycle. Thus, the *bit rate* in bits per second (bps) numerically equals the clock frequency in cycles per second (hertz, Hz).

bit time

The reciprocal of the bit rate is called the *bit time* and numerically equals the clock period in seconds (s). This amount of time is reserved on the serial data line (named SERDATA in the figure) for each bit that is transmitted. The time occupied by each bit is sometimes called a *bit cell*. The format of the actual signal that appears on the line during each bit cell depends on the *line code*. In the simplest line code, called *NRZ*, a 1 is transmitted by placing a 1 on the line for the entire bit cell, and a 0 is transmitted as a 0. However, more complex line codes have other rules, as discussed in the next subsection.

bit cell
line code
NRZ

Regardless of the line code, a serial data transmission or storage system needs some way of identifying the significance of each bit in the serial stream.

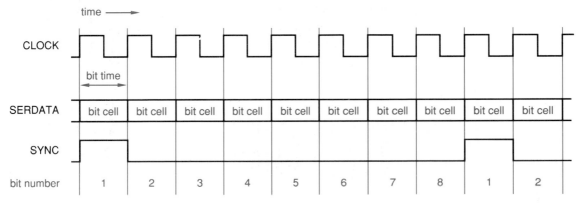

Figure 1–15   Basic concepts for serial data transmission.

For example, suppose that 8-bit bytes are transmitted serially. How can we tell which is the first bit of each byte? A *synchronization signal*, named **SYNC** in Figure 1–15, provides the necessary information; it is 1 for the first bit of each byte.

synchronization signal

Evidently, we need a minimum of three signals to recover a serial data stream: a clock to define the bit cells, a synchronization signal to define the word boundaries, and the serial data itself. In some applications, like the interconnection of modules in a computer or telecommunications system, a separate wire is used for each of these signals, since reducing the number of wires per connection from *n* to three is savings enough. We'll give an example of a 3-wire serial data system in Section 6.4.4.

In many applications, the cost of having three separate signals is still too high (e.g., three phone lines, three read/write heads). Such systems typically combine all three signals into a single serial data stream and use sophisticated analog and digital circuits to recover the clock and synchronization information from the data stream.

* 1.16.2   Serial Line Codes

The most commonly used line codes for serial data are illustrated in Figure 1–16. In the *Non-Return-to-Zero (NRZ)* code, each bit value is sent on the line for the entire bit cell. This is the simplest and most reliable coding scheme for short distance transmission. However, it generally requires a clock signal to be sent along with the data to define the bit cells. Otherwise, it is not possible for the receiver to determine how many 0s or 1s are represented by a continuous 0 or 1 level. Without a clock to define the bit cells, the NRZ waveform in Figure 1–16 might be erroneously interpreted as 01010.

Non-Return-to-Zero (NRZ)

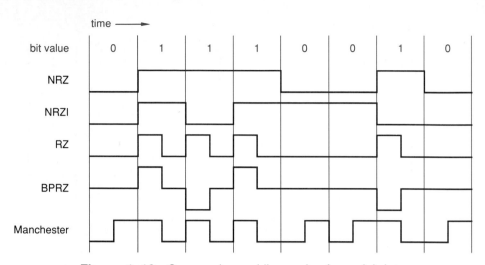

Figure 1–16   Commonly used line codes for serial data.

digital phase-locked loop (DPLL)

A *digital phase-locked loop (DPLL)* is an analog/digital circuit that can be used to recover a clock signal from a serial data stream. The DPLL works only if the serial data stream contains enough 0-to-1 and 1-to-0 transitions to give the DPLL "hints" about when the original clock transitions took place. With NRZ-coded data, the DPLL works only if the data does not contain any long, continuous streams of 1s or 0s.

transition-sensitive media

Some serial transmission and storage media are *transition sensitive*; they cannot transmit or store absolute 0 or 1 levels, only transitions between two discrete levels. For example, a magnetic disk or tape stores information by changing the polarity of the medium's magnetization in regions corresponding to the stored bits. When the information is recovered, it is not feasible to determine the absolute magnetization polarity of a region, only that the polarity changes between one region and the next.

Non-Return-to-Zero Invert-on-1s (NRZI)

Data stored in NRZ format on transition-sensitive media cannot be recovered unambiguously; the data in Figure 1–16 might be interpreted as 01110010 or 10001101. The *Non-Return-to-Zero Invert-on-1s (NRZI)* code overcomes this limitation by sending a 1 as the opposite of the level that was sent during the previous bit cell, and a 0 as the same level. A DPLL can recover the clock from NRZI-coded data as long as the data does not contain any long, continuous streams of 0s.

Return-to-Zero (RZ)

The *Return-to-Zero (RZ)* code is similar to NRZ except that, for a 1 bit, the 1 level is transmitted only for a fraction of the bit time, usually 1/2. With this code, data patterns that contain a lot of 1s create lots of transitions for a DPLL to use to recover the clock. However, as in the other line codes, a string of 0s has no transitions, and a long string of 0s makes clock recovery impossible.

All of the preceding codes transmit or store only two signal levels. The *Bipolar Return-to-Zero (BPRZ)* code transmits three signal levels: +1, 0, and −1. The code is like RZ except that 1s are alternately transmitted as +1 and −1; for this reason, the code is also known as *Alternate Mark Inversion (AMI)*.

Bipolar Return-to-Zero (BPRZ)
Alternate Mark Inversion (AMI)

The biggest advantage of BPRZ over the other codes is that for any data pattern, the average signal level of the BPRZ serial data stream is zero. This improves the noise immunity of the analog circuits that receive highly attenuated BPRZ signals sent over a long distance. In fact, the BPRZ code has been used in the telephone network for decades, where analog speech signals are converted into streams of 8000 8-bit digital samples per second that are transmitted in BPRZ format on 64 Kbps serial channels.

As with RZ, it is possible to recover the clock signal from a BPRZ stream as long as there aren't too many 0s in a row. Although TPC (The Phone Company) has no control over what you say (at least, not yet), they have a neat way of getting around this problem. If one of the 8-bit bytes that results from sampling your analog speech pattern is all 0s, they simply change one of its bits to 1! I'll bet you never noticed.

The last code in Figure 1–16 is called *Manchester* or *diphase* code. The major strength of this code is that, regardless of the transmitted data pattern, it provides at least one transition per bit cell, making it very easy to recover the clock. As shown in the figure, a 0 is encoded as a 0-to-1 transition in the middle of the bit cell, and a 1 is encoded as a 1-to-0 transition. The Manchester code's major strength is also its major weakness. Since it has more transitions per bit cell than other codes, it also requires more media bandwidth to transmit a given bit rate. Bandwidth is not a problem in coaxial cable, however, which is used in Ethernet local area networks to carry Manchester-coded serial data at the rate of 10 Mbps (megabits per second).

Manchester diphase

R E F E R E N C E S

The presentation in the first nine sections of this chapter is based on Chapter 4 of *Microcomputer Architecture and Programming*, by John F. Wakerly (Wiley, 1981). Precise, thorough, and entertaining discussions of these topics can also be found in Donald E. Knuth's *Seminumerical Algorithms* (Addison-Wesley, 1969). Mathematically inclined readers will find Knuth's analysis of the properties of number systems and arithmetic to be excellent, and all readers should enjoy the insights and history sprinkled throughout the text.

Descriptions of digital logic circuits for arithmetic operations, as well as an introduction to properties of various number systems, appear in *Computer Arithmetic* by Kai Hwang (Wiley, 1979). *Decimal Computation* by Hermann Schmid (Wiley, 1974) contains a thorough description of techniques for BCD arithmetic.

An introduction to algorithms for binary multiplication and division and to floating-point arithmetic appears in *Microcomputer Architecture and Programming: The 68000 Family* by John F. Wakerly (Wiley, 1989). A more thorough discussion of arithmetic techniques and floating-point number systems can be found in *Introduction to Arithmetic for Digital Systems Designers* by Shlomo Waser and Michael J. Flynn (Holt, Rinehart and Winston, 1982).

The classic book on error-detecting and error-correcting codes is *Error-Correcting Codes* by W. W. Peterson and E. J. Weldon, Jr. (MIT Press, 1972, 2nd ed.); however, this book is recommended only for mathematically sophisticated readers. A more accessible introduction can be found in *Error Control Coding: Fundamentals and Applications* by S. Lin and D. J. Costello, Jr. (Prentice-Hall, 1983). Another recent, communication-oriented introduction to coding theory can be found in *Error-Control Techniques for Digital Communication* by A. M. Michelson and A. H. Levesque (Wiley-Interscience, 1985). Hardware applications of codes in computer systems are discussed in *Error-Detecting Codes, Self-Checking Circuits, and Applications* by John F. Wakerly (Elsevier/North-Holland, 1978).

An introduction to coding techniques for serial data transmission, including mathematical analysis of the performance and bandwidth requirements of several codes, appears in *Introduction to Communications Engineering* by R. M. Gagliardi (Wiley-Interscience, 1988, 2nd ed.). A nice introduction to the serial codes used in magnetic disks and tapes is given in *Computer Storage Systems and Technology* by Richard Matick (Wiley-Interscience, 1977).

E X E R C I S E S

1.1   Perform the following number system conversions:

(a) $11010011_2 = ?_{16}$ (b) $174003_8 = ?_2$

(c) $10110111_2 = ?_{16}$ (d) $67.24_8 = ?_2$

(e) $10100.1101_2 = ?_{16}$ (f) $F3A5_{16} = ?_2$

(g) $11011001_2 = ?_8$ (h) $AB3D_{16} = ?_2$

(i) $101111.0111_2 = ?_8$ (j) $15C.38_{16} = ?_2$

1.2   Convert the following octal numbers into binary and hexadecimal:

(a) $1023_8 = ?_2 = ?_{16}$ (b) $761302_8 = ?_2 = ?_{16}$

(c) $163417_8 = ?_2 = ?_{16}$ (d) $552273_8 = ?_2 = ?_{16}$

(e) $5436.15_8 = ?_2 = ?_{16}$ (f) $13705.207_8 = ?_2 = ?_{16}$

1.3   Convert the following hexadecimal numbers into binary and octal:

(a) $1023_{16} = ?_2 = ?_8$ (b) $7E6A_{16} = ?_2 = ?_8$

(c) $BABE_{16} = ?_2 = ?_8$ (d) $C350_{16} = ?_2 = ?_8$

(e) $9E36.7A_{16} = ?_2 = ?_8$ (f) $DEAD.0AF_{16} = ?_2 = ?_8$

1.4   Convert the following numbers into decimal:

(a) $11010011_2 = ?_{10}$ (b) $174003_8 = ?_{10}$

(c) $10110111_2 = ?_{10}$ (d) $67.24_8 = ?_{10}$

(e) $10100.1101_2 = ?_{10}$ (f) $F3A5_{16} = ?_{10}$

(g) $12010_3 = ?_{10}$ (h) $AB3D_{16} = ?_{10}$

(i) $7156_8 = ?_{10}$ (j) $15C.38_{16} = ?_{10}$

1.5   Here's a problem to whet your appetite. What is the hexadecimal equivalent of 61453_{10}?

1.6   Perform the following number system conversions:

(a) $125_{10} = ?_2$ (b) $3489_{10} = ?_8$

(c) $209_{10} = ?_2$ (d) $97144_{10} = ?_8$

(e) $132_{10} = ?_2$ (f) $238511_{10} = ?_{16}$

(g) $727_{10} = ?_5$ (h) $57190_{10} = ?_{16}$

(i) $14435_{10} = ?_8$ (j) $65113_{10} = ?_{16}$

1.7   Each of the following arithmetic operations is correct in at least one number system. Determine possible radices of the numbers in each operation.

(a) $1234 + 5432 = 6666$ (b) $41/3 = 13$

(c) $33/3 = 11$ (d) $23 + 44 + 14 + 32 = 223$

(e) $302/20 = 12.1$ (f) $\sqrt{41} = 5$

1.8   The first expedition to Mars found only the ruins of a civilization. From the artifacts and pictures, the explorers deduced that the creatures who produced this civilization were four-legged beings with a tentacle that branched out at the end with a number of grasping "fingers." After much study, the explorers were able to translate Martian mathematics. They found the following equation:

$$5x^2 - 50x + 125 = 0$$

with the indicated solutions $x = 5$ and $x = 8$. The value $x = 5$ seemed legitimate enough, but $x = 8$ required some explanation. Then the explorers reflected on the way in which Earth's number system developed, and found evidence that the Martian system had a similar history. How many fingers would you say the Martians had? (From *The Bent of Tau Beta Pi*, February, 1956.)

1.9   Add the following pairs of binary numbers, showing all carries:

(a)   110101   (b)   101110   (c)   11011101   (d)   1110010
     + 11001   + 100101   + 1100011   + 1101101

1.10 Repeat Exercise 1.9 using subtraction instead of addition, and showing borrows instead of carries.

1.11 Add the following pairs of octal numbers:

(a)	1372	(b)	47135	(c)	175214	(d)	110321
	+ 4631		+ 5125		+ 152405		+ 56573

1.12 Add the following pairs of hexadecimal numbers:

(a)	1372	(b)	4F1A5	(c)	F35B	(d)	1B90F
	+ 4631		+ B8D5		+ 27E6		+ C44E

1.13 Write the 8-bit signed-magnitude, two's-complement, and ones'-complement representations for each of these decimal numbers: +18, +115, +79, −49, −3, −100.

1.14 Suppose a $4n$-bit number B is represented by an n-digit hexadecimal number H. Prove that the two's complement of B is represented by the 16's complement of H. Make and prove true a similar statement for octal representation.

1.15 Repeat Exercise 1.14 using the ones' complement of B and the 15s' complement of H.

1.16 Given an integer x in the range $-2^{n-1} \leq x \leq 2^{n-1} - 1$, we define $[x]$ to be the two's-complement representation of x, expressed as a positive number: $[x] = x$ if $x \geq 0$ and $[x] = 2^n - |x|$ if $x < 0$, where $|x|$ is the absolute value of x. Let y be another integer in the same range as x. Prove that the two's-complement addition rules given in Section 1.6 are correct by proving that the following equation is always true:

$$[x + y] \;=\; [x] + [y] \text{ modulo } 2^n$$

(*Hints:* Consider four cases based on the signs of x and y. Without loss of generality, you may assume that $|x| \geq |y|$.)

1.17 Repeat Exercise 1.16 using appropriate expressions and rules for ones'-complement addition.

1.18 Indicate whether or not overflow occurs when adding the following 8-bit two's-complement numbers:

(a)	11010100	(b)	10111001	(c)	01011101	(d)	00100110
	+ 10101011		+ 11010110		+ 00100001		+ 01011010

1.19 State an overflow rule for addition of two's-complement numbers in terms of counting operations in the modular representation of Figure 1–3.

sign extension

1.20 Show that a two's-complement number can be converted to a representation with more bits by *sign extension*. That is, given an n-bit two's-complement number X, show that the m-bit two's-complement representation of X, where $m > n$, can be obtained by appending $m - n$ copies of X's sign bit to the left of the n-bit representation of X.

1.21 Show that a two's-complement number can be converted to a representation with fewer bits by removing higher-order bits. That is, given an n-bit two's-complement number X, show that the m-bit two's-complement number Y obtained by discarding the d leftmost bits of X represents the same number as X if and only if the discarded bits all equal the sign bit of Y.

1.22 Why is the punctuation of "two's complement" and "ones' complement" inconsistent? (See the first two citations in the References.)

1.23 A n-bit binary adder can be used to perform an n-bit unsigned subtraction operation $X - Y$, by performing the operation $X + \overline{Y} + 1$, where X and Y are n-bit unsigned numbers and $\overline{Y}$ represents the bit-by-bit complement of Y. Demonstrate this fact as follows. First, prove that $(X - Y) = (X + \overline{Y} + 1) - 2^n$. Second, prove that the carry out of the n-bit adder is the opposite of the borrow from the n-bit subtraction. That is, show that the operation $X - Y$ produces a borrow out of the MSB position if and only if the operation $X + \overline{Y} + 1$ *does not* produce a carry out of the MSB position.

1.24 In most cases, the product of two n-bit two's-complement numbers requires $2n - 1$ or fewer bits to represent. In fact, there is only one case in which $2n$ bits are needed—find it.

1.25 Prove that a two's-complement number can be multiplied by 2 by shifting it one bit position to the left, with a carry of 0 into the least significant bit position and disregarding any carry out of the most significant bit position, assuming no overflow. State the rule for detecting overflow.

1.26 State and prove correct a technique similar to the one described in Exercise 1.25, for multiplying a ones'-complement number by 2.

1.27 Show how to subtract BCD numbers, by stating the rules for generating borrows and applying a correction factor. Show how your rules apply to each of the following subtractions: $9 - 3$, $5 - 7$, $4 - 9$, $1 - 8$.

1.28 How many different 3-bit binary state encodings are possible for the traffic-light controller of Table 1–12?

1.29 List all of the "bad" boundaries in the mechanical encoding disc of Figure 1–5, where an incorrect position may be sensed.

1.30 As a function of n, how many "bad" boundaries are there in a mechanical encoding disc that uses an n-bit binary code?

1.31 On-board altitude transponders on commercial and private aircraft use Gray code to encode the altitude readings that are transmitted to air traffic controllers. Why?

1.32 An incandescent light bulb is stressed every time it is turned on, so in some applications the lifetime of the bulb is limited by the number of on/off cycles rather than the total time it is illuminated. Use your knowledge of codes to suggest a way to double the lifetime of 3-way bulbs in such applications.

1.33   As a function of n, how many different distinct subcubes of an n-cube are there?

1.34   Find a way to draw a 3-cube on a sheet of paper (or other two-dimensional object) so that none of the lines cross, or prove that it's impossible.

1.35   Repeat Exercise 1.34 for a 4-cube.

1.36   Write a formula that gives the number of m-subcubes of an n-cube for a specific value of m. (Your answer should be a function of n and m.)

1.37   How many errors can be detected by a code with minimum distance d?

1.38   Define parity groups for a distance-3 Hamming code with 11 information bits.

1.39   Write the code words of a Hamming code with one information bit.

1.40   Exhibit the pattern for a 3-bit error that is not detected if the "corner" parity bits are not included in the two-dimensional codes of Figure 1–14.

1.41   What is the minimum number of parity bits required to obtain a distance-4, two-dimensional code with n information bits?

code rate   1.42   The *rate of a code* is the ratio of the number of information bits to the total number of bits in a code word. High rates, approaching 1, are desirable for efficient transmission of information. Construct a graph comparing the rates of distance-2 parity codes and distance-3 and -4 Hamming codes for up to 100 information bits.

1.43   Which type of distance-4 code has a higher rate—a two-dimensional code or a Hamming code? Support your answer with a table in the style of Table 1–15, including the rate as well as the number of parity and information bits of each code for up to 100 information bits.

1.44   Show how to construct a distance-6 code with four information bits. Write a list of its code words.

1.45   In the style of Figure 1–16, draw the waveforms for the bit pattern 10101110 when sent serially using the NRZ, NRZI, RZ, BPRZ, and Manchester codes, assuming that the bits are transmitted in order from left to right.

2

DIGITAL
CIRCUITS

Marketing hype notwithstanding, we live in an analog world, not a digital one. Voltages, currents, and other physical quantities in real circuits take on values that are infinitely variable, depending on properties of the real devices that comprise the circuits. Because real values are continuously variable, we could use a physical quantity such as a signal voltage in a circuit to represent a real number (e.g., 3.14159265358979 volts represents the mathematical constant *pi* to 14 decimal digits of precision).

Unfortunately, stability and accuracy in physical quantities are difficult to obtain in real circuits. They are affected by manufacturing tolerances, temperature, power-supply voltage, cosmic rays, and noise created by other circuits, among other things. If we used an analog voltage to represent *pi*, we might find that instead of being an absolute mathematical constant, *pi* varied over a range of 10% or more.

Also, many mathematical and logical operations are difficult to impossible to perform with analog quantities. While it is possible with some cleverness to build an analog circuit whose output voltage is the square root of its input voltage, no one has ever built a 100-input, 100-output analog circuit whose outputs are a set of voltages identical to the set of input voltages, but sorted arithmetically.

Table 2–1   Physical states representing bits in different computer logic and memory technologies.

Technology	State representing bit	
	0	1
Relay logic	Circuit open	Circuit closed
Transistor-transistor logic (TTL)	0–0.8 V	2.0–5.0 V
Complementary metal-oxide semiconductor (CMOS) logic	0–2.0 V	3.0–5.0 V
Fiber optics	Light off	Light on
Dynamic memory	Capacitor discharged	Capacitor charged
Nonvolatile, erasable memory	Electrons trapped	Electrons released
Bipolar read-only memory	Fuse blown	Fuse intact
Charge-coupled device memory	No charge	Charge present
Bubble memory	No magnetic bubble	Bubble present
Magnetic tape or disk	No flux reversal	Flux reversal
Polymer memory	Molecule in state A	Molecule in state B

2.1 LOGIC SIGNALS AND GATES

digital logic

binary digit
bit

Digital logic hides the pitfalls of the analog world by mapping the infinite set of real values for a physical quantity into two subsets corresponding to just two possible numbers—0 and 1. These two numbers, 0 and 1, are called *binary digits* or *bits*. If an application requires more than two discrete values, additional bits may be used, with a set of n bits representing 2^n different values.

Examples of the physical phenomena used to represent bits in some modern (and not-so-modern) digital technologies are given in Table 2–1. With most phenomena, there is an undefined region between the 0 and 1 states (e.g., voltage = 1.5, dim light, capacitor slightly charged, etc.). This undefined region, often

noise margin

called *noise margin*, is necessary so that the 0 and 1 states can be unambiguously defined and reliably detected. Noise can more easily corrupt results if the boundaries separating the 0 and 1 states are too close.

Because a wide range of physical values represent the same binary value, digital logic is highly immune to component and power supply variations and noise. Furthermore, *buffer amplifier* circuits can be used to regenerate "weak"

buffer amplifier

values into "strong" ones, so that digital signals can be transmitted over arbitrary distances without loss of information. For example, a buffer amplifier for CMOS logic converts any input voltage above 2.5 V into a 5 V output, and any smaller input voltage into a 0 V output.

Since the inputs of a logic circuit take on only discrete 0 and 1 values, the operation of a logic circuit can be described by a table that ignores physical values and lists only discrete logic values. A logic circuit whose outputs depend

Table 2–2
Truth table for a combinational
logic circuit.

X	Y	Z	F
0	0	0	0
0	0	1	1
0	1	0	0
0	1	1	0
1	0	0	0
1	0	1	0
1	1	0	1
1	1	1	1

only on its current inputs is called a *combinational circuit*; its operation is fully described by a *truth table* that lists all combinations of input values and the output value(s) produced by each one. Table 2–2 is the truth table for a logic circuit with three inputs X, Y, and Z and a single output F. *combinational circuit* *truth table*

As we'll show in Section 3.1, just three basic logic elements, AND, OR, and NOT, can be used to build any combinational digital logic circuit. Figure 2–1 shows the truth tables and symbols for these elements. The symbols and truth tables for AND and OR may be extended to gates with any number of inputs. An *AND gate* produces a 1 output only if all of its inputs are 1; an *OR gate* produces a 1 if any one or more of its inputs are 1. A *NOT gate*, often called a *complement gate* or *inverter*, produces an output value that is the opposite of its input value. Figure 2–2 is a logic circuit using AND, OR, and NOT gates that functions according to the truth table of Table 2–2. In Chapter 3 you'll learn how to go from a truth table to a logic circuit, and vice versa. *AND gate* *OR gate* *NOT gate* *complement gate* *inverter*

At this point, you should have the idea that digital logic circuits can be handled functionally, using tables and other abstract means to describe the operation of well-behaved 0s and 1s in a circuit. Even if you know nothing about analog electronics, you can understand the behavior of digital logic circuits. However, there comes a time in design and debugging when every digital logic designer

X	Y	X AND Y
0	0	0
0	1	0
1	0	0
1	1	1

X	Y	X OR Y
0	0	0
0	1	1
1	0	1
1	1	1

X	NOT X
0	1
1	0

Figure 2–1   Basic logic elements: (a) AND; (b) OR; (c) NOT.

Figure 2–2
Logic circuit with the truth table
of Table 2–2.

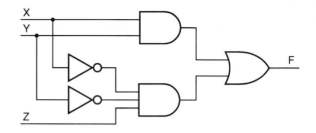

must temporarily throw out "the digital abstraction," and consider the analog phenomena that limit or disrupt digital performance. The purpose of the rest of this chapter is to prepare you for that day by discussing the electrical design of digital logic circuits, beginning with simple diode logic and working through to the most popular digital logic families that are used in industry today.[1]

2.2   DIODES

Many different physical devices, both electrical and nonelectrical, have been used to build logic gates, including mechanical relays, hydraulic valves, vacuum tubes, and cryotrons. The diode was the first semiconductor device used to build logic gates, and it is still used in some logic families today.

semiconductor diode
p-type material
n-type material
pn junction

A *semiconductor diode* is fabricated from two types of semiconductor material, called *p*-type and *n*-type, that are brought into contact with each other as shown in Figure 2–3(a). The point of contact between the *p* and *n* materials is called a *pn junction*. (Actually, a diode is normally fabricated from a single monolithic crystal of semiconductor material in which the two halves are "doped" with different impurities to give them *p*-type and *n*-type properties.)

The physical properties of a *pn* junction are such that positive current can easily flow from the *p*-type material to the *n*-type. Thus, if we build the circuit

---

[1]If all of this electrical "stuff" bothers you, don't worry, at least for now. The rest of this book is written to be as independent of this stuff as possible. But you'll need it later, if you ever expect to design digital systems in the real world.

Figure 2–3   Semiconductor diodes: (a) the *pn* junction; (b) forward-biased junction allowing current flow; (c) reverse-biased junction blocking current flow.

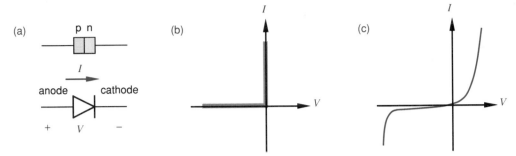

Figure 2-4   Diodes: (a) symbol; (b) transfer characteristic of an ideal diode; (c) transfer characteristic of a real diode.

shown in Figure 2–3(b), the *pn* junction acts almost like a short circuit. However, the physical properties also make it very difficult for positive current to flow in the opposite direction, from *n* to *p*. Thus, in the circuit of Figure 2–3(c), the *pn* junction behaves almost like an open circuit. This is called *diode action*.   *diode action*

Although it's possible to build vacuum tubes and other devices that exhibit diode action, modern systems use *pn* junctions—semiconductor diodes—which we'll henceforth call simply *diodes*. Figure 2–4(a) shows the schematic symbol   *diode* for a diode. As we've shown, in normal operation significant amounts of current can flow only in the direction indicated by the two arrows, from anode to cathode.[2] In effect, the diode acts like a short circuit as long as the voltage across the anode-to-cathode junction is nonnegative. If the anode-to-cathode voltage is negative, the diode acts like an open circuit and no current flows.

The transfer characteristic of an ideal diode shown in Figure 2–4(b) further illustrates this principle. If the anode-to-cathode voltage, *V*, is negative, the diode is said to be *reverse biased* and the current *I* through the diode is zero.   *reverse-biased diode* However, if *V* is nonnegative, the diode is said to be *forward biased* and *I* can   *forward-biased* be an arbitrarily large positive value. In fact, *V* can never get larger than zero,   *diode* because an ideal diode acts like a zero-resistance short circuit when forward biased.

A nonideal, real diode has a resistance that is less than infinity when reverse biased, and greater than zero when forward biased, so the transfer characteristic looks like Figure 2–4(c). When the diode is forward biased, the diode acts like a

[2]*Two* arrows, you ask? The second arrow is built into the diode symbol to help you remember the direction of current flow. Once you know this, there are many ways to remember which end is called the *anode* and which is the *cathode*. For example, if you're experienced with vacuum   *anode* tubes, you might remember that electrons travel from the hot cathode to the anode, and therefore   *cathode* positive current flow is from anode to cathode. If you were still watching "Sesame Street" when vacuum tubes went out of style, then you might like to think in terms of the alphabet—current flows alphabetically from A to C.

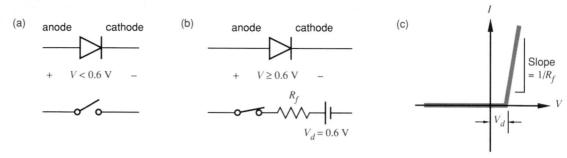

Figure 2–5   Model of a real diode: (a) reverse biased; (b) forward biased; (c) transfer characteristic of forward-biased diode.

small nonlinear resistance, in that the voltage drop increases as current increases, although not strictly proportionally. When the diode is reverse biased, a small

leakage current
diode breakdown

amount of negative *leakage current* flows. If the voltage is made too negative, the diode *breaks down*, and large amounts of negative current can flow; in most applications, this type of operation is avoided.[3]

A real diode can be modeled very simply as shown in Figure 2–5(a) and (b). When the diode is reverse biased, it acts like an open circuit; we ignore leakage current. When the diode is forward biased, it acts like a small resistance,

forward resistance
diode-drop

R_f, in series with V_d, a small voltage source. R_f is called the *forward resistance* of the diode, and V_d is called a *diode-drop*.

Careful choice of values for R_f and V_d yields a reasonable piecewise-linear approximation to the real diode transfer characteristic, as shown in Figure 2–5(c). In a typical small-signal diode such as a 1N914, the forward resistance R_f is about 25 Ω and the diode-drop V_d is about 0.6 V.

As you get a feel for diodes, it is important to remember that a real diode does not actually contain the 0.6 V source that appears in the model. It's just that, due to the nonlinearity of the real diode's transfer characteristic, significant amounts of current do not begin to flow until the diode's forward voltage V has reached about 0.6 V. Also note that in typical applications, the 25 Ω forward resistance of the diode is small compared to other resistances in the circuit, so that very little additional voltage drop occurs across the forward-biased diode once V has reached 0.6 V. Thus, for practical purposes, a forward-biased diode may be considered to have a fixed drop of 0.6 V or so.

zener diode

[3]*Zener diodes* take advantage of this effect, in particular the steepness of the *V–I* slope in the breakdown region. A Zener diode can function as a voltage regulator when used with a resistor to limit the breakdown current. A wide variety of Zeners with different specified breakdown voltages are produced for voltage-regulator applications.

	Signal level	Designation	Binary logic value
Table 2–3	0–2 volts	LOW	0
Logic levels in a simple diode	2–3 volts	noise margin	undefined
logic system.	3–5 volts	HIGH	1

2.3 DIODE LOGIC

Diode action can be exploited to perform logical operations. Consider a logic system with a 5 V power supply and the characteristics shown in Table 2–3. Within the 5-volt range, signal voltages are partitioned into two ranges, LOW and HIGH, with a 1-volt noise margin between. A voltage in the LOW range is considered to be a logic 0, and a voltage in the HIGH range is a logic 1.

LOW
HIGH

With these definitions, a *diode AND gate* can be constructed as shown in Figure 2–6(a). In this circuit, suppose that both inputs X and Y are connected to HIGH voltage sources, say 4 V, so that V_X and V_Y are both 4 V as in Figure 2–6(b). Then both diodes are forward biased, and the output voltage V_Z is one diode-drop above 4 V, or about 4.6 V. A small amount of current, determined by the value of R, flows from the 5 V supply through the two diodes and into the 4 V sources. The colored arrows in the figure show the path of this current flow.

diode AND gate

Now suppose that V_X drops to 1 V as in Figure 2–6(c). In the diode AND gate, the output voltage equals the *lower* of the two input voltages plus a diode-drop. Thus V_Z drops to 1.6 V, and diode *D2* is reverse biased (the anode is at 1.6 V and the cathode is still at 4 V). The single LOW input "pulls down" the output of the diode AND gate to a LOW value. Obviously, two LOW inputs create a LOW output as well. This functional operation is summarized in (d) and is repeated in terms of binary logic values in (e); clearly, this is an AND gate.

Figure 2–7(a) shows a logic circuit with two AND gates connected together; Figure 2–7(b) shows the equivalent electrical circuit with a particular set of input values. This example shows the necessity of diodes in the AND circuit: *D3* allows the output Z of the first AND gate to remain HIGH while the output C of the second AND gate is being pulled LOW by input signal B through *D4*.

Also notice in Figure 2–7(b) that a leakage current, I_{leak}, flows through the reverse-biased diode *D3*. Suppose that *R1* is 1000 Ω and I_{leak} is 100 μA (microamperes), both plausible values for real circuits. Then the leakage current creates a 100 mV drop across resistor *R1*, even if the X and Y inputs to the gate are left unconnected. This is no problem, given the noise margin in the circuit. But now suppose that twenty other gate inputs, not just one, are connected to output Z. Then the combined leakage currents create a voltage drop of 20·100 mV = 2 V, and the voltage at point Z drops to 3 V, the minimum logic 1 value in our system.

) leakage current

Figure 2–6 Diode AND gate: (a) electrical circuit; (b) both inputs HIGH; (c) one input HIGH, one LOW; (d) function table; (e) truth table.

fanout

So we see that real logic gates have a maximum *fanout*—the maximum number of gate inputs that a gate output can drive before logic levels are degraded beyond worst-case specifications.

Figure 2–7(c) shows the same circuit with a different set of input values. In this case, both gate outputs Z and C are LOW. Notice that since the logic LOW voltage of a gate is one diode-drop higher than its lowest input voltage, the logic LOW voltage creeps up as gates are cascaded. That is, an input voltage source of 1.0 V on input Y creates an output of 1.6 V at point Z, which in turn creates an output of 2.2 V at point C. This is outside the range we defined for valid logic LOW signals! To get a valid signal, we must either start out with a lower voltage at input Y, or insert a buffer amplifier on signal line Z to restore the logic level to a really low LOW voltage. In fact, all modern logic families contain such a buffer amplifier before the output of each gate.

A diode OR gate is shown in Figure 2–8. Here resistor R pulls down the output to a LOW level as long as both inputs are LOW. However, if either input

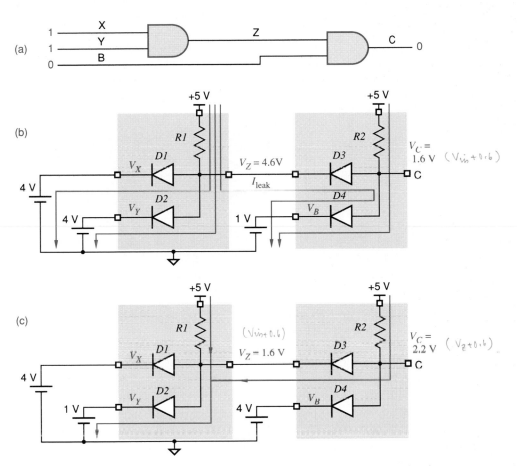

Figure 2–7 Two AND gates: (a) logic diagram; (b) electrical circuit; (c) same circuit with X,Y,B = 1,0,1.

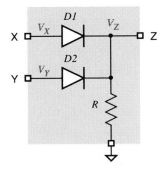

Figure 2–8
Diode OR gate circuit.

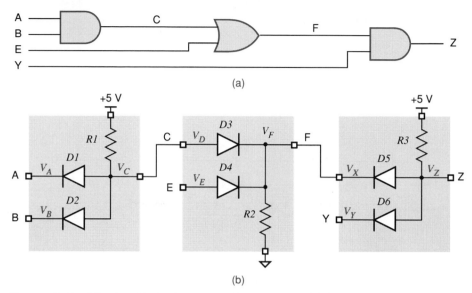

Figure 2–9 Circuit with both AND and OR gates: (a) logic circuit; (b) passive diode circuit equivalent.

is connected to a HIGH voltage source, the output is pulled HIGH, to within one diode-drop of the input.

As with AND gates, the logic levels of diode OR gate outputs suffer when they drive too many inputs or gates are cascaded without buffer amplifiers. Worse, diode AND gates and OR gates cannot normally be mixed together without intervening buffer amplifiers. For example, try to find a set of resistor values such that the passive AND/OR/AND circuit in Figure 2–9(a) behaves properly for all possible input combinations and valid signal voltages.

Diode logic forms a good theoretical basis for understanding logic gates, but it is seldom used in practice today because of the wide availability, high performance, and low cost of logic families with active buffer amplifiers in every gate. However, logic designers are occasionally tempted to use diode logic under special circumstances; for example, see Exercise 2.20.

2.4 BIPOLAR JUNCTION TRANSISTORS

bipolar junction transistor

base

emitter

collector

A *bipolar junction transistor* is a three-terminal device that, in most logic circuits, acts like a current-controlled switch. If we put a small current into one of the terminals, called the *base*, then the switch is "on"—current may flow between the other two terminals, called the *emitter* and the *collector*. If no current is put into the base, then the switch is "off"—no current flows between the emitter and the collector.

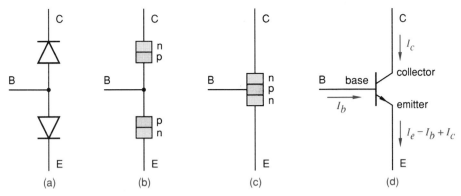

Figure 2–10 Development of an *npn* transistor: (a) back-to-back diodes; (b) equivalent *pn* junctions; (c) structure of an *npn* transistor; (d) *npn* transistor
symbol.

To study the operation of a transistor, we first consider the operation of a
pair of diodes connected as shown in Figure 2–10(a). In this circuit, current can
flow from node B to node C or node E, when the appropriate diode is forward
biased. However, no current can flow from C to E or vice versa, since for any
choice of voltages on nodes B, C, and E, one or both diodes will be reverse
biased. The *pn* junctions of the two diodes in this circuit are shown in (b).

Now suppose that we fabricate the back-to-back diodes so that they share
a common *p*-type region, as shown in Figure 2–10(c). The resulting structure is
called an *npn transistor* and has an amazing property. (At least, the physicists   *npn transistor*
working on transistors back in the 1950s thought it was amazing!) If we put
current across the base-to-emitter *pn* junction, then current is also enabled to
flow across the collector-to-base *np* junction (which is normally impossible) and
from there to the emitter.

The circuit symbol for the *npn* transistor is shown in Figure 2–10(d). Notice
that the symbol contains a subtle arrow in the direction of positive current flow.
This also reminds us that the base-to-emitter junction is a *pn* junction, the same
as a diode whose symbol has an arrow pointing in the same direction.

It is also possible to fabricate a *pnp transistor*, as shown in Figure 2–11.   *pnp transistor*
However, such transistors are seldom used in digital circuits, so we won't discuss
them further.

The current I_e flowing out of the emitter of an *npn* transistor is the sum
of the currents I_b and I_c flowing into the base and the collector. A transistor
is often used as a signal *amplifier*, because over a certain operating range (the   *amplifier*
active region) the collector current is equal to a fixed constant times the base   *active region*
current ($I_c = \beta \cdot I_b$). However, in digital circuits, we normally use a transistor as
a simple switch that's always fully "on" or fully "off," as explained next.

Figure 2–11
A *pnp* transistor: (a) structure;
(b) symbol.

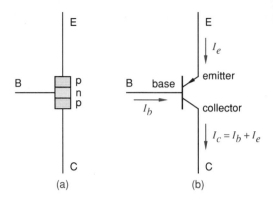

(a)   (b)

*common-emitter
configuration*

cut off (OFF)

Figure 2–12 shows the *common-emitter configuration* of an *npn* transistor, which is most often used in digital switching applications. This configuration uses two discrete resistors, *R1* and *R2*, in addition to a single *npn* transistor. In this circuit, if V_{IN} is 0 or negative, then the base-to-emitter diode junction is reverse biased, and no base current (I_b) can flow. If no base current flows, then no collector current (I_c) can flow, and the transistor is said to be *cut off (OFF)*.

Since the base-to-emitter junction is a *real* diode, as opposed to an ideal one, V_{IN} must reach at least +0.6 V (one diode-drop) before any base current can flow. Once this happens, Ohm's Law tells us that

$$I_b \;=\; (V_{IN} - 0.6)/R1$$

(We have ignored the forward resistance R_f of the forward-biased base-to-emitter junction, which is normally insignificant compared to the base resistor *R1*.) When base current flows, then collector current can flow in an amount proportional to I_b, that is,

$$I_c \;=\; \beta \cdot I_b$$

Figure 2–12
Common-emitter configuration of
an *npn* transistor.

The constant of proportionality, β, is called the *gain* of the transistor, and is in
the range of 10 to 100 for typical transistors.

Although the base current I_b controls the collector current flow I_c, it also indirectly controls the voltage V_{CE} across the collector-to-emitter junction, since V_{CE} is just the supply voltage V_{CC} minus the voltage drop across resistor *R2*:

$$
\begin{aligned}
V_{CE} &= V_{CC} - I_c \cdot R2 \\
&= V_{CC} - \beta \cdot I_b \cdot R2 \\
&= V_{CC} - \beta \cdot (V_{IN} - 0.6) \cdot R2/R1
\end{aligned}
$$

However, in an ideal transistor V_{CE} can never be less than zero (the transistor cannot create a negative potential out of thin air), and in a real transistor V_{CE} can never be less than $V_{CE(sat)}$, a transistor parameter that is typically about 0.2 V.

If the values of V_{IN}, β, *R1*, and *R2* in a given circuit configuration are such that the preceding equation predicts a value of V_{CE} that is less than $V_{CE(sat)}$, then the transistor cannot be operating in the active region and the equation does not apply. Instead, the transistor is operating in the *saturation region*, and is said to be *saturated (ON)*. No matter how much current I_b we put into the base, V_{CE} cannot drop below $V_{CE(sat)}$, and the collector current I_c is determined mainly by the load resistor *R2*:

$$
I_c = (V_{CC} - V_{CE(sat)})/(R2 + R_{CE(sat)})
$$

Here, $R_{CE(sat)}$ is the *saturation resistance* of the transistor. Typically, $R_{CE(sat)}$ is 50 Ω or less and is insignificant compared with *R2*.

Computer scientists might like to imagine the transistor as a device that continuously looks at its environment and executes the program in Table 2–4.

2.5 LOGIC INVERTERS

Figure 2–13 shows that we can make a logic inverter from an *npn* transistor in the common-emitter configuration. When the input voltage is LOW, the output voltage is HIGH, and vice versa.

In digital switching applications, transistors are operated so they are always either cut off or saturated. That is, digital circuits like the inverter in Figure 2–13 are designed so that their transistors are always (well, almost always) in one of the states depicted in Figure 2–14. When the input voltage V_{IN} is LOW, it is low enough that I_b is zero and the transistor is cut off; the collector-emitter junction looks like an open circuit. When V_{IN} is HIGH, it is high enough (and *R1* is low enough and β is high enough) that the transistor will be saturated for any reasonable value of *R2*; the collector-emitter junction looks almost like a short circuit. Input voltages in the undefined region between LOW and HIGH are

Table 2–4   A Pascal program that simulates the function of an *npn* transistor in the common-emitter configuration.

```
PROGRAM CommonEmitterNPNtransistor;
CONST {Transistor parameters}
   diodedrop = 0.6; {volts}
   beta = 10;
   VCEsat = 0.2; {volts}
   RCEsat = 50; {ohms}
VAR
   Vcc, Vin, R1, R2: real; {circuit parameters}
   Ib, Ic, Vce: real; {circuit conditions}
BEGIN
   IF (Vin < diodedrop) THEN BEGIN {cut off}
       Ib := 0.0;
       Ic := 0.0;
       Vce := Vcc;
     END
   ELSE BEGIN {active or saturated}
       Ib := (Vin - diodedrop) / R1;
       IF (Vcc - ((beta * Ib) * R2)) >= VCEsat THEN {active}
         BEGIN
           Ic := beta * Ib;
           Vce := Vcc - (Ic * R2);
         END
   END
   ELSE BEGIN {saturated}
       Vce := VCEsat;
       Ic := (Vcc - Vce) / (R2 + RCEsat);
     END;
END.
```

Figure 2–13   Transistor inverter: (a) logic symbol; (b) circuit diagram; (c) transfer characteristic.

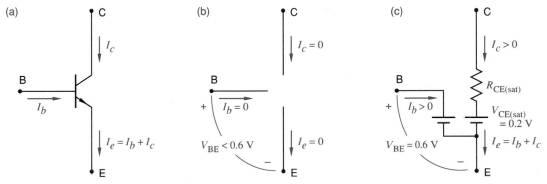

Figure 2–14 Normal states of an *npn* transistor in a digital switching circuit: (a) transistor symbol and currents; (b) equivalent circuit for a cut-off (OFF) transistor; (c) equivalent circuit for a saturated (ON) transistor.

not allowed, except during transitions. This undefined region corresponds to the noise margin that we discussed in conjunction with Table 2–1.

Another way to visualize the operation of a transistor inverter is shown in Figure 2–15. When V_{IN} is HIGH, the transistor switch is closed, and the output terminal is connected to ground, definitely a LOW voltage. When V_{IN} is LOW, the transistor switch is open and the output terminal is pulled to +5 V through a resistor; the output voltage is HIGH unless the output terminal is too heavily loaded (i.e., improperly connected through a low impedance to ground).

2.6 ELECTRICAL DESIGN OF A LOGIC INVERTER

Most digital designers use inverters and other logic elements whose transistor-level circuits have already been designed and fabricated by an integrated-circuit (IC) manufacturer. Still, it is useful to study some of the issues that an IC

Figure 2–15
Switch model for a transistor inverter.

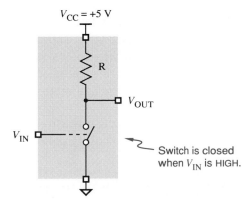

designer faces, because they relate to issues that a digital designer must consider when interconnecting ICs. We'll study a very simple, hypothetical logic family, whose basic logic element is the transistor inverter of Figure 2–13.

2.6.1   Power Supply and Logic Levels

The first choices made by an IC designer are the power-supply voltage and logic levels for the logic system. We'll make the following choices:

V_{CC} 5.0 V. This voltage is used by the industry-standard TTL and CMOS logic families that we'll study later, so we might as well get used to it.

LOW   0–0.5 V. For the transistor inverter, a LOW input level must be less than 0.6 V to guarantee that $Q1$ is cut off, so we specify 0.5 V just to be safe.

HIGH   2.0–5.0 V. We specify a minimum of 2.0 V in order to provide a 1.5-volt margin between the LOW and HIGH levels.

2.6.2   Resistor Choices

Given the preceding specifications, how would an IC designer choose the values of $R1$ and $R2$ in Figure 2–13? Let's start with $R2$. When $Q1$ is ON (V_{OUT} is LOW), the voltage drop across $R2$ is almost equal to V_{CC}, 5 V in our system. Suppose that $R2$ is 1 Ω. Then $R2$ and $Q1$'s collector have almost 5 amps of current flowing through them, and $R2$ dissipates almost 25 watts! Clearly, low values of $R2$ are impractical at worst, and wasteful at best.

On the other hand, suppose that $R2$ is very large, say 1 MΩ. Then $R2$ dissipates very little power when V_{OUT} is LOW, but when $Q1$ is OFF, $R2$ provides only a very weak pull-up to the HIGH level. In fact, the OFF resistance of $Q1$ is not infinite, and may be as little as 20 kΩ, in which case V_{OUT} for the "HIGH" level would be less than 0.5 V! Clearly, we must strike a compromise. Let's use a 2 kΩ resistor and see where that leads us.

$$I = \frac{V}{R} = \frac{5 - 0.2}{2 \times 10^3} = 2.4\,mA$$

With $R2 = 2$ kΩ, $Q1$'s collector and $R2$ have about 2.4 mA of current flowing through them when V_{OUT} is LOW, and $R2$ dissipates about 12 mW. So far, so good. Now, what value should we choose for $R1$? To make this choice, we need to make another assumption:

β Let us assume that transistor $Q1$ has a β of at least 10. This number is based on manufacturing parameters for real transistors.

At this point, we know that $Q1$ has about 2.4 mA of collector current when it is ON, and its β is at least 10. Therefore, if $Q1$ has at least 0.24 mA of base current, it is guaranteed to be ON (saturated), which is what we want. To obtain 0.24 mA of base current (I_b), we must have

$$0.24 \text{ mA} \ \leq \ I_b = (V_{\text{IN(HIGH,min)}} - 0.6 \text{ V})/R1$$

Here, $V_{\text{IN(HIGH,min)}}$ is minimum value of V_{IN} for a HIGH input, or 2.0 V. Solving, we compute

$$0.24 \text{ mA} \ \leq \ (2.0 \text{ V} - 0.6 \text{ V})/R1$$
$$R1 \ \leq \ (2.0 \text{ V} - 0.6 \text{ V})/(0.24 \text{ mA})$$
$$R1 \ \leq \ 5.833 \text{ k}\Omega$$

Since small values of $R1$ will produce more base current than we really need, we'll pick a value at the high end of the acceptable range, $R1 = 5.6 \text{ k}\Omega$.

2.6.3 Fanout

Now we must consider what happens when the output of one logic element drives the input of another. For example, Figure 2–16 shows a very simple circuit with two of our transistor inverters connected in series. As shown in (b), when input voltage V_A is HIGH, transistor $Q1A$ is ON, V_B is LOW, transistor $Q1B$ is OFF, and V_Z is HIGH. Each component of this analysis matches our discussion in the preceding subsection.

Now let us consider the case shown in part (c) of the figure, when V_A is LOW. Clearly, transistor $Q1A$ is OFF, but what voltage is V_B? Let us assume that the OFF resistance of $Q1A$ is high enough to be negligible. Then $R2A$ and $R1B$ form a voltage divider between 5.0 V and the base of $Q1B$ which, if $Q1B$ is ON, is at 0.6 V. Therefore,

$$
\begin{aligned}
V_B &= 5.0 \text{ V} - \frac{R2A}{R2A + R1B} \cdot (5.0 - 0.6) \text{ V} \\
&= 5.0 \text{ V} - \frac{2 \text{ k}\Omega}{2 \text{ k}\Omega + 5.6 \text{ k}\Omega} \cdot 4.4 \text{ V} \\
&= 3.84 \text{ V}
\end{aligned}
$$

This is well within the range that we defined and analyzed for a HIGH level, and so $Q1B$ is indeed ON, and V_Z is LOW.

Next, consider what happens if one inverter output drives three other inputs, as shown in Figure 2–17. The first inverter is said to have a *fanout* of 3 because it drives three inputs. When V_A is HIGH, the analysis is almost identical to the previous example. However, when V_A is LOW, the voltage divider that

fanout

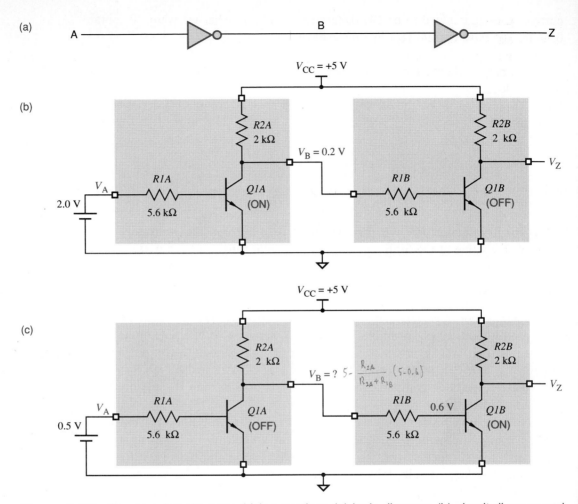

Figure 2–16   One transistor inverter driving another: (a) logic diagram; (b) circuit diagram and conditions when V_A is HIGH; (c) circuit diagram and conditions when V_A is LOW.

determines V_B has $R2A$ in series with the parallel combination of $R1B$, $R1C$, and $R1D$, so V_B is different:

$$V_B = 5.0 \text{ V} - \frac{R2A}{R2A + (R1B \parallel R1C \parallel R1D)} \cdot (5.0 - 0.6) \text{ V}$$

$$= 5.0 \text{ V} - \frac{2 \text{ k}\Omega}{(2 \text{ k}\Omega + 5.6/3 \text{ k}\Omega)} \cdot 4.4 \text{ V}$$

$$= 2.72 \text{ V}$$

So the HIGH value of V_B decreases when we increase the fanout. For some

Figure 2–17
One transistor inverter driving
three others: (a) logic diagram;
(b) circuit diagram and conditions
when V_A is LOW.

fanout value, V_B will be lower than what we allow for a valid HIGH signal in
our system. We can predict this fanout value as follows. We know that an
inverter input consumes about 0.24 mA of base current when the input voltage is
2.0 V, the lowest HIGH input that we allow. If an inverter's output is connected
to n inverter inputs, it must supply $n \cdot 0.24$ mA without dropping below 2.0 V.
Therefore, the voltage drop across $R2$ cannot be more than 3.0 V:

$$n \cdot 0.24 \text{ mA} \cdot R2 \leq 3.0 \text{ V}$$

$$n \leq \frac{3.0 \text{ V}}{0.24 \text{ mA} \cdot 2\text{k}\Omega}$$

$$n \leq 6.25$$

Thus the maximum fanout of one of our transistor inverter outputs driving other inputs is 6. In practice, we would probably specify a maximum fanout of only 4 or 5 to allow for less-than-perfect tolerances in our system's components (e.g., if $R1$ is less than 5.6 kΩ, then an input requires more than 0.24 mA of current at the HIGH level).

electrical specification spec

Fortunately, when we use a commercial logic family, we don't have to study the internals of each logic element and go through all of these calculations. Instead, the manufacturer provides an *electrical specification* (or *spec*) for the external characteristics of the logic element's inputs and outputs. For example, a manufacturer of the hypothetical inverter of this subsection might give customers two sets of numbers:

- $I_{OHmax} = 1.2$ mA at $V_{OHmin} = 2.0$ V. This means that an output in the HIGH state can provide up to 1.2 mA of current while maintaining an output voltage no lower than 2.0 V.

- $I_{IHmax} = 0.3$ mA at $V_{IHmin} = 2.0$ V. This means that an input recognizes 2.0 V or greater as being the HIGH state, and will consume no more than 0.3 mA of current in this state.

With this information, we can state that the maximum fanout for our hypothetical inverter is 4. Note that the specifications are more conservative than required by our analysis, in order to provide adequate manufacturing margins. We'll present a more detailed set of specifications for real logic elements when we study TTL and CMOS logic families later in this chapter.

2.6.4 Transition Time

transition time

The amount of time that the output of a logic circuit takes to change from one state to another is called the *transition time*. Figure 2–18(a) shows how we might like outputs to change state—in zero time. However, real outputs cannot change instantaneously, because they need time to charge the stray capacitance of the wires and other components that they drive. A more realistic view of a circuit's output is shown in (b). An output takes a certain time, called the *rise time* (t_r), to change from LOW to HIGH, and a possibly different time, called the *fall time* (t_f), to change from HIGH to LOW.

rise time (t_r)
fall time (t_f)

Even Figure 2–18(b) is not quite accurate, because the rate of change of the output voltage does not change instantaneously, either. Instead, the beginning

Figure 2–18
Transition times: (a) ideal case of zero-time switching; (b) a more realistic approximation; (c) actual timing, showing rise and fall times.

(a)

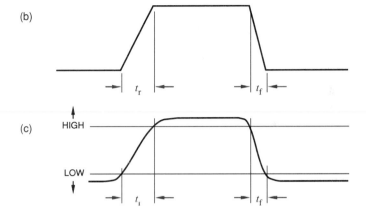

(b)

(c)

and the end of a transition are smooth, as shown in (c). Therefore, to avoid difficulties in defining the endpoints, rise and fall times are normally measured at the boundaries of the valid logic levels as indicated in the figure. With this convention, the rise and fall times indicate how long an output voltage takes to pass through the "undefined" region between LOW and HIGH.

For the transistor inverter of the previous section, the output's rise and fall times are quite different from each other. Figure 2–19 shows an equivalent circuit that we can use to analyze the transition times. Here, transistor $Q1$ is modeled as a switch that is open when V_{IN} is LOW, and closed when V_{IN} is HIGH. When the transistor switch is closed, the output terminal is connected to ground through the transistor's equivalent $R_{CE(sat)} \approx 50 \ \Omega$ and $V_{CE(sat)} \approx 0.2$ V. When the switch is open, the output is pulled up to +5 V through resistor $R2$.

An *equivalent load circuit* is used in Figure 2–19 to represent the inputs driven by the inverter output. The load circuit has three components:

equivalent load circuit

R_L, V_L These two components represent the *DC load* and determine the voltages and currents that are present when the output has settled into a stable HIGH or LOW state. The DC load doesn't have too much effect on transition times when the output changes states.

DC load

C_L This capacitance represents the *AC load* and determines the voltages and currents that are present while the output is changing, and how long it takes to change from one state to the other.

AC load

stray capacitance

A large AC load is undesirable because it slows down transition times; thus, the situations where a logic designer explicitly connects a capacitor to a logic circuit's output are rare. However, an AC load is still created by the *stray capacitance* present in every circuit; stray capacitance comes from at least three sources:

(1) Output circuits, including a gate's output transistor(s), internal wiring, and packaging, have some capacitance associated with them, on the order of several picofarads (pF) in typical technologies.

(2) The wiring that connects an output to other inputs has capacitance, about 1 pF per inch or more, depending on the wiring technology.

(3) Input circuits, including diodes, transistors, internal wiring, and packaging, have capacitance, from 2 to 15 pF per input in typical technologies.

The rise and fall times of V_{OUT} in Figure 2–19 depend on how long it takes to charge or discharge the stray capacitance C_L. As shown in Appendix A, the time required to charge or discharge C_L follows an exponential law. To simplify the analysis, let us ignore R_L and V_L, that is, let $R_L = \infty$ and $V_L = 0$ (but see Exercise 2.7). Figure 2–20(a) shows the timing for a HIGH-to-LOW transition, assuming that $C_L = 100$ pF, a typical value for a moderately loaded output. The switch $Q1$ is initially open and voltage V_{OUT} across C_L is 5.0 V. When the switch is closed at $t = 0$, the capacitive load C_L is discharged to 0.2 V through $R_{CE(sat)}$, and V_{OUT} may be written as a function of t:[4]

$$
\begin{aligned}
(t \leq 0) \qquad V_{OUT} &= 5.0 \text{ V} \\
(t \geq 0) \qquad V_{OUT} &= 0.2 \text{ V} + 4.8 e^{-t/R_{CE(sat)}C_L} \text{ V} \\
&= 0.2 \text{ V} + 4.8 e^{-t/(50 \cdot 100 \cdot 10^{-12})} \text{ V} \\
&= 0.2 \text{ V} + 4.8 e^{-t/(5 \cdot 10^{-9})} \text{ V}
\end{aligned}
$$

The *RC* time constant for HIGH-to-LOW transitions is $R_{CE(sat)} \cdot C_L$, which evaluates to $5 \cdot 10^{-9}$ second or 5 nanoseconds (ns). To measure transition times, recall that 0.5 V and 2.0 V are the defined boundaries for LOW and HIGH levels for our transistor inverter. To obtain the fall time, we must solve the foregoing equation for $V_{OUT} = 2.0$ and $V_{OUT} = 0.5$, yielding

$$
\begin{aligned}
t_{2.0} &= 4.9 \text{ ns} \\
t_{0.5} &= 13.9 \text{ ns}
\end{aligned}
$$

The fall time t_f is the difference between these two numbers, or 9 ns.

---

[4]We've ignored *R2* since it's so large compared to $R_{CE(sat)}$; for purists, the exact Thévenin equivalent of the discharging circuit is 0.317 V in series with 48.78 Ω.

Figure 2–19
Equivalent circuit for analyzing
transition times in a transistor
inverter.

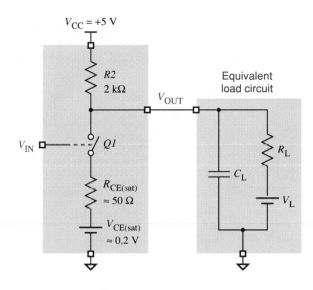

Figure 2–20
Transition timing for a transistor
inverter with $R_L = \infty$ and $V_L = 0$:
(a) HIGH-to-LOW; (b) LOW-to-
HIGH.

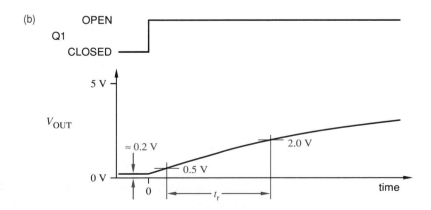

As shown in Figure 2–20(b), the rise time for a LOW-to-HIGH transition is much longer. (Some digital designers call such slow transitions *ooze*.) The switch $Q1$ is initially closed and voltage V_{OUT} across C_L is about 0.2 V. When the switch is opened at $t = 0$, C_L is charged to 5.0 V through $R2$, which is much larger than the $R_{CE(sat)}$ of the previous case:

$$(t \le 0) \qquad V_{OUT} = 0.2 \text{ V}$$
$$(t \ge 0) \qquad V_{OUT} = 5.0 \text{ V} - 4.8e^{-t/R2 \cdot C_L} \text{ V}$$
$$= 5.0 \text{ V} - 4.8e^{-t/(2000 \cdot 100 \cdot 10^{-12})} \text{ V}$$
$$= 5.0 \text{ V} - 4.8e^{-t/(200 \cdot 10^{-9})} \text{ V}$$

The RC time constant for this case is a whopping 200 ns. Solving the equation for the 0.5- and 2.0-volt crossings, we obtain

$$t_{0.5} = 12.9 \text{ ns}$$
$$t_{2.0} = 94.0 \text{ ns}$$

The rise time t_r is the difference between these two numbers, or 81.1 ns, almost an order of magnitude slower than t_f.

We could decrease the inverter output's rise time by using a smaller value of $R2$, but this would increase power consumption. We'll learn a more efficient way to improve rise time when we study TTL logic in Section 2.9.

2.6.5 Propagation Delay

Rise and fall times only partially describe the dynamic behavior of a logic element; we need additional parameters to relate output timing to input timing. A *signal path* is the electrical path from a particular input signal to a particular output signal of a logic element. The *propagation delay* t_p of a signal path is the amount of time that it takes for a change in the input signal to produce a change in the output signal.

A complex logic element with multiple inputs and outputs may specify a different value of t_p for each different signal path. Also, different values may be specified for a particular signal path, depending on the direction of the output change. Ignoring rise and fall times, Figure 2–21(a) shows two different propagation delays for the input-to-output signal path of a transistor inverter, depending on the direction of the output change:

t_{pHL} The time between an input change and the corresponding output change when the output is changing from HIGH to LOW.

t_{pLH} The time between an input change and the corresponding output change when the output is changing from LOW to HIGH.

Figure 2–21
Propagation delays for a
transistor inverter: (a) ignoring
rise and fall times; (b) measured
at midpoints of transitions.

Several factors lead to nonzero propagation delays, even when input and output rise and fall times are zero, as in Figure 2–21(a). For example, when the input to a saturated transistor is changed, the output does not change immediately; it takes extra time, called the *storage time*, to come out of saturation. Multi-   *storage time* transistor logic families such as TTL may require several internal transistors to change state before the output can change state. And even when the output begins to change state, with nonzero rise and fall times it takes quite some time to cross the region between states, as we showed in Figure 2–20. All of these factors are included in propagation delay.

To factor out the effect of rise and fall times, manufacturers usually specify propagation delays at the midpoints of input and output transitions, as shown in Figure 2–21(b). However, sometimes the delays are specified at the logic-level boundary points, especially if the device's operation may be adversely affected by slow rise and fall times. For example, Figure 2–22 shows how the minimum input pulse width for an S-R latch (discussed in Section 5.2.1) might be specified.

In addition, a manufacturer may specify absolute maximum input rise and fall times that must be satisfied to guarantee proper operation. Some components, especially high-speed TTL and CMOS circuits discussed later, may consume excessive current or oscillate if their input transitions are too slow.

Figure 2–22
Worst-case timing specified using
logic-level boundary points.

Figure 2–23   Three-input RTL NOR gate: (a) circuit diagram; (b) electrical function table; (c) truth table; (d) equivalent logic diagram; (e) logic symbol.

2.7   RESISTOR-TRANSISTOR LOGIC (RTL)

resistor-transistor logic (RTL)

bipolar logic family

direct-coupled transistor logic (DCTL)

Two or more transistor inverters of the type shown in Section 2.5 may be placed in parallel with a common pull-up resistor, as shown in Figure 2–23(a). Gates with this circuit structure belong to the *resistor-transistor logic (RTL)* family. Now obsolete, RTL was one of the first *bipolar logic families*—families whose gates use bipolar junction transistors. (The first family was *direct-coupled transistor logic (DCTL)*, which omitted the base resistors $R1$–$R3$.)

A HIGH input to an RTL gate turns on the corresponding transistor. The output is LOW if any one of its transistors is ON. Thus, the RTL output is LOW if any input is HIGH, as shown in the function table in (b). Associating LOW with logic 0 and HIGH with logic 1, we obtain the truth table in (c). The output values are just the opposite of what we would expect from an OR gate; thus, the RTL gate is equivalent to an OR gate followed by an inverter, as shown in (d). This is called an *OR-NOT* or *NOR gate*, and has the symbol shown in (e).

OR-NOT gate

NOR gate

Figure 2–24
Switch model for an RTL NOR
gate.

Figure 2–24 shows an easy way to visualize the operation of an RTL NOR
gate. When any input is HIGH, the corresponding transistor switch is closed, and
the output terminal is connected to a LOW voltage. If all of the inputs are LOW,
then all of the transistor switches are open, and the output terminal is pulled to
a HIGH voltage through a resistor.

We can also visualize an RTL NOR gate as three transistor inverters with
their outputs tied together, as shown in Figure 2–25(a). When transistor-inverter
outputs are connected together in this way, a *wired AND* function is produced
(that's *wired*, not *weird*). That is, if any transistor-inverter output is LOW, it
pulls the connection point LOW. The connection point is HIGH only if all of
the transistor inverters are calling for a HIGH output (i.e., the transistors are all
OFF). The wired-AND concept is important not just for RTL, but also for today's
popular TTL logic families, which we discuss later.

wired AND

A logic symbol corresponding to the wired-AND interpretation of the RTL
NOR gate is shown in Figure 2–25(b). The NOR function is equivalent to an
AND in which each input is first inverted. We'll show in Section 3.1 that the
NOR-gate symbols in Figure 2–23(d) and Figure 2–25(b) are mathematically and
functionally equivalent.

(a)

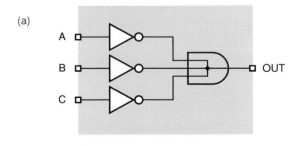

Figure 2–25
Wired-AND interpretation of RTL
NOR gate: (a) logic diagram;
(b) alternate NOR symbol.

(b)

2.7.1   Fan-in

fan-in

Since RTL technology is obsolete, we won't spend much time discussing its electrical characteristics. RTL does, however, provide a good starting point for discussing the concept of "fan-in." The *fan-in* of a logic family is the maximum number of inputs that a logic gate may have. The RTL NOR gate of Figure 2–23 has three inputs. Can we make an RTL NOR gate with 10 inputs?

Our initial answer to this question might be "yes." We simply place 10 transistors in parallel to make a 10-input RTL NOR. Any one of the transistors can pull the output LOW when its input is HIGH, and all of them are OFF when all the inputs are LOW. The output in this case is HIGH. Or is it?

Our fanout analysis of transistor inverters in Section 2.6.3 assumed that the OFF resistance of an *npn* transistor is high enough to be negligible, say 20 kΩ. However, the equivalent resistance of 10 parallel OFF transistors is only 2 kΩ. Therefore, the output voltage of a 10-input RTL NOR gate with $R0 = 2$ kΩ and $V_{CC} = 5.0$ V is 2.5 V, assuming that there is no load on the output. If we use this output to drive the inputs of just two other RTL gates, and assuming that $R1 = 5.6$ kΩ in the driven gates, the output voltage drops to 2.0 V. With more fanout the output drops even more, below an acceptable HIGH level.

So, we see that increasing the fan-in of an RTL gate reduces its fanout (see Exercise 2.9). The commercial RTL logic family used a different power supply and different resistor values than we assumed in the foregoing analysis, but it still had to limit fan-in to 4. Most other logic families have similar constraints that limit fan-in.

2.8   DIODE-TRANSISTOR LOGIC (DTL)

diode-transistor logic (DTL)

A diode AND gate of the type described in Section 2.3 may be combined with a transistor inverter as shown in Figure 2–26. This is called *diode-transistor logic (DTL)*. Historically, the DTL family followed RTL, and led to the development of today's popular TTL family, discussed in the next section. In fact, the DTL diode-input structure is still used in many TTL gates.

As shown in the function table in Figure 2–26(b), the output of the DTL gate is LOW only if both of its inputs are HIGH; we'll examine the circuit's operation in detail momentarily. Associating LOW with logic 0 and HIGH with logic 1, we obtain the truth table in (c). The output values are just the opposite of what we would expect from an AND gate; thus, the DTL gate is equivalent to an AND gate followed by an inverter, as shown in (d). This is called an *AND-NOT* or *NAND gate*, and has the symbol shown in (e).

AND-NOT gate
NAND gate

Diodes *D1* and *D2* in the DTL gate perform the AND function. The voltage divider consisting of resistors *R1* and *R2* ensures that enough base current flows

(a)

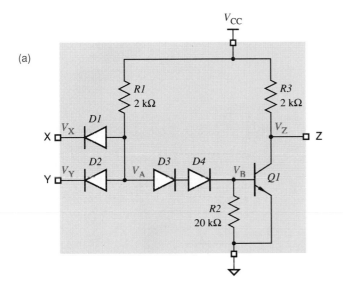

Figure 2–26
Two-input DTL NAND gate:
(a) circuit diagram; (b) electrical
function table; (c) truth table;
(d) equivalent logic diagram;
(e) logic symbol.

(b)

X Y	V_A	V_B	$Q1$	Z
L L	<1.8	<0.6	off	H
L H	<1.8	<0.6	off	H
H L	<1.8	<0.6	off	H
H H	1.8	0.6	on	L

(c)

X Y	Z
0 0	1
0 1	1
1 0	1
1 1	0

(d)

(e)

into $Q1$ to saturate it when X and Y are both HIGH. Diodes $D3$ and $D4$ perform no logic; they merely serve to increase the value of a LOW voltage that may appear on X or Y without causing transistor $Q1$ to turn on. That is, when $Q1$ is ON, V_B is 0.6 V and V_A is two diode-drops higher, or 1.8 V. This is true as long as V_X and V_Y are both 1.2 V or higher. Any lower voltage on input X or Y pulls V_A and V_B lower and cuts off $Q1$. Thus, DTL tolerates a higher voltage for a LOW input (almost 1.2 V, theoretically) than RTL does.

A DTL gate also has much better fanout capability than an RTL gate does. When an RTL output is in the HIGH state, current must flow out of the output and into the inputs that it drives. The output is said to be *sourcing current* in this case. According to our analysis in Section 2.6, an RTL output must source about 0.3 mA for each input that it drives. This current produces a voltage drop across the pull-up resistor $R0$ in the RTL gate, reducing the output voltage in the HIGH state.

The situation for a DTL gate driving another DTL gate is much different, as shown in Figure 2–27. In the HIGH state, the base current required to turn on $Q1B$ in the second gate is provided by resistor $R1B$ within the gate itself. The driving gate sources very little current, less than 100 µA of leakage current to

sourcing current

Figure 2–27 A DTL output driving a DTL input HIGH.

the reversed-biased (but nonideal) diode *D1B*. Thus, the reduction in the HIGH output level caused by the voltage drop across *R3A* in the driving gate is small and not a problem.

In the LOW state, the situation is also much different, as Figure 2–28 shows. Current flows *out of* the driven input pin and into the output pin of the gate that's driving it. The driving gate's output is said to be *sinking current* in this case. The amount of current is substantial, about 2 mA. But that's OK, because transistor *Q1A* is saturated, and can sink quite a bit of current without raising the LOW voltage very much.

sinking current

2.9 TRANSISTOR-TRANSISTOR LOGIC (TTL)

By far the most commonly used logic family is transistor-transistor logic (TTL). Actually, there are many different TTL families, with a range of speed, power consumption, and other characteristics. For simplicity's sake, we'll first describe the original but now obsolete *standard TTL* circuits, and then describe TTL's many variations.

standard TTL

2.9.1 Basic TTL NAND Gate

A basic two-input TTL NAND gate is shown in Figure 2–29. Comparing with the DTL NAND gate of the preceding section, we see the following changes:

Figure 2–28 A DTL output driving a DTL input LOW.

- The multiple input diodes $D1$ and $D2$ of the DTL NAND gate are replaced with $Q1$, a single multiple-emitter *npn* transistor with one emitter for each logic input.

- Reverse-biased diodes are provided from each logic input to ground. These diodes perform no logic, but improve logic signal quality, as explained later.

- The two series diodes $D3$ and $D4$ are replaced with $Q2$, a transistor in an "emitter follower" configuration that increases the base current available to the output transistor $Q3$ in the LOW state, thereby increasing the gate's LOW-state current-sinking capability.

- The relatively large (2 kΩ) pull-up resistor in the DTL output stage is replaced with a transistor $Q4$, diode $D1$, and a much smaller (130 Ω) pull-up, thereby increasing the gate's HIGH-state current-sourcing capability and decreasing the rise time t_r for LOW-to-HIGH transitions.

The *multiple-emitter transistor $Q1$* might be thought of as the "AND" of multiple single-emitter transistors. If any of its emitters is low with respect to the base, then current flows from the base to that emitter and from the collector to that emitter; thus the collector voltage may be pulled low. If all emitters are high, then the multiple-emitter transistor is cut off and the collector voltage may be pulled high.

*multiple-emitter
transistor*

(a)

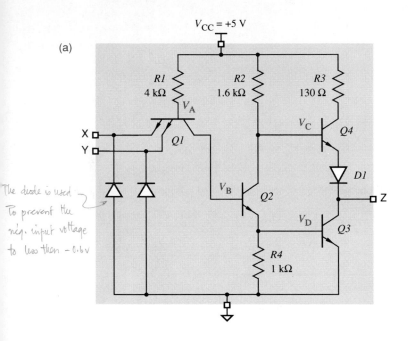

The diode is used —
To prevent the
neg. input voltage
to less than −0.6v

Figure 2–29
Two-input TTL NAND gate:
(a) circuit diagram; (b) function
table; (c) truth table; (d) logic
symbol.

(d)

(b)

X	Y	$Q1$	V_A	V_B	$Q2$	V_C	V_D	$Q3$	$Q4$	Z
L	L	on	≤1.4	≤1.0	off	≥3.6	≤0.4	off	on	H
L	H	on	≤1.4	≤1.0	off	≥3.6	≤0.4	off	on	H
H	L	on	≤1.4	≤1.0	off	≥3.6	≤0.4	off	on	H
H	H	off	1.8	1.2	on	0.8	0.6	on	off	L

(c)

X	Y	Z
0	0	1
0	1	1
1	0	1
1	1	0

TTL NAND gates can be designed with any desired number of inputs simply by changing the number of emitters on $Q1$ in Figure 2–29. Commercially available TTL NAND gates have as many as 13 inputs. A TTL inverter is designed as a 1-input NAND gate, and uses an ordinary single-emitter transistor for $Q1$.

We'll use the following definitions of LOW and HIGH in our analysis of TTL circuit behavior:

LOW 0–0.8 volts.

HIGH 2.0–5.0 volts.

Figure 2–30 shows the situation when all inputs of the TTL NAND gate are HIGH. In this case, almost no current flows in any base-to-emitter junction of $Q1$ (the leakage current in a 74-series TTL gate is specified to be no more than 40 μA per input in this case). However, the base-to-collector junction is forward biased by the V_{CC} potential through $R1$, an unusual but legitimate mode of operation for an NPN transistor (i.e., it's a hack). Thus, current flows from V_{CC} through

Figure 2–30
TTL NAND gate with both inputs
HIGH.

$R1$, through the base-to-collector junction of $Q1$ and into the base of $Q2$, so $Q2$ is ON. Current from the emitter of $Q2$ then provides base current for $Q3$, so $Q3$ is ON, too.

Each turned-on transistor in this situation has one diode-drop in its base-to-emitter junction, so V_D must be 0.6 V and V_B must be 1.2 V. There is also one diode-drop in the forward-biased base-to-collector junction of $Q1$, so V_A is 1.8 V. Recalling that the inputs X and Y are both at 2.0 V, we see that the base-to-emitter junctions of $Q1$ are indeed reverse-biased and no current flows in them, as we claimed at the beginning of this discussion.

The characteristics of the transistors and the resistor values in Figure 2–30 are such that $Q2$ and $Q3$ are saturated. Thus, V_C is about 0.2 V ($V_{CE(sat)}$) above V_D, or 0.8 V. In order to turn on $Q4$, V_C would have to be 1.4 V ($V_{CE(sat)(Q3)}$ + one diode-drop + $V_{BE(Q4)}$), and since it's not, $Q4$ must be OFF. Thus, when all of the TTL NAND gate's inputs are HIGH, $Q4$ is out of the picture and $Q3$ pulls the output LOW.

As shown in Figure 2–31, if any logic input is LOW, then the corresponding base-to-emitter junction is forward biased. Current flows from V_{CC} through $R1$, through the base-to-emitter junction of $Q1$, *out of* the input lead and into ground through whatever device is pulling that input LOW. (Thus, any device that wishes to drive a TTL input LOW must be prepared to *sink* current from the TTL input.)

Once again, the transistor characteristics and resistor values are chosen so that $Q1$ is saturated. Therefore, V_B is only about 0.2 V higher than the lowest

Figure 2–31
TTL NAND gate with one input LOW.

input voltage, or 1.0 V with the worst-case LOW voltage. Since about 1.2 V would be required at point B to turn on $Q2$ and $Q3$, they are OFF. Since $Q2$ is OFF, resistor $R2$ can pull V_C high, turning on $Q4$. Thus, when any of the TTL NAND gate's inputs is LOW, $Q3$ is out of the picture and $Q4$ pulls the output HIGH.

So far we have shown the input signals to a TTL gate as ideal voltage sources. Figure 2–32 shows the situation when a TTL input is driven LOW by the output of another TTL gate. Transistor $Q3A$ in the driving gate is ON, and thereby provides a path to ground for the current flowing out of the emitter of $Q1B$ in the driven gate. When current flows *into* a TTL output in the LOW state, as in this case, the output is said to be *sinking current*.

sinking current

Figure 2–33 shows the same circuit with a HIGH output. In this case, $Q4A$ in the driving gate is turned on enough to supply the small amount of leakage current required by $Q1B$ and reverse-biased protection diode in the driven gate. When current flows *out of* a TTL output in the HIGH state, the output is said to be *sourcing current*.

sourcing current

totem pole

The output structure consisting of $Q3$ and $Q4$ is called a *totem pole*. $Q4$ is sometimes replaced in this structure by a "Darlington transistor pair" to increase the HIGH output drive capability, as we'll see later.

Among the transistors in the TTL NAND gate, $Q4$ is the only one for which the component values are chosen so that it does *not* saturate in normal operation. Saturation is not necessary since the TTL HIGH level doesn't have to be all that

Figure 2–32
A TTL output driving
a TTL input LOW.

Figure 2–33
A TTL output driving
a TTL input HIGH.

high, and a nonsaturated transistor switches faster than a saturated one. We'll re-address these issues later, when we look at Schottky technology.

Since the totem-pole transistors *Q3* and *Q4* are normally complementary—one ON and the other OFF—you might question the purpose of the 130 Ω resistor *R3* in the totem pole. A value of 0 Ω would give even better driving capability in the HIGH state. This is certainly true from a DC point of view. However, when the TTL output is changing from HIGH to LOW or vice versa, there is a short time when both transistors may be on. The purpose of *R3* is to limit the amount of current that flows from V_{CC} to ground during this time.

current spike

Even with the 130 Ω resistor, higher-than-normal currents called *current spikes* flow when TTL totem-pole outputs are switched. Current spikes show up as noise on the power-supply and ground connections in a TTL circuit, especially when multiple outputs are switched simultaneously. For this reason, reliable TTL circuits require decoupling capacitors between V_{CC} and ground, distributed throughout the circuit so that there is a capacitor within an inch or so of each chip to supply the current needed during transitions. The large capacitors typically found in the power supply itself don't satisfy this requirement, because stray wiring inductance prevents them from supplying the current fast enough; hence the need for a *physically distributed* system of decoupling capacitors.

Since TTL input voltages are normally positive, the diodes between the input leads in Figure 2–31 and ground are normally reverse biased, and no current flows in them. However, if a negative voltage is applied, the diodes conduct and prevent the input voltage from becoming more negative than one diode-drop (approximately −0.6 V). Although one would never purposely apply a negative DC voltage to a TTL input, negative voltages can appear due to transmission-line effects during switching, and the diodes help to reduce these effects.

2.9.2   Logic Levels and Noise Margins

The input-output transfer characteristic of a TTL NAND gate or inverter under typical conditions is shown in Figure 2–34. This curve assumes that all but one of the inputs are tied to a "perfect" HIGH signal (e.g., 5.0 V), and the voltage on the remaining input is varied from 0 to 5 V, as shown on the X axis. The Y axis plots the output voltage of the gate under these conditions. If we believed this curve, we could define a TTL LOW level as any voltage under 1.2 V, and a HIGH level as anything over 1.5 V. Only when the input is between 1.2 and 1.5 V does the gate produce a nonlogic voltage under this definition.

Unfortunately, the typical transfer characteristic shown in Figure 2–34 is just that—typical, but not guaranteed. It varies greatly under different conditions of power supply voltage, temperature, and output loading. The transfer characteristic may even vary depending on when the device was fabricated. For example, after months of trying to figure out why gates made on some days were

Figure 2–34
Typical input-output transfer
characteristic of a TTL inverting
gate.

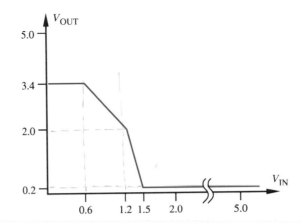

good and on other days were bad, one manufacturer discovered that the bad gates
were victims of airborne contamination by a particularly noxious perfume worn
by one of its production-line workers!

Sound engineering practice dictates that we make more conservative specifications for LOW and HIGH. The conservative specs for standard TTL are depicted
in Figure 2–35. The parameter V_{ILmax} denotes the maximum input voltage guaranteed to be recognized as a LOW, and is 0.8 V for standard TTL. V_{IHmin}, the
minimum input voltage guaranteed to be recognized as a HIGH, is 2.0 V.

    $V_{IL.max}$
    V_{IHmin}

Analysis of the TTL gate's output structure yields two more parameters.
V_{OLmax} is the maximum output voltage in the LOW state, 0.4 V for standard
TTL. V_{OHmin}, the minimum output voltage in the HIGH state, is 2.4 V. All of the
parameters in Figure 2–35 are guaranteed by TTL manufacturers over a range of
temperature, power supply voltage, and output loading.

    V_{OLmax}
    V_{OHmin}

At 0.8 V, V_{ILmax} exceeds V_{OLmax} by 0.4 V, so TTL has a *DC noise margin*
of 0.4 V in the LOW state. That is, it takes at least 0.4 V of noise to corrupt a
worst-case LOW output into a voltage that is not guaranteed to be recognizable
as a LOW input. There is a DC noise margin of 0.4 V in the HIGH state also.

    DC noise margin

Figure 2–35
TTL logic levels.

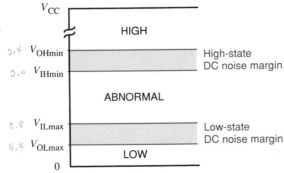

2.9.3 Fanout

fanout

Fanout is a measure of the number of gate inputs that are connected to (and driven by) a single gate output. In TTL circuits, fanout is limited by the ability of an output to supply current to the inputs that it drives, as we'll now explain.

current flow

By convention, the *current flow* in a TTL input or output lead is defined to be positive if the current actually flows *into* the lead, and negative if current flows *out of* the lead. As a result, when a TTL output is connected to one or more TTL inputs, the algebraic sum of all the input and output currents is 0.

I_{ILmax}

In the LOW state, the current required by a TTL input is guaranteed to be no more than I_{ILmax}, -1.6 mA for standard TTL. Recall from the discussion of Figure 2–32 that positive current is actually flowing from V_{CC}, through $R1$, through the emitter of $Q1$, *out of* the input lead (hence the minus sign), through the driving circuit, and into ground. I_{ILmax} is sometimes called a *LOW-state unit load*.

LOW-state unit load

I_{OLmax}

LOW-state fanout

A standard TTL output, on the other hand, is guaranteed to be capable of sinking I_{OLmax} of current while maintaining a valid LOW output level. For a standard TTL output, $I_{OLmax} = 16$ mA. Thus, a standard TTL output can drive up to 10 standard TTL inputs in the LOW state—the *LOW-state fanout* is 10. If the output is overloaded and is forced to sink more than 16 mA, the output voltage V_{OL} may rise above V_{OLmax} (0.4 V), because higher current means a higher voltage drop across $R_{CE(sat)}$. This reduces the LOW-state DC noise margin.

If a TTL output is forced to sink a lot more than I_{OLmax} (say, 50 mA or more), the device may be damaged, especially if high current is allowed to flow for more than a second or so. For example, suppose that a TTL output in the LOW state is short-circuited directly to the 5 V supply. The ON resistance, $R_{CE(sat)}$, of the saturated $Q3$ transistor in the TTL totem pole is about 10 Ω. Thus, $Q3$ must dissipate about $5^2 / 10$ or 2.5 watts. Don't try this yourself unless you're prepared to deal with the consequences! That's enough heat to destroy the device (and burn your finger) in a very short time.

I_{IHmax}

In the HIGH state, a TTL input requires no more than I_{IHmax} of current flowing into it, 40 µA for standard TTL. This is the leakage current flowing into the emitter of the cut-off transistor $Q1$ in Figure 2–33 and through the reverse-biased protection diode into ground. I_{IHmax} is sometimes called a *HIGH-state unit load*.

HIGH-state unit load

I_{OHmax}

HIGH-state fanout

A TTL output is guaranteed to be able to source at least I_{OHmax} of current in the HIGH state without V_{OH} falling below V_{OHmin}. The spec for standard TTL is $I_{OHmax} = -400$ µA; the *HIGH-state fanout* is 10, the same as LOW-state fanout.

Although the numerical fanouts for both states are equal, TTL has quite an asymmetry in current driving capability—a standard TTL output can sink 16 mA in the LOW state, but can source only 400 µA in the HIGH state. This asymmetry is no problem when TTL outputs drive other TTL inputs, because it is matched by a corresponding asymmetry in TTL input current requirements (I_{ILmax} is large, while I_{IHmax} is small). However, it *is* a problem when TTL is used to drive light-emitting diodes (LEDs), relays, solenoids, or other devices requiring large amounts of current (tens of milliamperes). Circuits using these devices must be designed so that current flows (and the driven device is "on") when the TTL output is in the LOW state, and little or no current flows in the HIGH state. Special TTL buffer/driver gates are made that can sink up to 60 mA in the LOW state, but that still have a rather puny current sourcing capability in the HIGH state (2.4 mA).

Other TTL families have different output voltage levels and input and output currents, as described later. When dealing with a single family, you can count unit loads to determine whether or not fanout capacity has been exceeded. However, when you interconnect gates from different families, you must add the input current requirements and compare with output drive capability in both states (HIGH and LOW) to check for fanout violations.

2.9.4   Effects of Loading

Loading an output with more than its rated fanout has several effects:

- In the LOW state, the output voltage (V_{OL}) may increase beyond its rated maximum.

- In the HIGH state, the output voltage (V_{OH}) may fall below its rated minimum.

- The propagation delay from input to output may increase beyond its rated maximum.

- The rise and fall times of LOW-to-HIGH and HIGH-to-LOW transitions may increase beyond their rated maximums.

- The operating temperature of the device may increase, thereby reducing the reliability of the device and eventually causing device failure.

The first four effects reduce the DC noise margins and timing margins of the circuit. Thus, a slightly overloaded circuit may work properly in ideal conditions, but experience says that it will fail once it's out of the friendly environment of the engineering lab.

2.9.5   Unused Inputs

Sometimes not all of the inputs of a logic gate are used. In a real design problem, you may need an n-input gate but have only an $n+1$-input gate available. Tying together two inputs of the $n+1$-input gate gives it the functionality of an n-input

Figure 2–36 Unused NAND input: (a) tied to another input; (b) pulled up; (c) driven by constant-1 output.

gate. You can convince yourself of this fact intuitively now, or use switching algebra to prove it after you've studied Section 3.1. Figure 2–36(a) shows a NAND gate with its inputs tied together.

floating input

Analysis of the TTL input structure shows that unused inputs left unconnected (or *floating*) behave as if they have a HIGH voltage applied—they are pulled HIGH by base resistor $R1$ in Figure 2–29. However, $R1$'s pull-up is much weaker than that of a TTL output driving the input. As a result, a small amount of circuit noise, such as that produced by other gates when they switch, can make a floating input spuriously behave like it's LOW. Therefore, unused TTL inputs should be tied to a stable HIGH or LOW voltage source.

An excellent LOW source is ground—inputs that need to be LOW, such as unused OR- and NOR-gate inputs, can simply be grounded. For a HIGH source, a direct connection to the 5 V power supply is not recommended, since a transient of over 5.5 V can damage inputs in some TTL families. Instead, unused NAND and AND inputs are normally tied to the 5 V supply through a current-limiting resistor, typically in the range 1–5 kΩ, as shown in Figure 2–36(b). Another possibility is to use the output of a spare inverter or NOR gate whose input is grounded, as in (c).

2.10 OTHER TTL GATE TYPES

Although the NAND gate is the "workhorse" of the TTL family, other types of gates can be built with the same general circuit structure.

2.10.1 NOR Gates

A TTL NAND gate performs the AND function in its multiple-emitter input transistor, and follows the AND with a transistor inverter. The multiple emitters perform logic just like the multiple input diodes in a DTL gate. However, it is also possible to adapt the RTL idea of using parallel transistors to perform logic in a TTL gate. The result is the TTL NOR gate shown in Figure 2–37.

(a)

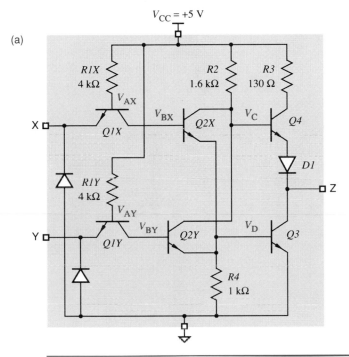

Figure 2–37
Two-input TTL NOR gate:
(a) circuit diagram; (b) function
table; (c) truth table; (d) logic
symbol.

(d)

(b)

X Y	Q1X	V_{AX}	V_{BX}	Q2X	Q1Y	V_{AY}	V_{BY}	Q2Y	V_C	V_D	Q3	Q4	Z
L L	on	≤1.4	≤1.0	off	on	≤1.4	≤1.0	off	≥3.6	≤0.4	off	on	H
L H	on	≤1.4	≤1.0	off	off	1.8	1.2	on	0.8	0.6	on	off	L
H I	off	1.8	1.2	on	on	≤1.4	≤1.0	off	0.8	0.6	on	off	L
H H	off	1.8	1.2	on	off	1.8	1.2	on	0.8	0.6	on	off	L

(c)

X Y	Z
0 0	1
0 1	0
1 0	0
1 1	0

The TTL NOR gate contains a pair of $Q1$ and $Q2$ transistors for each input (e.g., $Q1X$ and Q2X for input X). The $Q2$ transistors for all of the inputs are connected in parallel. Each $Q2$ transistor is ON if its corresponding input is HIGH, and the effect on the rest of the circuit is the same if one or more $Q2$ transistors are ON—an OR function. The output totem pole is identical to that of a TTL NAND gate. Thus, the gate's output is LOW if any $Q2$ is ON, that is, if any input is HIGH; a NOR function is performed.

The speed, input, and output characteristics of a TTL NOR gate are comparable to those of a TTL NAND. However, an n-input NOR gate uses more transistors and resistors and is thus more expensive in silicon area than an n-input NAND. Also, internal leakage current limits the number of $Q2$ transistors that can be placed in parallel, so NOR gates have poor fan-in. (The largest discrete TTL NOR gate has only 5 inputs, compared with a 13-input NAND.) As a result, NOR gates are less commonly used than NAND gates in TTL designs.

(a)

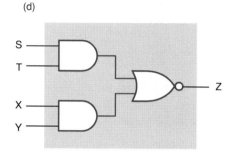

Figure 2–38
Two-wide, two-input TTL AOI gate: (a) circuit diagram; (b) function table; (c) truth table; (d) logic symbol.

(d)

(b)

S	T	X	Y	$Q1S$	V_{AS}	V_{BS}	$Q2S$	$Q1X$	V_{AX}	V_{BX}	$Q2X$	V_C	V_D	$Q3$	$Q4$	Z
L	x	L	x	on	≤1.4	≤1.0	off	on	≤1.4	≤1.0	off	≥3.6	≤0.4	off	on	H
L	x	x	L	on	≤1.4	≤1.0	off	on	≤1.4	≤1.0	off	≥3.6	≤0.4	off	on	H
x	L	L	x	on	≤1.4	≤1.0	off	on	≤1.4	≤1.0	off	≥3.6	≤0.4	off	on	H
x	L	x	L	on	≤1.4	≤1.0	off	on	≤1.4	≤1.0	off	≥3.6	≤0.4	off	on	H
L	x	H	H	on	≤1.4	≤1.0	off	off	1.8	1.2	on	0.8	0.6	on	off	L
x	L	H	H	on	≤1.4	≤1.0	off	off	1.8	1.2	on	0.8	0.6	on	off	L
H	H	L	x	off	1.8	1.2	on	on	≤1.4	≤1.0	off	0.8	0.6	on	off	L
H	H	x	L	off	1.8	1.2	on	on	≤1.4	≤1.0	off	0.8	0.6	on	off	L
H	H	H	H	off	1.8	1.2	on	off	1.8	1.2	on	0.8	0.6	on	off	L

(c)

S	T	X	Y	Z
0	x	0	x	1
0	x	x	0	1
x	0	0	x	1
x	0	x	0	1
0	x	1	1	0
x	0	1	1	0
1	1	0	x	0
1	1	x	0	0
1	1	1	1	0

2.10.2 AND-OR-INVERT Gates

AND-OR-INVERT (AOI) gate

The AND function of a multiple-emitter transistor and the OR function of parallel transistors are combined in a TTL *AND-OR-INVERT (AOI) gate*. Figure 2–38(a) shows the circuit for a two-wide, two-input AOI gate. It is identical to a 2-input NOR, except that each $Q1$ has been changed to a multiple-emitter transistor to perform the AND function. The overall logic function of the AOI gate is equivalent to a pair of AND gates followed by a NOR, as shown in Figure 2–38(d).

The speed, input, and output characteristics of a TTL AOI gate are quite comparable to those of a TTL NAND or NOR. As a result, these gates are very appealing because they can perform two levels of logic (AND-OR) with just one level of delay. Although most logic designers don't bother with AOI gates in their discrete designs, most larger-scale TTL devices use these gates internally.

(a)

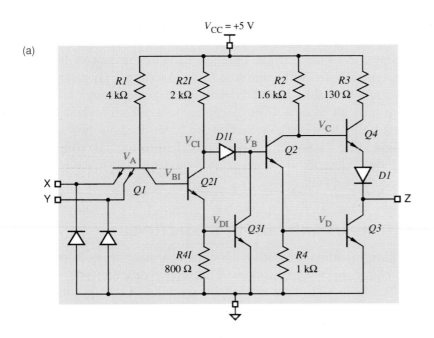

Figure 2–39
Two-input TTL AND gate:
(a) circuit diagram; (b) function
table; (c) truth table; (d) logic
symbol.

(d)

(b)

X Y	$Q1$	V_A	V_{BI}	$Q2I$	V_{CI}	V_{DI}	$Q3I$	V_B	$Q2$	V_C	V_D	$Q3$	$Q4$	Z
L L	on	≤1.4	≤1.0	off	1.8	≤0.4	off	1.2	on	0.8	0.6	on	off	L
L H	on	≤1.4	≤1.0	off	1.8	≤0.4	off	1.2	on	0.8	0.6	on	off	L
H L	on	≤1.4	≤1.0	off	1.8	≤0.4	off	1.2	on	0.8	0.6	on	off	L
H H	off	1.8	1.2	on	0.8	0.6	on	0.2	off	≥3.6	0	off	on	H

(c)

X Y	Z
0 0	0
0 1	0
1 0	0
1 1	1

For example, if you examine the internal logic diagrams of larger-scale TTL
devices in Chapter 4, you'll find that the AOI structure appears quite often.

2.10.3 Noninverting Gates

All of the RTL, DTL, and TTL gates that we've studied so far are *inverting*
gates—they produce the logical complement of an AND, OR, or AND-OR function.
Logical complementation is a natural byproduct of transistor amplifiers used in
these gates. The diode gates of Section 2.3 were *noninverting gates*—simple
ANDs and ORs—but they contained no buffer amplifiers to restore weak logic
levels to strong ones.

    In TTL and most other logic families, to obtain a noninverting gate that
includes a buffer amplifier, we must use an inverting gate followed by an inverter.
(As we'll show formally in Section 3.1, inverting the output of an AND or OR
function twice is equivalent to not inverting it at all.) Commercial TTL AND and
OR gates insert this extra level of inversion between a normal TTL input stage

inverting gate

noninverting gate

and output stage. For example, Figure 2–39 shows a two-input TTL AND gate. Components $Q2I$, $Q3I$, $D1I$, $R2I$, and $R4I$ form the extra inverter.

The extra inverter in a noninverting TTL gate increases its silicon area and propagation delay, typically by 50% compared to an inverting TTL gate. Thus, inverting gates are more desirable than noninverting ones, even though inverting gates may be more difficult to deal with conceptually, especially when lots of them are interconnected. However, in Section 3.3.2, we'll show how certain designs using noninverting gates can easily be modified to use inverting ones, and in Section 4.1 we'll describe documentations standards that make inverted signals a lot easier to understand.

2.11   OTHER TTL INPUT AND OUTPUT STRUCTURES

Circuit designers have modified the basic TTL circuit in many ways to produce gates that are tailored for specific applications. This section describes some of the more common variations in TTL input and output structures. Later, Section 2.12 describes different TTL families whose design parameters and technology have evolved to increase the speed and reduce the power requirements of the basic TTL circuit and its derivatives.

2.11.1   Schmitt-Trigger Inputs

Schmitt-trigger input

The input-output transfer characteristic for a typical TTL gate was shown in Figure 2–34. The corresponding transfer characteristic for a gate with *Schmitt-trigger inputs* is shown in Figure 2–40(a). A Schmitt trigger is a special analog circuit that shifts the switching threshold depending on whether the input is changing from LOW to HIGH or from HIGH to LOW.

For example, suppose the input of a Schmitt-trigger inverter is initially at 0 V, a solid LOW. Then the output is HIGH, typically about 3.4 V. If the input

Figure 2–40

A Schmitt-trigger inverter:
(a) input-output transfer characteristic; (b) logic symbol.

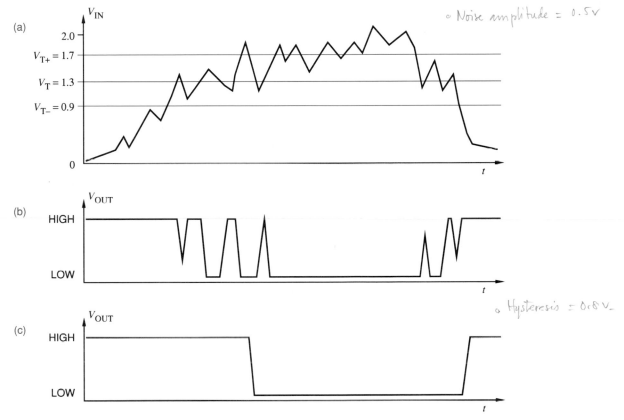

o Noise amplitude = 0.5v

o Hysteresis = 0.8 V

Figure 2–41   Device operation with slowly changing inputs: (a) a noisy, slowly changing input; (b) output produced by an ordinary inverter; (c) output produced
by an inverter with 0.8 V of hysteresis.

voltage is increased, the output will not go LOW until the input voltage reaches
about 1.7 V. However, once the output is LOW, it will not go HIGH again until the
input is decreased to about 0.9 V. Thus, the switching threshold for positive-going
input changes, denoted V_{T+}, is about 1.7 V, and for negative-going input changes,
denoted V_{T-}, is about 0.9 V. The difference between the two thresholds is called
hysteresis. The Schmitt-trigger inverter provides about 0.8 V of hysteresis.   *hysteresis*

    Schmitt-trigger inputs have better noise immunity than ordinary gate inputs
for signals that contain transmission-line reflections, discussed in Section 8.4, or
that have long rise and fall times. Such signals typically occur in physically long
connections, such as input-output buses and computer interface cables. Noise
immunity is important in these applications because long signal lines are more
likely to have reflections or to pick up noise from adjacent signal lines, circuits,
and appliances.

To demonstrate the Schmitt-trigger input's noise immunity, Figure 2–41(a) shows an input signal with long rise and fall times and about 0.5 V of noise on it. An ordinary inverter, without hysteresis, has the same switching threshold for both positive-going and negative-going transitions, $V_T \approx 1.3$ V. Thus, the ordinary inverter responds to the noise as shown in (b), producing multiple output changes each time the noisy input voltage crosses the switching threshold. However, as shown in (c), a Schmitt-trigger inverter does not respond to the noise, because its hysteresis is greater than the noise amplitude.

2.11.2 Darlington Outputs

Darlington transistor pair

Figure 2–42 shows a TTL gate in which the top transistor in the totem pole, $Q4$, has been replaced with a *Darlington transistor pair* consisting of $Q4$ and $Q5$. Transistor $Q4$ provides more base current for $Q5$, reducing its "on" resistance in the HIGH state; in some configurations, $Q5$ is actually driven to saturation. The resistance of $R3$ is also reduced. All of this means that a Darlington output provides more current-sourcing capability than a standard TTL output in the HIGH state, 1200 µA in the circuit shown.

The improved HIGH-state driving capability has two benefits. First, lowering the resistance of $Q5$ and $R3$ decreases t_r, the rise time of LOW-to-HIGH transitions, to a value almost as low as t_f, the fall time of HIGH-to-LOW transitions. Second, the HIGH-state fanout is increased, to 30 for the circuit shown.

What good is increased fanout in the HIGH state only? When two inputs of

Figure 2–42
TTL gate with a Darlington output.

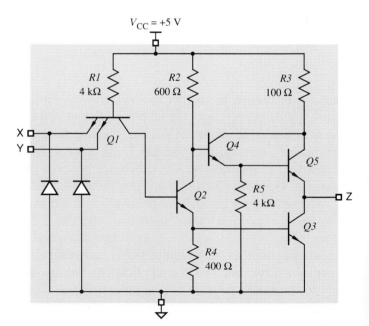

the same gate are tied together, as in Figure 2–36(a), the LOW-state input current is only −1.6 mA and the two tied inputs look like only one LOW-state unit load. However, in the HIGH state, the leakage currents add, and the tied inputs look like two unit loads. Thus, increased HIGH-state fanout allows inputs of the same gate to be tied together without reducing the total number of distinct gates that can be driven.

Although the original TTL family provided Darlington outputs in only a few gate types, most devices in the latest TTL logic families have Darlington outputs because of their benefits.

2.11.3 Open-Collector Outputs

The top transistor, $Q4$, in the TTL totem-pole output structure is said to provide *active pull-up*, since it actively pulls up the output voltage on a LOW-to-HIGH transition. This transistor is omitted in gates with *open-collector outputs*, as shown in Figure 2–43. An external *pull-up resistor* provides *passive pull-up* to the HIGH level, as in the RTL and DTL gates that we studied previously.

Because I_{OLmax} is limited, a relatively large external pull-up resistor (a few hundred ohms or greater) must be used with open-collector TTL outputs. As a result, LOW-to-HIGH output transitions are slower for open-collector gates than for gates with active pull-up. So why use open-collector outputs? They are

active pull-up
open-collector
 output
pull-up resistor
passive pull-up

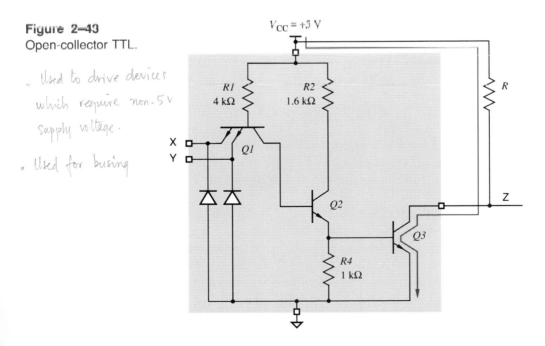

Figure 2–43
Open-collector TTL.

· Used to drive devices which require non-5 V supply voltage.

◦ Used for busing

Figure 2–44

Driving an LED with an open-collector output.

X or Y lo → Q3 OFF → LED OFF

X and Y hi → Q3 ON → LED ON

useful in at least three applications: driving light-emitting diodes (LEDs), relays, and other devices; busing; and performing wired logic.

An open-collector output can drive an LED as shown in Figure 2–44. If either input X or Y is LOW, $Q3$ is turned off and the LED is off. When X and Y are both HIGH, $Q3$ is turned on, the output Z is LOW, and the LED is turned on. The value of the pull-up resistor R is chosen so that the proper amount of current flows through the LED in the "on" state. For example, if the LED requires 10 mA for normal brightness and has a voltage drop of 1.6 V, we choose

$$R = (5.0 - 1.6 - 0.4)\,\text{V} \,/\, 10\,\text{mA} = 300\ \Omega$$

Note that we assumed an open-collector output voltage of 0.4 V, the worst-case specification. If the output can sink 10 mA and maintain a lower voltage, say 0.2 V, then a little more current will flow and the LED will be a little brighter than normal.

Open-collector outputs are very useful for driving relays, solenoids, and other devices that may use a non-5-volt supply. For example, Figure 2–45 shows how an open-collector output can be used to control a 12-volt relay. When the gate inputs are HIGH, $Q3$ is on and sinks current just as it would for an LED, resistor, or any other device. However, when one of the gate inputs is LOW, $Q3$ is off and no current flows. Therefore, the voltage at the Z output pin rises to the relay supply voltage, 12 V.

Figure 2–45
Driving a 12-volt relay with an
open-collector output.

It is important for $Q3$ to be able to withstand the higher potential of the
relay supply voltage without damage. In fact, TTL open-collector gates with
high-voltage outputs are rated to withstand potentials up to 30 V. However, as   *high-voltage output*
shown in the figure, one normally places a reversed-biased protection diode in
parallel with the relay coil to prevent an even higher voltage from appearing
at the output. As $Q3$ is turned off, a high voltage would otherwise be created
temporarily by the *inductive kick*, a transient that occurs when the magnetic field   *inductive kick*
of the relay coil suddenly collapses and its stored energy has nowhere to go.

2.11.4 Wired Logic

If the outputs of several open-collector gates are tied together with a single pull-
up resistor, then *wired logic* is performed. An AND function is obtained since the   *wired logic*
wired output is HIGH if and only if all of the individual gate outputs are HIGH;
any output going LOW is sufficient to pull the wired output LOW.

Note that wired logic cannot be performed using gates with totem-pole outputs. Two totem poles wired together and trying to maintain opposite logic values
result in a very high current flow and an abnormal output voltage. Figure 2–46
shows this situation, which is sometimes called *fighting*. The exact output volt-   *fighting*
age depends on the relative "strengths" of the fighting transistors, but is typically
about 1–1.5 V, a nonlogic voltage to be sure.

Open-collector outputs can be tied together to allow several devices to share   *open-collector bus*

Figure 2–46
Two totem-pole outputs trying to maintain opposite logic values on the same line.

a common bus. At any time all but one of the outputs on the bus are in their HIGH state. The remaining gate either stays in the HIGH state or pulls the bus LOW, depending on whether it wants to transmit a logical 1 or 0 on the bus.

Figure 2–47
Four open-collector outputs
driving two inputs in the LOW
state.

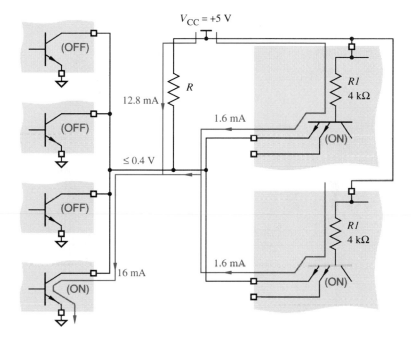

Control circuitry selects the particular device that is allowed to drive the bus at
any time.

A proper choice of value for the pull-up resistor must be made in open-
collector applications. Two calculations are made to bracket the allowable values
of R as follows:

*pull-up resistor
calculation*

Minimum The sum of the current through R in the LOW state and the LOW-
state input currents of the gates driven by the wired outputs must not
exceed the LOW-state driving capability of the active output, 16 mA
for standard TTL.

Maximum The voltage drop across R in the HIGH state must not reduce the output
voltage below V_{OH}, the guaranteed minimum HIGH output voltage
(2.4 V for standard TTL). This drop is produced by the HIGH-state
output leakage current of the wired outputs and the HIGH-state input
currents of the driven gates.

For example, suppose that four open-collector outputs are wired together
and drive two standard TTL inputs. As shown in Figure 2–47, to stay within the
I_{OLmax} spec of 16 mA, the current through R may be no more than

$$I_R = (16 - 2 \cdot 1.6) = 12.8 \, \text{mA}$$

Figure 2–48
Four open-collector outputs driving two inputs in the HIGH state.

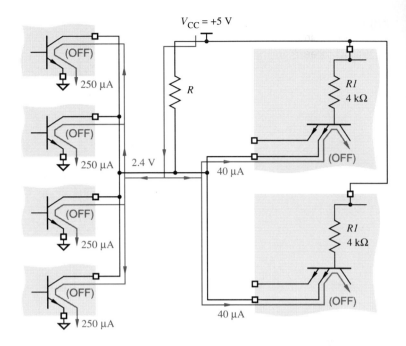

Hence the minimum value of R is given by

$$R = (5.0 - 0.0)/I_R = 391 \ \Omega$$

We assumed an output voltage of 0.0 V rather than 0.4 V (V_{OLmax}) in order to obtain a worst-case result.[5]

In the HIGH state, typical open-collector outputs have a leakage current of 250 μA. Hence, the HIGH-state current requirement as shown in Figure 2–48 is $((4 \cdot 250) + 2 \cdot 40) = 1080$ μA. This current must produce a voltage drop of no more than $(5 - 2.4) = 2.6$ V; thus the maximum value of R is $2.6/0.00108 = 2.407$ kΩ. Hence any value of R between 391 Ω and 2.407 kΩ may be used. Higher values reduce power consumption and improve the LOW-state noise margin, while lower values increase power consumption and improve both the HIGH-state noise margin and the speed of LOW-to-HIGH output transitions.

---

[5]That is, even if the open-collector output is so strong that it can pull the output voltage all the way down to 0.0 V (it's only required to pull down to 0.4 V), we'll never allow it to sink more than 16 mA, so it doesn't get overstressed. Some designers prefer to use 0.4 V in this calculation, figuring that if the output is so good that it can pull lower than 0.4 V, a little bit of excess sink current beyond 16 mA won't hurt.

Figure 2-49
Conceptual circuit for a NAND
gate with three-state output.

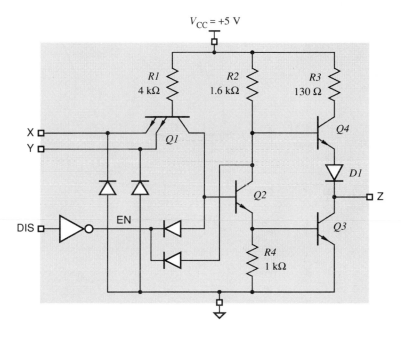

2.11.5 Three-State Outputs

Although open-collector devices may be used for busing as described in the preceding section, much more popular are devices with *three-state outputs*, also called *tri-state outputs*.[6] Three-state devices have an extra input, usually called "output enable" or "output disable," for placing the device's output(s) in a high-impedance state. Thus, an output can have one of three states—logic 0, logic 1, and *high impedance* (*Hi-Z* or *floating*). In the Hi-Z state, the output behaves as if it isn't even connected to the circuit, except for a small leakage current that may flow into or out of the output pin.

three-state outputs
tri-state outputs

*high-impedance
  state*
Hi-Z state
floating state
three-state bus

    A *three-state bus* is created by wiring several three-state outputs together. Control circuitry for the "output enables" must ensure that at most one output is enabled (not in its Hi-Z state) at any time. The single enabled device can transmit logic levels (HIGH and LOW) on the bus.

    Figure 2–49 shows roughly how three-state control of a NAND gate totem-pole output could be accomplished. When the DIS (disable) input is LOW, the gate functions normally. When DIS is HIGH, EN is LOW and pulls down the bases of *Q2* and *Q4*. This cuts off both of these transistors, and since *Q2* is cut off, so is *Q3*. Thus, both transistors in the totem pole are off, and the output is in the Hi-Z state.

    The circuit in Figure 2–50 is a little closer to a real circuit with three-state

---

[6]Tri-state is a trademark of National Semiconductor Corporation.

Figure 2–50

Three-state NAND gate.

control of a NAND gate totem-pole output. It uses three extra transistors $Q5$, $Q6$, and $Q7$, that act as the inverter for DIS. Instead of pulling down the base of $Q2$ directly, this circuit uses an extra emitter on $Q1$ to ensure that $Q2$ is off if DIS is HIGH.

In practice, the three-state control circuit may be more complex than we have shown, in order to provide adequate speed and noise margins on the control input, and proper dynamic behavior of the totem-pole output during transitions to and from the Hi-Z state. In particular, devices with three-state outputs are normally designed so that the output enable delay (Hi-Z to LOW or HIGH) is somewhat longer than the output disable delay (LOW or HIGH to Hi-Z). Thus if a control circuit activates one device's output-enable input at the same time that it deactivates a second's, the second device is guaranteed to enter the Hi-Z state before the first places a HIGH or LOW level on the bus.

If two three-state outputs on the same bus are enabled at the same time and try to maintain opposite states, the situation is similar to tying standard totem-pole outputs together—a nonlogic voltage is produced on the bus. The devices probably will not be damaged, but the large current drain through the tied outputs can produce noise pulses that affect circuit behavior elsewhere in the system.

There is a leakage current of typically 40 µA associated with a three-state output in its Hi-Z state. This current, as well as the input currents of receiving gates, must be taken into account when calculating the maximum number of devices that can be placed on a three-state bus.

In the LOW state, an enabled three-state output must be capable of sinking 40 μA of leakage current for every other three-state output on the bus, as well as sinking 1.6 mA for every standard TTL input on the bus. In the HIGH state, an active output must be capable of sourcing 40 μA of leakage current for every other three-state output on the bus, as well as 40 μA for every standard TTL input on the bus. As with standard TTL, separate LOW-state and HIGH-state calculations must be made to ensure that the fanout requirements of a particular circuit configuration are met.

2.12 TTL FAMILIES

TTL families have evolved over the years in response to the demands of digital designers for better performance. As a result, three TTL families have come and gone, and today's designers have at least four newer families from which to choose. All of the TTL families are compatible in that they use the same power supply voltage and logic levels, but each family has its own advantages in terms of speed, power consumption, and cost.

2.12.1 Early TTL Families

The original TTL family of logic gates was introduced by Sylvania in 1963. It was popularized by Texas Instruments, whose part numbers for gates and other TTL components have become an industry standard. For example, four 2-input TTL NAND gates of the type shown in Figure 2-29 are contained in a single package with the part number 7400.

Devices in a given TTL family have part numbers of the form 74FAM*nn*, where "FAM" is an alphabetic family mnemonic and *nn* is a numeric function designator. Devices in different families with the same value of *nn* perform the same function. In the original TTL family, "FAM" is null and the family is called *74-series TTL*.

74-series TTL

Resistor values in the basic TTL circuit can be changed to obtain gates with different performance characteristics. For example, lowering the resistor values reduces the time constants associated with signal transitions and therefore speeds up the circuit. By doing this, the *74H (High-speed TTL)* family achieves about a 33% reduction in the propagation delay of typical NAND gates (which have part number 74H00). Unfortunately, lowering the resistor values also increases the gate's power consumption, more than doubling it.

74H (High-speed TTL)

Going in the other direction, the *74L (Low-power TTL)* family increases resistor values to achieve a 90% reduction in power consumption. Speed suffers, however; the propagation delay of 74L TTL is almost four times longer than that of 74-series TTL.

74L (Low-power TTL)

The availability of three TTL families allowed digital designers in the 1970s to make a choice between high speed and low power consumption for their circuits. However, like many people in the 1970s, they wanted to "have it all, now." The development of Schottky transistors provided this opportunity, and made 74, 74H, and 74L TTL obsolete.

2.12.2 Schottky Transistors

Schottky diode

Schottky-clamped transistor

Schottky transistor

Transistors *Q1*, *Q2*, and *Q3* in the standard TTL gate, Figure 2–29, saturate in normal operation. Storage time, the time it takes for a transistor to come out of saturation, accounts for much of the propagation delay of standard TTL. Saturation can be eliminated by placing a *Schottky diode* between the base and collector of each saturating transistor, as shown in Figure 2–51. The resulting transistors, which do not saturate, are called *Schottky-clamped transistors* or *Schottky transistors* for short.

When forward biased, a Schottky diode's voltage drop is much less than a standard diode's, 0.25 V vs. 0.6 V. In a standard saturated transistor, the base-to-collector voltage is 0.4 V, as shown in Figure 2–52(a). In a Schottky transistor, the Schottky diode shunts current from the base into the collector before the transistor goes into saturation, as shown in (b).

An undesirable side effect of using Schottky-clamped transistors is that the collector-to-emitter voltage of the "almost saturated" transistor is 0.35 V, which

Figure 2–51
Schottky-clamped transistor:
(a) circuit; (b) symbol.

Figure 2–52
Operation of a transistor with large base current: (a) standard saturated transistor; (b) transistor with Schottky diode to prevent saturation.

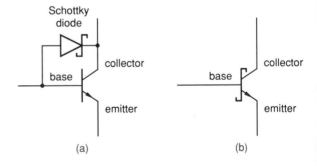

is 0.15 V higher than 0.2 V drop of a standard saturated transistor. Thus, the LOW-state noise margin provided by a Schottky output is less than that provided by a standard TTL output.

2.12.3 Schottky TTL Families

Historically, the first family to make use of Schottky transistors was *74S (Schottky TTL)*. With Schottky transistors and low resistor values, this family has much higher speed, but higher power consumption, than 74-series TTL.

74S (Schottky TTL)

Probably the most widely used TTL family is *74LS (Low-power Schottky TTL)*, introduced shortly after 74S. By combining Schottky transistors with higher resistor values, 74LS TTL matches the speed of 74-series TTL but has about one-fifth of its power consumption. Thus, 74LS TTL is a preferred logic family for new designs.

74LS (Low-power Schottky TTL)

Recently, IC processing and circuit innovations have produced two more Schottky logic families. The *74AS (Advanced Schottky TTL)* family offers speeds approximately twice as fast as 74S with approximately the same power consumption. The *74F (Fast TTL)* family from Fairchild Semiconductor is quite similar to 74AS. The *74ALS (Advanced Low-power Schottky TTL)* family offers both lower power and higher speeds than 74LS, and rivals 74LS in popularity for new designs.

74AS (Advanced Schottky TTL)
74F (Fast TTL)
74ALS (Advanced Low-power Schottky TTL)

2.12.4 Characteristics of TTL Families

Important characteristics of the TTL families are summarized in Table 2–5. The first three families in the table are obsolete, while the last four are widely used in contemporary systems.

The first two rows of the table list the propagation delay (in nanoseconds) and the power consumption (in milliwatts) of a typical 2-input NAND gate in each family. One figure of merit of a logic family is its *speed-power product* listed in the third row. This is simply the product of the propagation delay and power consumption of a typical gate; the result is measured in picojoules (pJ). Recall from physics that the joule is a unit of energy, so the speed-power product measures a sort of efficiency—how much energy a logic gate uses to switch its output once. In this day and age, it's obvious that the lower the energy usage, the better.[7]

speed-power product

The Schottky transistors used in 74S and 74LS TTL provide a better speed-power product than corresponding 74H and 74 devices, and Advanced Schottky

---

[7]There are practical as well as geopolitical reasons for saving energy. Lower energy consumption means lower cost of power supplies and cooling systems. Also, a digital system's reliability is improved more by running it cooler than by any other single reliability improvement strategy.

Table 2–5   Characteristics of TTL families.

Description	Symbol	Family 74	74L	74H	74S	74LS	74AS	74ALS
Typical propagation delay (ns)		9	33	6	3	9	1.6	5
Power consumption per gate (mW)		10	1	22	20	2	20	1.3
Speed-power product (pJ)		90	33	132	60	18	32	6.5
LOW-level input voltage (V)	V_{ILmax}	0.8	0.7	0.8	0.8	0.8	0.8	0.8
LOW-level output voltage (V)	V_{OLmax}	0.4	0.4	0.4	0.5	0.5	0.5	0.5
HIGH-level input voltage (V)	V_{IHmin}	2.0	2.0	2.0	2.0	2.0	2.0	2.0
HIGH-level output voltage (V)	V_{OHmin}	2.4	2.4	2.4	2.7	2.7	2.7	2.7
LOW-level input current (mA)	I_{ILmax}	−1.6	−0.18	−2.0	−2.0	−0.4	−2.0	−0.2
LOW-level output current (mA)	I_{OLmax}	16	3.6	20	20	8	20	8
HIGH-level input current (µA)	I_{IHmax}	40	10	50	50	20	20	20
HIGH-level output current (µA)	I_{OHmax}	−400	−200	−500	−1000	−400	−2000	−400

[handwritten margin notes:]
To compute noise margins

To compute fan out

$$\text{Fan out} = \frac{I_{OH\,max}}{I_{IH\,max}}$$

$$NM_L = V_{IL} - V_{OL}$$

$$NM_H = V_{OH\,min} - V_{IH\,min}$$

cost

processing improves 74AS and 74ALS devices even more. However, notice that the speed-power product is a function not only of the basic transistor technology, but also of the resistor values used in each family. For example, even though 74, 74H, and 74L TTL use the same transistors, their speed-power products are different. The fastest family, 74H, has the worst speed-power product. The same observation applies to 74S-74LS and 74AS-74ALS TTL. Evidently, for a given transistor technology, we must sacrifice efficiency to obtain the ultimate in performance.

Another figure of merit of a logic family, especially for accountants, is its *cost*. The advanced families with their better speed-power products have higher costs, which explains why they are not used universally. Ironically, the obsolete families, with their pathetic speed-power products, have *very* high costs (oops, prices), because they're used nowadays only in obsolete military equipment. But that's another story.

The remaining rows in Table 2–5 describe the input and output parameters of typical TTL gates in each of the families. Using this information, you can analyze the external behavior of TTL signals without knowing the details of the internal TTL circuit design. In other words, you can forget a lot of what you learned in the preceding three sections! The parameters are defined as follows:

V_{ILmax} The maximum voltage that an input is guaranteed to recognize as LOW.

V_{OLmax} The maximum voltage that a LOW output is guaranteed to produce (as long as I_{OLmax} is not exceeded).

V_{IHmin} The minimum voltage that an input is guaranteed to recognize as HIGH.

V_{OHmin} The minimum voltage that a HIGH output is guaranteed to produce (as long as I_{OHmax} is not exceeded).

I_{ILmax} The maximum current that an input requires to pull it LOW. Since this is a negative value, current flows *out of* the input pin.

I_{OLmax} The maximum current that an output can supply in the LOW state while maintaining a voltage less than or equal to V_{OLmax}. Since this is a positive value, current flows *into* the output pin.

I_{IHmax} The maximum current that an input requires to pull it HIGH. Since this is a positive value, current flows *into* the input pin.

I_{OHmax} The maximum current that an output can supply in the HIGH state while maintaining a voltage greater than or equal to V_{OHmin}. Since this is a negative value, current flows *out of* the output pin.

The voltage parameters in Table 2–5 determine DC noise margins. Since the parameters for all four Schottky families are the same, so are their margins. The LOW-state DC noise margin is the difference between V_{OLmax} and V_{ILmax}, or 0.3 V. The HIGH-state DC noise margin is the difference between V_{OHmin} and V_{IHmin}, or 0.7 V. In general, when different families are interconnected, you have to compare V_{OLmax} and V_{OHmin} of the driving gate with V_{ILmax} and V_{IHmin} of all the driven gates to determine the worst-case noise margins.

The current parameters in Table 2–5 determine fanout capability, and are especially important when an output drives inputs in one or more different families. Two calculations must be performed to determine whether an output is operating within its rated fanout capability:

HIGH-state fanout The I_{IHmax} values for all of the driven inputs are added. The sum must be less than I_{OHmax} of the driving output.

LOW-state fanout The I_{ILmax} values for all of the driven inputs are added, except that multiple inputs of the *same* NAND gate are counted only once. The sum must be less than I_{OLmax} of the driving output.

Note that the input and output characteristics of specific components may vary from the representative values given in Table 2–5, so you must always consult the manufacturer's data book when analyzing a real design.

Figure 2–53 shows the format of a typical manufacturer's data sheet for the 74LS00. The 54LS00 listed in the data sheet is identical to the 74LS00, except

SN54LS00, SN74LS00 QUADRUPLE 2-INPUT POSITIVE-NAND GATES

absolute maximum ratings over operating free-air temperature

Supply voltage, V_{CC} .... 7 V
Input voltage .... 7 V
Operating free-air temperature range: SN54 .... $-55°$ C to $125°$ C
                                       SN74 .... $0°$ C to $70°$ C
Storage temperature range .... $-65°$ C to $150°$ C

recommended operating conditions

		54LS00			74LS00			UNIT
		MIN	NOM	MAX	MIN	NOM	MAX	
V_{CC}	Supply voltage	4.5	5	5.5	4.75	5	5.25	V
V_{IH}	High-level input voltage	2			2			V
V_{IL}	Low-level input voltage			0.7			0.8	V
I_{OH}	High-level output current			-0.4			-0.4	mA
I_{OL}	Low-level output current			4			8	mA
T_A	Operating free-air temperature	-55		125	0		70	°C

electrical characteristics over recommended free-air temperature range ‡

PARAMETER	TEST CONDITIONS †	54LS00			74LS00			UNIT
		MIN	TYP‡	MAX	MIN	TYP‡	MAX	
V_{OH}	$V_{CC} = $ MIN, $V_{IL} = $ MAX, $I_{OH} = -0.4$ mA	2.5	3.4		2.7	3.4		V
V_{OL}	$V_{CC} = $ MIN, $V_{IH} = 2$ V, $I_{OL} = 4$ mA		0.25	0.4		0.25	0.4	V
	$V_{CC} = $ MIN, $V_{IH} = 2$ V, $I_{OL} = 8$ mA					0.35	0.5	V
I_I	$V_{CC} = $ MAX $V_I = 7$ V			0.1			0.1	mA
I_{IH}	$V_{CC} = $ MAX $V_I = 2.7$ V			20			20	µA
I_{IL}	$V_{CC} = $ MAX $V_I = 0.4$V			-0.4			-0.4	mA
I_{OS} §	$V_{CC} = $ MAX	-20		-100	-20		-100	mA
I_{CCH}	$V_{CC} = $ MAX $V_I = 0$ V		0.8	1.6		0.8	1.6	mA
I_{CCL}	$V_{CC} = $ MAX $V_I = 4.5$ V		2.4	4.4		2.4	4.4	mA

† For conditions shown as MIN or MAX, use the appropriate value specified under recommended operating conditions.
‡ Except typical values, which are at $V_{CC} = 5$ V, $T_A = 25$ °C.
§ Not more than one output should be shorted at a time, and the duration of the short should not exceed one second.

switching characteristics, $V_{CC} = 5$ V, $T_A = 25$ °C

PARAMETER	FROM (INPUT)	TO (OUTPUT)	TEST CONDITIONS	MIN	TYP	MAX	UNIT
t_{PLH}	X or Y	Z	$R_L = 2$K ohms, $C_L = 15$ pF		9	12	ns
t_{PHL}	X or Y	Z	$R_L = 2$K ohms, $C_L = 15$ pF		10	15	ns

Figure 2–53 Typical manufacturer's data sheet for 74LS00.

that it is specified to operate over the full "military" temperature and voltage range, and it costs more. Most 74LS, 74S, 74ALS, and 74AS parts have corresponding 54-series versions. The data sheet has four sections:

- *Absolute maximum ratings* indicate the worst-case conditions for operating or storing the device without damage.

  absolute maximum ratings

- *Recommended operating conditions* specify power-supply, input voltage, DC output loading, and temperature values under which the device is normally operated.

  recommended operating conditions

- *Electrical characteristics* specify additional DC voltages and currents that are observed at the device inputs and output when it is operated under the recommended conditions:

  electrical characteristics

  I_I Maximum input current for a very high HIGH input voltage.

  I_{OS} Output current with HIGH output shorted to ground.

  I_{CCH} Power-supply current when all inputs (on four NAND gates) are HIGH.

  I_{CCL} Power-supply current when all inputs (on four NAND gates) are LOW. (The number given is for the entire package, which contains four NAND gates, so the current per gate is one-fourth of the specified amount.)

- *Switching characteristics* give maximum and typical propagation delays under "typical" operating conditions of $V_{CC} = 5$ V and $T_A = 25°C$. A conservative designer must increase these delays by 5–10% to account for different power-supply voltages and temperatures, and even more under heavy loading conditions.

  switching characteristics

A data book also shows test circuits that are used to measure the parameters when the device is manufactured, and graphs that show how the typical parameters vary with operating conditions such as power-supply voltage (V_{CC}), ambient temperature (T_A), and load (R_L, C_L).

2.13 CURRENT-MODE LOGIC (CML)

The key to reducing propagation delay in a bipolar logic family is to prevent a gate's transistors from saturating. In the preceding section, we learned how Schottky diodes prevent saturation in TTL gates. However, it is also possible to prevent saturation by using a radically different circuit structure, called *current-mode logic (CML)*.

current-mode logic (CML)

Unlike the other logic families in this chapter, CML does not produce a large voltage swing between the LOW and HIGH levels. Instead, it has a small voltage swing, less than a volt, and it internally switches current between two possible paths, depending on the output state.

The first CML logic family was introduced by General Electric in 1961. The concept was refined by Motorola and others to produce today's 10K and 100K *emitter-coupled logic (ECL)* families. These ECL families are extremely fast, offering propagation delays as short as 1 ns. In fact, throughout the evolution of digital circuit technology, some type of CML has always been the fastest commercial logic family.

emitter-coupled logic (ECL)

Still, commercial ECL families aren't nearly as popular as TTL and CMOS, mainly because they consume much more power. In fact, high power consumption has made the design of ECL supercomputers, such as the Cray-1 and Cray-2, as much of a challenge in cooling technology as in digital design. Also, ECL has a poor speed-power product, does not provide a high level of integration, and is not directly compatible with TTL and CMOS. Nevertheless, proponents of ECL believe that, as the various digital techologies evolve, some kind of ECL will continue to be the family of choice for those who require maximum performance regardless of cost.

2.13.1 Basic CML Circuit

The basic idea of current-mode logic is illustrated by the inverter/buffer circuit in Figure 2–54. This circuit has both an inverting output (OUT1) and a noninverting output (OUT2). Two transistors are connected as a *differential amplifier* with a common emitter resistor. The supply voltages for this example are $V_{CC} = 5.0$,

differential amplifier

Figure 2–54
Basic CML inverter/buffer circuit with input HIGH.

Figure 2–55
Basic CML inverter/buffer circuit with input LOW.

V_{BB} = 4.0, and V_{EE} = 0 V, and the input LOW and HIGH levels are defined to be 3.6 and 4.4 V. This circuit produces output LOW and HIGH levels that are 0.6 V higher (4.2 and 5.0 V), but we'll deal with that problem later.

When V_{IN} is HIGH, as shown in the figure, transistor $Q1$ is on, but not saturated, and transistor $Q2$ is OFF. This is true because of a careful choice of resistor values and voltage levels, as shown in the following analysis. Since $Q1$ is on, V_E is one diode-drop lower than V_{IN}, or 3.8 V. Therefore, the current through $R3$ is 3.8/1.3 kΩ = 2.92 mA. If $Q1$ has a β of about 10, then about 2.65 mA of this current comes through the collector and $R1$, so the voltage drop across $R1$ is about 0.8 V, and V_{OUT1} is about 4.2 V (LOW). Since the collector-to-emitter voltage of $Q1$ (4.2 − 3.8 = 0.4 V) is greater than $V_{CE(sat)}$, $Q1$ is not saturated. $Q2$ is OFF because its base-to-emitter voltage (4.0 − 3.8 = 0.2 V) is less than 0.6 V. Thus, V_{OUT2} is pulled to 5.0 V (HIGH) through $R2$.

When V_{IN} is LOW, as shown in Figure 2–55, transistor $Q2$ is on, but not saturated, and transistor $Q1$ is OFF, as we'll show by an analysis similar to the foregoing. Since $Q2$ is on, V_E is one diode-drop lower than V_{BB}, or 3.4 V. Therefore, the current through $R3$ is 3.4/1.3 kΩ = 2.62 mA. If $Q2$ has a β of about 10, then about 2.38 mA of this current comes through the collector and $R2$, so the voltage drop across $R2$ is about 0.8 V, and V_{OUT2} is about 4.2 V. Since the collector-to-emitter voltage of $Q2$ (4.2 − 3.4 = 0.8 V) is greater than $V_{CE(sat)}$, $Q2$ is not saturated. $Q1$ is OFF because its base-to-emitter voltage (3.6 − 3.4 = 0.2 V) is less than 0.6 V. Thus, V_{OUT1} is pulled to 5.0 V through $R1$.

To perform logic with the basic circuit of Figure 2–55, we simply place

(a)

Figure 2–56
CML 2-input OR/NOR gate:
(a) circuit diagram; (b) function
table; (c) truth table; (d) logic
symbol.

(d)

(b)

X	Y	V_X	V_Y	$Q1$	$Q2$	$Q3$	V_E	V_{OUT1}	V_{OUT2}	OUT1	OUT2
L	L	3.6	3.6	OFF	OFF	on	3.4	5.0	4.2	H	L
L	H	3.6	4.4	OFF	on	OFF	3.8	4.2	5.0	L	H
H	L	4.4	3.6	on	OFF	OFF	3.8	4.2	5.0	L	H
H	H	4.4	4.4	on	on	OFF	3.8	4.2	5.0	L	H

(c)

X	Y	OUT1	OUT2
0	0	1	0
0	1	0	1
1	0	0	1
1	1	0	1

additional transistors in parallel with $Q1$. For example, Figure 2–56 shows a 2-input OR/NOR gate. If any input is HIGH, the corresponding input transistor is active, and V_{OUT1} is LOW (NOR output). At the same time, $Q3$ is OFF, and V_{OUT2} is HIGH (OR output).

Recall that the input levels for the inverter/buffer are defined to be 3.6 and 4.4 V, while the output levels that it produces are 4.2 and 5.0 V. This is obviously a problem. We could put a diode in series with each output to lower it by 0.6 V to match the input levels, but that still leaves another problem—the outputs have poor fanout. A HIGH output must supply base current to the inputs that it drives, and this current creates an additional voltage drop across $R1$ or $R2$, reducing the output voltage (and we don't have much margin to work with). These problems are solved in commercial ECL families, such as the 10K family that we describe next.

2.13.2 ECL 10K Family

The gates in today's most popular ECL family have 5-digit part numbers of the form "10xxx" (e.g., 10102, 10181, 10209), so the family is called *ECL 10K*. This family has the following improvements over the basic ECL circuit described previously: *ECL 10K family*

- An emitter-follower output stage shifts the output levels to match the input levels and provides very high current-driving capability, up to 50 mA per output. It is also responsible for the family's name, "emitter-coupled" logic.

- An internal, temperature- and voltage-compensated bias network provides V_{BB} without the need for a separate, external power supply.

- The family is designed to operate with $V_{CC} = 0$ (ground) and $V_{EE} = -5.2$ V. This improves the family's immunity to power-supply noise, because noise on V_{EE} is a "common-mode" signal that is rejected by the input structure's differential amplifier.

Figure 2–57(a) is the circuit for an ECL OR/NOR gate, one section of a quad OR/NOR gate with part number 10102. A pull-down resistor on each input ensures that if the input is left unconnected, it is treated as LOW. The bias network has component values selected to generate $V_{BB} = -1.29$ V for proper operation of the differential amplifier. Each output transistor, using the emitter-follower configuration, maintains its emitter voltage at one diode-drop below its base voltage, thereby achieving the required output level shift. Figure 2–57(b) summarizes the electrical operation of the gate.

The emitter-follower outputs used in ECL 10K require external pull-down resistors, as shown in the figure. The 10K family is designed to use external rather than internal pull-down resistors for good reason. The rise and fall times of ECL output transitions are so fast (typically 2 ns) that any connection longer than a few inches must be treated as a transmission line, and must be terminated as discussed in Section 8.4. Rather than waste power with an internal pull-down resistor, ECL 10K allows the designer to select an external resistor that satisfies both pull-down and transmission-line termination requirements. The simplest termination, sufficient for short connections, is to connect a resistor in the range of 270 Ω to 2 kΩ from each output to V_{EE}.

Logic LOW and HIGH levels are defined in the ECL 10K family as shown in Figure 2–58. Note that even though the power supply is negative, ECL assigns the names LOW and HIGH to the *algebraically* lower and higher voltages, respectively.

DC noise margins in ECL 10K are much less than in TTL, only 0.155 V in the LOW state and 0.125 V in the HIGH state. However, ECL gates do not need as much noise margin as TTL. Unlike TTL, an ECL gate generates very

(a)

(b)

X	Y	V_X	V_Y	$Q1$	$Q2$	$Q3$	V_E	V_{C2}	V_{C3}	V_{OUT1}	V_{OUT2}	OUT1	OUT2
L	L	−1.8	−1.8	OFF	OFF	on	−1.9	−0.2	−1.2	−0.9	−1.8	H	L
L	H	−1.8	−0.9	OFF	on	OFF	−1.5	−1.2	−0.2	−1.8	−0.9	L	H
H	L	−0.9	−1.8	on	OFF	OFF	−1.5	−1.2	−0.2	−1.8	−0.9	L	H
H	H	−0.9	−0.9	on	on	OFF	−1.5	−1.2	−0.2	−1.8	−0.9	L	H

(c)

X	Y	OUT1	OUT2
0	0	1	0
0	1	0	1
1	0	0	1
1	1	0	1

(d)

Figure 2–57 Two-input 10K ECL OR/NOR gate: (a) circuit diagram; (b) function table; (c) truth table; (d) logic symbol.

Figure 2–58
ECL 10K logic levels.

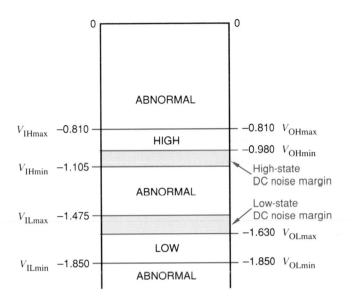

little power-supply and ground noise when it changes state, because its current requirement remains constant as it merely steers current from one path to another. Also, ECL's emitter-follower outputs have very low impedance in either state, and it is difficult to couple noise from an external source into a signal line driven by such a low-impedance output.

A typical ECL 10K gate has a propagation delay of 2 ns, comparable to 74AS TTL. With its outputs left unconnected, a 10K gate consumes about 26 mW of power, also comparable to a 74AS TTL gate, which consumes about 20 mW. However, the termination required by ECL 10K also consumes power, from 10 to 150 mW per output depending on the termination circuit configuration. A 74AS TTL output may or may not require a power-consuming termination circuit, depending on the physical characteristics of the application.

2.13.3 ECL 100K Family

Members of the *ECL 100K family* have 6-digit part numbers of the form "100xxx"   *ECL 100K family*
(e.g., 100101, 100117, 100170), but in general have functions different than 10K parts with similar numbers. The 100K family has the following major differences from the 10K family:

- Reduced power-supply voltage, $V_{EE} = -4.5$ V.

- Shorter propagation delays, typically 0.75 ns.

- Shorter transition times, typically 0.70 ns.

- Higher power consumption, typically 40 mW per gate.

2.14 METAL-OXIDE SEMICONDUCTOR (MOS) LOGIC

In bipolar logic families like TTL and ECL, the number of gates that can be fabricated in a single, integrated circuit is ultimately limited by the power consumed and dissipated by the circuit. That is, if we tried to build a circuit with 10,000 74ALS TTL gates in a 1 cm × 1 cm chip, the circuit would consume about 13 watts of power, most of which would be converted into heat that must be physically dissipated by the chip. Without extraordinary cooling measures (like the refrigeration systems used in Cray supercomputers), this heat would raise the circuit's temperature to 100°C or higher, causing improper operation and ultimately self-destruction.

metal-oxide semiconductor (MOS) transistor

Logic circuits that use *metal-oxide semiconductor (MOS) transistors* have much lower power consumption and dissipation per gate, and are therefore a natural choice for highly integrated chips. In this section, we'll introduce MOS transistors and logic circuits, and then in the next section we'll describe today's popular MOS (actually, CMOS) logic families.

2.14.1 MOS Transistors

An MOS transistor can be modeled as a 3-terminal device that acts like a voltage-controlled resistance. In digital switching applications, an MOS transistor is operated so its resistance is always either very high ("off") or very low ("on").

n-channel MOS (NMOS) transistor
gate
source
drain

The circuit symbol for an *n-channel MOS (NMOS) transistor* is shown in Figure 2–59. The terminals are called *gate*, *source*, and *drain*. (Note that the "gate" of an MOS transistor has nothing to do with a "logic gate.") As you might guess from the orientation of the circuit symbol, the drain is normally at a higher voltage than the source.

The voltage from gate to source (V_{gs}) in an NMOS transistor is normally zero or positive. If $V_{gs} = 0$, then the resistance from drain to source (R_{ds}) is very high, on the order of megohms. As we increase V_{gs} (i.e., increase the voltage on the gate), R_{ds} decreases to a very low value, as low as a few tens of ohms.

p-channel MOS (PMOS) transistor

There are also *p-channel MOS (PMOS) transistors*; their circuit symbol is shown in Figure 2–60. Operation is analogous to that of an NMOS transistor, except that the source is normally at a higher voltage than the drain, and V_{gs} is normally zero or negative. If V_{gs} is zero, then the resistance from source to drain (R_{ds}) is very high. As we algebraically decrease V_{gs}, (i.e., *decrease* the voltage on the gate), R_{ds} decreases to a very low value.

Figure 2–59
Circuit symbol for an *n-channel* MOS (NMOS) transistor.

gate

drain

source

V_{gs}

Voltage-controlled resistance: increase V_{gs} ==> decrease R_{ds}

Note: normally, $V_{gs} \geq 0$

Figure 2–60
Circuit symbol for a *p*-channel
MOS (PMOS) transistor.

The gate of an MOS transistor has a very high impedance. The gate voltage creates an electric field that enhances or retards the flow of current between source and drain. (In fact, MOS transistors are called *field-effect transistors (FETs)* or *MOSFETs.*) Regardless of gate voltage, no current flows from the gate to source, or from the gate to drain for that matter. You might have already guessed this from the MOS transistor symbols—there is no connection between the gate and the other two terminals of the device!

*field-effect transistor
(FET)*

MOSFET

2.14.2   NMOS Inverters and Gates

We can use an NMOS transistor and a resistor to create a logic inverter as shown in Figure 2–61. The main difference between this circuit and the bipolar transistor inverter of Figure 2–13 is that the NMOS inverter consumes no current on its V_{IN} terminal in either logic state. The bipolar transistor inverter requires base current to be supplied to it when V_{IN} is HIGH. Both inverters consume power through the pull-up resistor when V_{OUT} is LOW.

The pull-up resistor $R1$ in an NMOS inverter can be larger (and less power-consuming) than in a bipolar inverter, since the NMOS output does not have to supply current to drive another NMOS input in the HIGH state. However, current flows through $R1$ during LOW-to-HIGH output transitions, to charge the stray capacitance driven by the output. As we discussed in Section 2.6.4 for bipolar inverters, the charging resistance $R1$ and the load capacitance C_L determine the RC time constant. Hence, $R1$ can be made high to save power, but it must be low enough to produce acceptable transition times.

(a)

(b)

Figure 2–61
NMOS inverter: (a) circuit diagram; (b) transfer characteristic.

Figure 2–62
NMOS inverter with pull-up
transistor.

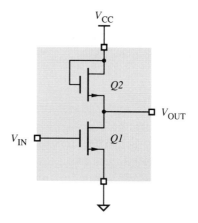

An integrated-circuit NMOS inverter does not use a discrete pull-up resistor
R1. Instead, a second NMOS transistor is used with its gate tied to V_{CC}, as shown
in Figure 2–62. This transistor is fabricated so that it is always only partially
"on," creating a pull-up resistance of the desired value.

An NMOS NOR gate can be fabricated much like an RTL NOR gate, by
putting additional NMOS transistors in parallel with *Q1*. If any input is HIGH,
the output is pulled LOW.

It is also possible to make NMOS NAND gates by putting additional transistors in series with *Q1*. All of the inputs must be HIGH to pull the output LOW.
However, "on" resistances of the transistors in the series configuration add, so
the individual "on" resistances must be smaller than in the parallel configuration,
in order to pull the output LOW. An MOS transistor with a lower "on" resistance requires more silicon area and, as a result, NOR gates (which use parallel
transistors) are preferred over NANDs in this technology.

Historically, PMOS transistors were actually the first MOS devices to be
used extensively in logic circuits, because of their ease of fabrication. However, PMOS circuits typically required unusual power supply voltages and had
much poorer performance than TTL. Therefore, they were soon supplanted by
NMOS circuits with improved performance and TTL compatibility. Most recently, "CMOS" circuits that combine both types of MOS transistors have been
developed to provide the best performance, as described in the next section.

2.15 COMPLEMENTARY MOS (CMOS) LOGIC

Although NMOS circuits dissipate less power per gate than bipolar circuits,
outputs in the LOW state still waste some power, and LOW-to-HIGH transitions
are slow. "CMOS" circuits use something like the TTL totem-pole structure to
eliminate both of these problems.

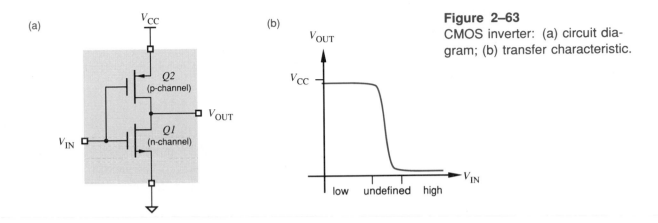

(a)

(b)

2.15.1 Basic CMOS Circuit

NMOS and PMOS transistors are used together to form *complementary MOS (CMOS) logic.* The circuit diagram and transfer characteristic of a CMOS inverter are shown in Figure 2–63. Like the top transistor in a TTL totem-pole output, the top, p-channel transistor in the CMOS circuit provides fast, active pull-up on LOW-to-HIGH transitions. Like the NMOS inverter, the CMOS inverter consumes almost no current on its V_{IN} input, only a tiny leakage current of 1–10 μA in either state.

complementary MOS (CMOS) logic

The CMOS inverter's biggest advantage is that its output structure consumes no current in either state, HIGH or LOW. As shown in Figure 2–64, only the top transistor is "on" when V_{IN} is LOW, and only the bottom transistor is "on" when V_{IN} is HIGH. (This figure uses different symbols for the p- and n-channel transistors that reflect this logical behavior.) In either state, one of the transistors is in the high-impedance "off" state.

The "on" resistances of the top and bottom transistors in a typical CMOS

Figure 2–64
CMOS inverter logical operation.

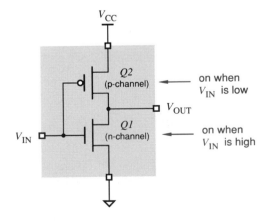

on when
V_{IN} is low

on when
V_{IN} is high

inverter are about the same, so the CMOS circuit has a symmetrical drive capability. That is, unlike a TTL output, a CMOS output sources current in the HIGH state as well as it sinks it in the LOW state. In advanced CMOS families, the "on" resistance of a transistor is as little as 50 Ω, while the "off" resistance is greater than 5 MΩ.

2.15.2   Logic Levels and Noise Margins

Since the DC load on a CMOS output driving CMOS inputs is practically nil, the output voltage in the HIGH state is practically equal to V_{CC}, and in the LOW state it's practically 0. However, a CMOS gate accepts much more relaxed input levels. Typically, any voltage between 0% and 30% of V_{CC} is recognized as LOW, and any voltage between 70% and 100% of V_{CC} is recognized as HIGH, as shown in Figure 2–65. Consequently, when CMOS outputs drive CMOS inputs, the DC noise margins are excellent, almost 30% of V_{CC}, compared to only 0.3 V or 6% of V_{CC} for 74LS TTL in the LOW state.

When a DC load, such as the input of a TTL gate, is connected to a CMOS output, then the output may have to sink significant current in the LOW state or source current in the HIGH state. This has the effect of raising the LOW-state and lowering the HIGH-state output voltage, but is acceptable as long as adequate noise margins are maintained. The amount of current that a CMOS output can source or sink while maintaining an acceptable output voltage is given in the specifications for different CMOS families, as we'll show in Section 2.15.7.

2.15.3   Power Consumption

A CMOS circuit consumes significant power only during switching. One source of this power consumption is a partial short-circuiting of the output structure. When the input voltage is between the LOW and HIGH levels, both the top and bottom output transistors may be partially "on," creating a series resistance as low as 600 Ω. In this case, current flows through the transistors from V_{CC} to ground, just as it does in a TTL totem-pole output during switching. If transitions are fast enough, this power consumption is insignificant. It is a problem only if an input is held between the LOW and HIGH levels for a long time.

Figure 2–65
Typical noise margins of CMOS
outputs driving CMOS inputs.

The second, and usually most significant, source of power consumption is the capacitive load (C_L) on the output. During a LOW-to-HIGH transition, current flows through the top transistor to charge C_L. Likewise, during a HIGH-to-LOW transition, current flows through the bottom transistor to discharge C_L. In each case, power is dissipated in the "on" resistance of the transistor. Because this power is dissipated only during transitions, it is called *dynamic power dissipation*.

dynamic power dissipation

The dynamic power dissipation of a CMOS output is an energy usage per unit time, which numerically equals the number of transitions per second times the energy used by one transition. The number of transitions per second (or *transition frequency*) depends on the circuit's application; a circuit operating at a higher frequency consumes more power. The energy for one transition could be determined by calculating the current through the charging transistor as a function of time (using the RC time constant as in Section 2.6.4), multiplying by the "on" resistance of the charging transistor, and integrating over time. However, there's an easier way, described below.

transition frequency

During a transition, the voltage across the load capacitance C_L changes by V_{CC}. According to the definition of capacitance (as in Appendix A), the total amount of charge that must flow to make a voltage change of V_{CC} across C_L is $C_L \cdot V_{CC}$. The total amount of energy used in one transition is charge times voltage, or $C_L \cdot V_{CC}^2$. If there are f transitions per second, the total dynamic power dissipation is

$$P_D \;=\; C_L \cdot V_{CC}^2 \cdot f$$

Based on this formula, dynamic power dissipation is often called CV^2f *power*. In most applications of CMOS circuits, CV^2f power is the main type of power consumption. Note that CV^2f power is also consumed by bipolar logic circuits, but at low to moderate frequencies it is insignificant compared to the static (DC) power consumption of bipolar circuits.

CV^2f *power*

2.15.4 Fanout

CMOS inverters and other CMOS gates consume almost no current on their logic inputs. Thus, from a DC point of view, the fanout of a CMOS gate driving other CMOS gates is virtually unlimited. From an AC point of view, however, the picture is not quite so rosy.

A CMOS input typically has more stray capacitance (5–15 pF) than a TTL or ECL input (3–5 pF), so the total capacitive load seen by each output tends to be higher in a system using CMOS logic. Worse, in all but the very highest-performance CMOS families, MOS transistors have significantly higher "on" resistances than the bipolar transistors used in TTL and ECL gates.

As a result, the RC time constants associated with charging and discharging stray capacitance are larger, and the rise and fall times are longer than in TTL and ECL. This increases the total propagation delay of a CMOS gate, especially when the gate drives a heavy capacitive load. In the design of gate arrays and other semicustom CMOS circuits, logic designers must constantly consider the speed degradation associated with high fanout, and provide extra buffers for high-fanout signals that also require high speed.

2.15.5 CMOS NAND and NOR Gates

Both NAND and NOR gates can be constructed using CMOS. A k-input gate uses k p-channel and k n-channel transistors. Figure 2–66 shows a CMOS NAND gate. If either input is LOW, the output Z has a low-impedance connection to V_{CC} through the corresponding "on" p-channel transistor, and the path to ground is blocked by the corresponding "off" n-channel transistor. If both inputs are HIGH, the path to V_{CC} is blocked, and Z has a low-impedance connection to ground.

Figure 2–67 shows a CMOS NOR gate. If both inputs are LOW, the output Z has a low-impedance connection to V_{CC} through the "on" p-channel transistors, and the path to ground is blocked by the "off" n-channel transistor. If either input is HIGH, the path to V_{CC} is blocked, and Z has a low-impedance connection to ground.

CMOS NAND and NOR gates do not have identical performance. For a given silicon area, an n-channel transistor has lower "on" resistance than a p-channel transistor. Therefore, when transistors are put in series, k n-channel transistors have lower "on" resistance than do k p-channel ones. As a result, a k-input NAND gate is generally faster than and preferred over a k-input NOR gate.

(a)

(b)

A B	$Q1$	$Q2$	$Q3$	$Q4$	Z
L L	off	on	off	on	H
L H	off	on	on	off	H
H L	on	off	off	on	H
H H	on	off	on	off	L

(c)

Figure 2–66
CMOS NAND gate: (a) circuit diagram; (b) function table; (c) logic symbol.

Figure 2–67
CMOS NOR gate: (a) circuit diagram; (b) function table; (c) logic symbol.

A	B	Q1	Q2	Q3	Q4	Z
L	L	off	on	off	on	H
L	H	off	on	on	off	L
H	L	on	off	off	on	L
H	H	on	off	on	off	L

2.15.6 Fan-In

The additive "on" resistance of series transistors limits the fan-in of CMOS gates, typically to 4 for NOR gates and 6 for NAND gates. As the number of inputs is increased, CMOS circuit designers sometimes compensate by increasing the size of the series transistors to reduce their resistance and the corresponding switching delay. However, at some point this becomes inefficient or impractical. Gates with a large number of inputs can be made faster and smaller by cascading gates with fewer inputs. For example, Figure 2–68 shows the logical structure of an 8-input CMOS NAND gate. The total delay through a 4-input NAND, a 2-input NOR, and an inverter is typically less than the delay of a one-level 8-input NAND circuit.

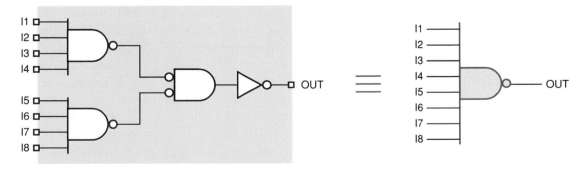

Figure 2–68   Logic diagram equivalent to the internal structure of an 8-input CMOS NAND gate.

Figure 2–69
Transfer characteristics of 74HC
and 74HCT circuits under typical
conditions.

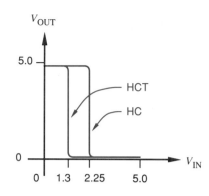

2.15.7   CMOS Families

4000-series CMOS   The first commercially successful CMOS family was *4000-series CMOS*. Although 4000-series circuits offered the benefit of low power consumption, they were fairly slow and were not easy to interface with TTL. Thus, the 4000 series has been supplanted in most applications by the more capable CMOS families discussed in this section. Devices in all of these families have part numbers of the form "74FAM*nn*," because they perform the same functions as like-numbered TTL parts.

74HC (High-speed CMOS)
74HCT (High-speed CMOS, TTL compatible)

The first two families are *74HC (High-speed CMOS)* and *74HCT (High-speed CMOS, TTL compatible)*. Compared with the original 4000 family, both of these families have higher speed and better current sinking and sourcing capability. The 74HC family is intended for use in systems that use CMOS logic exclusively. In particular, 74HC circuits are designed to recognize the CMOS input levels depicted in Figure 2–65, which are higher than TTL levels. For TTL-compatible applications, the inputs of 74HCT circuits are designed to recognize TTL levels. This is accomplished in the fabrication process by making transistors with different switching thresholds, producing the different transfer characteristics shown in Figure 2–69.

74AC (Advanced CMOS)
74ACT (Advanced CMOS, TTL compatible)

Introduced in the mid-1980s, the most advanced CMOS families are aptly named—*74AC (Advanced CMOS)* and *74ACT (Advanced CMOS, TTL compatible)*. These families are very fast, comparable to 74ALS TTL, and they can source or sink gobs of current, more than most 74AS TTL circuits can. Like 74HC and 74HCT, the 74AC and 74ACT families differ only in the input levels that they recognize; their output characteristics are the same.

Unlike TTL, typical CMOS circuits need not use a 5 V power supply for V_{CC}. The original 4000-series family allowed V_{CC} to be as high as 18 V for faster performance, and as low as 3 V for reduced power consumption. The newer families allow a range of V_{CC} from 2 to 6 V. In typical applications, lowering V_{CC} reduces power consumption more than proportionally, because

Table 2–6 Characteristics of CMOS families operating with V_{CC} between 4.5 and 5.5 V.

Description	Symbol	Condition	Family 74HC	74HCT	74AC	74ACT
Typical propagation delay (ns)			18	18	5.25	4.75
Quiescent power dissipation per gate (mW)		$V_{in} = 0$ or V_{CC}	0.0025	0.0025	0.005	0.005
		$V_{in} = 2.4$ V	n/a	12	n/a	7.5
Power dissipation capacitance (pF)	C_{PD}		24	24	30	30
Dynamic power dissipation per gate (mW/MHz)			0.60	0.60	0.75	0.75
Total power dissipation per gate (mW)		$f = 100$ kHz	0.0625	0.0625	0.080	0.080
		$f = 1$ MHz	0.6025	0.6025	0.755	0.755
		$f = 10$ MHz	6.0025	6.0025	7.505	7.505
Speed-power product (pJ)		$f = 100$ kHz	1.1	1.1	0.4	0.4
		$f = 1$ MHz	10.8	10.8	3.9	3.6
		$f = 10$ MHz	108	108	39	36
Input leakage current (μA)	I_{Imax}	$V_{in} = $ any	±1	±1	±1	±1
LOW-level input voltage (V)	V_{ILmax}		1.35	0.8	1.35	0.8
HIGH-level input voltage (V)	V_{IHmin}		3.85	2.0	3.85	2.0
LOW-level output current (mA)	I_{OLmaxC}	CMOS load	0.02	0.02	0.05	0.05
	I_{OLmaxT}	TTL load	4.0	4.0	24.0	24.0
LOW-level output voltage (V)	V_{OLmaxC}	$I_{out} \leq I_{OLmaxC}$	0.1	0.1	0.1	0.1
	V_{OLmaxT}	$I_{out} \leq I_{OLmaxT}$	0.33	0.33	0.37	0.37
HIGH-level output current (mA)	I_{OHmaxC}	CMOS load	−0.02	−0.02	−0.05	−0.05
	I_{OHmaxT}	TTL load	−4.0	−4.0	−24.0	−24.0
HIGH-level output voltage (V)	V_{OHminC}	$|I_{out}| \leq |I_{OHmaxC}|$	4.4	4.4	4.4	4.4
	V_{OHminT}	$|I_{out}| \leq |I_{OHmaxT}|$	3.84	3.84	3.76	3.76

dynamic power dissipation is given by CV^2f. However, a 5 V supply is still the most popular, especially when CMOS circuits are interfaced with TTL.

Electrical characteristics of CMOS families when used with a 5 V power supply are summarized in Table 2–6. (Some specifications are actually given with $V_{CC} = 4.5$ V or 5.5 V to allow a 10% power-supply margin.) From the first row of the table, and referring back to Table 2–5, you can see that 74HC and 74HCT are somewhat slower than 74LS TTL, but that 74AC and 74ACT have about the same speed as 74ALS.

quiescent power dissipation

A bigger distinction between CMOS and TTL is in power dissipation. A TTL gate dissipates power in resistors and ON transistors whether or not the gate's output is changing. In CMOS, the *quiescent power dissipation*—the power used when the output is not changing—is practically nil, a few μW.[8] The dynamic (CV^2f) power dissipation is much more significant in most applications.

The dynamic power dissipation of a CMOS gate depends on the voltage swing of the output (usually V_{CC}), the number of output transitions per second (f), and the capacitance that is being charged and discharged on each transition. To help designers estimate the power dissipation of CMOS circuits, manufacturers specify an effective *power dissipation capacitance, C_{PD}* for each device. This number represents the stray capacitances that must be charged or discharged when an output changes, including internal capacitances and pin and package capacitances, but excluding any load on the circuit. Thus, the total dynamic power dissipation is given by the formula

power dissipation capacitance, C_{PD}

$$P_D = (C_{load} + C_{PD}) \cdot V_{CC}^2 \cdot f$$

Here, C_{load} is the capacitance of the load, if any, attached to the CMOS output in a given application. As we discussed in Section 2.6.4, gate inputs may have a capacitance of 2–15 pF, and wiring may have a capacitance of 1 pF/inch or more.

Two rows of Table 2–6 shows typical values of C_{PD} for a CMOS gate and the resulting dynamic power dissipation per gate as a function of the transition frequency f, assuming that $C_{load} = 0$. The next two rows show the corresponding total power dissipation and speed-power product for several frequencies. At high transition frequencies (large f), CMOS families actually use more power than TTL.

For example, compare 74HCT CMOS at $f = 10$ MHz with 74LS TTL (Table 2–5); a CMOS gate uses three times as much power as a TTL gate at this frequency. Both 74HCT and 74LS may be used in systems with maximum "clock" frequencies of up to about 20 MHz, so you might think that CMOS is not so good for high-speed systems. However, the transition frequencies of most outputs in typical systems are much less than the maximum frequency present in

---

[8]This is strictly true only if the inputs have CMOS levels—0 V for LOW and V_{CC} for HIGH. 74HCT and 74ACT circuits can also accept TTL levels, with a HIGH input being as low as 2.4 V. The output specs are still met in this condition, since the switching threshold is well below 2.4 volts, and the bottom transistor is fully "on." However, the top transistor is also partially "on," resulting in additional current flow across the output transistors and power dissipation in the amount listed for the "$V_{in} = 2.4$ V" line in the table. Fortunately, TTL outputs are normally able to drive CMOS inputs to a HIGH level well above the 2.4-volt V_{OHmin} spec, usually to 3.5 V or more, so this extra power dissipation is usually not nearly as bad as the worst-case spec given in the table.

the system.[9] Thus, typical CMOS systems have a lower total power dissipation than they would have if they were built with TTL.

The I_{Imax} row of Table 2–6 shows that the input current for a CMOS input is 1 μA or less flowing into or out of the input, for any value of input voltage. In other words, CMOS inputs create almost no DC load on the circuits that drive them.

The next two rows of the table specify the input voltage levels that are recognized by CMOS circuits. As we explained previously, 74HC and 74AC families recognize CMOS levels. For worst-case operation, these levels are set at 30% of 4.5 V (V_{CCmin}) and 70% of 5.5 V (V_{CCmax}). The 74HCT and 74ACT families recognize input levels that are the same as in TTL families.

The remaining rows of the table specify the output characteristics of the CMOS families. Either of two sets of specifications may be used, depending on how an output is loaded. A *CMOS load* is one that requires the output to sink and source very little DC current, 20 μA for 74HC/HCT and 50 μA for 74AC/ACT. This is, of course, the case when the CMOS outputs drive only CMOS inputs. With CMOS loads, CMOS outputs maintain an output voltage within 0.1 V of the supply rails, 0 and V_{CC}. (Recall that a worst-case $V_{\text{CC}} = 4.5$ V is used for some table entries; hence $V_{\text{OHminC}} = 4.4$ V.)

CMOS load

A *TTL load* requires much more sink and source current, up to 4 mA for 74HC/HCT and 24 mA for 74AC/ACT. In this case, a higher voltage drop occurs across the "on" transistor in the CMOS output circuit, but the output voltage is still guaranteed to be well within the normal range of TTL output levels, especially in the HIGH state.

TTL load

2.15.0 CMOS/TTL Interfacing

A digital designer selects a "default" logic family to use in a system, based on general requirements of speed, power, cost, and so on. However, the designer may select devices from other families in some cases because of availability or other special requirements. (For example, not all 74LS part numbers are available in 74HCT, and vice versa.) Thus, it's important for a designer to understand the implications of connecting TTL outputs to CMOS inputs, and vice versa.

There are several factors to consider in TTL/CMOS interfacing, and the first is noise margin. The LOW-state DC noise margin depends on V_{OLmax} of the driving output and V_{ILmax} of the driven input, and equals $V_{\text{ILmax}} - V_{\text{OLmax}}$. Similarly, the HIGH-state DC noise margin equals $V_{\text{OHmin}} - V_{\text{IHmin}}$. Figure 2–70 shows the relevant numbers for TTL and CMOS families.

---

[9]As an example, consider an 8-bit binary counter (Section 6.3) clocked at 8 MHz. The least significant bit has a transition rate of 8 MHz, but the most significant bit has a transition rate of only 62.5 KHz.

Figure 2–70
Output and input levels for
interfacing TTL and CMOS
families. (Note that HC and AC
inputs are not TTL-compatible.)

For example, the LOW-state DC noise margin of HC or HCT driving TTL is $0.8 - 0.33 = 0.47$ V. On the other hand, the HIGH-state DC noise margin of TTL driving HC or AC is $2.7 - 3.85 = -1.15$ V. In other words, TTL driving HC or AC doesn't work unless the TTL HIGH output happens to be higher and the CMOS HIGH threshold happens to be lower by a total of 1.15 V compared to their worst-case specs.

The next factor to consider is fanout. As discussed in Section 2.9.3 for TTL, a designer must sum the input current requirements of devices driven by an output and compare with the output's capabilities in both states. Fanout is not a problem when TTL drives CMOS, since CMOS inputs require almost no current in either state. On the other hand, TTL inputs, especially in the LOW state, require substantial current, especially compared to 74HC and 74HCT output capabilities. For example, a 74HC or 74HCT output can drive 10 74LS or only two 74S TTL inputs.

The last factor is capacitive loading. We've seen that load capacitance increases both the delay and the power dissipation of logic circuits. Increases in delay are especially noticeable with 74HC and 74HCT outputs, whose transition times increase about 1 ns for each 5 pF of load capacitance. The transistors in 74AC and 74ACT outputs have very low "on" resistances, so their transition times increase only about 0.1 ns for each 5 pF of load capacitance.

For a given load capacitance, power-supply voltage, and application, all of the CMOS families have similar dynamic power dissipation, since each variable in the CV^2f equation is the same. On the other hand, TTL outputs have substantially lower dynamic power dissipation, since the voltage swing between TTL HIGH and LOW levels is smaller.

Figure 2–71
An integrated-circuit (IC) chip,
the National Semiconductor
NS32532, containing 370,000
transistors. (Photo courtesy of
National Semiconductor Corporation.)

pads

Unlike TTL inputs, CMOS inputs should never be left floating. The TTL input structure contains a resistor that provides some pull-up to the HIGH state. The CMOS input structure contains no such resistor and has extremely high impedance. Therefore, a very small amount of noise coupled into a CMOS input can make the device behave as if the input were HIGH or LOW. Worse, if the noise voltage is halfway between 0 and 1, both the top and bottom transistors may be "on," and excessive current may flow. In some cases, this current is sufficient to affect the power supply voltage seen by the device, and may actually create a sustained high-frequency oscillation at the device output. Thus, unused CMOS inputs must always be tied to V_{CC}, ground, or another HIGH or LOW signal source appropriate to the device's function.[10]

2.16 INTEGRATED CIRCUITS

One or more gates are fabricated on a single silicon chip, called an *integrated circuit (IC)*. Figure 2–71 shows a very large integrated-circuit chip, whose actual size is about half an inch wide. The relatively large squares on the periphery of

integrated circuit (IC)

---

[10]Well, almost always. Some inputs of larger-scale CMOS devices have internal pull-up or pull-down resistors that force the input to a particular state when no signal is connected. Because they consume power, these resistors are *not* included in most CMOS devices.

Figure 2–72
An NS32532 chip mounted
in a ceramic pin-grid array
package. (Photo courtesy of National
Semiconductor Corporation.)

package
(upside-down)

pins

chip

bonding wires
(not visible)

pad

the chip are called *pads*. The pads are the electrical contact points for the power supply and the input and output signals; the example chip has about 175 pads.

An IC chip is housed in a ceramic or plastic package. Figure 2–72 shows our example chip mounted in a ceramic "pin-grid-array" package with pins on a 0.1-inch grid. Tiny wires, called *bonding wires*, connect the pads on the chip to the pins of the package. Normally, the package is sealed and the IC chip inside is not visible.

bonding wires

wafer
die

When an IC chip is fabricated, it has lots of company. As shown in Figure 2–73, a single IC chip is replicated dozens or hundreds of times on a single *wafer* of silicon with a diameter as large as 8 inches. Each replica of the IC chip is called a *die*. After the wafer is fabricated, all of the dice are tested, and the defective ones are marked. Then the wafer is sliced up to produce the individual dice. Each unmarked die is mounted in a package, its pads are connected to the package pins, and the packaged IC is subjected to a final test and shipped to a customer.

Some people use the term "IC" to refer to a silicon die. Some use "chip" to refer to the same thing. Still others use "IC" or "chip" to refer to the combination of a silicon die and its package. Logic designers tend to use the two terms interchangeably, and they really don't care what they're talking about. They don't require a precise definition, since they're only looking at the functional and electrical behavior of these things. In the balance of this text, we'll use the term *IC* to refer to a packaged die.

IC

Figure 2–73
A wafer full of dice.
(Photo courtesy of
Motorola, Inc.)

one die

An IC is normally inserted in a *printed-circuit board (PCB)* [sometimes called a *printed-wiring board (PWB)*] that connects it to other ICs in a system. The multilayer PCBs used in typical digital systems have copper wiring etched on multiple, thin layers of fiberglass that are laminated into a single board about 1/16 inch thick. Figure 2–74 shows a PCB with many IC packages and other components.

printed-circuit board (PCB)

printed-wiring board (PWB)

ICs are classified by how many gates they contain. The simplest type of commercially available ICs are called *small-scale integration (SSI)*, and contain the equivalent of 1 to 20 gates. SSI ICs typically contain a handful of gates or flip-flops, the basic building blocks of digital design. One might also define SSI ICs in terms of cost: inflation willing, an SSI IC costs about 25 cents.

small-scale integration (SSI)

Figure 2–75 shows the pin diagrams for commonly used SSI ICs. These ICs use a 14-pin *dual in-line package (DIP)*, where the spacing between pins in a column is 0.1 inch and the spacing between columns is 0.3 inch, as shown in Figure 2–76(a). All 7400-series ICs with the same part number have the same assignment of gates to pin numbers (called the *pinout*), even if the alphabetic portion of the part name is different. For example, a 74LS00 and a 74S00 have the same pinout.

dual in-line package (DIP)

pinout

Note that pin diagrams given in Figure 2–75 are used only for mechanical analog components   120-pin ASIC   28-pin DIP   20-pin DIP   40-pin microprocessor

Figure 2–74   A PCB with ICs, analog components, and connectors. (Courtesy of DAVID Systems, Inc.)

reference, when a designer needs to determine the pin numbers for a particular IC. In the schematic diagram for a digital circuit, pin diagrams are not used. Instead, the various gates are grouped functionally, as we'll show in Section 4.1.

Although SSI ICs are still widely used as "glue" to tie together larger-scale elements in complex systems, they are gradually being supplanted by *programmable logic devices (PLDs)*, which we'll discuss in Chapter 7.

programmable logic device (PLD)
medium-scale integration (MSI)

The next larger commercially available ICs are called *medium-scale integration (MSI)*, and contain the equivalent of about 20 to 200 gates. An MSI IC typically contains a functional building block, such as a decoder, multiplexer, register, or counter, as we'll show in Chapters 4 and 6. One might also define MSI ICs in terms of cost: the Federal Reserve willing, an MSI IC costs under a dollar.

large-scale integration (LSI)

Large-scale integration (LSI) ICs are bigger still, containing the equivalent of 200 to 20,000 gates or more. LSI parts include memories, microprocessors, and programmable logic devices; their costs can range from a dollar to over $10.

Figure 2–75 Commonly used 7400-series SSI ICs.

Figure 2–76
Dual in-line packages (DIPs):
(a) 14-pin; (b) 20-pin; (c) 28-pin.

very large-scale integration (VLSI)

The dividing line between LSI and *very large-scale integration (VLSI)* is fuzzy, and tends to be stated in terms of transistor count rather than gate count. Any IC with over 500,000 transistors is definitely VLSI. Below that, the name depends on how well the marketing folks have promoted the product. Examples of VLSI devices include 1-Mbit memories and 32-bit microprocessors. Prices of VLSI devices can range from under $10 (for 1-Mbit memories) to hundreds of dollars (for the 1.2-million-transistor Intel 80486 microprocessor.) Perhaps the next level of integration will be called RLSI—ridiculously large-scale integration.

2.17 APPLICATION-SPECIFIC ICS (ASICS)

Probably the most exciting developments in IC technology for the average logic designer are not the ever-increasing chip sizes, but the ever-increasing opportunities to "design your own chip." Chips designed for a particular, limited product or application are called *application-specific ICs (ASICs)*. ASICs generally reduce the total component and manufacturing cost of a product by reducing chip count, PCB area, and power consumption, and they often provide higher performance.

application-specific IC (ASIC)

The *non-recurring engineering (NRE) cost* for designing an ASIC can exceed the cost of a discrete design by $5,000 to $250,000 or more. NRE charges are paid to the IC manufacturer and others who are responsible for designing the internal structure of the chip, creating tooling such as the metal masks for manufacturing the chips, developing tests for the manufactured chips, and actually making the first few sample chips. An ASIC design normally makes sense only when the NRE cost can be amortized by the per-unit savings over the expected sales volume of the product.

non-recurring engineering (NRE) cost

custom LSI

The NRE cost to design a *custom LSI* chip—a chip whose functions, internal architecture, and detailed transistor-level design is tailored for a specific customer—is very high, $250,000 or more. Thus, full custom LSI design is done only for chips that have general commercial application or that will enjoy very high sales volume in a specific application (e.g., a digital watch chip).

To reduce NRE cost, IC manufacturers have developed libraries of *standard cells* including commonly used "MSI" functions such as decoders, registers, and counters, and commonly used "LSI" functions such as memories, programmable logic arrays, and microprocessors. In a *standard-cell design*, the logic designer interconnects functions in much the same way as in a board-level design. Custom cells are created (at added cost, of course) only if absolutely necessary. All of the cells are then laid out on the chip, optimizing the layout to reduce propagation delays and minimize the size of the chip. Minimizing the chip size reduces the per-unit cost of the chip, since it increases the number of chips that can be fabricated on a single wafer. The NRE cost for a standard-cell design is typically on the order of $100,000.

standard cells

standard-cell design

Well, $100,000 is still a lot of money for most folks, so IC manufacturers have gone one step further to bring ASIC design capability to the masses. A *gate array* is an IC whose internal structure is an array of gates whose interconnections are initially unspecified.[11] The logic designer specifies the gate types and interconnections. Even though the chip design is ultimately specified at this very low level, the designer typically works with "macrocells," the same high-level functions used in board-level and standard-cell designs; software expands the high-level design into a low-level one.

gate array

The main difference between standard-cell and gate-array design is that the macrocells and the chip layout of a gate array are not as highly optimized as those in a standard-cell design, so the chip may be 25% or more larger. Also, there is no opportunity to create custom cells in the gate-array approach. On the other hand, a gate-array design can be completed faster and at lower NRE cost, ranging from about $5000 (what the marketing folks tell you initially) to $50,000 (what you find you've spent when you're all done).

One of the most recent developments in ASIC technology allows an ASIC to be created overnight (or even during the day, if that's when you like to work). In all of the ASIC technologies that we've described so far, gate-level interconnections are established when the device is manufactured. A *programmable gate array* contains nonvolatile memory cells that allow the interconnection pattern to be loaded and changed *after* the device is manufactured.

programmable gate array

Since a programmable gate array is a standard, in-stock part, its major appeal is that it can be programmed and tested in an application as soon as the design is completed, compared with 2–6 weeks to manufacture a fixed gate array. For designers who make mistakes, an even bigger appeal is that modifications can be made with very little time and cost, compared with many weeks and

---

[11] Actually, most gate-array technologies use an array of uncommitted CMOS transistors. Different gate types (NAND, NOR, etc.) and flip-flops (D, J-K, etc.) are created by interconnecting the transistors in different ways.

big bucks in the fixed gate-array and standard-cell approaches. The downside is that programmable gate arrays are slower and more costly than their fixed-interconnect cousins, but this may change with improvements in the technology.

The basic digital design methods that you'll study in the next five chapters apply very well to the functional design of ASICs. However, there are additional opportunities, constraints, and steps in ASIC design, which usually depend on the particular ASIC vendor and design environment.

REFERENCES

If you're interested in history, a nice introduction to all of the early bipolar logic families can be *Logic Design with Integrated Circuits*, by William E. Wickes (Wiley-Interscience, 1968). The classic introduction to TTL electrical characteristics appeared in *The TTL Applications Handbook*, edited by Peter Alfke and Ib Larsen (Fairchild Semiconductor, 1973).

For another perspective on the electronics material in this chapter, you can consult almost any modern electronics text. Many contain a much more analytical discussion of digital circuit operation; for example, see *Microelectronics* by J. Millman and A. Grabel (McGraw-Hill, 1987, 2nd ed.). Another good introduction to ICs and important logic families can be found in *VLSI System Design* by Saburo Muroga (Wiley, 1982). For NMOS and CMOS circuits in particular, two good books are *Introduction to VLSI Systems* by Carver Mead and Lynn Conway (Addison-Wesley, 1980) and *CMOS Digital Circuit Technology* by Masakazu Shoji (Prentice-Hall, 1988).

Characteristics of today's logic families can be found in the data books published by the device manufacturers. Texas Instruments' data books for TTL and CMOS devices are listed in Table 4–20 on page 337. Motorola provides an excellent introduction to the HC and HCT families in *High-Speed CMOS Logic Data* (publ. DL129, rev. 3, 1988), and to the more recent AC and ACT families in *FACT Data* (publ. DL138, 1988). ECL parts are covered in Motorola's *MECL Data* book (publ. DL122, 1987).

EXERCISES

2.1   If both V_Z and V_B in Figure 2–7(b) are 4.6 V, can we get V_C = 5.2 V? Explain.

2.2   Find a set of resistor values such that the passive AND/OR/AND circuit in Figure 2–9 behaves properly. Assume that a LOW input signal is 1 V, a HIGH input is 4 V. For all input combinations, the voltages at points C, F, and Z must be in the valid range for 1s and 0s defined in Table 2–3.

2.3   Repeat Exercise 2.2, assuming that a LOW input signal may be any voltage between 0 and 1 V, and a HIGH input is between 4 and 5 V.

2.4   Modify the program in Table 2–4 to account for leakage current in the OFF state.

2.5   A circuit designer was asked how to modify the transistor inverter design in Section 2.6 so that it has a best-case fanout of at least 10 instead of 6, and suggested changing the value of $R2$ from 2 kΩ to 1 kΩ. What value of $R1$ is required, and what is the fanout of the resulting inverter?

2.6   What single parameter in the transistor inverter design of Section 2.6 has the greatest effect on fanout? Explain.

2.7   Analyze the fall time of the transistor inverter output of Section 2.6, assuming that the output is driving four other inverter inputs. You must perform a piecewise analysis, since R_L and V_L are different for $V_{OUT} < 0.6$ V and $V_{OUT} \geq 0.6$ V. Also, note that the starting value for V_{OUT} is different because of the DC load. Compare your answer with the results of Section 2.6.4 and explain.

2.8   Repeat Exercise 2.7 for rise time.

2.9   Assuming that $R0 = 2$ kΩ, that $R1, R2, \ldots = 5.6$ kΩ, and that the OFF resistance of an *npn* transistor is 20 kΩ, compute the maximum fanout of RTL NOR gates with the following number of inputs: 2, 4, 8, 16.

2.10   Assuming "ideal" conditions, what is the minimum voltage that will be recognized as a HIGH in the TTL NAND gate in Figure 2–30?

2.11   What is the purpose of diode $D1$ in a TTL NAND gate?

2.12   Assuming "ideal" conditions, what is the maximum voltage that will be recognized as a LOW in the TTL NAND gate in Figure 2–31?

2.13   Find a commercial TTL part that can source 40 mA in the HIGH state. What is its application?

2.14   Show that a TTL totem-pole output could be used to drive an LED, with exactly the same connection as in Figure 2–44.

2.15   What happens if you try to drive an LED with its cathode grounded and its anode connected to a TTL totem-pole output?

2.16   What happens if you try to drive a 12-volt relay with a TTL totem-pole output?

2.17   Suppose that a single pull-up resistor to +5 V is used to provide a constant-1 logic source to 15 different 74LS00 inputs. What is the maximum value of this resistor? How much HIGH-state DC noise margin are you providing in this case?

2.18   The circuit in Figure X2.18 uses open-collector NAND gates to perform "wired logic." Write a truth table for output signal F and, if you've read Section 3.2, a logic expression for F as a function of the circuit inputs.

Figure X2.18

2.19 What is the maximum allowable value for *R1* in Figure X2.18? Assume that a 0.7 V HIGH-state noise margin is required. The 74LS01 has the specs shown in the 74LS column of Table 2–5, except that I_{OHmax} is 100 µA, a leakage current that flows *into* the output in the HIGH state.

2.20 A logic designer found a problem in a certain circuit's function after the circuit had been released to production and 1000 copies of it built. A portion of the circuit is shown in Figure X2.20 in black; all of the gates are 74LS00 NAND gates. The logic designer fixed the problem by adding the two diodes shown in color. What do the diodes do? Describe both the logical effects of this change on the circuit's function and the electrical effects on the circuit's noise margins.

Figure X2.20

Thévenin termination

2.21 A *Thévenin termination* for an open-collector or three-state bus has the structure shown in Figure X2.21(a). The idea is that, by selecting appropriate values of *R1* and *R2*, a designer can obtain a circuit equivalent to the termination in (b) for any desired values of *V* and *R*. The value of *V* determines the voltage on the bus when no device is driving it, and the value of R is selected to match the characteristic impedance of the bus for transmission-line purposes (Section 8.4). For each of the following pairs of V and R, determine the required values of $R1$ and $R2$.

(a) $V = 2.75$, $R = 148.5$ (b) $V = 2.7$, $R = 180$
(c) $V = 3.0$, $R = 130$ (d) $V = 2.5$, $R = 75$

Figure X2.21

$$R = \frac{R_1 R_2}{R_1 + R_2}$$

$$V = 5V \times \frac{R_2}{R_1 + R_2}$$

2.22 For each of the $R1$ and $R2$ pairs in Exercise 2.21, determine whether the termination can be properly driven by a three-state output in each of the following logic families: 74LS, 74S, 74ACT. For proper operation, the family's I_{OL} and I_{OH} specs must not be exceeded when $V_{OL} = V_{OLmax}$ and $V_{OH} = V_{OHmin}$, respectively.

2.23 Suppose that the output signal Γ in Figure X2.18 drives the inputs of two 74S04 inverters. Compute the minimum and maximum allowable values of $R2$, assuming that a 0.7 V HIGH-state noise margin is required.

2.24 Compute the maximum fanout for each of the following cases of a TTL output driving multiple inputs in a different logic family. Also indicate how much "excess" driving capability is available in the LOW or HIGH state for each case.

(a)   74LS driving 74S   (b)   74S driving 74LS
(c)   74LS driving 74ALS   (d)   74S driving 74AS
(e)   74ALS driving 74AS   (f)   74ALS driving 74LS

2.25 A 74LS125 is a buffer with a three-state output. When enabled, the output can sink 24 mA in the LOW state and source 2.6 mA in the HIGH state. When disabled, the output has a leakage current of ±20 μA (the sign depends on the output voltage— plus if the output is pulled HIGH by other devices, minus if it's LOW). Suppose a system is designed with multiple modules connected to a bus, where each module has a single 74LS125 to drive the bus, and one 74LS04 to receive information on the bus. What is the maximum number of modules that can be connected to the bus without exceeding the 74LS125's specs?

2.26   Repeat Exercise 2.25, this time assuming that a single pull-up resistor is connected from the bus to +5 V to guarantee that the bus is HIGH when no device is driving it. Calculate the maximum possible value of the pull-up resistor, and the number of modules that can be connected to the bus.

2.27   Find the circuit design in a TTL data book for an actual three-state gate, and explain how it works.

2.28   Using the graphs in a TTL data book, develop some rules of thumb for derating the maximum propagation delay specification of LS-TTL under nonoptimal conditions of power-supply voltage, temperature, and loading.

2.29   Determine the total power dissipation of the circuit in Figure X2.29 as function of transition frequency for two realizations: (a) using 74LS gates; (b) using 74HC gates. Assume that input capacitance is 3 pF for a TTL gate and 7 pF for a CMOS gate, that a 74LS gate has an internal power dissipation capacitance of 20 pF, and that there is an additional 20 pF of stray wiring capacitance in the circuit. Also assume that the W, X, Y, and Z inputs are always HIGH. Other information that you need for this problem can be found in Tables 2–5 and 2–6. State any other assumptions that you make. At what frequency does the TTL circuit dissipate less power than the CMOS circuit?

Figure X2.29

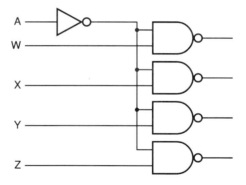

2.30   It is possible to operate 74AC CMOS devices with a 3.3-volt power supply. How much power does this save, compared to 5-volt operation?

open-drain output

2.31   The equivalent of a TTL open-collector output in CMOS or NMOS is called an *open-drain output*. Draw a circuit diagram of an inverter with an open-drain output.

2.32   Referring to a CMOS data book, look up the phenomonem of "latch up" and explain the circumstances under which it occurs.

2.33   Compute the LOW-state and HIGH-state DC noise margins for each of the following cases of a TTL output driving a TTL-compatible CMOS input or vice versa.

   (a)   74HCT driving 74LS   (b)   74ACT driving 74AS

   (c)   74LS driving 74HCT   (d)   74S driving 74ACT

2.34 Compute the maximum fanout for each of the following cases of a TTL-compatible CMOS output driving multiple inputs in a TTL logic family. Also indicate how much "excess" driving capability is available in the LOW or HIGH state for each case.

(a) 74HCT driving 74LS   (b) 74HCT driving 74S
(c) 74ACT driving 74AS   (d) 74ACT driving 74LS

2.35 Like TTL, CMOS circuits can perform two levels of logic with just a single "level" of transistors. For example, the circuit in Figure X2.35(a) is a CMOS *AND-OR-INVERT gate*. Write a function table for this circuit in the style of Figure 2–67(b), and a corresponding logic diagram using AND and OR gates and inverters.

AND-OR-INVERT gate

Figure X2.35

(a)

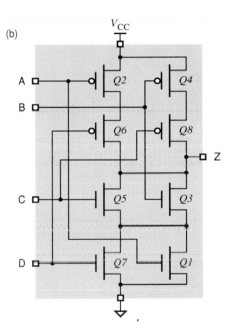
(b)

2.36 The circuit in Figure X2.35(b) is a CMOS *OR-AND-INVERT gate*. Write function table for this circuit in the style of Figure 2–67(b), and a corresponding logic diagram using AND and OR gates and inverters.

OR-AND-INVERT gate

2.37 Design a CMOS circuit that has the functional behavior shown in Figure X2.37. (*Hint:* Only six transistors are required.)

Figure X2.37

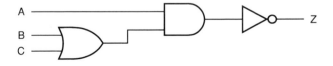

2.38 Design a CMOS circuit that has the functional behavior shown in Figure X2.38. (*Hint:* Only six transistors are required.)

Figure X2.38

transmission gate   2.39 The CMOS circuit in Figure X2.39(a) is called a *transmission gate*. In normal operation, the input signals EN and /EN are at opposite levels. When EN is HIGH and /EN is LOW, there is a low-impedance connection between points A and B. When EN is LOW and /EN is HIGH, points A and B are disconnected. Write a function table for the CMOS circuit in (b). What is its function, in words?

Figure X2.39

(a)

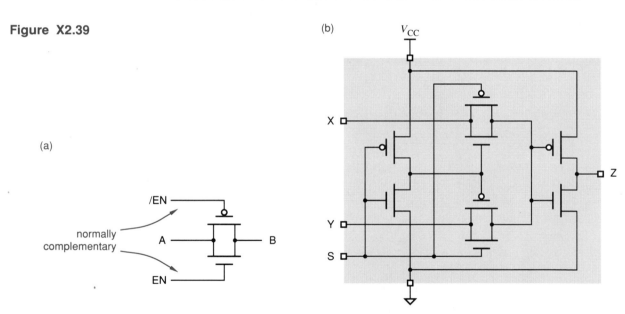

2.40 It is possible to drive one or more 74AC or 74HC inputs reliably with a 74LS TTL output by providing an external resistor to pull the TTL output all the way up to V_{CC} in the HIGH state. What are the design issues in choosing a value for this pull-up resistor?

2.41 For a given load capacitance and transition rate, which logic family in this chapter has the lowest dynamic power dissipation?

3

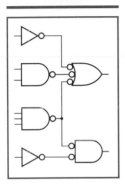

COMBINATIONAL LOGIC DESIGN PRINCIPLES

Logic circuits are classified into two types, "combinational" and "sequential." A *combinational logic circuit* is one whose outputs depend only on its current inputs. The rotary channel selector knob on an inexpensive TV is like a combinational circuit—its "output" selects a channel based only on the current position of the knob ("input").

combinational logic circuit

The outputs of a *sequential logic circuit* depend not only on the current inputs, but also on the past sequence of inputs, possibly arbitrarily far back in time. The channel selector controlled by the up and down pushbuttons on a fancy TV or VCR is a sequential circuit—the channel selection depends on the past sequence of up/down pushes, at least since when you started viewing 10 hours before, and perhaps as far back as when you first powered-up the device. Sequential circuits are discussed in Chapters 5 and 6.

sequential logic circuit

A combinational circuit may contain an arbitrary number of logic gates and inverters but no feedback loops. A *feedback loop* is a signal path of a circuit that allows the output of a gate to propagate back to the input of that same gate; such a loop generally creates sequential circuit behavior.

feedback loop

In combinational circuit *analysis*, we start with a logic diagram, and proceed to a formal description of the function performed by that circuit, such as a truth table or a logical expression. In *synthesis*, we do the reverse, starting with a formal description and proceeding to a logic diagram.

analysis

synthesis

design

    Design is a superset of synthesis, since in a real design problem we usually start out with an informal (word or thought) description of the circuit. Often the most challenging and creative part of design is to formalize the circuit description, defining the circuit's input and output signals and specifying its functional behavior by means of truth tables and equations. Once we've created the formal circuit description, we can usually follow a "turn-the-crank" synthesis procedure to obtain a logic diagram for a circuit with the required functional behavior.

    Combinational circuits may have one or more outputs. Most analysis and synthesis techniques can be extended in an obvious way from single-output to multiple-output circuits (e.g., "Repeat these steps for each output"). However, some techniques can also be extended in a not-so-obvious way for improved effectiveness in the multiple-output case. For example, in multiple-output synthesis, common subexpressions can sometimes be shared among different outputs. We will revisit these ideas later, but we will not discuss formal multiple-output synthesis techniques in detail (see References).

    Before launching into a discussion of combinational logic circuits, we must introduce switching algebra, the fundamental tool for analyzing and designing logic circuits of all types.

3.1 SWITCHING ALGEBRA

Formal analysis techniques for digital circuits have their roots in the work of an English mathematician, George Boole. In 1854, he invented a two-valued

Boolean algebra

algebraic system, now called *Boolean algebra*, to "give expression ... to the fundamental laws of reasoning in the symbolic language of a Calculus." Using this system, a philosopher, logician, or inhabitant of the planet Vulcan can formulate propositions that are true or false, combine them to make new propositions, and determine the truth or falsehood of the new propositions. For example, if we agree that "People who haven't studied this material are either failures or not nerds," and "No computer designer is a failure," then we can answer questions like "If you're a nerdy computer designer, then have you already studied this?"

    Long after Boole, in 1938, Bell Labs researcher Claude E. Shannon showed how to adapt Boolean algebra to analyze and describe the behavior of circuits built from relays, the most commonly used digital logic elements of that time.

switching algebra

In Shannon's *switching algebra*, the condition of a relay contact, open or closed, is represented by a variable X that can have one of two possible values, 0 or 1. In today's logic technologies, these values correspond to a wide variety of physical conditions—light off or on, capacitor discharged or charged, fuse blown or intact, and so on—as we detailed in Table 2–1 on page 54.

    In the remainder of this section, we develop the switching algebra directly, using "first principles" and what we already know about the behavior of logic elements (gates and inverters). For more historical and/or mathematical treatments of this material, consult the References.

3.1.1 Axioms

In switching algebra we use a symbolic variable, such as X, to represent the condition of a logic signal. A logic signal is in one of two possible conditions—low or high, off or on, and so on, depending on the technology. We say that X has the value "0" for one of these conditions and "1" for the other.

For example, with the TTL and CMOS logic circuits in Chapter 2, the *positive-logic convention* dictates that we associate the value "0" with a LOW voltage and "1" with a HIGH voltage. The *negative-logic convention* makes the opposite association: 0 = HIGH and 1 = LOW. However, the choice of positive or negative logic has no effect on our ability to develop a consistent algebraic description of circuit behavior; it only affects details of the physical-to-algebraic abstraction, as we'll explain later in our discussion of "duality." For the moment, we may ignore the physical realities of logic circuits and pretend that they operate directly on the logic symbols 0 and 1.

positive-logic convention

negative-logic convention

The *axioms* (or *postulates*) of a mathematical system are a minimal set of basic definitions that we assume to be true, from which all other information about the system can be derived. The first two axioms of switching algebra embody the "digital abstraction" by formally stating that a variable X can take on only one of two values:

axiom

postulate

$$(A1) \quad X = 0 \quad \text{if } X \neq 1 \qquad (A1') \quad X = 1 \quad \text{if } X \neq 0$$

Notice that we stated these axioms as a pair, with the only difference between A1 and A1′ being the interchange of the symbols 0 and 1. This is a characteristic of all the axioms of switching algebra, and is the basis of the "duality" principle that we'll study later.

In Section 2.5 we showed the design of an inverter, a logic circuit whose output signal level is the opposite (or *complement*) of its input signal level. We use a *prime (′)* to denote an inverter's function. That is, if a variable X denotes an inverter's input signal, then X′ denotes the value of a signal on the inverter's output.[1] This notation is formally specified in the second pair of axioms:

complement

prime (′)

$$(A2) \quad \text{If } X = 0, \text{ then } X' = 1 \qquad (A2') \quad \text{If } X = 1, \text{ then } X' = 0$$

---

[1] The notations $\overline{X}$, $\sim X$, and $\neg X$ are also used by some authors to denote the complement of X. The overbar notation is probably the most widely used and the best looking typographically. However, we use the prime notation to get you used to writing logical expressions on a single text line without the more graphical overbar, and to force you to parenthesize complex complemented subexpressions—because this is what you'll have to do when you start using computer-aided logic design languages and systems.

Figure 3–1
Signal naming and algebraic
notation for an inverter.

As shown in Figure 3–1, the output of an inverter with input signal X may have an arbitrary signal name, say Y. However, algebraically, we write Y = X′ to say "signal Y always has the opposite value as signal X." The prime (′) is an

algebraic operator
expression
NOT operation

algebraic operator, and X′ is an *expression*, which you can read as "X prime" or "NOT X." This usage is analogous to what you've learned in programming languages, where if J is an integer variable, then −J is an expression whose value is 0 − J. Although this may seem like a small point, you'll learn that the distinction between signal names (X, Y), expressions (X′), and equations (Y = X′) is very important when we study documentation standards and logic design tools. In the logic diagrams in this book, we maintain this distinction by writing signal names in black and expressions in color.

In Section 2.3 we showed how to build a 2-input diode AND gate, a circuit whose output is 1 if both of its inputs are 1. The function of a 2-input AND gate

logical
  multiplication
multiplication dot (·)

is sometimes called *logical multiplication* and is symbolized algebraically by a *multiplication dot (·)*. That is, an AND gate with inputs X and Y has an output signal whose value is X·Y, as shown in Figure 3–2(a). Some authors, especially mathematicians and logicians, denote logical multiplication with a wedge (X ∧ Y). We follow standard engineering practice by using the dot (X·Y), and we also understand simple juxtaposition (XY) to denote logical multiplication.[2]

We also showed in Section 2.3 how to build a 2-input diode OR gate, a circuit whose output is 1 if either of its inputs is 1. The function of a 2-input

logical addition

OR gate is sometimes called *logical addition* and is symbolized algebraically by a plus sign (+). An OR gate with inputs X and Y has an output signal whose value is X + Y, as shown in Figure 3–2(b). Some authors denote logical

---

[2]Most texts use juxtapostion extensively, but we don't. In general, juxtaposition is a clear notation only when signal names are limited to a single character. Otherwise, is XY a logical product or a 2-character signal name? One-character variable names are common in algebra, but in real digital design problems, we prefer to use multicharacter signal names that mean something. Thus, we need a separator between names, and the separator might just as well be a multiplication dot rather than a space. In summary, we limit our use of juxtaposition to expressions where it creates no ambiguity. For example, we may write X(Y + Z) instead of X·(Y + Z).

(a)   (b)

Figure 3–2   Signal naming and algebraic notation: (a) AND gate; (b) OR gate.

addition with a vee $(X \vee Y)$, but we follow the standard engineering practice of using the plus sign $(X + Y)$. By convention, in a logic expression involving both multiplication and addition, multiplication has *precedence*, just as in integer expressions in conventional programming languages. That is, the expression $W \cdot X + Y \cdot Z$ is equivalent to $(W \cdot X) + (Y \cdot Z)$.

precedence

The last three pairs of axioms state the formal definitions of the AND and OR operations by listing the output produced by each gate for each possible input combination:

AND operation
OR operation

(A3)	$0 \cdot 0 = 0$	(A3′)	$1 + 1 = 1$
(A4)	$1 \cdot 1 = 1$	(A4′)	$0 + 0 = 0$
(A5)	$0 \cdot 1 = 1 \cdot 0 = 0$	(A5′)	$1 + 0 = 0 + 1 = 1$

The five pairs of axioms, A1–A5 and A1′–A5′, completely define switching algebra. All other facts about the system can be proved using these axioms as a starting point.

3.1.2 Single-Variable Theorems

During the analysis or synthesis of logic circuits, we often write algebraic expressions that characterize a circuit's actual or desired behavior. Switching-algebra *theorems* are statements, known to be always true, that allow us to manipulate algebraic expressions to allow simpler analysis or more efficient synthesis of the corresponding circuits. For example, the theorem $X + 0 = X$ allows us to substitute every occurrence of $X + 0$ in an expression with X.

theorem

Table 3–1 lists switching-algebra theorems involving a single variable X. How do we know that these theorems are true? We can either prove them ourselves or take the word of someone who has. OK, we're in college now, let's learn how to prove them.

Most theorems in switching algebra are exceedingly simple to prove using a technique called *perfect induction*. Axiom A1 gives us the key to this technique— since a switching variable can take on only two different values, 0 and 1, we can prove a theorem involving a single variable X by proving that it is true for both $X = 0$ and $X = 1$. For example, to prove theorem T1, we make two substitutions:

perfect induction

Table 3–1
Switching-algebra theorems with one variable.

(T1)	$X + 0 = X$	(T1′)	$X \cdot 1 = X$	(Identities)
(T2)	$X + 1 = 1$	(T2′)	$X \cdot 0 = 0$	(Null elements)
(T3)	$X + X = X$	(T3′)	$X \cdot X = X$	(Idempotency)
(T4)	$(X')' = X$			(Involution)
(T5)	$X + X' = 1$	(T5′)	$X \cdot X' = 0$	(Complements)

$$(X = 0) \qquad 0 + 0 = 0 \qquad \text{true, according to axiom A4}'$$

$$(X = 1) \qquad 1 + 0 = 1 \qquad \text{true, according to axiom A5}'$$

All of the theorems in Table 3–1 can be proved using perfect induction, as requested in the Exercises 3.3 and 3.4.

3.1.3 Two- and Three-Variable Theorems

Switching-algebra theorems with two or three variables are listed in Table 3–2. Each of these theorems is easily proved by perfect induction, by evaluating the theorem statement for the four possible combinations of X and Y, or the eight possible combinations of X, Y, and Z.

The first two theorem pairs are identical to the commutative and associative laws for addition and multiplication of integers and reals. Taken together, they indicate that the parenthesization or order of terms in a logical sum or logical product is irrelevant. For example, from a strictly algebraic point of view, an expression such as $W \cdot X \cdot Y \cdot Z$ is ambiguous; it should be written as $(W \cdot (X \cdot (Y \cdot Z)))$ or $(((W \cdot X) \cdot Y) \cdot Z)$ or $(W \cdot X) \cdot (Y \cdot Z)$ (see Exercise 3.11). But the theorems tell us that the ambiguous form of the expression is OK, because we get the same results in any case. We even could have rearranged the order of the variables (e.g., $X \cdot Z \cdot Y \cdot W$) and gotten the same results.

As trivial as this discussion may seem, it is very important because it forms the theoretical basis for using logic gates with more than two inputs. We defined *binary operator* · and + as *binary operators*—operators that combine *two* variables. Yet we use 3-input, 4-input, and larger AND and OR gates in practice. The theorems tell us we can connect gate inputs in any order (some printed-circuit-board layout programs actually take advantage of this). We can use either one n-input gate or $(n - 1)$ 2-input gates interchangeably, though propagation delay and cost are likely to be higher with 2-input gates.

Theorem T8 is identical to the distributive law for integers and reals—that is, logical multiplication distributes over logical addition. Hence, we can

Table 3–2 Switching-algebra theorems with two or three variables.

(T6)	$X + Y = Y + X$	(T6')	$X \cdot Y = Y \cdot X$	(Commutativity)
(T7)	$(X + Y) + Z = X + (Y + Z)$	(T7')	$(X \cdot Y) \cdot Z = X \cdot (Y \cdot Z)$	(Associativity)
(T8)	$X \cdot Y + X \cdot Z = X \cdot (Y + Z)$	(T8')	$(X + Y) \cdot (X + Z) = X + Y \cdot Z$	(Distributivity)
(T9)	$X + X \cdot Y = X$	(T9')	$X \cdot (X + Y) = X$	(Covering)
(T10)	$X \cdot Y + X \cdot Y' = X$	(T10')	$(X + Y) \cdot (X + Y') = X$	(Combining)
(T11)	$X \cdot Y + X' \cdot Z + Y \cdot Z = X \cdot Y + X' \cdot Z$			(Consensus)
(T11')	$(X + Y) \cdot (X' + Z) \cdot (Y + Z) = (X + Y) \cdot (X' + Z)$			

"multiply out" an expression to obtain a sum-of-products form, as in the example below:

$$V \cdot (W + X) \cdot (Y + Z) = V \cdot W \cdot Y + V \cdot W \cdot Z + V \cdot X \cdot Y + V \cdot X \cdot Z$$

However, switching algebra also has the unfamiliar property that the reverse is true—logical addition distributes over logical multiplication—as demonstrated by theorem T8′. Thus, we can also "add out" an expression to obtain a product-of-sums form:

$$(V \cdot W \cdot X) + (Y \cdot Z) = (V + Y) \cdot (V + Z) \cdot (W + Y) \cdot (W + Z) \cdot (X + Y) \cdot (X + Z)$$

Theorems T9 and T10 are used extensively in the minimization of logic functions. For example, if the subexpression $X + X \cdot Y$ appears in a logical expression, the *covering theorem* T9 says that we need only include X in the expression; X is said to *cover* $X \cdot Y$. The *combining theorem* T10 says that if the subexpression $X \cdot Y + X \cdot Y'$ appears in an expression, we can replace it with X. Since Y must be 0 or 1, either way the original subexpression is 1 if and only if X is 1.

covering theorem
cover
combining theorem

Although we could easily prove T9 by perfect induction, the truth of T9 is more obvious if we prove it using the other theorems that we've proved so far:

$$
\begin{aligned}
X + X \cdot Y &= X \cdot 1 + X \cdot Y && \text{(according to T1′)} \\
&= X \cdot (1 + Y) && \text{(according to T8)} \\
&= X \cdot 1 && \text{(according to T2)} \\
&= X && \text{(according to T1′)}
\end{aligned}
$$

Likewise, the other theorems can be used to prove T10, where the key step is to use T8 to rewrite the left-hand side as $X \cdot (Y + Y')$.

Theorem T11 is known as the *consensus theorem*. The $Y \cdot Z$ term is called the *consensus* of $X \cdot Y$ and $X' \cdot Z$. The idea is that if the $Y \cdot Z$ term is 1, then either $X \cdot Y$ or $X' \cdot Z$ must also be 1, so the $Y \cdot Z$ term is redundant and may be dropped. The consensus theorem has two important applications. It forms the basis of the iterative-consensus method of finding prime implicants, as we'll show in Section 3.4.3. And it can also be used to eliminate certain timing hazards in combinational logic circuits, as we'll see in Section 3.5.

consensus theorem
consensus

In all of the theorems, it is possible to replace each variable with an arbitrary logic expression. A simple replacement is to complement one or more variables:

$$(X + Y') + Z' = X + (Y' + Z') \qquad \text{(based on T7)}$$

But more complex expressions may be substituted as well:

$$(V' + X) \cdot (W \cdot (Y' + Z)) + (V' + X) \cdot (W \cdot (Y' + Z))' = V' + X \qquad \text{(based on T10)}$$

Table 3–3   Switching-algebra theorems with n variables.

(T12)	$X + X + \cdots + X = X$	(Generalized idempotency)
(T12′)	$X \cdot X \cdot \ \cdots \ \cdot X = X$	
(T13)	$(X_1 \cdot X_2 \cdot \ \cdots \ \cdot X_n)' = X_1' + X_2' + \cdots + X_n'$	(DeMorgan's theorems)
(T13′)	$(X_1 + X_2 + \cdots + X_n)' = X_1' \cdot X_2' \cdot \ \cdots \ \cdot X_n'$	
(T14)	$[F(X_1, X_2, \ldots, X_n, +, \cdot)]' = F(X_1', X_2', \ldots, X_n', \cdot, +)$	(Generalized DeMorgan's theorem)
(T15)	$F(X_1, X_2, \ldots, X_n) = X_1 \cdot F(1, X_2, \ldots, X_n) + X_1' \cdot F(0, X_2, \ldots, X_n)$	(Shannon's expansion theorems)
(T15′)	$F(X_1, X_2, \ldots, X_n) = [X_1 + F(0, X_2, \ldots, X_n)] \cdot [X_1' + F(1, X_2, \ldots, X_n)]$	

3.1.4 n-Variable Theorems

finite induction
basis step
induction step

Several important theorems, listed in Table 3–3, are true for an arbitrary number of variables, n. Most of these theorems can be proven using a two-step method called *finite induction*—first proving that the theorem is true for $n = 2$ (the *basis step*) and second proving that if the theorem is true for $n = i$, then it is also true for $n = i + 1$ (the *induction step*). For example, consider the generalized idempotency theorem T12. For $n = 2$, T12 is equivalent to T3 and is therefore true. If it is true for a logical sum of i X's, then it is also true for a sum of $i + 1$ X's, according to the following reasoning:

$$
\begin{aligned}
X + X + X + \cdots + X &= X + (X + X + \cdots + X) \quad (i + 1 \text{ X's on either side}) \\
&= X + (X) \quad (\text{if T12 is true for } n = i) \\
&= X \quad (\text{according to T3})
\end{aligned}
$$

Thus, the theorem is true for all finite values of n.

DeMorgan's theorems

DeMorgan's theorems (T13 and T13′) are probably the most commonly used of all the theorems of switching algebra. Theorem T13 says that an n-input AND gate whose output is complemented is equivalent to an n-input OR gate whose inputs are complemented. That is, the circuits of Figure 3–3(a) and (b) are equivalent.

In Section 2.8 we showed how to build a DTL NAND gate as a diode AND gate followed by a transistor inverter (NOT gate). The NAND gate's logic symbol, based on this circuit realization, is shown in Figure 3–3(c). However, theorem T13 tells us that the logic symbol in Figure 3–3(d) denotes the same logic function (bubbles on the OR-gate inputs indicate logical inversion). By observing the inputs and outputs of a NAND gate, it is impossible to determine whether it has been built internally as an AND gate followed by an inverter or as inverters followed by an OR gate, *because both circuits perform precisely the same logic function.* Although the choice of symbol has no bearing on the functionality of a circuit, we'll show in Section 4.1 that the proper choice can make the circuit's function much easier to understand.

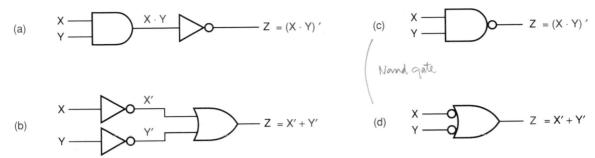

Figure 3–3   Equivalent circuits according to DeMorgan's theorem T13: (a) AND-NOT; (b) NOT-OR; (c) logic symbol for a NAND gate; (d) equivalent symbol for a NAND gate.

A similar symbolic equivalence can be inferred from theorem T13′. As shown in Figure 3–4, a NOR gate may be realized as an OR gate followed by an inverter, or as inverters followed by an AND gate.

Theorems T13 and T13′ are just special cases of a *generalized DeMorgan's theorem*, T14, that applies to an arbitrary logic expression F. By definition, the *complement of a logic expression*, denoted (F)′, is an expression whose value is the opposite of F's for every possible input combination. Theorem T14 is very important because it gives us a way to manipulate and simplify the complement of an expression.

Theorem T14 states that, given any *n*-variable logic expression, its complement can be obtained by swapping + and · and complementing all variables. For example, suppose that we have

generalized DeMorgan's theorem

complement of a logic expression

$$F(W,X,Y,Z) = (W' \cdot X) + (X \cdot Y) + (W \cdot (X' + Z'))$$
$$= ((W)' \cdot X) + (X \cdot Y) + (W \cdot ((X)' + (Z)'))$$

In the second line we have enclosed complemented variables in parentheses to

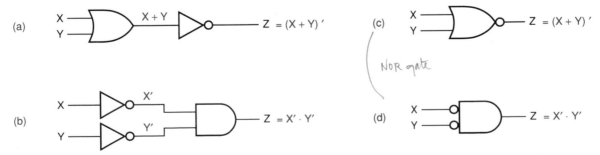

Figure 3–4   Equivalent circuits according to DeMorgan's theorem T13′: (a) OR-NOT; (b) NOT-AND; (c) logic symbol for a NOR gate; (d) equivalent symbol for a NOR gate.

remind you that the $'$ is an operator, not part of the variable name. Applying theorem T14, we obtain

$$[F(W,X,Y,Z)]' \;=\; ((W')' + X') \cdot (X' + Y') \cdot (W' + ((X')' \cdot (Z')'))$$

Using theorem T4, this can be simplified to

$$[F(W,X,Y,Z)]' \;=\; (W + X') \cdot (X' + Y') \cdot (W' + (X \cdot Z))$$

In general, we can apply theorem T14 to a parenthesized expression by swapping + and $\cdot$, complementing all uncomplemented variables, and uncomplementing all complemented ones.

The generalized DeMorgan's theorem T14 can be proved by showing that all logic functions can be written as either a sum or a product of subfunctions, and then applying T13 and T13$'$ recursively. However, a much more enlightening and satisfying proof can be based on the principle of duality, explained next.

3.1.5 Duality

We stated all of the axioms of switching algebra in pairs. The primed version of each axiom (e.g., A5$'$) is obtained from the unprimed version (e.g., A5) by simply swapping 0 and 1 and, if present, $\cdot$ and +. As a result, we can state the following *metatheorem*, a theorem about theorems:

metatheorem

> *Principle of Duality* Any theorem or identity in switching algebra remains true if 0 and 1 are swapped and $\cdot$ and + are swapped throughout.

The metatheorem is true because the duals of all the axioms are true, so duals of all switching-algebra theorems can be proved using duals of the axioms.

After all, what's in a name, or in a symbol for that matter? If the software that was used to typeset this book had a bug, one that swapped $0 \leftrightarrow 1$ and $\cdot \leftrightarrow +$ throughout this chapter, you still would have learned exactly the same switching algebra; only the nomenclature would have been a little weird, using words like "product" to describe an operation that uses the symbol "+".

Duality is important because it doubles the usefulness of everything that you learn about switching algebra and manipulation of switching functions. Stated more practically, from a student's point of view, it halves the amount that you have to learn! For example, once you learn how to synthesize two-stage AND-OR logic circuits from sum-of-products expressions, you automatically know a dual technique to synthesize OR-AND circuits from product-of-sums expressions.

There is just one convention in switching algebra where we did not treat ·
and + identically, so duality does not necessarily hold true—can you figure out
what it is? Consider the following statement of theorem T9 and its "dual":

$$X + X \cdot Y = X \qquad \text{(theorem T9)}$$
$$X \cdot X + Y = X \qquad \text{(after applying the principle of duality)}$$
$$X + Y = X \qquad \text{(after applying theorem T3')}$$

Clearly the last line above is false—where did we go wrong? The problem is
in operator precedence. We were able to write the left-hand side of the first line
without parentheses because of our convention that · has precedence. However,
once we applied the principle of duality, we should have given precedence to
+ instead, or written the second line as $X \cdot (X + Y) = X$. The best way to avoid
problems like this is to parenthesize an expression fully before taking its dual.

Let us formally define the *dual of a logic expression*. If $F(X_1, X_2, \ldots, X_n, +, \cdot, ')$ *dual of a logic*
is a fully parenthesized logic expression involving the variables $X_1, X_2, \ldots, X_n$ and   *expression*
the operators +, ·, and ', then the dual of F, written F^D, is the same expression
with + and · swapped:

$$F^D(X_1, X_2, \ldots, X_n, +, \cdot, ') = F(X_1, X_2, \ldots, X_n, \cdot, +, ')$$

You already knew this, of course, but we wrote the definition in this way just to
highlight the similarity between duality and the generalized DeMorgan's theorem
T14, which may now be restated as follows:

$$[F(X_1, X_2, \ldots, X_n)]' = F^D(X_1', X_2', \ldots, X_n')$$

Let's examine this statement in terms of a physical network.

Figure 3–5(a) shows the electrical function table for a logic element that
we'll simply call a "type-1" gate. Under the positive-logic convention (LOW = 0
and HIGH = 1), this is an AND gate, but under the negative-logic convention
(LOW = 1 and HIGH = 0), it is an OR gate, as shown in Figure 3–5(b) and (c).
We can also imagine a "type-2" gate, shown in Figure 3–6, that is a positive-
logic OR or a negative-logic AND. Similar tables can be developed for gates with
more than two inputs.

Suppose that we are given an arbitrary logic expression, $F(X_1, X_2, \ldots, X_n)$.
Following the positive-logic convention, we can build a circuit corresponding
to this expression using inverters for NOT operations, type-1 gates for AND,
and type-2 gates for OR, as shown in Figure 3–7. Now suppose that, without
changing this circuit, we simply change the logic convention from positive to
negative. Then we should redraw the circuit as shown in Figure 3–8. Clearly,
for every possible combination of input voltages (HIGH and LOW), the circuit
still produces the same output voltage. However, from the point of view of
switching algebra, the output value—0 or 1—is the opposite of what it was

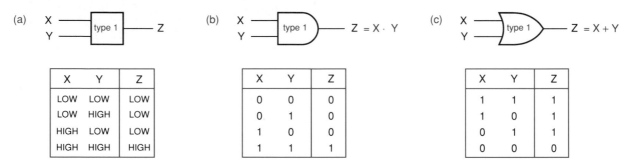

Figure 3–5   A "type-1" logic gate: (a) electrical function table; (b) logic function table and symbol under a positive-logic convention; (c) logic function table and symbol under a negative-logic convention.

under the positive-logic convention. Likewise, each input value is the opposite of what it was. Therefore, for each possible input combination to the circuit in Figure 3–7, the output is the opposite of that produced by the opposite input combination applied to the circuit in Figure 3–8:

$$F(X_1,X_2,...,X_n) = [F^D(X_1',X_2',...,X_n')]'$$

By complementing both sides, we get the generalized DeMorgan's theorem:

$$[F(X_1,X_2,...,X_n)]' = F^D(X_1',X_2',...,X_n')$$

Amazing!

So, we have seen that duality is the basis for the generalized DeMorgan's theorem, and we have promised that it will halve the number of methods you must learn to manipulate and simplify logic functions. Some authors exploit duality even further, allowing the logic convention (positive or negative) to be changed at arbitrary points in a circuit, so that AND gates can be used as OR

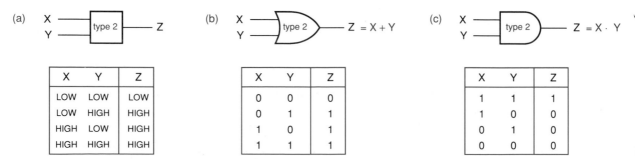

Figure 3–6   A "type-2" logic gate: (a) electrical function table; (b) logic function table and symbol under a positive-logic convention; (c) logic function table and symbol under a negative-logic convention.

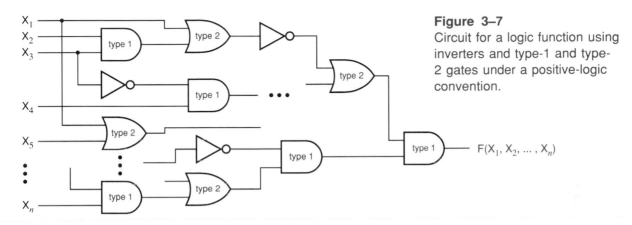

Figure 3-7
Circuit for a logic function using inverters and type-1 and type-2 gates under a positive-logic convention.

gates, and vice versa. The goal of this system, called *mixed logic*, is to promote   *mixed logic*
more effective use of available gates and more understandable logic diagrams.

Getting correct results with mixed logic requires proper identification of the convention-change points, careful naming of signals, and using inverters as nonlogic elements called *oops functions*. In this book, we take advantage of   *oops function*
duality in a more customary and less confusing way. We use the positive-logic convention throughout this book, but as you'll see in Section 4.1, we choose signal names and gate symbols in a way that promotes better understanding of circuit functionality.

3.1.6 Standard Representations of Logic Functions

Before moving on to analysis and synthesis of combinational logic functions we'll introduce some necessary nomenclature and notation.

The most basic representation of a logic function is the *truth table*. Similar   *truth table*
in philosophy to the perfect-induction proof method, this brute-force representa-

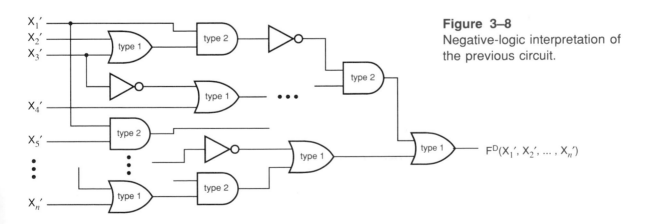

Figure 3-8
Negative-logic interpretation of the previous circuit.

Table 3–4
General truth table structure for a
3-variable logic function, F(X,Y,Z).

Row	X	Y	Z	F
0	0	0	0	F(0,0,0)
1	0	0	1	F(0,0,1)
2	0	1	0	F(0,1,0)
3	0	1	1	F(0,1,1)
4	1	0	0	F(1,0,0)
5	1	0	1	F(1,0,1)
6	1	1	0	F(1,1,0)
7	1	1	1	F(1,1,1)

tion simply lists the output of the circuit for every possible input combination. Traditionally, the input combinations are arranged in rows in ascending binary counting order, and the corresponding output values are written in a column next to the rows. The general structure of a 3-variable truth table is shown in Table 3–4. The rows are numbered 0–7 corresponding to the binary input combinations, but this numbering is not an essential part of the truth table. The truth table for a particular 3-variable logic function is shown in Table 3–5. Each distinct pattern of 0s and 1s in the output column yields a different logic function; there are 2^8 such patterns. Thus, the logic function in Table 3–5 is one of 2^8 different logic functions of three variables.

The truth table for an n-variable logic function has 2^n rows. Obviously, truth tables are practical to write only for logic functions with a small number of variables, say, 10 for students and about 5 for everyone else.

The information contained in a truth table can also be conveyed algebraically. To do so, we first need some definitions:

literal

- A *literal* is a variable or the complement of a variable. Examples: X, Y, X', Y'.

Table 3–5
Truth table for a particular 3-variable logic function, F(X,Y,Z).

[handwritten annotations:]
○ logic function :

○ minterms → $\Sigma (0,3,4,6,7)$

○ maxterms → $\Pi (1,2,5)$

Row	X	Y	Z	F
0	0	0	0	1
1	0	0	1	0
2	0	1	0	0
3	0	1	1	1
4	1	0	0	1
5	1	0	1	0
6	1	1	0	1
7	1	1	1	1

Table 3–6
Minterms and maxterms for a
3-variable logic function, $F(X,Y,Z)$.

Row	X	Y	Z	F	Minterm	Maxterm
0	0	0	0	$F(0,0,0)$	$X'\cdot Y'\cdot Z'$	$X + Y + Z$
1	0	0	1	$F(0,0,1)$	$X'\cdot Y'\cdot Z$	$X + Y + Z'$
2	0	1	0	$F(0,1,0)$	$X'\cdot Y\cdot Z'$	$X + Y' + Z$
3	0	1	1	$F(0,1,1)$	$X'\cdot Y\cdot Z$	$X + Y' + Z'$
4	1	0	0	$F(1,0,0)$	$X\cdot Y'\cdot Z'$	$X' + Y + Z$
5	1	0	1	$F(1,0,1)$	$X\cdot Y'\cdot Z$	$X' + Y + Z'$
6	1	1	0	$F(1,1,0)$	$X\cdot Y\cdot Z'$	$X' + Y' + Z$
7	1	1	1	$F(1,1,1)$	$X\cdot Y\cdot Z$	$X' + Y' + Z'$

- A *product term* is a single literal or a logical product of two or more literals. Examples: Z', $W\cdot X\cdot Y$, $X\cdot Y'\cdot Z$, $W'\cdot Y'\cdot Z$. *product term*

- A *sum-of-products expression* is a logical sum of product terms. Example: $Z' + W\cdot X\cdot Y + X\cdot Y'\cdot Z + W'\cdot Y'\cdot Z$. *sum-of-products expression*

- A *sum term* is single literal or a logical sum of two or more literals. Examples: Z', $W + X + Y$, $X + Y' + Z$, $W' + Y' + Z$. *sum term*

- A *product-of-sums expression* is a logical product of sum terms. Example: $Z'\cdot(W + X + Y)\cdot(X + Y' + Z)\cdot(W' + Y' + Z)$. *product-of-sums expression*

- A *normal term* is a product or sum term in which no variable appears more than once. A nonnormal term can always be simplified to a constant or a normal term using one of theorems T3, T3′, T5, or T5′. Examples of nonnormal terms: $W\cdot X\cdot X\cdot Y'$, $W + W + X' + Y$, $X\cdot X'\cdot Y$. Examples of normal terms: $W\cdot X\cdot Y'$, $W + X' + Y$. *normal term*

- An *n*-variable *minterm* is a normal product term with n literals. There are 2^n such product terms. Examples of 4-variable minterms: $W'\cdot X'\cdot Y'\cdot Z'$, $W'\cdot X'\cdot Y'\cdot Z$, $W'\cdot X'\cdot Y\cdot Z'$. *minterm*

- An *n*-variable *maxterm* is a normal sum term with n literals. There are 2^n such sum terms. Examples of 4-variable maxterms: $W' + X' + Y' + Z'$, $W' + X' + Y' + Z$, $W' + X' + Y + Z'$. *maxterm*

There is a close correspondence between the truth table and minterms and maxterms. A minterm can be defined as a product term that is 1 in exactly one row of the truth table. Likewise, a maxterm can be defined as a sum term that is 0 in exactly one row of the truth table. Table 3–6 shows this correspondence for a 3-variable truth table.

An *n*-variable minterm can be represented by an *n*-bit integer, the *minterm* *minterm number*
number. We'll use the name *minterm i* to denote the minterm corresponding to *minterm i*

maxterm i

row i of the truth table. In minterm i, a particular variable appears complemented if the corresponding bit in the binary representation of i is 0; otherwise, it is uncomplemented. For example, row 5 has binary representation 101 and the corresponding minterm is $X \cdot Y' \cdot Z$. As you might expect, the correspondence for maxterms is just the opposite: in *maxterm i*, a variable appears complemented if the corresponding bit in the binary representation of i is 1. Thus, maxterm 5 (101) is $X' + Y + Z'$.

canonical sum

Based on the correspondence between the truth table and minterms, we can easily create an algebraic representation of a logic function from its truth table. The *canonical sum* of a logic function is a sum of the minterms corresponding to truth-table rows (input combinations) for which the function produces a 1 output. For example, the canonical sum for the logic function in Table 3–5 is

$$F = \Sigma_{X,Y,Z}(0,3,4,6,7)$$
$$= X' \cdot Y' \cdot Z' + X' \cdot Y \cdot Z + X \cdot Y' \cdot Z' + X \cdot Y \cdot Z' + X \cdot Y \cdot Z$$

minterm list

Here, the notation $\Sigma_{X,Y,Z}(0,3,4,6,7)$ is a *minterm list* and means "the sum of minterms 0, 3, 4, 6, and 7 with variables X, Y, and Z." You can visualize that each minterm "turns on" the output for exactly one input combination. Any logic function can be written as a canonical sum.

canonical product

The *canonical product* of a logic function is a product of the maxterms corresponding to input combinations for which the function produces a 0 output. For example, the canonical product for the logic function in Table 3–5 is

$$F = \Pi_{X,Y,Z}(1,2,5)$$
$$= (X + Y + Z') \cdot (X + Y' + Z) \cdot (X' + Y + Z')$$

maxterm list

Here, the notation $\Pi_{X,Y,Z}(1,2,5)$ is a *maxterm list* and means "the product of maxterms 1, 2, and 5 with variables X, Y, and Z." You can visualize that each maxterm "turns off" the output for exactly one input combination. Any logic function can be written as a canonical product.[3]

It's easy to convert between a minterm list and a maxterm list. For a function of n variables, the possible minterm and maxterm numbers are in the set $\{0, 1, \ldots, 2^n - 1\}$; a minterm or maxterm list contains a subset of these numbers. To switch between list types, take the set complement, for example,

$$\Sigma_{A,B,C}(0,1,2,3) = \Pi_{A,B,C}(4,5,6,7)$$
$$\Sigma_{X,Y}(1) = \Pi_{X,Y}(0,2,3)$$
$$\Sigma_{W,X,Y,Z}(0,1,2,3,5,7,11,13) = \Pi_{W,X,Y,Z}(4,6,8,9,10,12,14,15)$$

---

[3] You can tell from the parallelism in the preceding paragraphs that duality not only halves what you have to learn. It also halved the amount of original material that I had to write!

We have now learned five possible representations for a combinational logic function:

(1) A truth table.

(2) An algebraic sum of minterms, the canonical sum.

(3) A minterm list using the Σ notation.

(4) An algebraic product of maxterms, the canonical product.

(5) A maxterm list using the Π notation.

Each one of these representations specifies exactly the same information; given any one of them, we can derive the other four using a simple mechanical process.

3.2 COMBINATIONAL CIRCUIT ANALYSIS

We analyze a combinational logic circuit by obtaining a formal description of its logic function. Once we have a description of the logic function, a number of other operations are possible:

- We can determine the behavior of the circuit for various input combinations.

- We can manipulate an algebraic description to suggest different circuit structures for the logic function.

- We can transform an algebraic description into a standard form corresponding to a readily available circuit structure. For example, a sum-of-products expression can be transferred directly to a device called a "programmable logic array."

- We can use an algebraic description of the circuit's functional behavior in the analysis of a larger system that includes the circuit.

Given a logic diagram for a combinational circuit, such as Figure 3–9, there are a number of ways to obtain a formal description of the circuit's function. The most primitive functional description is the truth table.

Using only the basic axioms of switching algebra, we can obtain the truth table of an n-input circuit by working our way through all 2^n input combinations. For each input combination, we determine all of the gate outputs produced by that input, propagating information from the circuit inputs to the circuit outputs. Figure 3–10 applies this "exhaustive" technique to our example circuit. Written on each signal line in the circuit is a sequence of eight logic values, the values present on that line when the circuit inputs XYZ are 000, 001, ..., 111. The truth table can be written by transcribing the output sequence of the final OR gate, as

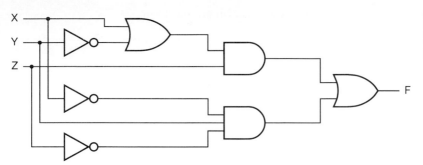

Figure 3–9
A three-input, one-output logic circuit.

shown in Table 3–7. Once we have the truth table for the circuit, we can also directly write a logic expression—the canonical sum or product—if we wish.

The number of input combinations of a logic circuit grows exponentially with the number of inputs, so the exhaustive approach can quickly become exhausting. Instead, we normally use an algebraic approach whose complexity is more linearly proportional to the size of the circuit. The method is simple—we build up a parenthesized logic expression corresponding to the logic operators and structure of the circuit. We start at the circuit inputs and propagate expressions through gates towards the output. Using the theorems of switching algebra,

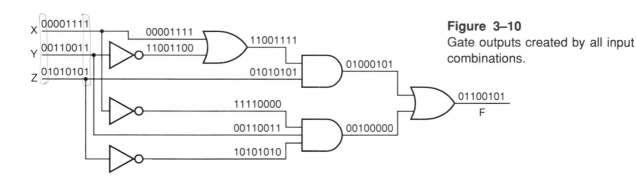

Figure 3–10
Gate outputs created by all input combinations.

Table 3–7
Truth table for the logic circuit of Figure 3–10.

Row	X	Y	Z	F
0	0	0	0	0
1	0	0	1	1
2	0	1	0	1
3	0	1	1	0
4	1	0	0	0
5	1	0	1	1
6	1	1	0	0
7	1	1	1	1

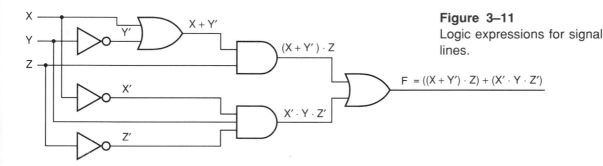

Figure 3–11
Logic expressions for signal lines.

we may simplify the expressions as we go, or we may defer all algebraic manipulations until an output expression is obtained.

Figure 3–11 applies the algebraic technique to our example circuit. The output function is given on the output of the final OR gate:

$$F = ((X + Y') \cdot Z) + (X' \cdot Y \cdot Z')$$

No switching-algebra theorems were used in obtaining this expression. However, we can use theorems to transform this expression into another form. For example, a sum of products can be obtained by "multiplying out":

$$F = X \cdot Z + Y' \cdot Z + X' \cdot Y \cdot Z'$$

The new expression corresponds to a different circuit for the same logic function, as shown in Figure 3–12. Similarly, we can "add out" the original expression to obtain a product of sums:

$$
\begin{aligned}
F &= ((X + Y') \cdot Z) + (X' \cdot Y \cdot Z') \\
&= (X + Y' + X') \cdot (X + Y' + Y) \cdot (X + Y' + Z') \cdot (Z + X') \cdot (Z + Y) \cdot (Z + Z') \\
&= 1 \cdot 1 \cdot (X + Y' + Z') \cdot (X' + Z) \cdot (Y + Z) \cdot 1 \\
&= (X + Y' + Z') \cdot (X' + Z) \cdot (Y + Z)
\end{aligned}
$$

The corresponding logic circuit is shown in Figure 3–13.

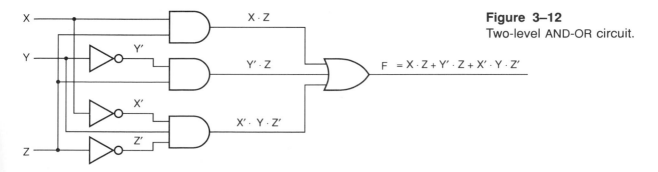

Figure 3–12
Two-level AND-OR circuit.

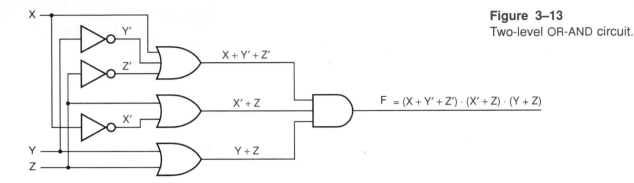

Figure 3–13
Two-level OR-AND circuit.

Our next example of algebraic analysis uses a circuit with NAND and NOR gates, shown in Figure 3–14. This analysis is a little messier than the previous example, because each gate produces a complemented subexpression, not just a simple sum or product. However, the output expression can be simplified by repeated application of the generalized DeMorgan's theorem:

$$
\begin{aligned}
F &= [((W \cdot X')' \cdot Y)' + (W' + X + Y')' + (W + Z)']' \\
&= ((W' + X)' + Y')' \cdot (W \cdot X' \cdot Y)' \cdot (W' \cdot Z')' \\
&= ((W \cdot X')' \cdot Y) \cdot (W' + X + Y') \cdot (W + Z) \\
&= ((W' + X) \cdot Y) \cdot (W' + X + Y') \cdot (W + Z)
\end{aligned}
$$

Quite often, DeMorgan's theorem can be applied *graphically* to simplify algebraic analysis. Recall from Figures 3–3 and 3–4 that NAND and NOR gates each have two equivalent symbols. By judiciously redrawing Figure 3–14, we make it possible to cancel out some of the inversions during the analysis by using theorem T4 [(X')' = X], as shown in Figure 3–15. This manipulation leads us to a simplified output expression directly:

$$
F = ((W' + X) \cdot Y) \cdot (W' + X + Y') \cdot (W + Z)
$$

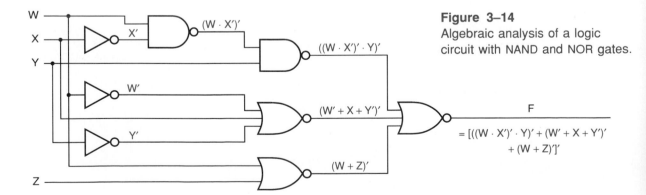

Figure 3–14
Algebraic analysis of a logic circuit with NAND and NOR gates.

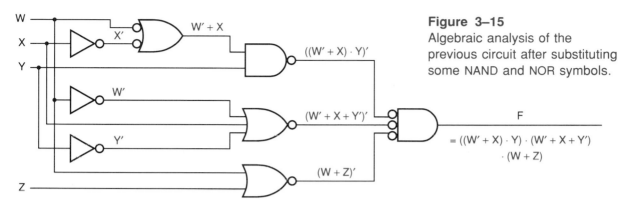

Figure 3–15
Algebraic analysis of the previous circuit after substituting some NAND and NOR symbols.

Figures 3–14 and 3–15 were just two different ways of drawing the same physical logic circuit. However, when we simplify a logic expression using the theorems of switching algebra, we get an expression corresponding to a different physical circuit. For example, the simplified expression above corresponds to the circuit of Figure 3–16, which is physically different from the one in the previous two figures. Furthermore, we could multiply out and add out the expression to obtain sum-of-products and product-of-sums expressions corresponding to two more physically different circuits for the same logic function.

Although we used logic expressions above to convey information about the physical structure of a circuit, we don't always do this. For example, we might use the expression $G(W,X,Y,Z) = W \cdot X \cdot Y + Y \cdot Z$ to describe any one of the circuits in Figure 3–17. Normally, the only sure way to determine a circuit's structure is to look at its schematic drawing. However, for certain restricted classes of circuits, structural information can be inferred from logic expressions. For example, the circuit in Figure 3–17(a) could be described without reference to the drawing if we said that it is a "two-level AND-OR circuit for $W \cdot X \cdot Y + Y \cdot Z$," while the circuit in Figure 3–17(b) could be described as a "two-level NAND-NAND circuit for $W \cdot X \cdot Y + Y \cdot Z$."

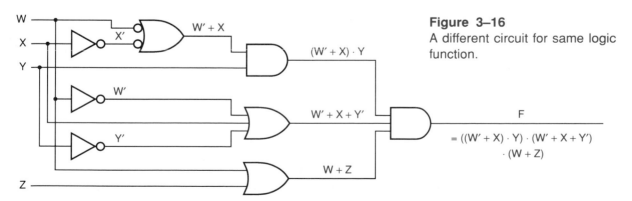

Figure 3–16
A different circuit for same logic function.

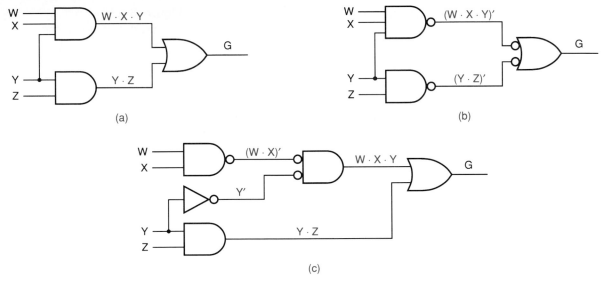

Figure 3–17 Three circuits for $G(W,X,Y,Z) = W \cdot X \cdot Y + Y \cdot Z$: (a) two-level AND-OR; (b) two-level NAND-NAND; (c) ad hoc.

3.3 COMBINATIONAL CIRCUIT SYNTHESIS

3.3.1 Circuit Descriptions and Designs

What is the starting point for designing combinational logic circuits? Usually, we are given a word description of a problem or we develop one ourselves. Occasionally, the description might be a list of input combinations for which a signal should be on or off, the verbal equivalent of a truth table or the Σ or Π notation introduced in the preceding section. For example, the description of a 4-bit prime-number detector might be, "Given a 4-bit input combination $N = N_3N_2N_1N_0$, this function produces a 1 output for $N = 1,2,3,5,7,11$, and 13, and 0 otherwise." A logic function described in this way can be designed directly from the canonical sum or product expression. For the prime-number detector, we have

$$
\begin{aligned}
F &= \Sigma_{N_3,N_2,N_1,N_0}(1,2,3,5,7,11,13) \\
&= N_3{'}N_2{'}N_1{'}N_0 + N_3{'}N_2{'}N_1N_0{'} + N_3{'}N_2{'}N_1N_0 + N_3{'}N_2N_1{'}N_0 \\
&\quad + N_3{'}N_2N_1N_0 + N_3N_2{'}N_1N_0 + N_3N_2N_1{'}N_0
\end{aligned}
$$

The corresponding circuit is shown in Figure 3–18.

More often, we describe a logic function using the English-language connectives "and," "or," and "not." For example, we might describe an alarm circuit by saying, "The ALARM output is 1 if the PANIC input is 1, or if the ENABLE input is 1, the EXITING input is 0, and the house is not secure; the house is secure

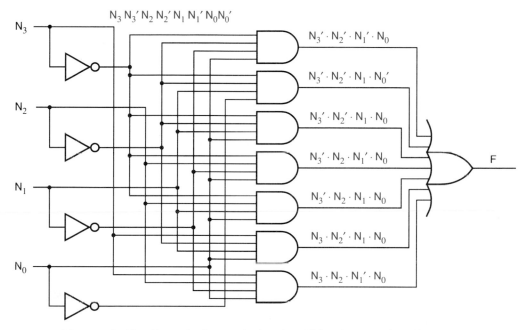

Figure 3–18 Canonical-sum design for 4-bit prime-number detector.

if the WINDOW, DOOR, and GARAGE inputs are all 1." Such a description can be translated directly into algebraic expressions:

$$\text{ALARM} = \text{PANIC} + \text{ENABLE} \cdot \text{EXITING}' \cdot \text{SECURE}'$$
$$\text{SECURE} = \text{WINDOW} \cdot \text{DOOR} \cdot \text{GARAGE}$$
$$\text{ALARM} = \text{PANIC} + \text{ENABLE} \cdot \text{EXITING}' \cdot (\text{WINDOW} \cdot \text{DOOR} \cdot \text{GARAGE})'$$

Notice that we used the same method in switching algebra as in ordinary algebra to formulate a complicated expression—we defined an auxiliary variable SECURE to simplify the first equation, developed an expression for SECURE, and used substitution to arrive at the final expression. We can easily draw a circuit using AND, OR, and NOT gates that realizes the final expression, as shown in Figure 3–19. (We say that a circuit *realizes* ("makes real") an expression if *realize* its output function equals that expression, and the circuit is called a *realization* *realization* of the function.)

Once we have an expression, any expression, for a logic function, we can do other things besides building a circuit directly from the expression. We can manipulate the expression to get different circuits. For example, the ALARM expression above can be multiplied out to get the sum-of-products circuit in Figure 3–20. Or, if the number of variables is not too large, we can construct the truth table for the expression and use any of the synthesis methods that apply

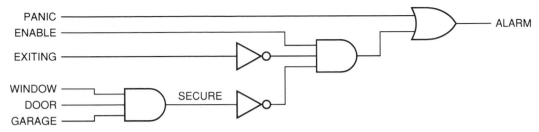

Figure 3–19   Alarm circuit derived from logic expression.

to truth tables, including the canonical sum or product method described earlier and the minimization methods described later.

In general, it's easier to describe a circuit in words using logical connectives and to write the corresponding logic expressions than it is to write a complete truth table, especially if the number of variables is large. However, sometimes we have to work with imprecise word descriptions of logic functions, for example, "The ERROR output should be 1 if the GEARUP, GEARDOWN, and GEARCHECK inputs are inconsistent." In this situation, the truth-table approach makes the most sense because it allows us to decide what the output should be for every input combination, based on our knowledge and understanding of the problem environment.

3.3.2   Circuit Manipulations

The design methods that we've described so far use AND, OR, and NOT gates. We might like to use NAND and NOR gates, too—they're faster than ANDs and ORs in most technologies. However, most people don't develop logical propositions in terms of NAND and NOR connectives. That is, you probably wouldn't say, "I won't go out with you if you're not clean and nonthreatening and you're not smart and not obnoxious." It would be more natural for you to say, "I'll go out with you if you're clean and nonthreatening, or if you're smart and not

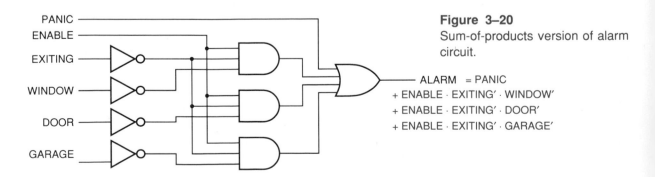

Figure 3–20
Sum-of-products version of alarm circuit.

ALARM = PANIC
 + ENABLE · EXITING′ · WINDOW′
 + ENABLE · EXITING′ · DOOR′
 + ENABLE · EXITING′ · GARAGE′

Figure 3–21
Alternative sum-of-products
realizations: (a) AND-OR;
(b) AND-OR with extra inverter
pairs; (c) NAND-NAND.

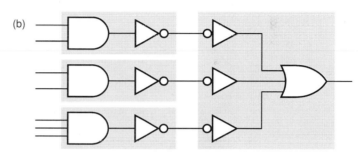

obnoxious." So, given a "natural" logic expression, we need ways to translate it into other forms.

We can translate any logic expression into an equivalent sum-of-products expression, simply by multiplying it out. As shown in Figure 3–21(a), such an expression may be realized directly with AND and OR gates. The inverters required for complemented inputs are not shown.

As shown in Figure 3–21(b), we may insert a pair of inverters between each AND-gate output and the corresponding OR-gate input in a two-level AND-OR circuit. According to theorem T4, these inverters have no effect on the output function of the circuit. In fact, we've drawn the second inverter of each pair with its inversion bubble on its input to provide a graphical reminder that the inverters cancel. However, if these inverters are absorbed into the AND and OR gates, we wind up with AND-NOT gates at the first level and a NOT-OR gate at the second level. These are just two different symbols for the same type of gate—a NAND gate. Thus, a two-level *AND-OR circuit* may be converted to a two-level *NAND-NAND circuit* simply by substituting gates.

AND-OR circuit
NAND-NAND circuit

If any of the product terms in the sum-of-products expression contain just a single literal, then we may gain or lose inverters in the transformation from AND-OR to NAND-NAND. For example, Figure 3–22 is an example where an inverter on the W input is no longer needed, but an inverter must be added to the Z input.

We have shown that any sum-of-products expression can be realized in ei-

Figure 3–22
Another two-level sum-of-products circuit: (a) AND-OR; (b) AND-OR with extra inverter pairs; (c) NAND-NAND.

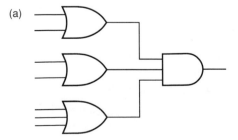

ther of two ways—as an AND-OR circuit or as a NAND-NAND circuit. The dual of this statement is also true: any product-of-sums expression can be realized as an *OR-AND* circuit or as a *NOR-NOR* circuit. Figure 3–23 shows an example. Any logic expression can be translated into an equivalent product-of-sums expression by adding it out, and hence has both OR-AND and NOR-NOR circuit realizations. The same kind of manipulations can be applied to arbitrary logic circuits.

OR-AND circuit
NOR-NOR circuit

Figure 3–23
Realizations of a product-of-sums expression: (a) OR-AND; (b) OR-AND with extra inverter pairs; (c) NOR-NOR.

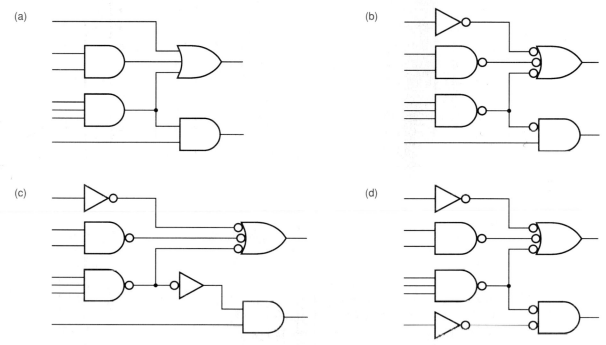

Figure 3–24 Manipulations of logic symbols: (a) original circuit; (b) transformation including a non-standard gate; (c) inverter used to eliminate nonstandard gate; (d) preferred inverter placement.

For example, Figure 3–24(a) shows a circuit built from AND and OR gates. After adding pairs of inverters, we obtain the circuit in Figure 3–24(b). However, one of the gates, a 2-input AND gate with a single inverted input, is not a standard type. We can use a discrete inverter as shown in Figure 3–24(c) to obtain a circuit that uses only standard gate types—NAND, AND, and inverters. Actually, a better way to use the inverter is shown in Figure 3–24(d); one level of gate delay is eliminated, and the bottom gate becomes a NOR instead of AND. In most logic technologies, inverting gates like NAND and NOR are faster than noninverting gates like AND and OR.

3.3.3 Combinational Circuit Minimization

It's often uneconomical to build logic circuits directly from the first logic expression that pops into our heads. Canonical sum and product expressions are especially expensive, because the number of possible minterms or maxterms (and hence gates) grows exponentially with the number of variables. We *minimize* a   *minimize* combinational circuit by reducing the number and size of gates that are needed to build it.

The traditional combinational circuit minimization methods that we'll study have as their starting point a truth table or, equivalently, a minterm list or maxterm list. If we are given a logic function that is not expressed in this form, then we must convert it to an appropriate form before using these methods. For example, if we are given an arbitrary logic expression, then we can evaluate it for every input combination to construct the truth table. The minimization methods reduce the cost of a two-level AND-OR, OR-AND, NAND-NAND, or NOR-NOR circuit in three ways:

(1) By minimizing the number of first-level gates.

(2) By minimizing the number of inputs on each first-level gate.

(3) By minimizing the number of inputs on the second-level gate. This is actually a side effect of the first reduction.

However, the minimization methods do not consider the cost of input inverters; they assume that both true and complemented versions of all input variables are available. While this is not the case in SSI-based design, it's quite appropriate for modern circuits built with programmable logic devices; such devices have both true and complemented versions of all input variables available "for free."

Most minimization methods are based on a generalization of the combining theorems, T10 and T10':

$$\text{given product term} \cdot Y + \text{given product term} \cdot Y' = \text{given product term}$$
$$(\text{given sum term} + Y) \cdot (\text{given sum term} + Y') = \text{given sum term}$$

That is, if two product or sum terms differ only in the complementing or not of one variable, we can combine them into a single term with one less variable. So we save one gate and the remaining gate has one fewer input.

We can apply this algebraic method repeatedly to combine minterms 1, 3, 5, and 7 of the prime-number detector that was shown in Figure 3–18:

$$
\begin{aligned}
F &= \Sigma_{N_3,N_2,N_1,N_0}(1,3,5,7,2,11,13) \\
&= N_3'N_2'N_1'N_0 + N_3'N_2'N_1N_0 + N_3'N_2N_1'N_0 + N_3'N_2N_1N_0 + \ldots \\
&= (N_3'N_2'N_1'N_0 + N_3'N_2'N_1N_0) + (N_3'N_2N_1'N_0 + N_3'N_2N_1N_0) + \ldots \\
&= N_3'N_2'N_0 + N_3'N_2N_0 + \ldots \\
&= N_3'N_0 + \ldots
\end{aligned}
$$

The resulting circuit is shown in Figure 3–25; it has three fewer gates and one of the remaining gates has only two inputs.

If we had worked a little harder on the preceding expression, we could have saved a couple more first-level gate inputs, though not any gates. It's difficult

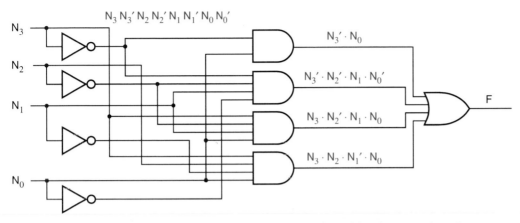

Figure 3–25   Simplified sum-of-products realization for 4-bit prime-number detector.

to find terms that can be combined in a jumble of algebraic symbols. In the next subsection, we'll begin to explore a minimization method that is more fit for human consumption. Our starting point will be the graphical equivalent of a truth table.

3.3.4 Karnaugh Maps

A *Karnaugh map* is a graphical representation of a logic function's truth table.   *Karnaugh map*
Figure 3–26 shows Karnaugh maps for logic functions of 2, 3, and 4 variables. The map for an n-input logic function is an array with 2^n cells, one for each possible input combination or minterm.

The rows and columns of a Karnaugh map are labeled so that the input combination for any cell is easily determined from the row and column headings for that cell. The small number inside each cell is the corresponding minterm

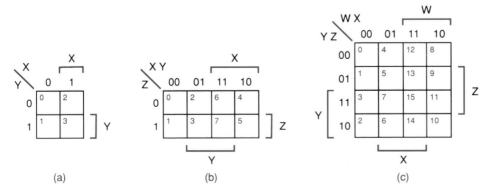

Figure 3–26   Karnaugh maps: (a) 2-variable; (b) 3-variable; (c) 4-variable.

Figure 3–27
Truth table and Karnaugh map
for $F = \Sigma_{X,Y,Z}(1,2,5,7)$.

X	Y	Z	F
0	0	0	0
0	0	1	1
0	1	0	1
0	1	1	0
1	0	0	0
1	0	1	1
1	1	0	0
1	1	1	1

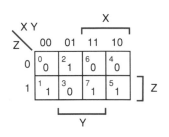

number in the truth table, assuming that the truth table inputs are labeled alpha-betically from left to right (e.g., X, Y, Z) and the rows are numbered in binary counting order, like all the examples in this text. For example, cell 13 in the 4-variable map corresponds to the truth-table row in which W X Y Z = 1101.

When we draw the Karnaugh map for a given function, each cell of the map contains the information from the like-numbered row of the function's truth table—a 0 if the function is 0 for that input combination, a 1 otherwise.

In this text, we use two redundant labelings for map rows and columns. For example, consider the 4-variable map in Figure 3–26(c). The columns are labeled with the four possible combinations of W and X, W X = 00, 01, 11, and 10. Similarly, the rows are labeled with the Y Z combinations. These labels give us all the information we need. However, we also use brackets to associate four regions of the map with the four variables. Each bracketed region is the part of the map in which the indicated variable is 1. Obviously, the brackets convey the same information that is given by the row and column labels.

When we draw a map by hand, it is much easier to draw the brackets than to write out all of the labels. So feel free to do so in your homework assignments. However, we retain the labels in the text's Karnaugh maps as an additional aid to understanding. In any case, you must be sure to label the rows and columns in the proper order to preserve the correspondence between map cells and truth table row numbers shown in Figure 3–26.

To represent a logic function on a Karnaugh map, we simply copy 1s and 0s from the truth table or equivalent to the corresponding cells of the map. Figure 3–27 shows the truth table and Karnaugh map for a logic function that we analyzed (beat to death?) in Section 3.2. From now on, we'll reduce the clutter in maps by copying only the 1s or the 0s, not both.

3.3.5   Minimizing Sums of Products

By now you must be wondering about the "strange" ordering of the row and column numbers in a Karnaugh map. There is a very important reason for this ordering—each cell corresponds to an input combination that differs from each

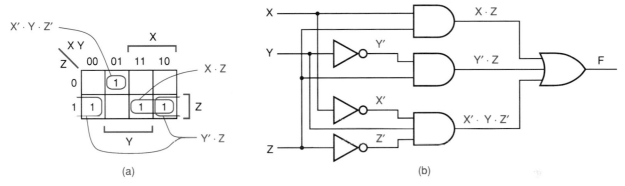

Figure 3–28 $F = \Sigma_{X,Y,Z}(1,2,5,7)$: (a) Karnaugh map combining adjacent 1-cells; (b) AND-OR circuit.

of its immediately adjacent neighbors in only one variable. For example, cells 5 and 13 in the 4-variable map differ only in the value of **W**. In the 3- and 4-variable maps, corresponding cells on the left/right or top/bottom borders are less obvious neighbors; for example, cells 12 and 14 in the 4-variable map are adjacent because they differ only in the value of Y.

Each input combination with a "1" in the truth table corresponds to a minterm in the logic function's canonical sum. Since pairs of adjacent "1" cells in the Karnaugh map have minterms that differ in only one variable, the minterm pairs can be combined into a single product term using the generalization of theorem T10, $term \cdot Y + term \cdot Y' = term$. Thus, we can use a Karnaugh map to simplify the canonical sum of a logic function. For example, consider cells 5 and 7 in Figure 3–27, and their contribution to the canonical sum for this function:

$$
\begin{aligned}
F &= \ldots + X \cdot Y' \cdot Z + X \cdot Y \cdot Z \\
&= \ldots + (X \cdot Z) \cdot Y' + (X \cdot Z) \cdot Y \\
&= \ldots + X \cdot Z
\end{aligned}
$$

Remembering wraparound, we see that cells 1 and 5 in Figure 3–27 are also adjacent and can be combined:

$$
\begin{aligned}
F &= X' \cdot Y' \cdot Z + X \cdot Y' \cdot Z + \ldots \\
&= X' \cdot (Y' \cdot Z) + X \cdot (Y' \cdot Z) + \ldots \\
&= Y' \cdot Z + \ldots
\end{aligned}
$$

In general, we can simplify a logic function by combining pairs of adjacent 1-cells (minterms) whenever possible, and writing a sum of product terms that cover all of the 1-cells. Figure 3–28(a) shows the result for our example logic function. We circle a pair of 1s to indicate that the corresponding minterms are combined into a single product term. The corresponding AND/OR circuit is shown in (b).

In many logic functions, the cell-combining procedure can be extended to combine more than two 1-cells into a single product term. For example, consider the canonical sum for the logic function $F = \Sigma_{X,Y,Z}(0,1,4,5,6)$. We can use the algebraic manipulations of the previous examples iteratively to combine four of the five minterms:

$$
\begin{aligned}
F &= X' \cdot Y' \cdot Z' + X' \cdot Y' \cdot Z + X \cdot Y' \cdot Z' + X \cdot Y' \cdot Z + X \cdot Y \cdot Z' \\
&= [(X' \cdot Y') \cdot Z' + (X' \cdot Y') \cdot Z] + [(X \cdot Y') \cdot Z' + (X \cdot Y') \cdot Z] + X \cdot Y \cdot Z' \\
&= X' \cdot Y' + X \cdot Y' + X \cdot Y \cdot Z' \\
&= [X' \cdot (Y') + X \cdot (Y')] + X \cdot Y \cdot Z' \\
&= Y' + X \cdot Y \cdot Z'
\end{aligned}
$$

In general, 2^i 1-cells may be combined to form a product term containing $n - i$ literals, where n is the number of variables in the function.

A precise mathematical rule determines how 1-cells may be combined and the form of the corresponding product term:

A set of 2^i 1-cells may be combined if there are i variables of the logic function that take on all 2^i possible combinations within that set, while the remaining $n - i$ variables have the same value throughout that set. The corresponding product term has $n - i$ literals, where a variable is complemented if it appears as 0 in all of the 1-cells, and uncomplemented if it appears as 1.

rectangular sets of 1s

Graphically, this rule means that we can circle *rectangular* sets of 2^n 1s, literally as well as figuratively stretching the definition of rectangular to account for wraparound at the edges of the map. We can determine the literals of the corresponding product terms directly from the map; for each variable we make the following determination:

- If a circle covers only areas of the map where the variable is 0, then the variable is complemented in the product term.

- If a circle covers only areas of the map where the variable is 1, then the variable is uncomplemented in the product term.

- If a circle covers both areas of the map where the variable is 0 and areas where it is 1, then the variable does not appear in the product term.

A sum-of-products expression for a function must contain product terms (circled sets of 1-cells) that cover all of the 1s and none of the 0s on the map.

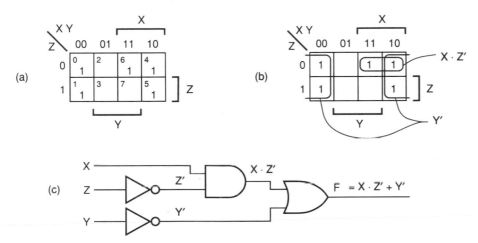

Figure 3–29 $F = \Sigma_{X,Y,Z}(0,1,4,5,6)$: (a) initial Karnaugh map; (b) Karnaugh map with circled product terms; (c) AND/OR circuit.

The Karnaugh map for our most recent example, $F = \Sigma_{X,Y,Z}(0,1,4,5,6)$, is shown in Figure 3–29(a) and (b).[4] We have circled one set of four 1s, corresponding to the product term Y', and a set of two 1s corresponding to the product term $X \cdot Z'$. Notice that the second product term has one less literal than the corresponding product term in our algebraic solution ($X \cdot Y \cdot Z'$). By circling the largest possible set of 1s containing cell 6, we have found a less expensive realization of the logic function, since a 2-input AND gate should cost less than a 3-input one. The fact that two different product terms now cover the same 1-cell (4) does not affect the logic function, since for logical addition $1 + 1 = 1$, not 2! The corresponding two-level AND/OR circuit is shown in Figure 3–29(c).

A Karnaugh map and minimized circuit for our prime-number detector are shown in Figure 3–30. As mentioned earlier, some of the gates in the minimal solution have fewer inputs than we (or at least I) could come up with in the algebraic approach.

At this point, we need some more definitions to clarify what we're doing:

- A *minimal sum* of a logic function $F(X_1,\ldots,X_n)$ is a sum-of-products ex-   *minimal sum*
  pression for F such that no sum-of-products expression for F has fewer
  product terms, and any sum-of-products expression with the same number
  of product terms has at least as many literals.

That is, the minimal sum has the fewest possible product terms (first-level gates and second-level gate inputs) and, within that constraint, the fewest possible

---

[4]Cell numbers are shown in Figure 3–29(a) to show you the correspondence between cell positions and minterm numbers, while they are omitted in (b) to reduce clutter.

$$F = \Sigma_{N3N2N1N0}(1,2,3,5,7,11,13)$$

$$F = N_3' \cdot N_0 + N_3' \cdot N_2' \cdot N_1 + N_2' \cdot N_1 \cdot N_0 + N_2 \cdot N_1' \cdot N_0$$

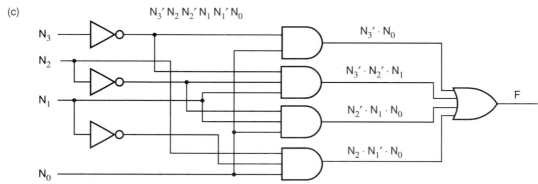

Figure 3–30 Prime-number detector: (a) initial Karnaugh map; (b) Karnaugh map with circled product terms; (c) minimized circuit.

literals (first-level gate inputs). Thus, among our three prime-number detector circuits, only the one in Figure 3–30 realizes a minimal sum.

implies
- A logic function $P(X_1,...,X_n)$ *implies* a logic function $F(X_1,...,X_n)$ if for every input combination such that $P = 1$, then $F = 1$ also.

includes
covers

That is, F is 1 for every input combination that P is 1, and maybe some more. We may write the shorthand $P \Rightarrow F$. We may also say that "F *includes* P," or that "F *covers* P."

prime implicant
- A *prime implicant* of a logic function $F(X_1,...,X_n)$ is a normal product term $P(X_1,...,X_n)$ that implies F, such that if any variable is removed from P, then the resulting product term does not imply F.

In terms of a Karnaugh map, a prime implicant of F is a circled set of 1-cells satisfying our combining rule, such that if we try to make it larger (covering twice as many cells), it covers one or more 0s.

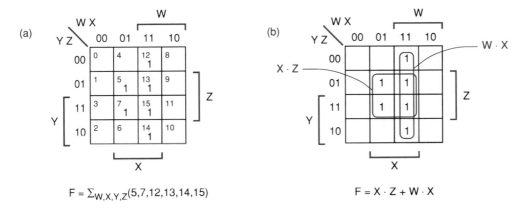

$$F = \Sigma_{W,X,Y,Z}(5,7,12,13,14,15) \qquad\qquad F = X \cdot Z + W \cdot X$$

Figure 3–31 $F = \Sigma_{W,X,Y,Z}(5,7,12,13,14,15)$: (a) Karnaugh map; (b) prime implicants.

Now comes the most important part, a theorem that limits how much work we must do to find a minimal sum for a logic function:

Prime-Implicant Theorem   A minimal sum is a sum of prime implicants.

That is, to find a minimal sum, we need not consider any product terms that are not prime implicants. This theorem is easily proved by contradiction. Suppose that a product term P in a "minimal" sum is *not* a prime implicant. Then according to the definition of prime implicant, if P is not one, it is possible to remove some literal from P to obtain a new product term P^* that still implies F. If we replace P with P^* in the "minimal" sum, the resulting sum still equals F but has one fewer literal. Therefore, the presumed "minimal" sum was not minimal after all.

Another minimization example, this time a 4-variable function, is shown in Figure 3–31. There are just two prime implicants, and it's quite obvious that both of them must be included in the minimal sum in order to cover all of the 1-cells on the map. We didn't draw the logic diagram for this example because you should know how to do that yourself by now.

The sum of all the prime implicants of a logic function is called the *complete*   *complete sum*
sum. Although the complete sum is always a legitimate way to realize a logic function, it's not always minimal. For example, consider the logic function shown in Figure 3–32. It has five prime implicants, but the minimal sum includes only three of them. So, how can we systematically determine which prime implicants to include and which to leave out? You guessed it—we need two more definitions:

- A *distinguished 1-cell* of a logic function is an input combination that is   *distinguished 1-cell*
  covered by only one prime implicant.

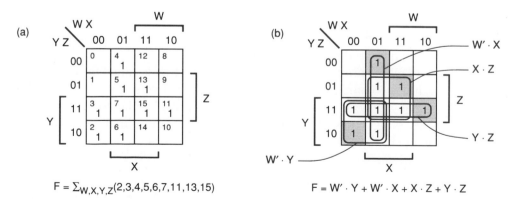

$$F = \Sigma_{W,X,Y,Z}(1,3,4,5,9,11,12,13,14,15)$$

$$F = X \cdot Y' + X' \cdot Z + W \cdot X$$

Figure 3–32 $F = \Sigma_{W,X,Y,Z}(1,3,4,5,9,11,12,13,14,15)$: (a) Karnaugh map; (b) prime implicants and distinguished 1-cells.

essential prime implicant

- An *essential prime implicant* of a logic function is a prime implicant that covers one or more distinguished 1-cells.

Since an essential prime implicant is the *only* prime implicant that covers some 1-cell, it *must* be included in every minimal sum for the logic function. So, the first step in the prime implicant selection process is simple—we identify distinguished 1-cells and the corresponding prime implicants, and include the essential prime implicants in the minimal sum. Then we need only determine how to cover the 1-cells, if any, that are not covered by the essential prime implicants. In the example of Figure 3–32, the three distinguished 1-cells are shaded, and the corresponding essential prime implicants are circled with heavier lines. All of the 1-cells in this example are covered by essential prime implicants, so we need go no further. Likewise, Figure 3–33 shows an example where all of the prime

$$F = \Sigma_{W,X,Y,Z}(2,3,4,5,6,7,11,13,15)$$

$$F = W' \cdot Y + W' \cdot X + X \cdot Z + Y \cdot Z$$

Figure 3–33 $F = \Sigma_{W,X,Y,Z}(2,3,4,5,6,7,11,13,15)$: (a) Karnaugh map; (b) prime implicants and distinguished 1-cells.

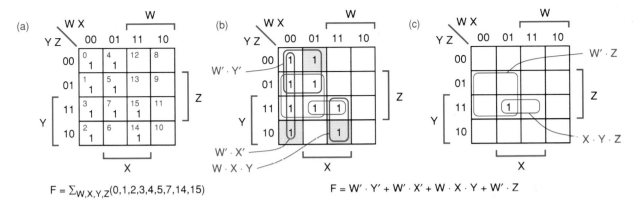

$$F = \Sigma_{W,X,Y,Z}(0,1,2,3,4,5,7,14,15)$$

$$F = W' \cdot Y' + W' \cdot X' + W \cdot X \cdot Y + W' \cdot Z$$

Figure 3–34 $F = \Sigma_{W,X,Y,Z}(0,1,2,3,4,5,7,14,15)$: (a) Karnaugh map; (b) prime implicants and distinguished 1-cells; (c) reduced map after removal of essential prime implicants and covered 1-cells.

implicants are essential, and so all are included in the minimal sum.

A logic function in which not all the 1-cells are covered by essential prime implicants is shown in Figure 3–34. By removing the essential prime implicants and the 1-cells they cover, we obtain a reduced map with only a single 1-cell and two prime implicants that cover it. The choice in this case is simple—we use the $W' \cdot Z$ product term because it has fewer inputs and therefore lower cost.

For more complex cases, we need yet another definition:

- Given two prime implicants P and Q in a reduced map, P is said to *eclipse* Q (written P $\supseteq$ Q) if P covers at least all the 1-cells covered by Q.[5]

eclipse

If P costs no more than Q and eclipses Q, then removing Q from consideration cannot prevent us from finding a minimal sum; that is, P is at least as good as Q. An example is shown in Figure 3–35. After removing essential prime implicants, we are left with two 1-cells, each of which is covered by two prime implicants. However, X·Y·Z eclipses the other two prime implicants, which therefore may be removed from consideration. The two 1-cells are then covered only by X·Y·Z, which is a *secondary essential prime implicant* that must be included in the minimal sum.

secondary essential prime implicant

Figure 3–36 shows a more difficult case—a logic function with no essential prime implicants. By trial and error we can find two different minimal sums for this function. We can also approach the problem systematically using the *branching method*. Starting with any 1-cell, we arbitrarily select one of the prime implicants that covers it, and include it as if it were essential. This simplifies the remaining problem, which we can complete in the usual way to find a tentative

branching method

---

[5]This is a combination of the properties of dominance and equality defined in other texts. The word "include" could also be substituted for "eclipse" in this context.

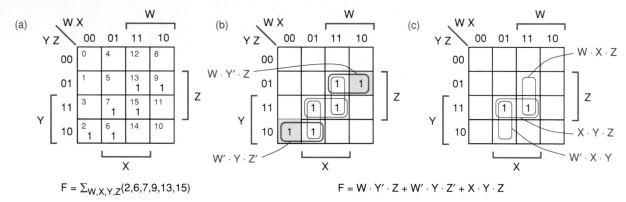

$$F = \Sigma_{W,X,Y,Z}(2,6,7,9,13,15)$$

$$F = W \cdot Y' \cdot Z + W' \cdot Y \cdot Z' + X \cdot Y \cdot Z$$

Figure 3–35 $F = \Sigma_{W,X,Y,Z}(2,6,7,9,13,15)$: (a) Karnaugh map; (b) prime implicants and distinguished 1-cells; (c) reduced map after removal of essential prime implicants and covered 1-cells.

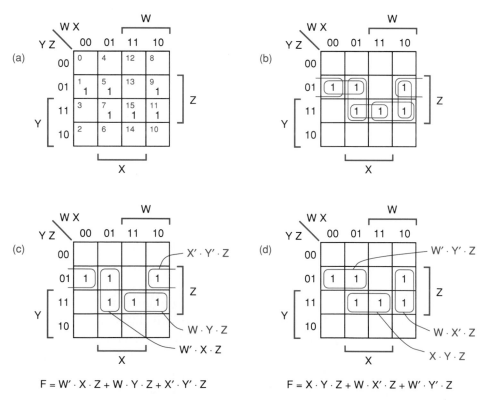

$$F = W' \cdot X \cdot Z + W \cdot Y \cdot Z + X' \cdot Y' \cdot Z$$

$$F = X \cdot Y \cdot Z + W \cdot X' \cdot Z + W' \cdot Y' \cdot Z$$

Figure 3–36 $F = \Sigma_{W,X,Y,Z}(1,5,7,9,11,15)$: (a) Karnaugh map; (b) prime implicants; (c) a minimal sum; (d) another minimal sum.

minimal sum. We repeat this process starting with all other prime implicants that cover the starting 1-cell, generating a different tentative minimal sum from each starting point. We may get stuck along the way and have to apply the branching method recursively. Finally, we examine all of the tentative minimal sums generated in this way and select one that is truly minimal.

Obviously, prime-implicant selection can be a fairly involved process. Automatic programs for logic circuit minimization, discussed in Section 3.4, apply additional optimizations and heuristics not discussed here. In real logic-design applications, you are likely to encounter only two kinds of minimization problems: functions of a few variables that you can "eyeball" using the methods of this section, and more complex, multiple-output functions that are hopeless without the use of a minimization program.

* 3.3.6 Simplifying Products of Sums

Using the principle of duality, we can also minimize product-of-sums expressions, by looking at the 0s on a Karnaugh map. Recall that each 0 on the map corresponds to a maxterm in the canonical product of the logic function. The 0s can be combined to form sum terms with a smaller number of literals using the following rule:

> A set of 2^i 0-cells may be combined if there are i variables of the logic function such that the i variables take on all 2^i possible combinations within that set, while the remaining $n - i$ variables take on the same combination within that set. The resulting sum term has $n - i$ literals, where a literal is complemented if its variable appears as 1 in all of the 0-cells, and uncomplemented if its variable appears as 0.

The rules for including variables in the sum terms are as follows:

- If a circle covers only areas of the map where the variable is 1, then the variable is complemented in the sum term.

- If a circle covers only areas of the map where the variable is 0, then the variable is uncomplemented in the sum term.

- If a circle covers both areas of the map where the variable is 1 and areas where it is 0, then the variable does not appear in the sum term.

Notice that these rules are the precise duals of our rules for combining 1-cells and including variables in the corresponding product terms when forming sums

---

[*]Throughout this book, optional sections are marked with an asterisk.

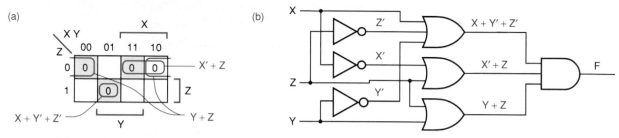

Figure 3–37 $F = \Sigma_{X,Y,Z}(1,2,5,7) = \Pi_{X,Y,Z}(0,3,4,6)$: (a) Karnaugh map; (b) OR/AND circuit.

of products. The circled sets of 0s are called "prime implicates" and are formally defined below:

prime implicate

- A *prime implicate* of a logic function $F(X_1,\ldots,X_n)$ is a normal sum term $S(X_1,\ldots,X_n)$ implied by F, such that if any variable is removed from S, then the resulting sum term is not implied by F.

In terms of a Karnaugh map, a prime implicate of F is a circled set of 0-cells satisfying our combining rule, such that if we try to make it larger (covering twice as many cells), it covers one or more 1s. Finally, we have a dual for the prime-implicant theorem:

Prime-Implicate Theorem A minimal product is a product of prime implicates.

Figure 3–37 shows the product-of-sums procedure for our favorite logic function, $F = \Sigma_{X,Y,Z}(1,2,5,7)$. Since this function is specified as a list of minterms, the first step is to write the corresponding maxterm list, $F = \Pi_{X,Y,Z}(0,3,4,6)$, so we can mark all of the function's 0s on the map. Next, we circle and choose prime implicates, using the same kind of rules as in the sum-of-products procedure. The only notable difference is in the way we form sum terms corresponding to circled sets of 0s—each sum term is precisely the complement of the product term we would have written if the 0s were 1s. Another example is shown in Figure 3–38.

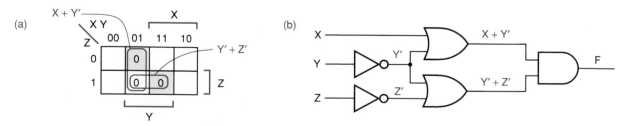

Figure 3–38 $F = \Sigma_{X,Y,Z}(0,1,4,5,6) = \Pi_{X,Y,Z}(2,3,7)$: (a) Karnaugh map; (b) OR/AND circuit.

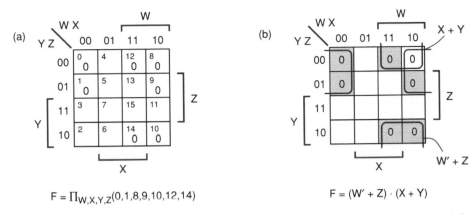

$$F = \Pi_{W,X,Y,Z}(0,1,8,9,10,12,14) \qquad\qquad F = (W' + Z) \cdot (X + Y)$$

Figure 3–39 $F = \Sigma_{W,X,Y,Z}(2,3,4,5,6,7,11,13,15) = \Pi_{W,X,Y,Z}(0,1,8,9,10,12,14)$:
(a) Karnaugh map; (b) prime implicates and distinguished 0-cells.

In general, to find the lowest-cost two-level realization of a logic function, we have to find both a minimal sum of products and a minimal product of sums, and compare them. Both product-of-sums examples above were worked previously as sums of products, in Figures 3–28 and 3–29. In the first example, the OR-AND circuit has exactly the same cost (number of gates and gate inputs) as our earlier AND-OR circuit; in the second case, the AND-OR circuit is cheaper. An example where the OR-AND circuit is cheaper is shown in Figure 3–39; the product of sums has just two sum terms. The corresponding sum of products, which we obtained in Figure 3–33, has four product terms.

If you find the product-of-sums minimization procedure to be a little bit disorienting, you can always use a simple trick. Complement the given function and find a minimal sum-of-products expression using the methods of the preceding subsection. Then complement the result using the generalized DeMorgan's theorem, to obtain a minimal product-of-sums expression for the original function. (Note that if you simply "add out" a minimal sum-of-products expression for the original function, the resulting product-of-sums expression is not necessarily minimal; for example, see Exercise 3.38.)

* 3.3.7 "Don't-Care" Input Combinations

Sometimes the specification of a combinational circuit is such that its output doesn't matter for certain input combinations, called *don't-cares*. This may *don't-care* be true because the outputs really don't matter when these input combinations occur, or because these input combinations never occur in normal operation. For example, suppose we wanted to build a prime-number detector whose 4-bit input $N = N_3 N_2 N_1 N_0$ is always a BCD digit; then minterms 10–15 should never occur. A prime BCD-digit detector function may therefore be written as follows:

$$F = \Sigma_{N3N2N1N0}(1,2,3,5,7) + d(10,11,12,13,14,15) \qquad F = N_3' \cdot N_0 + N_2' \cdot N_1$$

Figure 3–40   Prime BCD-digit detector: (a) initial Karnaugh map; (b) Karnaugh map with prime implicants and distinguished 1-cells.

$$F = \Sigma_{N_3,N_2,N_1,N_0}(1,2,3,5,7) + d(10,11,12,13,14,15)$$

The d(...) term lists the don't-care input combinations for the function. Here F must be 1 for input combinations corresponding to the minterm set (1,2,3,5,7), F can have any values for the don't-care set (10,11,12,13,14,15), and F must be 0 for all other input combinations.

Figure 3–40 shows how to find a minimal sum-of-products realization for the prime BCD-digit detector, including don't-cares. The d's in the map denote the don't-care input combinations. We modify the procedure for circling sets of 1s (prime implicants) as follows:

- Allow d's to be included when circling sets of 1s, to make the sets as large as possible. This reduces the number of variables in the corresponding prime implicants. Two such prime implicants (N_2N_0 and $N_2'N_1$) appear in the example.

- Do not circle any sets that contain only d's. Including the corresponding product term in the function would unnecessarily increase its cost. Two such product terms (N_3N_2 and N_3N_1) are circled with dashed lines in the example.

- Just a reminder: as usual, do not circle any 0s.

The remainder of the procedure is the same. In particular, we look for distinguished 1-cells and *not* distinguished d-cells, and we include only the corresponding essential prime implicants and any others that are needed to cover all the 1s on the map. In the example, the two essential prime implicants are sufficient to cover all of the 1s on the map. Two of the d's also happen to be

$$F = \Sigma_{W,X,Y,Z}(4,5,13,15) + d(2,3,7,9,14) \qquad F(W,X,Y,Z) = W' \cdot X \cdot Y' + X \cdot Z$$

Figure 3–41 $F = \Sigma_{W,X,Y,Z}(4,5,13,15) + d(2,3,7,9,14)$:   (a) Karnaugh map; (b) prime implicants and distinguished 1-cells.

covered, so F will be 1 for don't-care input combinations 10 and 11, and 0 for the other don't-cares.

Figure 3–41 shows another minimization problem with don't-cares. Once again, d's are included to make the circled sets as large as possible, and sets containing only d's are not circled. The single essential prime implicant $W' \cdot X \cdot Y'$ does not cover all the 1s. However, prime implicant $X \cdot Z$ eclipses the other two remaining prime implicants (considering only 1s, not d's), and is added to the first to obtain the minimal sum.

* 3.3.8   Multiple-Output Minimization

Most practical combinational logic circuits require more than one output. We can always handle a circuit with *n* outputs as *n* independent single-output design problems. However, in doing so, we may miss some opportunities for optimization. For example, consider the following two logic functions:

$$F = \Sigma_{X,Y,Z}(3,6,7)$$
$$G = \Sigma_{X,Y,Z}(0,1,3)$$

Figure 3–42 shows the design of F and G as two independent single-output functions. However, as shown in Figure 3–43, we can also find a pair of sum-of-products expressions that share a product term, such that the resulting circuit has one less gate than our original design.

When we design multiple-output combinational circuits using discrete gates, product-term sharing obviously reduces circuit size and cost. Of course, most modern designs don't use discrete gates; they use the larger-scale elements of Chapter 4 and the structured logic elements of Chapter 7. However, as we'll see

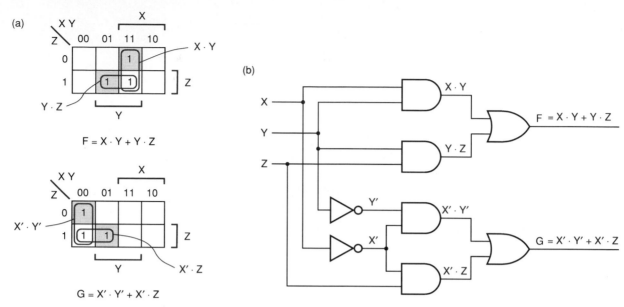

Figure 3–42   Treating a 2-output design as two independent single-output designs: (a) Karnaugh maps; (b) "minimal" circuit.

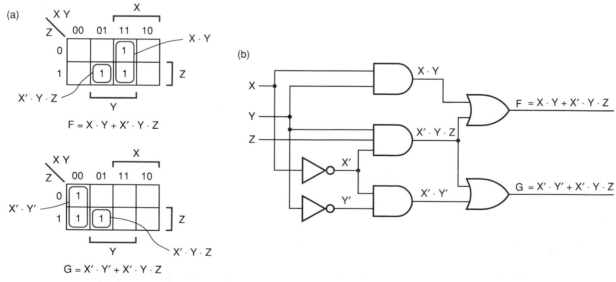

Figure 3–43   Multiple-output minimization for a 2-output circuit: (a) minimized maps including a shared term; (b) minimal multiple-output circuit.

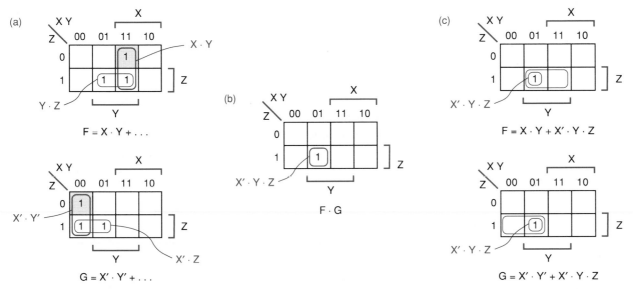

Figure 3-44 Karnaugh maps for a set of two functions: (a) maps for F and G; (b) 2-product map for F · G; (c) reduced maps for F and G after removal of essential prime implicants and covered 1-cells.

in Section 7.2, programmable logic devices (PLDs) contain precisely the sum-of-products structure that we've been learning how to minimize, and they allow product terms to be shared among multiple outputs. Thus, the ideas introduced in this subsection are used extensively in modern logic minimization programs for multiple-output PLDs.

You probably could have "eyeballed" the Karnaugh maps for F and G in Figure 3-43, and discovered the minimal solution. However, larger circuits, such as might be built with a PLD, can be mimimized only with a formal multiple-output minimization algorithm. We'll outline the ideas in such an algorithm here; details can be found in the References.

The key to successful multiple-output minimization of a set of n functions is to consider not only the n original single-output functions, but also "product functions." An *m-product function* of a set of n functions is the product of m of *m-product function* the functions, where $2 \leq m \leq n$. There are $2^n - n - 1$ such functions. Fortunately, $n = 2$ in our example and there is only one product function, F·G, to consider. The Karnaugh maps for F, G, and F·G are shown in Figure 3-44; in general, the map for an m-product function is obtained by ANDing the maps of its m components.

A *multiple-output prime implicant* of a set of n functions is a prime im-   *multiple-output* plicant of one of the n functions or of one of the product functions. The first   *prime implicant* step in multiple-output minimization is to find all of the multiple-output prime implicants. Each prime implicant of an m-product function is a possible term *distinguished 1-cell*

essential prime implicant

Quine-McCluskey algorithm

to include in the corresponding *m* outputs of the circuit. If we were trying to minimize a set of 8 functions, we would have to find the prime implicants for $2^8 - 8 - 1 = 247$ product functions as well as for the 8 given functions. Obviously, multiple-output minimization is not for the faint-hearted!

Once we have found the multiple-output prime implicants, we try to simplify the problem by identifying the essential ones. A *distinguished 1-cell* of a particular single-output function F is a 1-cell that is covered by exactly one prime implicant of F or of the product functions involving F. The distinguished 1-cells in Figure 3–44 are shaded. An *essential prime implicant* of a particular single-output function is one that contains a distinguished 1-cell. As in single-output minimization, the essential prime implicants must be included in a mimimum-cost solution. Only the 1-cells that are not covered by essential prime implicants are considered in the remainder of the algorithm.

The final step is to select a minimal set of prime implicants to cover the remaining 1-cells. In this step we must consider all *n* functions simultaneously, including the possibility of sharing; details of this procedure are discussed in the References. In the example of Figure 3–44(c), we see that there exists a single, shared product term that covers the remaining 1-cell in both F and G.

* 3.4 PROGRAMMED MINIMIZATION METHODS

Most "real" combinational logic functions are far too big to be minimized by the Karnaugh-map approach, but it's still important to minimize them in many situations. That's because, as we'll discuss in Section 7.2, "random" combinational logic functions are most often built nowadays using programmable logic devices (PLDs) that contain a two-level AND-OR structure internally. Minimization isn't just a matter of saving 10% or 20% of the gates in a design. The number of AND gates in a PLD is limited, and so minimization can often make the difference between whether or not a design "fits" at all in an available PLD.

We know that minimization operations can be performed visually for functions of a few variables using the Karnaugh-map method. We'll show in this section that the same operations can be performed for functions of an arbitrarily large number of variables (at least in principle) using a tabular method called the *Quine-McCluskey algorithm*. Like all algorithms, the Quine-McCluskey algorithm can be translated into a computer program. And like the map method, the algorithm has two steps: (a) finding all prime implicants of the function, and (b) selecting a minimal set of prime implicants that covers the function.

The Quine-McCluskey algorithm is often described in terms of handwritten tables and manual check-off procedures. However, since no logic designer

---

*This section and all of its subsections are optional.

ever performs these procedures manually, it's more appropriate for us to discuss the algorithm in terms of data structures and procedures in a high-level programming language. The goal of this section is to give you an appreciation for computational complexity involved in a large minimization problem, and to mention a few alternatives to the traditional algorithm that are used in practice. We consider only fully specified, single-output functions; don't-cares and multiple-output functions can be handled by fairly straightforward modifications to the single-output algorithms, as discussed in the References.

3.4.1 Represention of Product Terms

The starting point for the Quine-McCluskey minimization algorithm is the truth table or, equivalently, the minterm list of a function. If the function is specified differently, it must first be converted into this form. For example, an arbitrary n-variable logic expression can by multiplied out (perhaps using De-Morgan's theorem along the way) to obtain a sum-of-products expression. Once we have a sum-of-products expression, each p-variable product term produces 2^{n-p} minterms in the minterm list.

   We showed in Section 3.1.6 that a minterm of an n-variable logic function can be represented by an n-bit integer (the minterm number), where each bit indicates whether the corresponding variable is complemented or uncomplemented. However, a minimization algorithm must also deal with product terms that are not minterms, where some variables do not appear at all. Thus, we must represent three possibilities for each variable in a general product term:

1 Uncomplemented.

0 Complemented.

x Doesn't appear.

These possibilities are represented by a string of n of the above digits in the *cube representation* of a product term. For example, if we are working with product terms of up to eight variables, X7, X6, ..., X1, X0, we can write the following product terms and their cube representations:

$$X7' \cdot X6 \cdot X5 \cdot X4' \cdot X3 \cdot X2 \cdot X1 \cdot X0' \equiv 01101110$$
$$X3 \cdot X2 \cdot X1 \cdot X0' \equiv xxxx1110$$
$$X7 \cdot X5' \cdot X4 \cdot X3 \cdot X2' \cdot X1 \equiv 1x01101x$$
$$X6 \equiv x1xxxxxx$$

cube representation

Notice that for convenience, we named the variables just like the bit positions in n-bit binary integers.

In terms of the *n*-cube and *m*-subcube nomenclature of Section 1.14, the string 1x01101x represents a 2-subcube of an 8-cube, and the string 01101110 represents a 0-subcube of an 8-cube. However, in the miminimization literature, the maximum dimension *n* of a cube or subcube is usually implicit, and an *m*-subcube is simply called an "*m*-cube" or a "cube" for short; we'll follow this practice in this section.

To represent a product term in a computer program, we can use a data structure with *n* elements, each of which has three possible values. In Pascal, we might make the following declarations:

```
TYPE
  trit = (complemented, uncomplemented, doesntappear);
  cube = ARRAY [0..15] OF trit; {Represents a single product
                                    term with up to 16 variables.}
VAR
  P1, P2, P3: cube;  {Allocate three cubes for use by program.}
```

However, these declarations do not lead to a particularly efficient internal representation of cubes. As we'll see, cubes are easier to manipulate using conventional computer instructions if an *n*-variable product term is represented by two *n*-bit computer words, as suggested by the following declarations:

```
CONST
  maxVars = 16;      {Maximum # of variables in a product term.}
  maxCint = 65535;   {2**(maxVars)-1}
TYPE
  word = 0..maxCint;     {This is a maxVars-bit integer.}
  cube = RECORD t: word; {Bits 1 for uncomplemented variables.}
              f: word; {Bits 1 for complemented variables.}
         END;
VAR
  P1, P2, P3: cube;  {Allocate three cubes for use by program.}
```

Here, a `word` is a 16-bit integer, and a 16-variable product term is represented by a record with two `word`s, as shown in Figure 3–45(a). The first `word` in a `cube` has a 1 for each variable in the product term that appears uncomplemented (or "true," `.t`), and the second has a 1 for each variable that appears complemented (or "false," `.f`). If a particular bit position has 0s in both `word`s, then the corresponding variable does not appear, while the case of a particular bit position having 1s in both `word`s is not used. Thus, the program variable `P1` in (b) represents the product term P1 = X15·X12′·X10′·X9·X4′·X1·X0. If we wished to represent a logic function F of up to 16 variables, containing up to 100 product terms, we could declare an array of 100 `cube`s:

```
VAR
  F: ARRAY [1..100] OF cube; {Storage for a logic function
                              with up to 100 product terms.}
```

Figure 3–45
Internal representation
of 16-variable product
terms in a Pascal program:
(a) general format; (b) P1 =
X15·X12'·X10'·X9·X4'·X1·X0.

Using the foregoing cube representation, it is possible to write short, efficient procedures that manipulate product terms in useful ways. For example, the following Pascal function compares two cubes for equality:

```
FUNCTION EqualCubes(C1, C2: cube): boolean;
BEGIN
  EqualCubes := (C1.t = C2.t) AND (C1.f = C2.f);
END;
```

Figure 3–46 depicts how two cubes can be compared and combined if possible using theorem T10, term·X + term·X' – term. This theorem says that two product terms can be combined if they differ in only one variable that appears complemented in one term and uncomplemented in the other. Combining two m-cubes

Figure 3–46   Cube manipulations: (a) determining whether two cubes are combinable using theorem T10, term·X + term·X' = term; (b) combining cubes using theorem T10.

Table 3–8   A Pascal program that finds prime implicants using the Quine-McCluskey algorithm.

```
PROGRAM FindPIsQuineMcCluskey(input,output);
CONST
  maxVars = 16;          {Max # of variables in a product term.}
  maxCint = 65535;       {2**(maxVars)-1}
  maxCubes = 1000;       {Max # of cubes to allocate per level.}
TYPE
  word = 0..maxCint;     {This is a maxVars-bit integer.}
  cube = RECORD t: word; {Bits 1 for uncomplemented variables.}
                f: word; {Bits 1 for complemented variables.}
         END;
VAR
  cubes: ARRAY [0..maxVars, 1..maxCubes] OF cube;
  covered: ARRAY [0..maxVars, 1..maxCubes] OF boolean;
  numCubes: ARRAY [0..maxVars] OF 0..maxCubes;
  m: 0..maxVars;           {Value of m in an m-cube, i.e., ``level m.''}
  j, k, p: 1..maxCubes;   {Index into the cubes or covered array.}
  tempCube: cube;
  found: boolean;

{Procedure and function definitions should be placed here.}
```

yields an $(m + 1)$-cube. Using cube representation, we can apply the combining theorem to a few examples:

$$010 + 000 = 0x0$$
$$00111001 + 00111000 = 0011100x$$
$$101xx0x0 + 101xx1x0 = 101xxxx0$$
$$x111xx00110x000x + x111xx00010x000x = x111xx00x10x000x$$

In the Pascal algorithm in the next subsection, we assume the existence of a Pascal function and procedure that support the theorem:

```
FUNCTION Combinable(C1, C2: cube): boolean;
{Returns true if C1 and C2 differ in only one variable,
 which appears true in one and false in the other.}
```

```
PROCEDURE Combine(C1, C2: cube; VAR C3: cube);
{Combines C1 and C2 using theorem T10, and stores the
 result in C3.  Assumes Combinable(C1,C2) is true.}
```

These procedures can be written in a few lines of assembly language using the bit-by-bit "logical operations" provided in almost all computers and in the C language. Unfortunately, bit-by-bit logical operations are not provided in standard Pascal, so we have given a pictorial description of the procedures.

Table 3–8 (continued)

```
BEGIN
  {Initialize number of m-cubes at each level m.}
  FOR m := 0 to maxVars DO numCubes[m] := 0;

  {Read a list of minterms (0-cubes) supplied by the user, storing them
   in the cubes[0,j] subarray, setting covered[0,j] to false for each minterm,
   and setting numCubes[0] to the total number of minterms read.}
  ReadMinterms;

  FOR m := 0 to maxVars-1 DO {For all levels except the last...}
    FOR j := 1 TO numCubes[m] DO {For all cubes at this level...}
      FOR k := j+1 TO numCubes[m] DO {For other cubes at this level...}
        BEGIN
          IF Combinable(cubes[m,j], cubes[m,k]) THEN
            BEGIN
              {Mark the cubes as covered.}
              covered[m,j] := true; covered[m,k] := true;
              {Combine into an (m+1)-cube, store in tempCube.}
              Combine(cubes[m,j], cubes[m,k], tempCube);
              found := false; {See if we've generated this one before.}
              FOR p := 1 TO numCubes[m+1] DO
                IF EqualCubes(cubes[m+1,p],tempCube) THEN found := true;
              IF NOT found THEN {Add the new cube to the next level.}
                BEGIN
                  numCubes[m+1] := numCubes[m+1] + 1;
                  cubes[m+1,numCubes[m+1]] := tempCube;
                  covered[m+1,numCubes[m+1]] := false;
                END;
            END;
        END;
  FOR m := 0 to maxVars DO {For all levels...}
    FOR j := 1 TO numCubes[m] DO {For all cubes at this level...}
      {Print uncovered cubes -- these are the prime implicants.}
      IF NOT covered[m,j] THEN PrintCube(cubes[m,j]);
END.
```

3.4.2 Finding Prime Implicants by Combining Product Terms

The first step in the Quine-McCluskey algorithm is to determine all of the prime implicants of the logic function. With a Karnaugh map, we do this visually by identifying "largest possible rectangular sets of 1s." In the algorithm, this is done by systematic, repeated application of theorem T10 to combine minterms, then 1-cubes, 2-cubes, and so on, creating the largest possible cubes (smallest possible product terms) that cover only 1s of the function.

Table 3–8 is a Pascal program that applies the algorithm to functions with up to 16 variables. It declares two 2-dimensional arrays, `cubes[m,j]` and `covered[m,j]`, to keep track of m-cubes for $m = 0$ to `maxVars`. The 0-cubes (minterms) are supplied by the user. Starting with the 0-cubes, the program examines every pair of cubes at each level and combines them when possible into cubes at the next level. Cubes that are combined into a next-level cube are marked as "covered;" cubes that are not covered by a next-level cube are prime implicants.

Even though the program is short, an experienced programmer might become extremely pessimistic just looking at its structure. The inner `FOR` loop is nested four levels deep, and the number of times it might be executed is on the order of `maxVars` $\cdot$ `maxCubes`^{3}. That's right, that's an exponent, not a footnote. We picked the value `maxCubes` = 1000 somewhat arbitrarily (in fact, too optimistically for many functions), but if you believe this number, then the inner loop can be executed *billions and billions* of times.

The maximum number of minterms of an n-variable function is 2^n, of course, and so by all rights the program in Table 3–8 should declare `maxCubes` to be 2^{16}, at least to handle the maximum possible number of 0-cubes. Such a declaration would not be overly pessimistic. If an n-variable function has a product term equal to a single variable, then 2^{n-1} minterms are in fact needed to cover that product term.

For larger cubes, the situation is actually worse. The number of possible m-subcubes of an n-cube is $\binom{n}{m} \times 2^{n-m}$, where the binomial coefficient $\binom{n}{m}$ is the number of ways to choose m variables to be x's, and 2^{n-m} is the number of ways to assign 0s and 1s to the remaining variables. For 16-variable functions, the worst case occurs with $m = 5$; there are 8,945,664 possible 5-subcubes of a 16-cube. The total number of distinct m-subcubes of an n-cube, over all values of m, is 3^n. So a general minimization program might require a *lot* more memory than we've allocated in Table 3–8.

There *are* a few things that we can do to optimize the storage space and execution time required in Table 3–8 (see Exercises 3.60–3.63), but they are piddling compared to the overwhelming combinatorial complexity of the problem. Thus, even with today's fast computers and huge memories, direct application of the Quine-McCluskey algorithm for generating prime implicants is generally limited to functions with only a few variables (less than 15–20).

3.4.3 Finding Prime Implicants by Iterative Consensus

Programmable logic devices have a lot of inputs, typically 16 or more, and PLD-based functions often use most or all of the available inputs. Therefore, a minimization algorithm whose storage requirements and computational complexity grow with the number of inputs, such as the Quine-McCluskey algorithm of the preceding subsection, is not too useful in PLD-based design.

Figure 3–47   Determining whether one cube covers another cube using the covering theorem T9,
termA + termA·termB = termA.

On the other hand, the number of product terms available in a typical PLD is comparatively limited, typically 4–16 per PLD output. Therefore, if a function is going to fit in a PLD at all, it must have only a small number of terms in its minimal sum-of-products expression. Also, designers tend to specify logic functions using a small number of product terms to begin with, far fewer than a canonical sum would require.

Thus, logic functions intended for PLD realization can be minimized by an algorithm whose computational complexity grows with the number of product terms, rather than with the number of input variables. *Iterative consensus* is such an algorithm. The starting point for the iterative consensus algorithm is a sum-of-products expression, or equivalently, a list of cubes. The product terms *need not* be minterms or prime implicants, but *may* be either or anything in between. In other words, the cubes in the list may have any and all dimensions, from 0 to n in an n-variable function. Starting with the list of cubes, the iterative consensus algorithm generates a list of all the prime-implicant cubes of the function, without ever having to generate a full minterm list.

Before describing the iterative consensus algorithm, we'll revisit two theorems of switching algebra that it uses heavily. The first is the covering theorem T9, termA + termA·termB = termA, which can be applied to a pair of cubes as shown in Figure 3–47. The iterative consensus algorithm uses this theorem to eliminate covered cubes; only the larger cube must be retained.

The second theorem, as you must have guessed, is the consensus theorem T11, X·term1 + X′·term2 = X·term1 + X′·term2 + term1·term2. In general, given two product terms, termA and termB, their consensus is said to *exist* if there is a variable X such that we can write

$$termA = X·term1$$
$$termB = X′·term2$$

and if no variable other than X appears complemented in one term and uncomplemented in the other. The product term1·term2 is called the consensus of termA and termB.

iterative consensus algorithm

existence of consensus

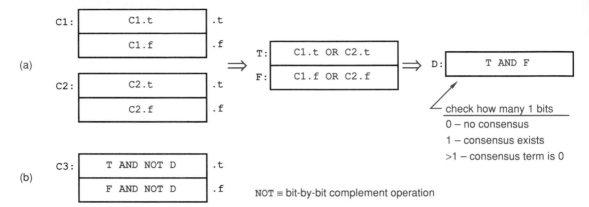

Figure 3–48   Consensus operations on cubes: (a) determining whether the consensus of two terms exists; (b) deriving the consensus term.

As shown in the References, it can be proved that starting with any sum-of-products expression, iteratively adding consensus terms and removing covered terms eventually results in a complete sum. That is, all of the prime implicants of the given function are generated, all other product terms are removed, and the procedure eventually terminates. One version of the iterative consensus algorithm, stated in words, is as follows:

(1) Make a list of the product terms in the given expression.

(2) Starting at the beginning of the list, compare each product term with every product term above it.

(3) If any product term is found to be covered by another, delete the covered term from the list.

(4) If the consensus of two product terms exists, compare the consensus term with all product terms in the list, and add it to the bottom of the list if it is not covered by any other term.

(5) Keep doing this until every product term in the list has been compared with all the ones above it. The remaining terms are all the prime implicants of the function.

Figure 3–48(a) shows how a program can determine whether the consensus exists for two product terms given in cube representation. The consensus term, if it exists, can be formed as shown in (b).

A Pascal program that embodies the iterative consensus algorithm is shown in Table 3–9. The program uses a 1-dimensional array, `cubes[j]`, to store the initial list of cubes, and it adds cubes at the end of the array as it goes along.

The program "deletes" a covered cube by simply marking it in a corresponding array, `covered[j]`; this array is checked at each step so that covered cubes are not considered.

Compared with the Quine-McCluskey program of Table 3–8, the iterative consensus program has two favorable characteristics. First, the number of iterations of the innermost FOR-loop is slightly less, on the order of $\mathtt{maxCubes}^3$ instead of $\mathtt{numVars} \cdot \mathtt{maxCubes}^3$. Second and more important, the number of cubes generated is typically far less; in fact, $\mathtt{maxCubes} = 100$ would work for the majority of practical design problems with PLD realizations.

As in the Quine-McCluskey algorithm, several enhancements can be made to the iterative consensus algorithm to improve its performance. One improvement has been used enough to have a name; it is called the *generalized consensus algorithm*. The basic idea is to execute a shorter procedure `maxVars` times, where execution i generates all possible consensus terms that result from input variable i appearing true in one product term and complemented in another. To be more specific, the generalized consensus algorithm consists of executing the following steps for each value of i:

generalized consensus algorithm

(1) Partition the `cubes` array into three parts, `C1`, `C0`, and `Cx`, according to whether each cube contains a 1, 0, or x for variable i.

(2) Compare every possible pair of cubes that results from taking one cube from `C1` and the other from `C0`. If a consensus exists add it to a list `CC` of consensus cubes.

(3) Compare each consensus cube in `CC` with all cubes in `CC` and `Cx`, and remove any consensus cubes that are covered by other cubes. (Consensus cubes cannot be covered by cubes in `C0` or `C1` [Exercise 3.64]).

(4) Compare the remaining consensus cubes in `CC` with the cubes in `C0`, `C1`, and `Cx`, removing any of the latter that are covered by a consensus cube.

(5) Combine all four arrays and iterate on the next variable.

The innermost loop of this procedure has a worst-case execution time on the order of $\mathtt{numVars} \cdot (.5s \cdot \mathtt{maxCubes})^2 \cdot \mathtt{maxCubes}$, where s is the fraction of all the cubes that go into the `C0` and `C1` arrays on each step. The maximum value of s is 1, but a more typical value for large circuits might be 0.5 or less.

3.4.4 Finding a Minimal Cover Using Prime-Implicant Tables

The second step in minimizing a combinational logic function, once we have a list of all its prime implicants, is to select a minimal subset of them to cover all the 1s of the function. The Quine-McCluskey algorithm uses a two-dimensional

Table 3-9   A Pascal program that finds prime implicants using iterative consensus.

```
PROGRAM FindPIsIterativeConsensus(input,output);
CONST
   maxVars = 16;          {Max # of variables in a product term.}
   maxCint = 65535;       {2**(maxVars)-1}
   maxCubes = 1000;       {Max # of cubes to allocate.}
TYPE
   word = 0..maxCint;     {This is a maxVars-bit integer.}
   cube = RECORD t: word; {Bits 1 for uncomplemented variables.}
                 f: word; {Bits 1 for complemented variables.}
          END;
VAR
   cubes: ARRAY [1..maxCubes] OF cube;
   covered: ARRAY [1..maxCubes] OF boolean;
   numCubes: 0..maxCubes; {Number of cubes actually used in arrays.}
   j, k, p: 1..maxCubes;  {Index into the cubes or covered array.}
   tempCube: cube;
   found: boolean;

{The following procedures operate as shown in the previous two figures.}

FUNCTION Covers(C1, C2: cube): boolean;
{Returns true if C1 covers C2, that is, if C2 has 1s
 everywhere that C1 does.  Note that it is possible for
 Covers(X,Y) and Covers(Y,X) both to return true only if X=Y.}

FUNCTION ConsensusExists(C1, C2: cube): boolean;
{Returns true if the consensus of C1 and C2 exists.}

PROCEDURE StoreConsensus(C1, C2: cube; VAR C3: cube);
{Computes the consensus of C1 and C2, and stores the
 result in C3.  Assumes ConsensusExists(C1,C2) is true.}
```

prime-implicant table

array called a *prime-implicant table* to do this. Figure 3–49(a) shows a small but representative prime-implicant table, corresponding to the Karnaugh-map minimization problem of Figure 3–35. There is one column for each minterm of the function, and one row for each prime implicant. Each entry is a bit that is 1 if and only if the prime implicant for that row covers the minterm for that column (shown in the figure as a check).

The steps for selecting prime implicants with the table are analogous to the steps that we used in Section 3.3.5 with Karnaugh maps:

(1) Identify distinguished 1-cells. These are easily identified in the table as columns with a single 1, as shown in Figure 3–49(b).

(2) Include all essential prime implicants in the minimal sum. A row that

Table 3–9   (continued)

```
BEGIN
  {Read a list of product terms (m-cubes) supplied by the user, storing them
   in the cubes[j] array, setting covered[j] to false for each product term,
   and setting numCubes to the total number of product terms read.}
  ReadProductTerms;

  j := 2;  {Start with the second cube in the array.}
  WHILE j <= numCubes DO    {For each cube...}
    BEGIN
        FOR k := 1 TO j-1 DO  {For all cubes before it in the array.}
          {Look only at pairs of uncovered cubes.}
          IF covered[j] OR covered[k] THEN BEGIN END {Don't bother.}
          {Else mark cubes that are covered, which are subsequently ignored.}
          ELSE IF Covers(cubes[j],cubes[k]) THEN covered[k] := true
          ELSE IF Covers(cubes[k],cubes[j]) THEN covered[j] := true
          {Else check for consensus.}
          ELSE IF ConsensusExists(cubes[j],cubes[k]) THEN
            BEGIN {Consensus exists, store it in tempCube}
              StoreConsensus(cubes[j],cubes[k],tempCube);
              found := false;            {Determine whether tempCube is   }
              FOR p := 1 TO numCubes DO  (  covered by an existing cube.  }
                IF Covers(cubes[p],tempCube) THEN found := true;
              IF NOT found THEN {Add tempCube to the end of array.}
                BEGIN
                  numCubes := numCubes + 1;
                  cubes[numCubes] := tempCube;
                  covered[numCubes] := false;
                END;
            END;
      j := j+1;
    END;
  FOR j := 1 TO numCubes DO {For all cubes...}
    {Print uncovered cubes -- these are the prime implicants.}
    IF NOT covered[j] THEN PrintCube(cubes[j]);
END.
```

contains a check in one or more distinguished-1-cell columns corresponds to an essential prime implicant.

(3) Remove from consideration the essential prime implicants and the 1-cells (minterms) that they cover. In the table, this is done by deleting the corresponding rows and columns, marked in color in Figure 3–49(b). If any rows have no checks remaining, they are also deleted; the corresponding prime implicants are *redundant*, that is, completely covered by essential prime implicants. This step leaves the reduced table shown in (c).

redundant prime implicant

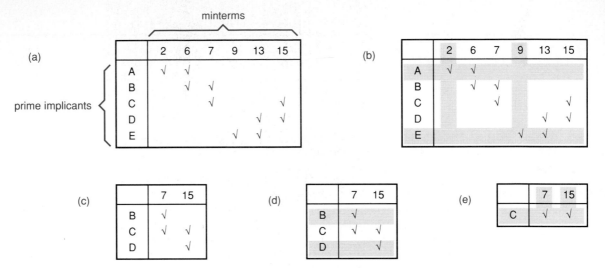

Figure 3–49 Prime-implicant tables: (a) original table; (b) showing distinguished 1-cells and essential prime implicants; (c) after removal of essential prime implicants; (d) showing eclipsed rows; (e) after removal of eclipsed rows, showing secondary essential prime implicant.

(4) Remove from consideration any prime implicants that are eclipsed by others with equal or lesser cost. In the table, this is done by deleting any rows whose checked columns are a proper subset of another row's, and deleting all but one of a set of rows with identical checked columns.[6] This is shown in color in (d), and leads to the further reduced table in (e).

(5) Identify distinguished 1-cells and include all secondary essential prime implicants in the minimal sum. As before, any row that contains a check in one or more distinguished-1-cell columns corresponds to a secondary essential prime implicant.

(6) If all remaining columns are covered by the secondary essential prime implicants, as in (e), we're done. Otherwise, if any secondary essential prime implicants were found in the previous step, we go back to step (3) and iterate. Otherwise, the branching method must be used, as described in Section 3.3.5. This involves picking rows one at a time, treating them as if they were essential, and recursing (and cursing) on steps (3)–(6).

[6]When a function is minimized for a PLD-based design, all of its prime implicants may be considered to have equal cost, because all of the AND gates in a PLD have all of the inputs available. Otherwise, the prime implicants must be sorted and selected according to the number of AND-gate inputs.

Although a prime-implicant table allows a fairly straightforward prime-implicant selection algorithm, the data structure required in a corresponding computer program is huge, since it requires on the order of $p \cdot 2^n$ bits, where p is the number of prime implicants and n is the number of input bits (assuming that the given function produces a 1 output for most input combinations). Worse, executing the steps that we so blithely described in a few sentences above requires a huge amount of computation.

3.4.5 Finding a Minimal Cover Using Iterative Consensus

As we've said before, practical PLD-based designs can have lots of inputs, so an applicable minimization algorithm must not generate or use a minterm list of the function; that rules out prime-implicant tables. However, the iterative consensus algorithm works at least for the first part of the minimization procedure, finding all prime implicants. With a little thought, we can design the second part of the procedure, finding a minimal cover, using iterative consensus as the basic operation, and hope thereby to obtain reasonable performance.

The first step in selecting prime implicants for the minimal cover is to identify essential prime implicants, as in steps (1) and (2) of the prime-implicant-table algorithm in the preceding subsection. And we must do this without generating a minterm list.

Let C denote the set of *all* the prime implicants of the given logic function, and let P denote a prime implicant that we would like to test for being essential. Suppose we temporarily delete P from C and run the iterative consensus algorithm on the remain terms in C. If P is not essential, then all of the minterms it covers are also covered by other terms that remain in C; therefore, the iterative consensus algorithm will recreate P. On the other hand, if P is essential, it will not reappear. This procedure can be applied to all of the prime implicants in C to find all of the essential ones. *essential-test procedure*

Let E be the set of essential prime implicants found above. Now suppose we run the iterative consensus algorithm on E; we'll generate some additional product terms that aren't necessarily prime implicants. However, any of the new terms that *are* prime implicants (i.e., that appear in C) are redundant, since they are completely covered by essential prime implicants. The essential and the redundant prime implicants are removed from C, corresponding to step (3) in the preceding subsection. *remove-redundant procedure*

At this point, if there are no prime implicants left in C, we're done; the function is completely covered by just the essential prime implicants. Otherwise, we must go on to find secondary essential prime implicants. To do this, we must first identify and remove from C any prime implicants that are eclipsed by others, corresponding to step (4) in the preceding subsection. We do this as follows. *remove-eclipsed procedure*

For each prime implicant P remaining in C, we want to know if P plus the terms in E completely cover any other equal- or lower-cost prime implicant Q in C. Any such Q may be deleted from C, since P covers at least as much of what's left to cover as Q does. We can perform this test for each P by temporarily adding P to E and running the iterative consensus algorithm on E. If any Q in C appears in the resulting set of product terms, it is eclipsed by P and is deleted from C, as in step (5) in the preceding subsection.

If the remove-eclipsed procedure removed one or more terms from C, then we have a chance of finding secondary essential prime implicants. So we go back to the essential-test procedure and iterate. Otherwise, we're stuck with the branching method, in which we curse and recurse.

3.4.6 Other Minimization Methods

Although the previous subsections form a good introduction to logic minimization algorithms, the methods they describe are by no means the latest and greatest. Spurred on by the ever increasing density of VLSI chips, many researchers have discovered more effective ways to minimize combinational logic functions. Their results fall roughly into three categories:

(1) *Computational improvements.* Improved algorithms typically use clever data structures or rearrange the order of the steps to reduce the memory requirements and execution time of the classical algorithms.

(2) *Heuristic methods.* Some minimization problems are just too big to be solved using an "exact" algorithm. These problems can be attacked using shortcuts and well-educated guesses to reduce memory size and execution time to a fraction of what an "exact" algorithm would require. However, instead of finding a provably minimal expression for a logic function, heuristic methods attempt to find an "almost minimal" expression. Even for problems that can be solved by an "exact" method, a heuristic method typically finds a good solution ten times faster. The most successful hueristic program, Espresso-II, does in fact produce minimal or near-minimal results for the majority of problems (within one or two product terms), including problems with dozens of inputs and hundreds of product terms.

(3) *Looking at things differently.* As we mentioned earlier, multiple-output minimization can be handled by straightforward, fairly mechanical modifications to single-output minimization methods. However, by looking at multiple-output minimization as a problem in multi-valued (nonbinary) logic, the designers of the Espresso-MV algorithm were able to make substantial performance improvements over Espresso II.

More information on these methods can be found in the References.

* 3.5 TIMING HAZARDS

The analysis methods that we developed in Section 3.2 ignore circuit delay and predict only the *steady-state behavior* of combinational logic circuits. That is, they predict a circuit's output as a function of its inputs under the assumption that the inputs have been stable for a long time, relative to the delays in the circuit's electronics. However, we showed in Section 2.6 that the actual delay from an input change to the corresponding output change in a real logic circuit is nonzero and depends on many factors.

steady-state behavior

Because of circuit delays, the *transient behavior* of a logic circuit may differ from what is predicted by a steady-state analysis. In particular, a circuit's output may produce a short pulse, often called a *glitch*, at a time when steady-state analysis predicts that the output should not change. A *hazard* is said to exist when a circuit has the possibility of producing such a glitch. Whether or not the glitch actually occurs depends on the exact delays and other electrical characteristics of the circuit. Since such parameters are difficult to control in production circuits, a logic designer must be prepared to eliminate hazards (the *possibility* of a glitch) even though a glitch may occur only under a worst-case combination of logical and electrical conditions.

transient behavior

glitch
hazard

Any combinational circuit can be analyzed for the presence of hazards. However, a well-designed digital system is structured so that hazard analysis is not needed for most of its circuits. In such a system, all of the inputs to a combinational circuit are changed at a particular time, and the outputs are not "looked at" until they have had time to settle to a steady-state value. Hazard analysis and elimination is typically needed only in the design of asynchronous sequential circuits, such as the feedback sequential circuits discussed in Section 5.5. You'll rarely have reason to design such a circuit, but if you do, an understanding of hazards will be absolutely essential for a reliable result.

3.5.1 Static Hazards

A *static-1 hazard* is the possibility of a circuit's output producing a 0 glitch when we would expect the output to remain at a nice steady 1 based on a static analysis of the circuit function. A formal definition is given below.

static-1 hazard

Definition: A static-1 hazard is a pair of input combinations that: (a) differ in only one input variable and (b) both give a 1 output; such that it is possible for a momentary 0 output to occur during a transition in the differing input variable.

For example, consider the logic circuit in Figure 3–50(a). Suppose that X and Y are both 1 and Z is changing from 1 to 0. Figure 3–50(b) shows the timing diagram assuming that the propagation delay through each gate or inverter is one

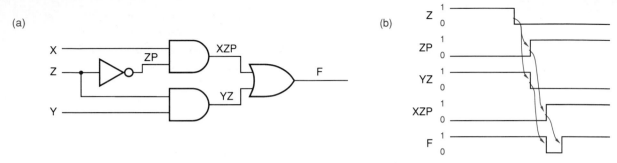

Figure 3–50   Circuit with a static-1 hazard: (a) logic diagram; (b) timing diagram.

static-0 hazard

unit time. Even though "static" analysis predicts that the output is 1 for both input combinations X,Y,Z = 111 and X,Y,Z = 110, the timing diagram shows that F goes to 0 for one unit time during a 1-0 transition on Z, because of the delay in the inverter that generates Z'.

A *static-0 hazard* is the possibility of a 1 glitch when we expect the circuit to have a steady 0 output:

Definition: A static-0 hazard is a pair of input combinations that: (a) differ in only one input variable and (b) both give a 0 output; such that it is possible for a momentary 1 output to occur during a transition in the differing input variable.

Since a static-0 hazard is just the dual of a static-1 hazard, an OR-AND circuit that is the dual of Figure 3–50(a) would have a static-0 hazard.

An OR-AND circuit with four static-0 hazards is shown in Figure 3–51(a). One of the hazards occurs when W,X,Y = 000 and Z is changed, as shown in (b). You should be able to find the other three hazards and eliminate all of them after studying the next subsection.

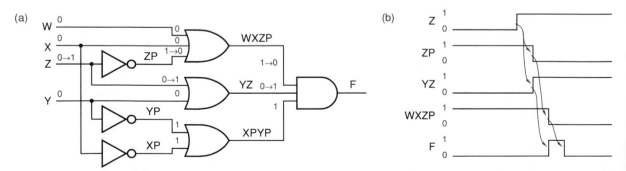

Figure 3–51   Circuit with static-0 hazards: (a) logic diagram; (b) timing diagram.

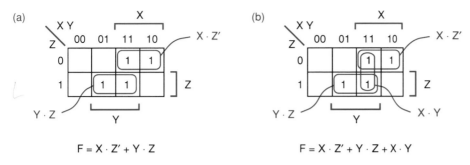

Figure 3–52 Karnaugh map for the circuit of Figure 3–50: (a) as originally designed; (b) with static-1 hazard eliminated.

3.5.2 Finding Static Hazards Using Maps

A Karnaugh map can be used to detect static hazards in a two-level sum-of-products or product-of-sums circuit. The existence or nonexistence of static hazards depends on the circuit design for a logic function.

A properly designed two-level sum-of-products (AND-OR) circuit has no static-0 hazards. A static-0 hazard would exist in such a circuit only if both a variable and its complement were connected to the same AND gate, which would be silly. However, the circuit *may* have static-1 hazards. Their existence can be predicted from a Karnaugh map where the product terms corresponding to the AND gates in the circuit are circled.

Figure 3–52(a) shows the Karnaugh map for the circuit of Figure 3–50. It is clear from the map that there is no single product term that covers both input combinations XYZ = 111 and XYZ = 110. Thus, intuitively, it is possible for the output to "glitch" momentarily to 0 if the AND gate output that covers one of the combinations goes to 0 before the AND gate output covering the other input combination goes to 1. The way to eliminate the hazard is also quite apparent: simply include an extra product term (AND gate) to cover the hazardous input pair, as shown in Figure 3–52(b). The extra product term, it turns out, is the *consensus* of the two original terms; in general, we must add consensus terms to eliminate hazards. The corresponding hazard-free circuit is shown in Figure 3–53.   *consensus*

Another example is shown in Figure 3–54. In this example, three product terms must be added to eliminate the static-1 hazards.

A properly designed two-level product-of-sums (OR-AND) circuit has no static-1 hazards. It *may* have static-0 hazards, however. These hazards can be detected and eliminated by studying the adjacent 0s in the Karnaugh map, in a manner dual to the foregoing.

Figure 3–53
Circuit with static-1 hazard
eliminated.

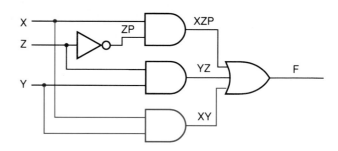

3.5.3 Dynamic Hazards

dynamic hazard

A *dynamic hazard* is the possibility of an output changing more than once as the result of a single input transition. Multiple output transitions can occur if there are multiple paths with different delays from the changing input to the changing output.

For example, consider the circuit in Figure 3–55; it has three different paths from input X to the output F. One of the paths goes through a slow OR gate, and

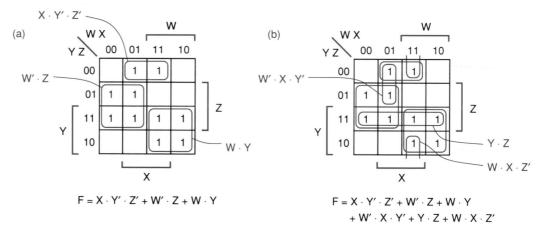

$$F = X \cdot Y' \cdot Z' + W' \cdot Z + W \cdot Y$$

$$F = X \cdot Y' \cdot Z' + W' \cdot Z + W \cdot Y$$
$$+ W' \cdot X \cdot Y' + Y \cdot Z + W \cdot X \cdot Z'$$

Figure 3–54   Karnaugh map for another sum-of-products circuit: (a) as originally designed; (b) with extra product terms to cover static-1 hazards.

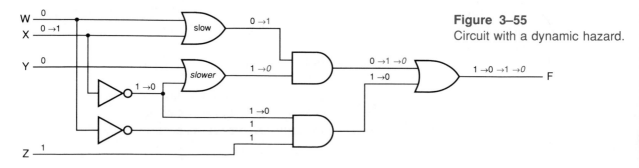

Figure 3–55
Circuit with a dynamic hazard.

another goes through an OR gate that is even slower. If the input to the circuit is W,X,Y,Z = 0,0,0,1, then the output will be 1, as shown. Now suppose we change the X input to 1. Assuming that all of the gates except the two marked "slow" and "slower" are very fast, the transitions shown in black occur next, and the output goes to 0. Eventually, the output of the "slow" OR gate changes, creating the transitions shown in nonitalic color, and the output goes to 1. Finally, the output of the "slower" OR gate changes, creating the transitions shown in italic color, and the output goes to its final state of 0.

Dynamic hazards do not occur in a properly designed two-level AND-OR or OR-AND circuit, that is, one in which no variable and its complement are connected to the same first-level gate. In multilevel circuits, dynamic hazards can be discovered using a method described in the References.

3.5.4   Designing Hazard-Free Circuits

As we mentioned earlier, there are a few situations, such as the design of feedback sequential circuits, that require the design of hazard-free combinational circuits. Techniques for finding hazards in arbitrary circuits, described in the references, are a pain to use. So, when you require a hazard-free design, it's best to use a circuit structure that is easy to analyze.

In particular, we have indicated that a properly designed two-level AND-OR circuit has no static-0 or dynamic hazards. Static-1 hazards may exist in such a circuit, but they can be found and eliminated using the map method described earlier. If cost is not a problem, then a brute-force method of obtaining a hazard-free realization is to use the complete sum—the sum of all of the prime implicants of the logic function (see Exercise 3.68). In a dual manner, a hazard-free two-level OR-AND circuit can be designed for any logic function. Finally, note that everything we've said about AND-OR circuits naturally applies to the corresponding NAND-NAND designs, and for OR-AND applies to NOR-NOR.

R E F E R E N C E S

A historical description of Boole's development of "the science of Logic" appears in *The Computer from Pascal to von Neumann* by Herman H. Goldstine (Princeton University Press, 1972). Claude E. Shannon showed how Boole's work could be applied to logic circuits in "A Symbolic Analysis of Relay and Switching Circuits" (*Trans. AIEE*, Vol. 57, 1938, pp. 713–723).

Although the two-valued Boolean algebra is the basis for switching algebra, a Boolean algebra need not have only two values. Boolean algebras with 2^n values exist for every integer n; for example, see *Discrete Mathematical*

Figure 3–56
A 4-variable Veitch diagram.

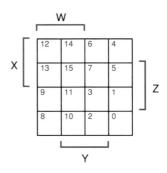

Huntington
postulates

Structures and Their Applications by Harold S. Stone (SRA, 1973). Such algebras may be formally defined using the so-called *Huntington postulates* devised by E. V. Huntington in 1907; for example, see *Digital Design* by M. Morris Mano (Prentice-Hall, 1984). A mathematician's development of Boolean algebra based on a more modern set of postulates appears in *Modern Applied Algebra*, by G. Birkhoff and T. C. Bartee (McGraw-Hill, 1970). Our engineering-style, "direct" development of switching algebra follows Edward J. McCluskey's, as in his *Introduction to the Theory of Switching Circuits* (McGraw-Hill, 1965) and *Logic Design Principles* (Prentice-Hall, 1986).

The prime implicant theorem was proved by W. V. Quine in "The Problem of Simplifying Truth Functions" (*Am. Math. Monthly*, Vol. 59, No. 8, 1952, pp. 521–531). In fact it is possible to prove a more general prime implicant theorem showing that there exists at least one minimal sum that is a sum of prime implicants even if we remove the constraint on the number of literals in our definition of "minimal."

Veitch diagram

A graphical method for simplifying Boolean functions was first proposed by E. W. Veitch in "A Chart Method for Simplifying Boolean Functions" (*Proc. ACM*, May 1952, pp. 127–133); the format of the original *Veitch diagram* is shown in Figure 3–56. The maps we use today differ from Veitch's charts only in the assignment of minterms to the cells, as presented by M. Karnaugh in "A Map Method for Synthesis of Combinational Logic Circuits" (*Trans. AIEE, Comm. and Electron.*, Vol. 72, Part I, November 1953, pp. 593–599). Any assignment in which only one variable changes between adjacent rows or columns will work, so why did Karnaugh bother to change Veitch's perfectly good charts? I'm not sure, but recall Dr. Leonard J. McCoy's disgusted statement when he boarded the renovated *Enterprise* in *Star Trek: The Motion Picture*, "I know engineers—they love to change things!"

Karnaugh maps, Veitch charts, or whatever, are only suitable for minimizing very small logic functions. For larger problems, we need a more systematic algorithm that can be translated into a computer program. Such an algorithm was described by W. V. Quine in "A Way to Simplify Truth Functions" (*Am. Math.*

Monthly, Vol. 62, No. 9, 1955, pp. 627–631) and modified by E. J. McCluskey in "Minimization of Boolean Functions" (*Bell Sys. Tech. J.*, Vol. 35, No. 5, November 1956, pp. 1417–1444). The "Quine-McCluskey" algorithm is fully described in McCluskey's books cited earlier.

McCluskey's 1965 book also covers the iterative consensus algorithm for finding prime implicants, and proves that it works. The iterative consensus algorithm was first published by T. H. Mott, Jr. in "Determination of the Irredundant Normal Forms of a Truth Function by Iterated Consensus of the Prime Implicants" (*IRE Trans. Electron. Computers*, Vol. EC-9, No. 2, 1960, pp. 245–252). The generalized consensus algorithm was published by Pierre Tison in "Generalization of Consensus Theory and Application to the Minimization of Boolean Functions" (*IEEE Trans. Electron. Computers*, Vol. EC-16, No. 4, 1967, pp. 446–456). All of these algorithms are described by Thomas Downs in *Logic Design with Pascal* (Van Nostrand Reinhold, 1988).

As we explained in Section 3.4.6, the huge number of prime implicants in some logic functions makes it impractical or impossible deterministically to find them all or select a minimal cover. However, efficient heuristic methods can find solutions that are close to minimal, at least close enough for government work. The Espresso-II method is described in *Logic Minimization Algorithms for VLSI Synthesis* by R. K. Brayton, C. McMullen, G. D. Hachtel, and A. Sangiovanni-Vincentelli (Kluwer Academic Publishers, 1984). The more recent Espresso-MV and Espresso-EXACT algorithms are described in "Multiple-Valued Minimization for PLA Optimization," by R. L. Rudell and A. Sangiovanni-Vincentelli (*IEEE Trans. CAD*, Vol. CAD-6, No. 5, 1987, pp. 727–750).

In this chapter we described a map method for finding static hazards in two-level AND-OR and OR-AND circuits, but any combinational circuit can be analyzed for hazards. In both his 1965 and 1986 books, McCluskey defines the *0-sets* and *1-sets* of a circuit and shows how they can be used to find static hazards. He also defines *P-sets* and *S-sets* and shows how they can be used to find dynamic hazards.

0-set
1-set
P-set
S-set

Many deeper and varied aspects of switching theory have been omitted from this book, but have been beaten to death in other books and literature. A good starting point for an academic study of classical switching theory is Zvi Kohavi's book, *Switching and Finite Automata Theory* (McGraw-Hill, 1978, 2nd ed.), which includes material on set theory, symmetric networks, functional decomposition, threshold logic, fault detection, and path sensitization. Another area of great academic interest (but little commercial activity) is nonbinary *multiple-valued logic*, in which each signal line can take on more than two values. In his 1986 book, McCluskey gives a good introduction to multiple-valued logic, explaining its pros and cons and why it has seen little commercial development.

multiple-valued logic

E X E R C I S E S

3.1   Design a non-trivial-looking logic circuit that contains a feedback loop but whose output depends only on its current input.

3.2   Using variables NERD, DESIGNER, FAILURE, and STUDIED, write a boolean expression that is 1 for successful designers who never studied and for nerds who studied all the time.

3.3   Prove theorems T2–T5 using perfect induction.

3.4   Prove theorems T1′–T3′ and T5′ using perfect induction.

3.5   Prove theorems T6–T9 using perfect induction.

3.6   Prove the combining theorem T10 without using perfect induction, but assuming that theorems T1–T9 and T1′–T9′ are true.

3.7   Show that the combining theorem, T10, is just a special case of consensus (T11) used with covering (T9).

3.8   Prove that $(X + Y') \cdot Y = X \cdot Y$ *without* using perfect induction. You may assume that theorems T1–T11 and T1′–T11' are true.

3.9   Prove that $(X + Y) \cdot (X' + Z) = X \cdot Z + X' \cdot Y$ *without* using perfect induction. You may assume that theorems T1–T11 and T1′–T11' are true.

3.10 Show that an n-input AND gate can be replaced by $n - 1$ 2-input AND gates. Can the same statement be made for NAND gates? Justify your answer.

3.11 How many physically different ways are there to realize $V \cdot W \cdot X \cdot Y \cdot Z$ using four 2-input AND gates (4/4 of a 74LS08)? Justify your answer.

3.12 Use switching algebra to prove that tying together two inputs of an $n + 1$-input AND or OR gate gives it the functionality of an n-input gate.

3.13 Prove DeMorgan's theorems (T13 and T13′) using finite induction.

3.14 Which logic symbol more closely approximates the internal realization of an RTL NOR gate, Figure 3–4(c) or (d)? Why?

3.15 According to DeMorgan's theorem, the complement of $X + Y \cdot Z$ is $X' \cdot Y' + Z'$. Yet both functions are 1 for XYZ = 110. How can both a function and its complement be 1 for the same input combination? What's wrong here?

3.16 Use the theorems of switching algebra to rewrite the following expression using as few inversions as possible (complemented parentheses are allowed).

$$B' \cdot C + A \cdot C \cdot D' + A' \cdot C + E \cdot B' + E \cdot (A + C) \cdot (A' + D')$$

3.17 Prove Shannon's expansion theorems. (*Hint:* Don't get carried away; it's easy.)

3.18 Shannon's expansion theorems can be generalized to "pull out" not just one but *i* variables, so that a logic function can be expressed as a sum or product of 2^i terms. State the generalized Shannon expansion theorems.

generalized Shannon expansion theorems

3.19 Show how the generalized Shannon expansion theorems lead to the canonical sum and canonical product representations of logic functions.

3.20 Prove or disprove the following propositions:

  (a) Let A and B be switching-algebra *variables*. Then $A \cdot B = 0$ and $A + B = 1$ implies that $A = B'$.

  (b) Let X and Y be switching-algebra *expressions*. Then $X \cdot Y = 0$ and $X + Y = 1$ implies that $X = Y'$.

3.21 An *Exclusive OR (XOR) gate* is a 2-input gate whose output is 1 if and only if exactly one of its inputs is 1. Write a truth table, sum-of-products expression, and corresponding AND-OR circuit for the Exclusive OR function.

Exclusive OR (XOR) gate

3.22 From the point of view of switching algebra, what is the function of a 2-input XOR gate whose inputs are tied together? How might the output behavior of a real XOR gate differ?

3.23 Any set of logic-gate types that can realize any logic function is called a *complete set* of logic gates. For example, 2-input AND gates, 2-input OR gates, and inverters are a complete set, because any logic function can be expressed as a sum of products of variables and their complements, and AND and OR gates with any number of inputs can be made from 2-input gates. Do 2-input NAND gates form a complete set of logic gates? Prove your answer.

complete set

3.24 Do 2-input NOR gates form a complete set of logic gates? Prove your answer.

3.25 Do 2-input XOR gates form a complete set of logic gates? Prove your answer.

3.26 Define a two-input gate, other than NAND, NOR, or XOR, that forms a complete set of logic gates if the constant inputs 0 and 1 are allowed. Prove your answer.

3.27 Some people think that there are *four* basic logic functions, AND, OR, NOT, and *BUT*. Figure X3.27 is a possible symbol for a 4-input, 2-output *BUT gate*. Invent a useful, nontrivial function for the BUT gate to perform. The function should have something to do with the name (BUT). Keep in mind that, due to the symmetry of the symbol, the function should be symmetric with respect to the A and B inputs of each section and with respect to sections 1 and 2. Describe your BUT's function and write its truth table.

BUT
BUT gate

Figure X3.27

3.28 Most students have no problem using theorem T8 to "multiply out" logic expressions, but many develop a mental block if they try to use theorem T8' to "add out" a logic expression. How can duality be used to overcome this problem?

3.29 How many different logic functions are there of n variables?

3.30 How many different 2-variable logic functions $F(X,Y)$ are there? Write a simplified algebraic expression for each of them.

self-dual logic
function

$\oplus$

3.31 A *self-dual logic function* is a function F such that $F = F^D$. Which of the following functions are self-dual? (The symbol $\oplus$ denotes the XOR operation.)

    (a) $F = X$ (b) $F = \Sigma_{X,Y,Z}(0,3,5,6)$

    (c) $F = X \cdot Y' + X' \cdot Y$ (d) $F = W \cdot (X \oplus Y \oplus Z) + W' \cdot (X \oplus Y \oplus Z)'$

    (e) A function F of 7 variables such   (f) A function F of 10 variables such
        that $F = 1$ if and only if 4 or more   that $F = 1$ if and only if 5 or more
        of the variables are 1   of the variables are 1

3.32 How many self-dual logic functions of n input variables are there? (*Hint:* Consider the structure of the truth table of a self-dual function.)

3.33 Prove that any n-input logic function $F(X_1,\ldots,X_n)$ that can be written in the form $F = X_1 \cdot G(X_2,\ldots,X_n) + X_1' \cdot G^D(X_2,\ldots,X_n)$ is self-dual.

3.34 Write the truth table for each of the following logic functions:

    (a) $F = X' \cdot Y + X' \cdot Y' \cdot Z$ (b) $F = W' \cdot X + Y' \cdot Z' + X' \cdot Z$
    (c) $F = W + X' \cdot (Y' + Z)$ (d) $F = A \cdot B + B' \cdot C + C' \cdot D + D' \cdot A$
    (e) $F = V \cdot W + X' \cdot Y' \cdot Z$ (f) $F = (A' + B' + C \cdot D) \cdot (B + C' + D' \cdot E')$
    (g) $F = (W \cdot X)' \cdot (Y' + Z')'$ (h) $F = (((A + B)' + C')' + D)'$
    (i) $F = (A' + B + C) \cdot (A + B' + D') \cdot (B + C' + D') \cdot (A + B + C + D)$

3.35 Write the canonical sum and product for each of the following logic functions:

    (a) $F = \Sigma_{X,Y}(1,2)$ (b) $F = \Pi_{A,B}(0,1,2)$
    (c) $F = \Sigma_{A,B,C}(2,4,6,7)$ (d) $F = \Pi_{W,X,Y}(0,1,3,4,5)$
    (e) $F = X + Y' \cdot Z'$ (f) $F = V' + (W' \cdot X)'$

3.36 Assuming that an inverting gate has a propagation delay of 5 ns, and a noninverting gate has a propagation delay of 8 ns, compare the speeds of the circuits in Figure 3–24(a), (c), and (d).

3.37 Using switching algebra, show that the logic functions for the AND-OR circuits in Figures 3–28 and 3–29 equal their OR-AND counterparts in Figures 3–37 and 3–38.

3.38 Determine whether the product-of-sums expressions obtained by "adding out" the minimal sums in Figure 3–28 and 3–29 are minimal.

3.39 Prove that the rule for combining 2^i 1-cells in a Karnaugh map is true, using the axioms and theorems of switching algebra.

3.40 An *irredundant sum* for a logic function F is a sum of prime implicants for F such *irredundant sum* that if any prime implicant is deleted, the sum no longer equals F. This sounds a lot like a minimal sum, but an irredundant sum is not necessarily minimal. For example, the minimal sum of the function in Figure 3–35 has only three product terms, but there is an irredundant sum with four product terms. Find the irredundant sum and draw a map of the function, circling only the prime implicants in the irredundant sum.

3.41 Find another logic function in Section 3.3 that has one or more non-minimal irredundant sums, and draw its map, circling only the prime implicants in the irredundant sum.

3.42 Why do you suppose that the cost of inverters is not included in the definition of "minimal" for logic minimization?

3.43 Find a minimal sum-of-products expression for each of the following logic functions:

(a) $F = \Sigma_{X,Y,Z}(1,3,5,6,7)$ (b) $F = \Sigma_{W,X,Y,Z}(1,4,5,6,7,9,14,15)$
(c) $F = \Pi_{W,X,Y}(0,1,3,4,5)$ (d) $F = \Sigma_{W,X,Y,Z}(0,2,5,7,8,10,13,15)$
(e) $F = \Pi_{A,B,C,D}(1,7,9,13,15)$ (f) $F = \Sigma_{A,B,C,D}(1,4,5,7,12,14,15)$

3.44 Find a minimal product-of-sums expression for each function in Exercise 3.43.

3.45 Derive the minimal product-of-sums expression for the prime BCD-digit detector function of Figure 3–40. Determine whether or not the expression algebraically equals the minimal sum-of-products expression and explain your result.

3.46 Draw a Karnaugh map and assign variables to the inputs of the AND-XOR circuit in Figure X3.46 so that its output is $F = \Sigma_{W,X,Y,Z}(6,7,12,13)$. Note that the output gate is a 2-input XOR rather than an OR.

Figure X3.46

3.47 Find a minimal sum-of-products expression for each of the following logic functions:

(a) $F = \Sigma_{W,X,Y,Z}(0,1,3,5,14) + d(8,15)$ (b) $F = \Sigma_{W,X,Y,Z}(0,1,2,8,11) + d(3,9,15)$
(c) $F = \Sigma_{A,B,C,D}(1,5,9,14,15) + d(11)$ (d) $F = \Sigma_{A,B,C,D}(1,5,6,7,9,13) + d(4,15)$
(e) $F = \Sigma_{W,X,Y,Z}(3,5,6,7,13) + d(1,2,4,12,15)$

3.48 Repeat Exercise 3.47, finding a minimal product-of-sums expression for each logic function.

3.49 For each logic function in the two preceding exercises, determine whether the minimal sum-of-products expression equals the minimal product-of-sums expression. Also compare the circuit cost for realizing each of the two expressions.

3.50 The text indicates that a truth table or equivalent is the starting point for traditional combinational minimization methods. A Karnaugh map itself contains the same information as a truth table. Given a sum-of-products expression, it is possible to write the 1s corresponding to each product term directly on the map without developing an explicit truth table or minterm list, and then proceed with the map minimization procedure. Find a minimal sum-of-products expression for each of the following logic functions in this way:

(a) $F = X' \cdot Z + X \cdot Y + X \cdot Y' \cdot Z$ (b) $F = A' \cdot C' \cdot D + B' \cdot C \cdot D + A \cdot C' \cdot D + B \cdot C \cdot D$
(c) $F = W \cdot X \cdot Z' + W \cdot X' \cdot Y \cdot Z + X \cdot Z$ (d) $F = (X' + Y') \cdot (W' + X' + Y) \cdot (W' + X + Z)$
(e) $F = A \cdot B \cdot C' \cdot D' + A' \cdot B \cdot C' + A \cdot B \cdot D + A' \cdot C \cdot D + B \cdot C \cdot D'$

3.51 Repeat Exercise 3.50, finding a minimal product-of-sums expression for each logic function.

5-variable Karnaugh map

3.52 A Karnaugh map for a 5-variable function can be drawn as shown in Figure X3.52. In such a map, cells that occupy the same relative position in the $V = 0$ and $V = 1$ submaps are considered to be adjacent. (Many worked examples of 5-variable Karnaugh maps appear in Sections 5.4.4 and 5.4.5.) Find a minimal sum-of-products expression for each of the following functions using a 5-variable map:

(a) $F = \Sigma_{V,W,X,Y,Z}(5,7,13,15,16,20,25,27,29,31)$
(b) $F = \Sigma_{V,W,X,Y,Z}(0,7,8,9,12,13,15,16,22,23,30,31)$
(c) $F = \Sigma_{V,W,X,Y,Z}(0,1,2,3,4,5,10,11,14,20,21,24,25,26,27,28,29,30)$
(d) $F = \Sigma_{V,W,X,Y,Z}(0,2,4,6,7,8,10,11,12,13,14,16,18,19,29,30)$
(e) $F = \Pi_{V,W,X,Y,Z}(4,5,10,12,13,16,17,21,25,26,27,29)$
(f) $F = \Sigma_{V,W,X,Y,Z}(4,6,7,9,11,12,13,14,15,20,22,25,27,28,30) + d(1,5,29,31)$

Figure X3.52

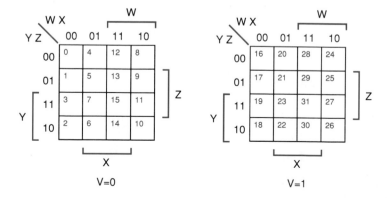

3.53 Repeat Exercise 3.52, finding a minimal product-of-sums expression for each logic function.

3.54 A Karnaugh map for a 6-variable function can be drawn as shown in Figure X3.54. *6-variable* In such a map, cells that occupy the same relative position in adjacent submaps *Karnaugh map* are considered to be adjacent. Minimize the following functions using 6-variable maps:

(a) $F = \Sigma_{U,V,W,X,Y,Z}(1,5,9,13,21,23,29,31,37,45,53,61)$

(b) $F = \Sigma_{U,V,W,X,Y,Z}(0,4,8,16,24,32,34,36,37,39,40,48,50,56)$

(c) $F = \Sigma_{U,V,W,X,Y,Z}(2,4,5,6,12-21,28-31,34,38,50,51,60-63)$

Figure X3.54

3.55 A 3-bit "comparator" circuit receives two 3-bit numbers, $P = P_2P_1P_0$ and $Q = Q_2Q_1Q_0$. Design a minimal sum-of-products circuit that produces a 1 output if and only if $P > Q$.

3.56 Find minimal multiple-output sum-of-products expressions for $F = \Sigma_{X,Y,Z}(0,1,2)$, $G = \Sigma_{X,Y,Z}(1,4,6)$, and $H = \Sigma_{X,Y,Z}(0,1,2,4,6)$.

3.57 Prove whether or not the following expression is a minimal sum. Do it the easiest way possible (algebraically, not using maps).

$$F = T' \cdot U \cdot V \cdot W \cdot X + T' \cdot U \cdot V' \cdot X \cdot Z + T' \cdot U \cdot W \cdot X \cdot Y' \cdot Z$$

3.58 There are $2n$ m-subcubes of an n-cube for the value $m = n - 1$. Show their text representations and the corresponding product terms. (You may use ellipses as required, e.g., $1, 2, \ldots, n$.)

3.59 There is just one m-subcube of an n-cube for the value $m = n$; its text representation is xx...xx. Write the product term corresponding to this cube.

3.60 The Pascal program in Table 3–8 uses memory inefficiently, because it allocates memory for a maximum number of cubes at each level, even if this maximum is never used. Redesign the program so that the cubes and used arrays are one-dimensional arrays, and each level uses only as many array entries as needed. *Hint:* You can still allocate cubes sequentially, but keep track of the starting point in the array for each level.

3.61 As a function of m, how many times is each distinct m-cube rediscovered in Table 3–8, only to be found in the inner loop and thrown away? Suggest some ways to eliminate this inefficiency.

3.62 The third FOR-loop in Table 3–8 tries to combine all m-cubes at a given level with all other m-cubes at that level. In fact, only m-cubes with x's in the same positions can be combined, so it is possible to reduce the number of loop iterations by using a more sophisticated data structure. Design a data structure that segregates the cubes at a given level according to the position of their x's, and determine the maximum size required for various elements of the data structure. Rewrite Table 3–8 accordingly.

3.63 Estimate whether the savings in inner-loop iterations achieved in Exercise 3.62 outweighs the overhead of maintaining a more complex data structure. Try to make reasonable assumptions about how cubes are distributed at each level, and indicate how your results are affected by these assumptions.

3.64 Explain why, in the generalized consensus algorithm, cubes in CC cannot be covered by cubes in C0 and C1.

3.65 Write a Pascal program that selects a minimal cover for a logic function using the iterative consensus method of Section 3.4.4. Your program should behave reasonably if it given a function for which the branching method is required. Reasonable behaviors include halting with an error message, picking a pseudo-essential prime implicant at random, and actually executing the branching method. Getting stuck in an infinite loop is not reasonable.

3.66 For each of the following logic expressions, find all of the static hazards in the corresponding two-level AND-OR or OR-AND circuit, and design a hazard-free circuit that realizes the same logic function.

(a) $F = W \cdot X + W' Y'$

(b) $F = W \cdot X' \cdot Y' + X \cdot Y' \cdot Z + X \cdot Y$

(c) $F = W' \cdot Y + X' \cdot Y' + W \cdot X \cdot Z$

(d) $F = W' \cdot X + Y' \cdot Z + W \cdot X \cdot Y \cdot Z + W \cdot X' \cdot Y \cdot Z'$

(e) $F = (W + X + Y) \cdot (X' + Z')$

(f) $F = (W + Y' + Z') \cdot (W' + X' + Z') \cdot (X' + Y + Z)$

(g) $F = (W + Y + Z') \cdot (W + X' + Y + Z) \cdot (X' + Y') \cdot (X + Z)$

3.67 *(Hamlet circuit.)* Complete the timing diagram and explain the function of the circuit in Figure X3.67. Where does the circuit get its name?

Figure X3.67

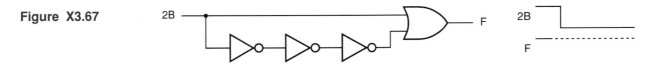

3.68 Prove that a two-level AND-OR circuit corresponding to the complete sum of a logic function is always hazard free.

3.69 Find a four-variable logic function whose minimal sum-of-products realization is not hazard free, but where there exists a hazard-free sum-of-products realization with fewer product terms than the complete sum.

4

74LS138

COMBINATIONAL LOGIC DESIGN PRACTICES

The preceding chapter described the theoretical principles used in combinational logic design. In this chapter, we'll build on that foundation and describe many of the devices, structures, and methods used by engineers to solve practical digital design problems.

A practical combinational circuit may have dozens of inputs and outputs and could require hundreds, thousands, even millions of terms to describe as a sum of products, and *billions and billions* of rows to describe in a truth table. Thus, most real combinational logic design problems are too large to solve by "brute-force" application of theoretical techniques.

But wait, you say, how could any human being conceive of such a complex logic circuit in the first place? The key is structured thinking. A complex circuit or system is conceived as a collection of smaller subsystems, each of which has a much simpler description.

In combinational logic design, there are several straightforward structures— decoders, multiplexers, comparators, and the like—that turn up quite regularly as building blocks in larger systems. The most important of these structures appear in this chapter. We describe each structure generally and then give examples and applications using ICs in the TTL logic family. But first, we start this chapter by presenting documentation standards that are used by digital designers to ensure that their designs are correct, manufacturable, and maintainable.

4.1 DOCUMENTATION STANDARDS

Good documentation is essential for correct design and efficient maintenance of digital systems. In addition to being accurate and complete, documentation must be somewhat instructive, so that a test engineer, maintenance technician, or even the original design engineer (six months after designing the circuit) can figure out how the system works just by reading the documentation.

Although the type of documentation depends on system complexity and the engineering and manufacturing environments, a documentation package should generally contain at least the following five items:

(1) A *block diagram* is an informal pictorial description of the system's major functional modules and their basic interconnections.

block diagram

(2) A *schematic diagram* is a formal specification of the electrical components of the system, their interconnections, and all of the details needed to construct the system, including IC types, reference designators, and pin numbers. We've been using the term *logic diagram* for an informal drawing that does not have quite this level of detail.

schematic diagram

logic diagram

(3) A *timing diagram* shows the values of various logic signals as a function of time, including the cause-and-effect delays between critical signals.

timing diagram

(4) A *structured logic description* describes the internal function of a structured logic device, such as a programmable logic array (PLA) or read-only memory (ROM). Such a description may be in the form of logic equations, state tables or diagrams, or program listings.

structured logic description

(5) A *circuit description* is a narrative text document that, in conjunction with the other documentation, describes concisely how the circuit is intended to be used and explains how it works internally. The circuit description should list any assumptions and potential pitfalls in the circuit's design and operation, and point out the use of any nonobvious design "tricks."

circuit description

You've probably already seen block diagrams in many contexts. We present a few rules for drawing them in the next subsection, and then we concentrate on schematics for combinational logic circuits in the rest of this section. Section 4.2.1 introduces timing diagrams. Structured logic descriptions, such as PLD equations, are covered in Chapter 7.

The last area of documentation, the circuit description, is very important in practice. Just as an experienced programmer creates a program design document before beginning to write code, an experienced logic designer starts writing the circuit description before drawing the schematic. Unfortunately, the circuit description is sometimes the last document to be created, and sometimes it's never written at all. A circuit without a description is difficult to debug, manufacture, test, maintain, modify, and enhance. Therefore, we now take a short break from our story to bring you this important message:

> In order to create great products, logic designers must develop their language and writing skills, especially in the area of *logical* outlining and organization. The most successful logic designers (and later, project leaders, system architects, company founders, and venture capitalists) are the ones who communicate their ideas, proposals, and decisions effectively to others. Even though it's a lot of fun to tinker in the digital design lab, don't use that as an excuse to shortchange your writing and communication oriented courses and projects.

4.1.1 Block Diagrams

block diagram

We now return to our regularly scheduled story. A *block diagram* shows the inputs, outputs, functional modules, internal data paths, and important control signals of a system. In general, it should not be so detailed that it occupies more than one page, yet it must not be too vague. A small block diagram may have three to six blocks, while a large one may have 10 to 15, depending on system complexity. In any case, the block diagram must show the most important system elements and how they "play together." Large systems may require additional block diagrams of individual subsystems, but there should always be a "top-level" diagram showing the entire system.

Figure 4–1 shows a sample block diagram. Each block is labeled with the function of the block, not the individual chips that comprise it. As another example, Figure 4–2(a) shows the block-diagram symbol for a 32-bit register. If the register is to be built using four 74LS377 8-bit registers, and this information is important to someone reading the diagram (e.g., for cost reasons), then it can be conveyed as shown in (b). However, splitting the block to show individual chips as in (c) is incorrect.

bus

A *bus* is a collection of two or more related signal lines. In a block diagram, buses are drawn with a double or heavy line. A slash and a number are often used to indicate how many individual signal lines are contained in a bus. Active levels (defined later) and inversion bubbles may or may not appear in block diagrams; in most cases, they are unimportant at this level of detail. However, important control signals and buses should have names, usually the same names that appear in the more detailed schematic.

The flow of control and data in a block diagram should be clearly indicated. Logic diagrams are generally drawn with signals flowing from left to right, but in block diagrams this ideal is more difficult to achieve. Inputs and outputs may be on any side of a block, and the direction of signal flow may be arbitrary. Arrowheads are used on buses and ordinary signal lines to eliminate ambiguity.

SHIFT-AND-ADD MULTIPLIER

Figure 4–1   Block diagram for a simple digital design project.

4.1.2 Gate Symbols

The symbol shapes for AND and OR gates and buffers are shown in Figure 4–3(a).
(Recall from Chapter 2 that a buffer is a circuit that simply converts "weak" logic
signals into "strong" ones.) To draw logic gates with more than about four inputs,
we expand the AND and OR symbols as shown in (b). A small circle, called an
inversion bubble, denotes logical complementation and is used in the symbols   *inversion bubble*
for NAND and NOR gates and inverters as shown in (c).

Using the generalized DeMorgan's theorem, we can manipulate the logic
expressions for gates with complemented outputs. For example, if X and Y are
the inputs of a NAND gate with output Z, then we can write

$$Z = (X \cdot Y)'$$
$$= X' + Y'$$

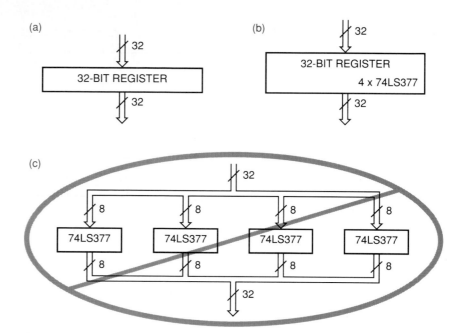

(a)

32

32-BIT REGISTER

32

(b)

32

32-BIT REGISTER

4 x 74LS377

32

(c)

32

8   8   8   8

74LS377   74LS377   74LS377   74LS377

8   8   8   8

32

Figure 4–2   A 32-bit register block: (a) realization unspecified; (b) chips specified; (c) too much detail.

This gives rise to two different but equally correct symbols for a NAND gate, as we demonstrated in Figure 3–3 on page 155. In fact, this sort of manipulation may be applied to gates with uncomplemented inputs, too. For example, consider the following equations for an AND gate:

$$Z = X \cdot Y$$
$$= ((X \cdot Y)')'$$
$$= (X' + Y')'$$

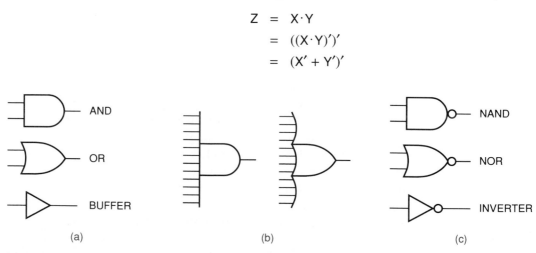

AND

OR

BUFFER

NAND

NOR

INVERTER

(a)   (b)   (c)

Figure 4–3   Shapes for basic logic gates: (a) AND, OR, and buffers; (b) expansion of inputs; (c) inversion bubbles.

Figure 4–4
Equivalent gate symbols under the generalized DeMorgan's theorem.

Thus, an AND gate may be symbolized as an OR gate with inversion bubbles on its inputs and output.

Equivalent symbols for standard gates that can be obtained by these manipulations are summarized in Figure 4–4. Even though both symbols in a pair represent the same logic function, the choice of one symbol or the other in a logic diagram is not arbitrary, at least not if we are adhering to good documentation standards. As we'll show in the next three subsections, proper choices of signal names and gate symbols can make logic diagrams much easier to use and understand.

4.1.3 Signal Names and Active Levels

Each input and output signal in a logic circuit should have a descriptive alphanumeric label, the signal's name. Most computer-aided design systems for drawing logic circuits also allow certain special characters, such as *, _, and !, to be included in signal names. In the analysis and synthesis examples in Chapter 3, we used mostly single-character signal names (X, Y, etc.), because the circuits didn't do much. However, in a real system, well-chosen signal names convey information to someone reading the logic diagram the same way that variable names in a software program do. The names usually indicate an action that is controlled (GO, PAUSE) or a condition that is detected (READY, ERROR) by the signal.

Although it is absolutely necessary to name only a circuit's main inputs and outputs, most logic designers find it useful to name internal signals as well. During circuit debugging, it's nice to have a name to use when pointing to an internal signal that's behaving strangely. Most computer-aided design systems automatically generate labels for unnamed signals, but a user-chosen name is preferable to a computer-generated one like XSIG1057.

Table 4–1
Each line shows a different naming convention for active levels.

Active low	Active high
READY–	READY+
ERROR.L	ERROR.H
ADDR15(L)	ADDR15(H)
RESET*	RESET
/TRANSMIT	TRANSMIT

active level
active high

active low
assert
negate
deassert

Each signal name should have an *active level* associated with it. A signal is *active high* if it performs the named action or denotes the named condition when it is HIGH or 1. (Under the positive-logic convention, which we use throughout this book, "HIGH" and "1" are equivalent.) A signal is *active low* if it performs the named action or denotes the named condition when it is LOW or 0. A signal is said to be *asserted* when it is at its active level. A signal is said to be *negated* (or, sometimes, *deasserted*) when it is not at its active level.

The active level of each signal in a circuit is normally specified as part of its name, according to some convention. Examples of several different active-level naming conventions are shown in Table 4–1. The choice of one of these or other signal naming conventions is sometimes just a matter of personal preference, but more often it is constrained by the engineering environment. Since the active-level designation is part of the signal name, the naming convention must be compatible with the input requirements of any computer-aided design tools, such as schematic editors and simulators, that will process the signal names. In this

active-level naming convention

text, we'll use the last convention in the table: an active-low signal name has a prefix of /, and an active-high signal has no prefix. The / prefix may be read as "not."

It's extremely important to understand the difference between a signal name and an equation that describes a signal's function in terms of other signals. A

signal name
signal equation

signal name is just a name—an alphanumeric label. A *signal equation* is an assignment of a logic expression to a signal name. A similar concept is used in programming languages, where the left-hand side of an assignment statement contains a variable name, and the right-hand side contains an expression whose value will be given to the named variable (e.g., $Z := -(X+Y)$ in Pascal). In a programming language, you can't put an expression on the left-hand side of an assignment statement. In logic design, you shouldn't use an expression as a signal name.

Logic signals may have names like X, READY, and /GO. The "/" in /GO is just part of the signal's name, like an underscore in a Pascal variable name. There is *no* signal whose name is READY′—this is an expression since ′ is an operator—but there *may* be another signal named /READY such that /READY = READY′ during normal operation of the circuit. We are very careful in this book to distinguish between signal names, which are always printed in black, and logic expressions, which are always printed in color when they are written near the corresponding signal lines.

In a computer-aided design system, you can associate *names*, not expressions, with all of the signals in a logic circuit, following conventions similar to what we describe in this section. Expressions that describe each signal in terms of other signals are a consequence of the circuit's structure. If you need to know these expressions, you can derive them by hand or, if you're fortunate, just give a command to your computer-aided design system to work them out for you.

4.1.4   Active Levels for Pins

When we draw the outline of an AND or OR symbol, or a rectangle representing a larger-scale logic element, we think of the given logic function as occurring *inside* that symbolic outline. In Figure 4–5(a), we show the logic symbols for an AND and OR gate and for a larger-scale element with an ENABLE input. The AND and OR gates have active-high inputs—they require 1s on the input to assert their outputs. Likewise, the larger-scale element has an active-high ENABLE input, which must be 1 to enable the element to do its thing. In (b), we show the same logic elements with active-low input and output pins. Exactly the same logic functions are performed *inside* the symbolic outlines, but the inversion bubbles indicate that 0s must now be applied to the input pins to activate the logic functions, and that the outputs are 0 when they are "doing their thing."

Thus, active levels may be associated with the input and output pins of gates and larger-scale logic elements. We use an inversion bubble to indicate an active-low pin and the absence of a bubble to indicate an active-high pin. For example, the AND gate in Figure 4–6(a) performs the logical AND of two active-high inputs and produces an active-high output: if both inputs are asserted (1), the output is asserted (1). The NAND gate in (b) also performs the AND function, but it produces an active-low output. Even a NOR or OR gate can be construed to perform the AND function using active-low inputs and outputs, as shown in

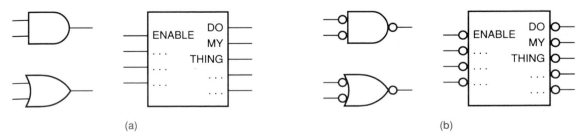

(a)   (b)

Figure 4–5   Logic symbols: (a) AND, OR, and a larger-scale logic element; (b) the same elements with active-low inputs and outputs.

Figure 4–6   Four ways of obtaining an AND function: (a) AND gate (7408); (b) NAND gate (7400); (c) NOR gate (7402); (d) OR gate (7432).

(c) and (d). All four gates in the figure can be said to perform the same function: the output of each gate is asserted if both of its inputs are asserted. Figure 4–7 shows the same idea for the OR function: the output of each gate is asserted if either of its inputs is asserted.

Sometimes a noninverting buffer is used simply to boost the fanout of a logic signal without changing its function. Figure 4–8 shows the possible logic symbols for both inverters and noninverting buffers. In terms of active levels, all of the symbols perform exactly the same function: each asserts its output signal if and only if its input is asserted.

4.1.5   Bubble-to-Bubble Logic Design

Experienced logic designers formulate their circuits in terms of the logic functions performed *inside* the symbolic outlines, and select discrete gates to fit the constraints of the design. Design constraints include input and output active levels of the selected larger-scale elements, speed, cost, and the availability of discrete gate types. However, it is still possible to choose logic symbols and descriptive signal names, including active-level designators, to make the function of a logic circuit quite easy to understand. Usually, this means choosing signal names and gate types and symbols so that most of the inversion bubbles "cancel out." In this way, simplified logic expressions using AND and OR operators may be read directly from the logic diagram.

Figure 4–7   Four ways of obtaining an OR function: (a) OR gate (7432); (b) NOR gate (7402); (c) NAND gate (7400); (d) AND gate (7408).

Figure 4–8   Alternate logic symbols: (a,b) inverters; (c,d) noninverting buffers.

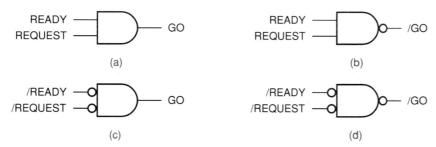

Figure 4–9 Many ways to GO: (a) active-high inputs and output; (b) active-high inputs, active-low output; (c) active-low inputs, active-high output; (d) active-low inputs and outputs.

For example, suppose we need to produce a signal that tells a device to "GO" when we are "READY" and we get a "REQUEST." Clearly, an AND function is required; in switching algebra, we can write GO = READY·REQUEST. However, we can use different gates to perform the AND function, depending on the active level required for the GO signal and the active levels of the available input signals. Figure 4–9(a) shows the simplest case, where GO must be active-high and the available input signals are also active-high; we use an AND gate. If, on the other hand, the device that we're controlling requires an active-low /GO signal, we can use a NAND gate as shown in (b). If the available input signals are active-low, we can use a NOR or OR gate as shown in (c) and (d).

The active levels of available signals don't always match the active levels of available gates. For example, suppose we are given input signals /READY (active-low) and REQUEST (active-high). Figure 4–10 shows two different ways to generate GO using an inverter to generate the appropriate active level for the AND function. The second way is generally preferred, since inverting gates like NOR are generally faster than noninverting ones like AND. We drew the inverter differently in each case to make the output's active level match its signal name.

The following rules are useful for performing bubble-to-bubble logic design:

- The signal name on a device's output should have the same active level as the device's output pin, that is, active-low if the device symbol has an inversion bubble on the output pin, active-high if not.

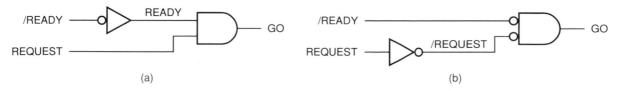

Figure 4–10 Two more ways to GO, with mixed input levels: (a) with an AND gate; (b) with a NOR gate.

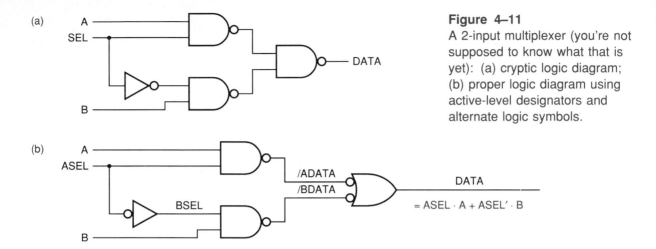

Figure 4–11
A 2-input multiplexer (you're not supposed to know what that is yet): (a) cryptic logic diagram; (b) proper logic diagram using active-level designators and alternate logic symbols.

- If the active level of an input signal is the same as that of the input pin to which it is connected, then the logic function inside the symbolic outline is activated when the signal is asserted. This should be the most common case in a logic diagram.

- If the active level of an input signal is the opposite of that of the input pin to which it is connected, then the logic function inside the symbolic outline is activated when the signal is negated. This case should be avoided whenever possible, because it forces us to keep track mentally of a logical negation to understand the circuit.

For example, consider the circuit in Figure 4–11(a). What does it do? In Section 3.2 we showed several ways to analyze such a circuit, and we could certainly obtain a logic expression for the DATA output using these techniques. However, when the circuit is redrawn in Figure 4–11(b), the output function can be read directly from the logic diagram. The DATA output is asserted when either /ADATA or /BDATA is asserted. If ASEL is asserted, then /ADATA is asserted if and only if A is asserted; that is, /ADATA is a copy of A. If ASEL is negated, BSEL is asserted and /BDATA is a copy of B. In other words, DATA is a copy of A if ASEL is asserted, and DATA is a copy of B if ASEL is negated. Even though there are five inversion bubbles in the logic diagram, we mentally had to perform only one negation to understand the circuit—that BSEL is asserted if ASEL is not.

    If we wish, we can write an algebraic expression for the DATA output. We use the technique of Section 3.2, simply propagating expressions through gates towards the output. In doing so, we can ignore pairs of inversion bubbles that cancel, and directly write the expression shown in color in the figure.

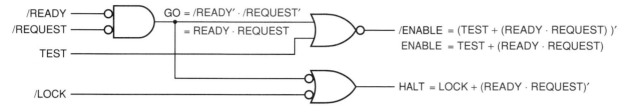

Figure 4-12 Another properly drawn logic diagram.

Another example is shown in Figure 4–12. Reading directly from the logic diagram, we see that /ENABLE is asserted if /READY and /REQUEST are asserted or if TEST is asserted. The HALT output is asserted if /READY and /REQUEST are not both asserted or if /LOCK is asserted. Once again, this example has only one place where a gate input's active level does not match the input signal level, and this is reflected in the verbal description of the circuit.

We can, if we wish, write algebraic equations for the /ENABLE and HALT outputs. As we propagate expressions through gates towards the output, we obtain expressions like /READY'·/REQUEST'. However, we can use our active-level naming convention to simplify terms like /READY'. The circuit contains no signal with the name READY; but if it did, it would satisfy the relationship READY = /READY' according to the naming convention. This allows us to write the /ENABLE and HALT equations as shown. Complementing both sides of the /ENABLE equation, we obtain an equation that describes a hypothetical active-high ENABLE output in terms of hypothetical active-high inputs.

We'll see more examples of bubble-to-bubble logic design in this and later chapters, especially as we begin to use larger-scale logic elements.

4.1.6   Drawing Layout

Logic diagrams and schematics should be drawn with gates in their "normal" orientation with inputs on the left and outputs on the right. The logic symbols for larger-scale logic elements are also normally drawn with inputs on the left and outputs on the right.

A complete schematic page should be drawn with system inputs on the left and outputs on the right, and the general flow of signals should be from left to right. If an input or output appears in the middle of the page, it should be extended to the left or right edge, respectively. In this way, a reader can find all inputs and outputs by looking at the edges of the page only. All signal paths on the page should be connected when possible; paths may be broken if the drawing gets crowded, but breaks should be flagged in both directions, as described later.

Sometimes block diagrams are drawn without crossing lines to give a neater appearance, but for logic diagrams this is not recommended. For logic diagrams

Figure 4–13
Line crossings and connections.

Hand drawn

Machine drawn   not allowed

crossing   connection   connection

it is best to allow lines to cross and to indicate connections clearly with a dot. However, some computer-aided design systems (and some designers) can't draw legible connection dots. To distinguish between crossing lines and connected lines, they adopt the convention that only "T" type connections are allowed, as shown in Figure 4–13. This a good convention to follow in any case.

Schematics that fit on a single page are the easiest to work with. The largest practical paper size for a schematic might be E-size (34×44 inches). Although its drawing capacity is great, such a large paper size is fairly unwieldy to work with. The best compromise of drawing capacity and practicality is B-size (11×17 inches) paper. It can be easily folded for storage and quick reference in standard 3-ring notebooks, and it can be copied on most office copiers (because of all the accountants with their B-size ledger paper). Some designers draw their

Figure 4–14 Flat schematic structure.

Figure 4–15
Hierarchical schematic structure.

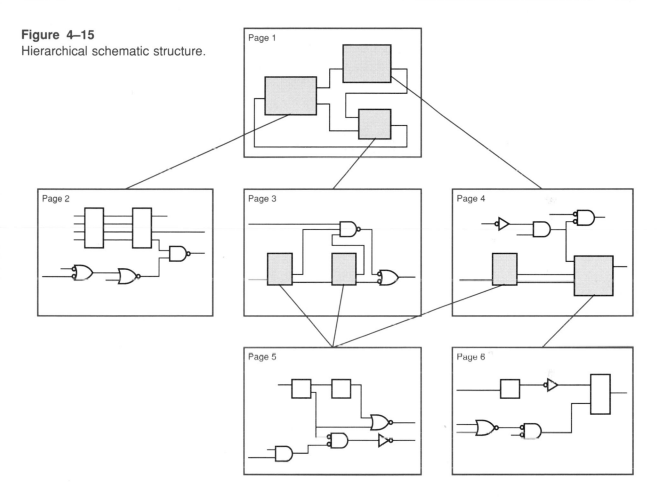

original schematics on D-size (22×34 inches) paper, and use a 50%-reduced B-size copy for reference. Regardless of paper size, schematics seem to come out best when the page is used in landscape format, that is, with its long dimension oriented from left to right, the direction of most signal flow.

Schematics that don't fit on a single page should be broken up into individual pages in a way that minimizes the connections (and confusion) between pages. They may also use a coordinate system, like that of a road map, to flag *signal flags* the source and destination of signals that travel from one page to another. An outgoing signal should have flags referring to all of the destinations of that signal, while an incoming signal should have a flag referring to the source only. That is, an incoming signal should be flagged to the place where it is generated, not to a place somewhere in the middle of a chain of destinations that use the signal.

A multiple-page schematic usually has a "flat" structure. As shown in *flat schematic* Figure 4–14, each page is carved out from the complete schematic and can con- *structure*

*hierarchical
schematic
structure*

nect to any other page as if all the pages were on one large sheet. However, much like programs, schematics can also be constructed hierarchically, as illustrated in Figure 4–15. In this approach, the "top-level" schematic is just a single page that may take the place of a block diagram. Typically, the top-level schematic contains no gates or other logic elements; it shows the major subsystems and their interconnections. The subsystems are in turn defined on lower-level pages, which may contain ordinary gate-level descriptions, or may themselves define even lower-level hierarchies. If a particular lower-level hierarchy is used more than once, it may be reused (or "called," in the programming sense) multiple times by the higher-level pages. Most computer-aided logic design systems support hierarchical schematics.

4.1.7 Buses

bus

A collection of two or more related signal lines is called a *bus*. For example, a microprocessor system might have an address bus containing 16 address lines, ADDR0–ADDR15, and a data bus with 8 data lines, DATA0–DATA7. The signal names in a bus are not necessarily related or ordered as in these first examples. For example, a microprocessor system might have a control bus containing five signals, ALE, MIO, /RD, /WR, and RDY.

Logic diagrams use special notation for buses in order to reduce the amount of drawing and to improve readability. As shown in Figure 4–16, a bus typically has its own descriptive name, such as ADDR[15–0], DATA[7–0], or CONTROL, and is drawn with a double line or a thicker line than ordinary signals. Individual signals may be put into or pulled out of the bus by connecting an ordinary signal line to the bus and writing its name on the signal line. Sometimes a special connection dot is also used, as shown in the example. A computer-aided design system keeps track of the individual signals in a bus. When it actually comes time to build a circuit from the schematic, signal lines in a bus are treated just as though they had all been drawn individually.

The symbols at the right-hand edge of Figure 4–16 are interpage signal flags. They indicate that the LA bus goes out to page 2, the DB bus is bidirectional and connects to page 2, and the CONTROL bus is bidirectional and connects to pages 2 and 3.

4.1.8 Additional Schematic Information

IC type

Complete schematic diagrams indicate IC types, reference designators, and pin numbers, as in Figure 4–17. The *IC type* is a part number identifying the integrated circuit that performs a given logic function. For example, a 2-input NAND gate might be identified as a 74LS00 or a 74LS37. Even when there is only one possible generic part number for a given gate type (e.g., 7408 for a 2-input

Figure 4–16 Examples of buses.

AND gate), the IC type identifies the logic family and speed for that gate (e.g., 74LS08, 74ALS08, 74F08, etc.).

The *reference designator* for an IC identifies a particular instance of that IC type installed in the system. In conjunction with the system's mechanical documentation, the reference designator allows a particular IC to be located during assembly, test, and maintenance of the system. Traditionally, reference designators for ICs begin with the letter U (for "unit").

reference designator

Once a particular IC is located, *pin numbers* are used to locate individual logic signals on its pins. Sometimes the manufacturers' catalogs contain pin diagrams for ICs in the form shown in Figure 4–18. However, logic diagrams and

pin number

Figure 4–17
Schematic diagram for a circuit using a 74LS00.

schematics should never contain these pin diagrams. Instead, the pin numbers are written near the corresponding inputs and outputs of the standard logic symbol, as shown in Figure 4–17.

In the rest of this book, just to make you comfortable with properly drawn schematics, we'll include reference designators and pin numbers for all of our logic circuit examples.

* 4.1.9 IEEE Standard Symbols

ANSI
IEEE

Together with the American National Standards Institute (ANSI), the Institute of Electrical and Electronic Engineers (IEEE) has developed a standard set of logic symbols. The most recent revision of the standard is ANSI/IEEE Std 91-1984,

*Throughout this book, optional sections are marked with an asterisk.

Figure 4–18
Pin diagram for a 74LS00 quad 2-input NAND gate.

Figure 4–19
Internal and external logic states.

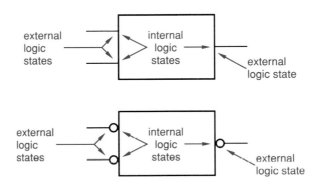

IEEE Standard Graphic Symbols for Logic Functions. It is compatible with standard 617 of the International Electrotechnical Commission (IEC), and must be used in all logic diagrams drawn for the U.S. Department of Defense.

The IEEE standard supports the notion of bubble-to-bubble logic design with the following definitions:

- An *internal logic state* is a logic state assumed to exist inside a symbol outline at an input or an output.

- An *external logic state* is a logic state assumed to exist outside a symbol outline either (1) on an input line prior to any external qualifying symbol at that input, or (2) on an output line beyond any external qualifying symbol at that output.

A *qualifying symbol* is graphics or text added to the basic outline of a device's logic symbol to describe the physical or logical characteristics of the device. The "external qualifying symbol" mentioned above is typically an inversion bubble, which denotes a "negated" input or output, for which the external 0-state corresponds to the internal 1-state.[1] This concept is illustrated in Figure 4–19. When the standard says that a signal is in its *internal 1-state*, we would say that the signal is asserted. Likewise, when the standard says that a signal is in its *internal 0-state*, we would say that the signal is negated.

The IEEE standard provides two different types of symbols for logic gates. One type, called *distinctive-shape symbols*, is what we've been using all along. The other type, called *rectangular-shape symbols*, uses the same shape for all the gates, along with an internal label to identify the type of gate. Figure 4–20 com-

internal logic state

external logic state

qualifying symbol

internal 1-state
internal 0-state

distinctive-shape symbols
rectangular-shape symbols

---

[1] The standard also allows an external, triangular "polarity symbol" to be used to specify active-low inputs and outputs, for which the external LOW level corresponds to the internal 1-state. However, under a positive-logic convention, the bubble and the triangular polarity symbol are equivalent, so we'll stick with the more traditional bubble.

Figure 4–20
Distinctive- and rectangular-shape logic symbols.

pares the two types. According to the IEEE standard, "the distinctive-shape symbol is not preferred." Some people think this statement means that rectangular-shape symbols *are* preferred. However, all the standard really says is that it gives no preference to distinctive-shape symbols compared to rectangular-shape symbols. On the other hand, since most digital designers, authors, and computer-aided design systems prefer the distinctive-shape symbols, that's what we use in this book.

Before the promulgation of the IEEE standard, logic symbols for larger-scale logic elements were drawn in an ad hoc manner; the only standard rule was to use rectangles with inputs on the left and outputs on the right. Although the logic symbol might contain a short description of the element (e.g., "3–8 decoder," "2–1 multiplexer"), it was usually necessary to refer to a separate table to determine the element's logic function. However, the IEEE standard contains a rich set of concepts, such as bit grouping, common control blocks, and dependency notation, that allow some or all of a larger-scale logic element's function to be displayed in the symbol itself. We'll introduce these concepts as appropriate when we discuss larger-scale logic elements later in this chapter.

4.2 CIRCUIT TIMING

4.2.1 Timing Diagrams

timing diagram

A *timing diagram* illustrates the logical behavior of signals in a digital circuit as a function of time. Timing diagrams are an important part of the documentation of any digital system. They can be used both to explain the timing relationships among signals within a system, and to define the timing requirements of external signals that are applied to the system.

Figure 4–21
Timing diagrams for combinational circuits: (a) causality and
propagation delay; (b) uncertain
or "don't-care" delays.

Figure 4–21(a) is a simple timing diagram that illustrates the delay between inputs and outputs of a combinational circuit. In each waveform, the upper line represents a logic 1, and the lower line a logic 0. Signal transitions are drawn as slanted lines, to remind us that they do not occur in zero time in real circuits. (Also, slanted lines look nicer than vertical ones.) Arrows may be drawn, especially in complex timing diagrams, to show *causality*—which input transitions cause which output transitions. In any case, the most important information provided by a timing diagram is a specification of the *delay* between transitions.

causality

delay

Delay in a real circuit is normally measured between the centerpoints of transitions, so the delays in a timing diagram are marked this way. A single timing diagram may contain many different delay specifications. Each different delay is marked with a different identifier, such as t_{RDY} and t_{OUT} in the figure. In large timing diagrams, the delay identifiers are usually numbered for easier reference (e.g., $t_1, t_2, \ldots, t_{42}$). In either case, the timing diagram is accompanied by a *timing table* that specifies each delay amount and the conditions under which it applies.

timing table

Since the delays of real digital components can vary depending on voltage, temperature, and manufacturing parameters, delay is seldom specified as a single number. Instead, a timing table may specify a range of values by giving *minimum*, *typical*, and *maximum* values for each delay. The idea of a range of delays is sometimes carried over into the timing diagram itself by showing the transitions to occur at uncertain times, as in Figure 4–21(b).

Figure 4–22
Timing diagrams for "data" signals: (a) certain and uncertain transitions; (b) sequence of values on an 8-bit bus.

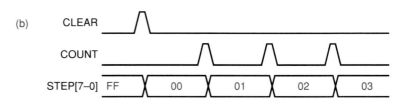

For some signals, the timing diagram needn't show whether the signal changes from 1 to 0 or from 0 to 1 at a particular time, only that a transition occurs at that time. Any signal that carries a bit of "data" has this characteristic—the actual value of the data bit varies according to circumstances but, regardless of value, the bit is transferred, stored, or processed at a particular time relative to "control" signals in the system. Figure 4–22(a) is a timing diagram that illustrates this concept. The "data" signal is normally at a steady 0 or 1 value, and transitions occur only at the times indicated. The idea of an uncertain delay time can also be used with "data" signals, as shown for the DATAOUT signal.

Quite often in digital systems, a group of data signals in a bus is processed by identical circuits. In this case, all signals in the bus have the same timing, and can be represented by a single line in the timing diagram and corresponding specifications in the timing table. If the bus bits are known to take on a particular combination at a particular time, this is sometimes shown in the timing diagram using binary, octal, or hexadecimal numbers, as in Figure 4–22(b).

* 4.2.2 Propagation Delay

propagation delay

In Section 2.6.5, we formally defined the *propagation delay* of a signal path as the time that it takes for a change at the input of the path to produce a change at the output of the path. A combinational circuit with many inputs and outputs has many different paths, and each one may have a different propagation delay. Also, the propagation delay when the output changes from LOW to HIGH (t_{pLH}) may be different from the delay when it changes from HIGH to LOW (t_{pHL}).

The manufacturer of a combinational-logic IC normally specifies all of these different propagation delays, or at least the delays that would be of interest in typical applications. A logic designer who combines ICs in a larger circuit uses the individual device specifications to analyze the overall circuit timing. The delay of a path through the overall circuit is the sum of the delays through subpaths in the individual devices.

The timing specification for an individual device may give minimum, typical, and maximum values for each propagation-delay path and transition direction:

- *Maximum.* This specification is the one that is most often used by experienced designers, since a path "never" has a propagation delay longer than the maximum. However, the definition of "never" varies among logic families and manufacturers. For example, "maximum" propagation delays of 74LS and 74S TTL devices are specified with $V_{CC} = 5$ V, $T_A = 25°C$, and almost no capacitative load. If the voltage or temperature is different, or if the capacitative load is more than 15 pF, the delay may be longer. On the other hand, a "maximum" propagation delay is specified for 74AC and 74ACT devices over the full operating voltage and temperature range, and with a heavier capacitative load of 50 pF.   *maximum delay*

- *Typical.* This specification is the one that is most often used by designers who don't expect to be around when their product leaves the friendly environment of the engineering lab and is shipped to customers. The "typical" delay is what you see from a device that was manufactured on a good day and is operating under near-ideal conditions. Most parts, perhaps 99%, really are manufactured on "good" days and exhibit near-typical delays. However, if you design a system that works only if all of its 100 ICs meet the "typical" timing specs, probability theory says that 63% $(1 - .99^{100})$ of the systems won't work.[2]   *typical delay*

- *Minimum.* This is the smallest propagation delay that a path will ever exhibit. Most well-designed circuits don't depend on this number; that is, they will work properly even if the delay is zero. That's good, because   *minimum delay*

---

[2]The failures aren't spread out evenly, though, since all ICs from a given batch tend to behave about the same. Therefore, everything works fine for a while, just long enough for the system to get into volume production and for everyone to become complacent and self-congratulatory. Then, unknown to the production department, a "slow" batch of some IC type arrives from a supplier and gets used everywhere, so that *nothing* works. The production engineers scurry around trying to analyze the problem (not easy, because the designer is long gone and didn't bother to write a circuit description), and in the meantime the company loses big bucks because it is unable to ship its product.

Table 4–2 Propagation delay in nanoseconds of selected TTL SSI parts.

	74LS				74S			
	Typical		Maximum		Typical		Maximum	
Part number	t_{pLH}	t_{pHL}	t_{pLH}	t_{pHL}	t_{pLH}	t_{pHL}	t_{pLH}	t_{pHL}
'00, '04, '10, '20	9	10	15	15	3	3	4.5	5
'02, '27	10	10	15	15	3.5	3.5	5.5	5.5
'08, '11, '20	8	10	15	20	4.5	5	7	7.5
'14	15	15	22	22				
'30	8	13	15	20	4	4.5	6	7
'32	14	14	22	22	4	4	7	7
'86 (2 levels)	12	10	23	17	7	6.5	10.5	10
'86 (3 levels)	20	13	30	22	7	6.5	10.5	10

manufacturers don't specify minimum delay in most moderate-speed logic families, including 74LS and 74S TTL. A conservative designer assumes a minimum delay of zero for these circuits.[3] However, in high-speed families, including ECL and 74AC and 74ACT CMOS, a nonzero minimum delay is specified to help the designer ensure that timing requirements of latches and flip-flops, discussed in Section 5.2, are met.

Table 4–2 lists the typical and maximum delays of several 74LS and 74S TTL gates. Table 4–3 does the same thing for most of the TTL MSI parts that are introduced later in this chapter. All inputs of a TTL gate have the same propagation delay, though the numbers for LOW-to-HIGH and HIGH-to-LOW transitions are usually different. Also note that two different sets of numbers are given for the 74LS86 XOR gate (discussed in Section 4.7.1). If one input is LOW, and the other is changed, the change propagates through two levels of logic internally, and we observe the first set of delays. If one input is HIGH, and the other is changed, the change propagates through *three* levels of logic internally, and we observe the second set of delays. Similar behavior is exhibited by the 74LS138 and 74LS139 in the MSI table.

To accurately analyze the timing of a circuit containing multiple SSI and MSI devices, a designer may have to study its logical behavior in excruciating

---

[3]Some designs won't work if the propagation delay actually goes to zero, but the cost of modifying the circuit to handle the zero-delay case may be unreasonable, especially since this case is expected never to occur. To obtain a design that always works under "reasonable" conditions, TTL designers often assume minimum delays of one-fourth to one-third of the published *typical* delays.

Table 4–3   Propagation delay in nanoseconds of selected TTL MSI parts.

| | | | 74LS | | | | 74S | | | |
| | | | Typical | | Maximum | | Typical | | Maximum | |
Part number	From	To	t_{pLH}	t_{pHL}	t_{pLH}	t_{pHL}	t_{pLH}	t_{pHL}	t_{pLH}	t_{pHL}
'138	any select	output (2)	11	18	20	41	4.5	7	7	10.5
	any select	output (3)	21	20	27	39	7.5	8	12	12
	/G2A, /G2B	output	12	20	18	32	5	7	8	11
	G1	output	14	13	26	38	7	7	11	11
'139	any select	output (2)	13	22	20	33	5	6.5	7.5	10
	any select	output (3)	18	25	29	38	7	8	12	12
	enable	output	16	21	24	32	5	6.5	8	10
'151	any select	Y	27	18	43	30	12	12	18	18
	any select	/Y	14	20	23	32	10	9	15	13.5
	any data	Y	20	16	32	26	8	8	12	12
	any data	/Y	13	12	21	20	4.5	4.5	7	7
	enable	Y	26	20	42	32	11	12	16.5	18
	enable	/Y	15	18	24	30	9	8.5	13	12
'153	any select	output	19	25	29	38	11.5	12	18	18
	any data	output	10	17	15	26	6	6	9	9
	enable	output	16	21	24	32	10	9	15	13.5
'157	select	output	15	18	23	27	9.5	9.5	15	15
	any data	output	9	9	14	14	5	4.5	7.5	6.5
	enable	output	13	14	21	23	8.5	7.5	12.5	12
'182	any /Gi or /Pi	C1, C2, C3					4.5	4.5	7	7
	any /Gi or /Pi	/G					5	7	7.5	10.5
	any /Pi	/P					4.5	6.5	6.5	10
	C0	C1, C2, C3					6.5	7	10	10.5
'280	any input	EVEN	33	29	50	45	14	11.5	21	18
	any input	ODD	23	31	35	50	14	11.5	21	18
'283	C0	any Si	16	15	24	24	11	12	18	18
	any Ai, Bi	any Si	15	15	24	24	12	11.5	18	18
	C0	C4	11	11	17	22	6	7.5	11	11
	any Ai, Bi	C4	11	12	17	17	7.5	8.5	12	12
'381	CIN	any Fi	18	14	27	21	10	10	17	17
	any Ai, Bi	/G	20	21	30	33	12	12	20	20
	any Ai, Bi	/P	21	33	23	33	11	18	11	18
	any Ai, Bi	any Fi	20	15	30	23	18	16	27	25
	function select	any Fi	35	34	53	51	18	18	30	30
	function select	/G, /P	31	32	47	48	18	18	30	30
'682	any Pi	/PEQQ	13	15	25	25				
	any Qi	/PEQQ	14	15	25	25				
	any Pi	/PGTQ	20	15	30	30				
	any Qi	/PGTQ	21	19	30	30				

detail. For example, when inverting gates (NAND, NOR, etc.) are placed in series, a LOW-to-HIGH change at one gate's output causes a HIGH-to-LOW change at the next one's, and so the differences between t_{pLH} and t_{pHL} tend to average out. On the other hand, when noninverting gates (AND, OR, etc.) are placed in series, a transition causes all outputs to change in the same direction, and so the gap between t_{pLH} and t_{pHL} tends to widen. The analysis gets more complicated if there are MSI devices in the delay path, or if there are multiple paths from a given input signal to a given output signal.

worst-case delay

Thus, in large circuits, analysis of all of the different delay paths and transition directions can be very complex. To permit a simplified "worst-case" analysis, designers often use a single *worst-case delay* specification that is the maximum of t_{pLH} and t_{pHL} specifications. The worst-case delay through a circuit is then computed as the sum of the worst-case delays through the individual components, independent of the transition direction and other circuit conditions. This may give an overly pessimistic view of the overall circuit delay, but it saves design time and it's guaranteed to work. Of course, designers who have access to sophisticated CAD software can simply run a timing analysis program to obtain true minimum, typical, and maximum delays for any desired circuit path.

4.3 DECODERS

decoder

A *decoder* is a multiple-input, multiple-output logic circuit that converts coded inputs into coded outputs, where the input and output codes are different. The input code generally has fewer bits than the output code, and there is a one-to-

one-to-one mapping

one mapping from input code words into output code words. In a *one-to-one mapping*, each input code word produces a different output code word.

The general structure of a decoder circuit is shown in Figure 4–23. The enable inputs, if present, must be asserted for the decoder to perform its normal mapping function. Otherwise, the decoder maps all input code words into a single, "disabled," output code word.

The most commonly used input code is an n-bit binary code, where an n-bit word represents one of 2^n different coded values, normally the integers from 0

Figure 4–23
Decoder circuit structure.

through $2^n - 1$. Sometimes an n-bit binary code is truncated to represent fewer than 2^n values. For example, in the BCD code, the 4-bit combinations 0000 through 1001 represent the decimal digits 0–9, and combinations 1010 through 1111 are not used.

The most commonly used output code is a 1-out-of-m code, which contains m bits, where one bit is asserted at any time. Thus, in a 1-out-of-4 code with active-high outputs, the code words are 0001, 0010, 0100, and 1000. With active-low outputs, the code words are 1110, 1101, 1011, and 0111.

4.3.1 Binary Decoders

The most commonly used decoder circuit is an n-to-2^n decoder or *binary decoder*.   *binary decoder*
Such a decoder has an n-bit binary input code and a 1-out-of-2^n output code. For example, Figure 4–24(a) shows the inputs and outputs and Table 4–4 is the truth table of a 2-to-4 decoder. The input code word $I1, I0$ represents an integer in the range 0–3. The output code word $Y3, Y2, Y1, Y0$ has Yi equal to 1 if and only if the input code word is the binary representation of i and the *enable input* EN   *enable input*

Figure 4–24
A 2-to-4 decoder: (a) inputs and outputs; (b) logic diagram.

one output asserted
to make true, pull 0.

(a)   (b)

Table 4–4
Truth table for a 2-to-4 decoder.

Inputs			Outputs			
EN	I1	I0	Y3	Y2	Y1	Y0
0	x	x	0	0	0	0
1	0	0	0	0	0	1
1	0	1	0	0	1	0
1	1	0	0	1	0	0
1	1	1	1	0	0	0

$Y_3 = 0001$
$Y_2 = 00010$
$Y_1 = 00100$
$Y_0 = 01000$

Table 4–5
Position encoding for a 3-bit mechanical encoding disk.

Disk position	I2	I1	I0	Binary decoder output
0°	0	0	0	Y0
45°	0	0	1	Y1
90°	0	1	1	Y3
135°	0	1	0	Y2
180°	1	1	0	Y6
225°	1	1	1	Y7
270°	1	0	1	Y5
315°	1	0	0	Y4

decode

is 1. If EN is 0, then all of the outputs are 0. A gate-level circuit for the 2-to-4 decoder is shown in Figure 4–24(b). Each AND gate *decodes* one combination of the input code word I1,I0.

The decoder's truth table introduces a "don't-care" notation for input combinations. If one or more input values do not affect the output values for some combination of the remaining inputs, they are marked with an "x" for that input combination. This convention can greatly reduce the number of rows in the truth table, as well as make the functions of the inputs more clear.

The input code of an n-bit binary decoder need not represent the integers from 0 through $2^n - 1$. For example, Table 4–5 shows the 3-bit Gray-code output of a mechanical encoding disk with eight positions. The eight disk positions can be decoded with a 3-bit binary decoder with the appropriate assignment of signals to the decoder outputs, as shown in Figure 4–25.

decimal decoder
BCD decoder

Also, it is not necessary to use all of the outputs of a decoder. For example, a *decimal* or *BCD decoder* can be obtained from a 4-to-16 decoder by using only the first ten outputs, Y0–Y9. Commercially available decimal decoders, such as the 74LS42, are just 4-to-16 decoders in which outputs Y10–Y15 have been omitted because of pin limitations.

Figure 4–25
Using a 3-to-8 binary decoder to decode a Gray code.

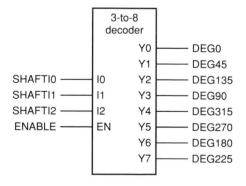

Figure 4–26
Logic symbol for one half of a
74LS139 dual 2-to-4 decoder:
(a) conventional symbol; (b) default signal names associated
with external pins.

(a)   (b)

4.3.2 Pin and Signal Names for Larger-Scale Elements

Before describing some commercially available TTL MSI decoders, we need to
remind you about active levels for pins and set up a consistent convention for
naming the internal signals and external pins of decoders and other MSI parts.

Larger-scale elements almost always have their signal names defined in
terms of the functions performed *inside* their symbolic outline, as explained in
Section 4.1.4. For example, Figure 4–26(a) shows the logic symbol for one
section of a 74LS139 dual 2-to-4 decoder, an MSI part that we'll fully describe
in the next subsection. When the G input is asserted, one of the outputs Y0–Y3 is
asserted, as selected by a 2-bit code applied to the A and B inputs. It is apparent
from the symbol that the G input pin and all of the output pins are active low.

When the '139 symbol appears in the logic diagram for a real application,
its inputs and outputs have signals connected to other devices, and each such
signal has a name that indicates its function in the application. However, when
we describe the '139 in isolation, we might still like to have a name for the
signal on each external pin. Figure 4–26(b) shows our naming convention in
this case. Active-high pins are given the same name as the internal signal, while
active-low pins have the internal signal name preceded by a /.

Unfortunately, not all IC manufacturers use consistent drawing and naming
conventions for internal signals, external pins, and active levels. For example,
the Texas Instruments *TTL Logic Data Book* (1988) properly draws the G input
in the 74LS139 symbol as active low, but improperly shows the Y0–Y3 outputs
as active high. Furthermore, it includes an overbar in the external names of some
active-low pins, including $\overline{G}$ in the '139, but fails to do so for others, such as
Y0–Y7 in the 74LS138. The result is that a digital designer still must study the
truth table and internal logic diagram of a larger-scale element to be absolutely
sure what it does.

4.3.3 The 74LS139 Dual 2-to-4 Decoder

In this and subsequent sections, we'll look at commonly used TTL MSI parts
that perform larger-scale functions. These parts are the most important building blocks in a digital designer's toolbox, because their level of functionality

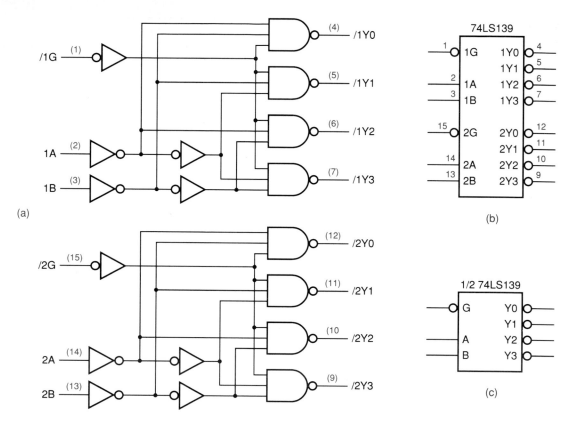

Figure 4–27 The 74LS139 dual 2-to-4 decoder: (a) logic diagram, including pin numbers for a standard 16-pin dual in-line package; (b) traditional logic symbol; (c) logic symbol for one decoder.

matches a designer's level of thinking when partitioning a large problem into smaller chunks. Even for ASIC designers, understanding these TTL MSI parts is important, because most ASIC environments provide standard cells that are very similar in function.

74LS139   Two independent and identical 2-to-4 decoders are contained in a single TTL MSI part, the *74LS139*.[4] The gate-level circuit diagram for this IC is shown in Figure 4–27(a). Notice that the outputs and the enable input of the '139 are active-low. Most commercially available decoders are designed with

---

[4] As of this writing, 74LS is the most inexpensive and commonly used logic family, but most logic gates and larger-scale elements are available in a variety of families. For example, the 74LS139, 74S139, 74ALS139, 74AS139, 74HCT139, 74ACT139, and 74AC139 are all dual 2-to-4 decoders with the same logic function and pinouts, but in electrically different TTL and CMOS families. Thus, throughout this text, we use "74LS" merely as a generic prefix. And we'll sometimes omit the prefix and write, for example, '139.

Table 4–6
Truth table for one half of a
74LS139 dual 2-to-4 decoder.

Inputs			Outputs			
/G	B	A	/Y3	/Y2	/Y1	/Y0
1	x	x	1	1	1	1
0	0	0	1	1	1	0
0	0	1	1	1	0	1
0	1	0	1	0	1	1
0	1	1	0	1	1	1

active-low outputs, since inverting gates are generally faster than noninverting
ones. Also notice that the '139 has extra inverters on its select inputs. Without
these inverters, each select input would present three unit loads instead of one,
consuming much more of the fanout budget of the device that drives it.

A logic symbol for the 74LS139 is shown in Figure 4–27(b). Notice that
all of the signal names inside the symbol outline are active-high (no /), and that
inversion bubbles indicate active-low inputs and outputs. Sometimes a schematic
may contain a generic symbol for just one decoder, one half of a '139, as shown
in (c). In this case, a computer-aided design system may defer the assignment
of the generic function to one half or the other of a particular '139 package until
the printed-circuit board is laid out, in much the same way as the assignment of
2-input NAND gates to 74LS00 packages can be deferred.

Table 4–6 is the truth table for a 74LS139-type decoder. The truth tables
in some manufacturers' data books use L and H to denote the input and output
signal voltage levels, so there can be no ambiguity about the electrical function
of the device; a truth table written this way is sometimes called a *function table*.   *function table*
However, since we use positive logic throughout this book, we can use 0 and 1
without ambiguity. In any case, the truth table gives the logic function in terms
of the *external pins* of the device. A truth table for the function performed *inside*
the symbol outline would look just like Table 4–4 except that the input signal
names would be G, B, A.

Some logic designers draw the symbol for 74LS139s and other logic functions without inversion bubbles. Instead, they use an overbar on signal names
inside the symbol outline to indicate negation, as shown in Figure 4–28(a). This
notation is self-consistent, but it is inconsistent with our drawing standards for
bubble-to-bubble logic design. The symbol shown in (b) is absolutely *incorrect*:
according to this symbol, a logic 1, not 0, must be applied to the enable pin to
enable the decoder.

Some manufacturers' data books have similar inconsistencies. For example,
the Texas Instruments *TTL Logic Data Book* uses names like 1$\overline{\text{G}}$ for the 139's
enable inputs, with the overbar indicating an active-low pin, but names like 1Y0

Figure 4-28
More ways to symbolize a
74LS139: (a) correct but to
be avoided; (b) incorrect because
of double negations.

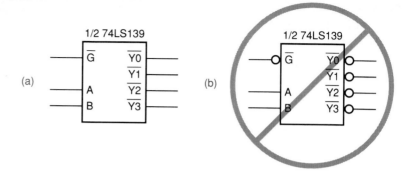

for active-low output pins. Worse, the same name, 1Y0 is used *inside* the symbol outline to denote the active-high internal signal.

Apparently, the TI data book really intends for 1Y0 to denote the external signal, since the columns of its '139 function table, which is written in terms of external signal levels, have headings like 1Y0. Unfortunately, the notation is not consistent among different parts, so you have to study the description of each part to know what's really going on. Be assured, however, that the signal names in this text *are* consistent.

4.3.4   The 74LS138 3-to-8 Decoder

74LS138

The *74LS138* is a commercially available MSI 3-to-8 decoder whose gate-level circuit diagram and symbol are shown in Figure 4-29; its truth table is given in Table 4-7. Like the 74LS139, the 74LS138 has active-low outputs, and it has three enable inputs (G1, /G2A, /G2B), all of which must be asserted for the selected output to be asserted.

The logic function of the '138 is straightforward—an output is asserted if and only if the decoder is enabled and the output is selected. Thus, we can easily write logic equations for an internal output signal such as Y5 in terms of the internal input signals:

$$Y5 = \underbrace{G1 \cdot G2A \cdot G2B}_{\text{enable}} \cdot \underbrace{C \cdot B' \cdot A}_{\text{select}}$$

However, because of the inversion bubbles, we have the following relations between internal and external signals:

$$G2A = /G2A'$$
$$G2B = /G2B'$$
$$Y5 = /Y5'$$

Figure 4–29   The 74LS138 3-to-8 decoder: (a) logic diagram, including pin numbers for a standard 16-pin dual in-line package, (b) traditional logic symbol.

Table 4–7   Truth table for a 74LS138 3-to-8 decoder.

G1	/G2A	/G2B	C	B	A	/Y7	/Y6	/Y5	/Y4	/Y3	/Y2	/Y1	/Y0
0	x	x	x	x	x	1	1	1	1	1	1	1	1
x	1	x	x	x	x	1	1	1	1	1	1	1	1
x	x	1	x	x	x	1	1	1	1	1	1	1	1
1	0	0	0	0	0	1	1	1	1	1	1	1	0
1	0	0	0	0	1	1	1	1	1	1	1	0	1
1	0	0	0	1	0	1	1	1	1	1	0	1	1
1	0	0	0	1	1	1	1	1	1	0	1	1	1
1	0	0	1	0	0	1	1	1	0	1	1	1	1
1	0	0	1	0	1	1	1	0	1	1	1	1	1
1	0	0	1	1	0	1	0	1	1	1	1	1	1
1	0	0	1	1	1	0	1	1	1	1	1	1	1

The table headers are grouped as **Inputs** (G1, /G2A, /G2B, C, B, A) and **Outputs** (/Y7, /Y6, /Y5, /Y4, /Y3, /Y2, /Y1, /Y0).

$$/y5 = (G1 \cdot /G2A' \cdot /G2B' \cdot C \cdot B' \cdot A)' = G1' + /G2A + /G2B + C' + B + A'$$

$$/y6 = (G1 \cdot /G2A' \cdot /G2B' \cdot C \cdot B \cdot A')' = G1' + /G2A + /G2B + C' + B' + A$$

Therefore, if we're interested, we can write the following equation for the external output signal /Y5 in terms of external input signals:

$$/Y5 = Y5' = (G1 \cdot /G2A' \cdot /G2B' \cdot C \cdot B' \cdot A)'$$
$$= G1' + /G2A + /G2B + C' + B + A'$$

On the surface, this equation doesn't resemble what we might expect for a decoder, since it is a logical sum rather than a product. However, if we practice bubble-to-bubble logic design, we don't have to worry about this; we simply give the output signal an active-low name and remember that it's active low when we connect it to other inputs.

4.3.5 Cascading Binary Decoders

Multiple binary decoders can be used to decode larger code words. Figure 4–30 shows how two 3-to-8 decoders can be combined to make a 4-to-16 decoder. The availability of both active-high and active-low enable inputs on the 74LS138 makes it possible to enable one or the other directly based on the state of the most significant input bit. <u>The top decoder (U1) is enabled when N3 is 0, and the bottom one (U2) is enabled when N3 is 1.</u>

Figure 4–30
Design of a 4-to-16 decoder using 74LS138s.

Figure 4–31
Design of a 5-to-32 decoder
using 74LS138s and a 74LS139.

For even larger code words, binary decoders can be cascaded hierarchically. Figure 4–31 shows how to use half of a 74LS139 to decode the two high-order bits of a 5-bit code word, thereby enabling one of four 74LS138s that decode the three low-order bits.

4.3.6 Binary Decoders as Minterm Generators

The outputs of an n-to-2^n binary decoder correspond to n-variable minterms since, when the decoder is enabled, each output is asserted for exactly one row of an n-variable truth table. If an n-input logic function is specified as a minterm list, a decoder can be used to build the function directly, as a canonical sum (sum of minterms).

For example, consider the function $F = \Sigma_{X,Y,Z}(0,2,3,5)$, whose Karnaugh map is shown in Figure 4–32(a). A traditional sum-of-products circuit using NAND gates is shown in (b). Alternatively, the canonical sum can be realized with a decoder and one NAND gate as shown in (c), simply by "picking off" and summing the appropriate minterms. (Note that X, Y, and Z must be connected to the decoder inputs in the proper order.) A designer might prefer one design or the other, depending on trade-offs in speed, cost, and other things. The decoder solution is particularly flexible, since the designer can change it as required to

Figure 4–32
Designing a circuit for the logic function $F = \Sigma_{X,Y,Z}(0,2,3,5)$:
(a) Karnaugh map; (b) NAND-based minimal sum of products;
(c) decoder-based canonical sum.

Figure 4–33 Using a decoder to realize a multiple-output function.

any canonical sum with four or fewer minterms, simply by connecting different decoder outputs to the NAND gate (see Exercise 4.16).

Using a decoder to realize the canonical sum is impractical for logic functions with more than a few inputs, of course, since the number of minterms and decoder outputs grows exponentially with the number of inputs. On the other hand, the decoder approach makes a lot of sense for *multiple-output* logic functions with a few inputs. For example, Figure 4–33 shows how a decoder and three NAND gates can be used to realize three different canonical sums of the same input variables.

Probably the most attractive aspect of designing with decoders is that it's so easy—mindless, if you will. Don't knock it! The decoder approach isn't used too much anymore, but not because it's mindless. Instead, "programmable logic devices" have expanded the approach, allowing logic designers to realize not only sums of minterms, but also sums of arbitrary product terms of a large number of input variables, simply by listing the desired terms. These devices are described fully in Chapter 7.

4.3.7 Seven-Segment Decoders

Look at your wrist and you'll probably see a *seven-segment display*. This type of display, which normally uses light-emitting diodes (LEDs) or liquid-crystal display (LCD) elements, is used in watches, calculators, and instruments to display decimal data. A digit is displayed by illuminating a subset of the seven line segments shown in Figure 4–34(a).

A *seven-segment decoder* has 4-bit BCD as its input code and the "seven-segment code," which is graphically depicted in Figure 4–34(b), as its output code. Figure 4–35 and Table 4–8 are the logic diagram and truth table for a *74LS49* seven-segment decoder. Except for the strange (clever?) connection

seven-segment display

seven-segment decoder

74LS49

Figure 4–34   Seven-segment display: (a) segment identification; (b) decimal digits.

of the "blanking input" /BI, each output of the 74LS49 is a minimal product-of-sums realization for the corresponding segment, assuming "don't-cares" for the nondecimal input combinations. The INVERT-OR-AND structure used for each output may seem a little strange, but it is equivalent under the generalized DeMorgan's theorem to an AND-OR-INVERT gate, which is a fairly fast and compact structure to build in TTL.

   Most modern seven-segment display elements have decoders built into them, so that a 4-bit BCD word can be applied directly to the device. Many of the older, discrete seven-segment decoders have special high-voltage or high-current outputs that are well suited for driving large, high-powered display elements.

Table 4–8
Truth table for a 74LS49 seven-segment decoder.

		Inputs					Outputs				
/BI	D	C	B	A	a	b	c	d	e	f	g
0	x	x	x	x	0	0	0	0	0	0	0
1	0	0	0	0	1	1	1	1	1	1	0
1	0	0	0	1	0	1	1	0	0	0	0
1	0	0	1	0	1	1	0	1	1	0	1
1	0	0	1	1	1	1	1	1	0	0	1
1	0	1	0	0	0	1	1	0	0	1	1
1	0	1	0	1	1	0	1	1	0	1	1
1	0	1	1	0	0	0	1	1	1	1	1
1	0	1	1	1	1	1	1	0	0	0	0
1	1	0	0	0	1	1	1	1	1	1	1
1	1	0	0	1	1	1	1	0	0	1	1
1	1	0	1	0	0	0	0	1	1	0	1
1	1	0	1	1	0	0	1	1	0	0	1
1	1	1	0	0	0	1	0	0	0	1	1
1	1	1	0	1	1	0	0	1	0	1	1
1	1	1	1	0	0	0	0	1	1	1	1
1	1	1	1	1	0	0	0	0	0	0	0

Figure 4–35
The 74LS49 seven-segment decoder: (a) logic diagram, including pin numbers; (b) traditional logic symbol.

Figure 4–36
IEEE standard symbols for
decoders: (a) 74LS138;
(b) 74LS139.

(a)

(b)

* 4.3.8   IEEE Standard Symbols

Up until now, we have used the "traditional" logic symbols for decoders. Although they show the active levels of the inputs and outputs, they do not indicate the logical function of the device—you already have to know what a 74LS138 or 74LS139 does when you read its symbol.

The IEEE standard for logic symbols, on the other hand, allows a decoder's logic function to be displayed as part of the symbol, shown in Figure 4–36. These symbols use several concepts of the standard:

internal qualifying symbol
qualifying label

- *Internal qualifying symbols.* Individual input and output signals may be labeled with qualifying symbols inside the logic-symbol outline to describe the signals' characteristics. In this book, we call such symbols *qualifying labels* for short.

General qualifying symbol
coder

- *General qualifying symbols.* The top of a logic symbol may contain an alphanumeric label to denote the general function performed by the device. Decoders and encoders (called *coders*) use the general qualifying symbol X/Y to indicate the type of coding performed, where X is the input code and Y is the output code. For example, a 3-to-8 decoder may be labeled BIN/1-OF-8.

internal value

- *Internal values.* Each input combination of a coder produces an "internal value" that is displayed by the coder's outputs. The internal values for a 3-to-8 decoder are 0–7.

input weight

- *Input weights.* The inputs of a coder may have qualifying labels indicating the numerical weights associated with those inputs. In this case, the internal value at any time is the sum of the weights of the asserted inputs. The input weights for a 3-to-8 decoder are 1, 2, and 4.

output value

- *Output values.* Each output may have a qualifying label listing the internal

Figure 4–37
A composite symbol with one or
more logic connections between
its elements.

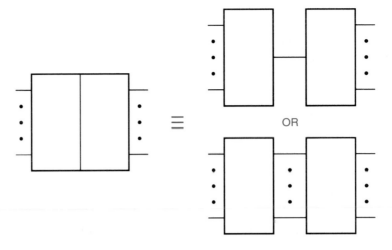

values that cause that output to be asserted. In a binary decoder, each
output is asserted for just one internal value.

- *Enable input.* An enable input has the qualifying label **EN** and permits   *enable input*
  action when asserted. When negated, an enable input imposes the external
  high-impedance state on three-state outputs, and the negated state on other
  outputs. The 74LS138 and 74LS139 have active-low outputs, which are
  therefore 1 when the enable input is negated.

- *Embedded and abutted elements.* The outlines of individual logic elements   *embedded element*
  may be embedded or abutted to form a larger composite symbol. There   *abutted element*
  is at least one common logic connection when the dividing line between
  two outlines is perpendicular to the direction of signal flow, as shown in
  Figure 4–37. For example, the 74LS138 symbol has an embedded 3-input
  AND gate that drives the internal **EN** input. There is no connection between
  the elements when the dividing line is in the direction of signal flow, as
  shown in Figure 4–38. For example, the 74LS139 contains two separate
  2-to-4 decoders.

The ability to embed individual logic elements in a larger symbol is probably
the most useful feature of the IEEE standard. For example, Figure 4–39 shows

Figure 4–38
A composite symbol with no
connection between its elements.

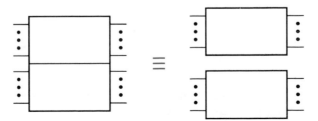

Figure 4–39
IEEE standard symbol for a
fictitious decoder, the 74JFW328.

74JFW328

the symbol for a 3-to-8 decoder with a different set of enable inputs than the 74LS138. The fictitious *74JFW328* decoder is enabled when pin 6 is 0 or both pins 4 and 5 are 1.

The absence of a standardized method of giving unique *names* to pins and displaying the names on the logic symbol is probably the biggest defect of the IEEE standard. For example, how do we name the internal signals on pins 4, 5, and 6 of the 74JFW328? Worse, how can we distinguish between the internal signals on pins 3 and 11, which are both labeled "4" on the logic symbol? Apparently, the standard makers expected us to refer to these signals by pin number only. But this is awkward and inconvenient, not only in textbooks, but also in design and debugging. It is far more natural to refer to a signal by a functional name than by a pin number. (Indeed, in ASIC design there are no pin numbers for internal signals!) As a result, while we will later introduce a few more aspects of the IEEE standard for larger-scale logic elements, we'll stick with traditional symbols in our design examples.

4.4 THREE-STATE BUFFERS

In Section 2.11.5, we showed the electrical design of TTL devices whose outputs may be in one of three states—0, 1, or Hi-Z. In this section, we'll show how to use them.

three-state buffer
three-state driver

three-state enable

The most basic three-state device is a *three-state buffer*, often called a *three-state driver*. The logic symbols for four physically different three-state buffers are shown in Figure 4–40. The basic symbol is that of a noninverting buffer (a,b) or an inverter (c,d). The extra signal at the top of the symbol is a *three-state enable* input, which may be active high (a,c) or active low (b,d). When the enable input is asserted, the device behaves like an ordinary buffer or inverter. When the enable input is negated, the device output "floats," that is, it goes to a high-impedance (Hi-Z), disconnected state and functionally behaves as if it weren't even there.

Figure 4–40 Various three-state buffers: (a) noninverting, active-high enable;
(b) noninverting, active-low enable; (c) inverting, active-high enable;
(d) inverting, active-low enable.

Three-state devices allow multiple sources to share a single "party line,"
as long as only one device "talks" on the line at a time. Figure 4–41 gives an
example of how this can be done. Three input bits, SSRC2–SSRC0, select one
of eight sources of data that may drive a single line, SDATA. A 3-to-8 decoder,
the 74LS138, ensures that only one of the eight SEL lines is asserted at a time,
enabling only one three-state buffer to drive SDATA. However, if not all of the
EN lines are asserted, then none of the three-state buffers is enabled. The logic
value on SDATA is undefined in this case.

Typical three-state devices are designed so that they go into the Hi-Z state
faster than they come out of the Hi-Z state. (In terms of the specifications in a

Figure 4–41
Eight sources sharing a three-state party line.

Figure 4–42
Timing diagram for the three-state party line.

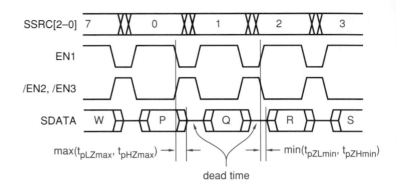

data book, t_{pLZ} and t_{pHZ} are both less than t_{pZL} and t_{pZH}; also see Section 2.11.5.) This means that if the outputs of two three-state devices are connected to the same party line, and we simultaneously disable one and enable the other, the first device will get off the party line before the second one gets on. This is important because, if both devices were to drive the party line at the same time, and if both were trying to maintain opposite output values (0 and 1), then excessive current would flow and create noise in the system, as discussed in Section 2.11.4. This is often called *fighting*.

fighting

Unfortunately, delays and timing skews in control circuits make it difficult to ensure that the enable inputs of different three-state devices change "simultaneously." Even when this is possible, a problem arises if three-state devices from different-speed logic families (or even different ICs manufactured on different days) are connected to the same party line. The turn-on time (t_{pZL} or t_{pZH}) of a "fast" device may be shorter than the turn-off time (t_{pLZ} or t_{pHZ}) of a "slow" one, and the outputs may still fight.

dead time

The only really safe way to use three-state devices is to design control logic that guarantees a *dead time* on the party line during which no one is driving it. The dead time must be long enough to account for the worst-case differences between turn-off and turn-on times of the devices and for skews in the three-state control signals. A timing diagram that illustrates this sort of operation for the party line of Figure 4–41 is shown in Figure 4–42. This timing diagram also illustrates a drawing convention for three-state signals—when in the Hi-Z state, they are shown at an "undefined" level halfway between 0 and 1.[5]

---

[5] The actual voltage level of a floating signal depends on circuit details, such as resistive and capacitive load, and may vary over time. Also, the interpretation of this level by other circuits depends on the input characteristics of those circuits, so it's best not to count on a floating signal as being anything other than "undefined." Sometimes a pull-up resistor is used on three-state party lines to ensure that a floating value is pulled to a HIGH voltage and interpreted as logic 1. This is especially important on party lines that drive CMOS devices, which may consume excessive current when their input voltage is halfway between logic 0 and 1.

Figure 4–43

Pin diagrams for the 74LS125 and 74LS126 three-state buffers.

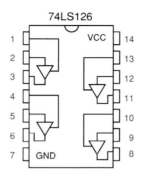

4.4.1 Standard SSI and MSI Three-State Buffers

Like logic gates, several independent three-state buffers may be packaged in a single SSI IC. For example, Figure 4–43 shows the pin diagrams for *74LS125* and *74LS126*, each of which contains four independent noninverting three-state buffers in a 14-pin package. The three-state enable inputs in the '125 are active low, and in the '126 they are active high.

    Most party-line applications use a bus with more than one bit of data. For example, in an 8-bit microprocessor system, the data bus is eight bits wide, and

74LS125

74LS126

Figure 4–44

The 74LS541 octal three-state buffer: (a) logic diagram, including pin numbers for a standard 20-pin dual in-line package; (b) traditional logic symbol.

Figure 4–45
Using a 74LS541 as a microprocessor input port.

74LS541
octal

hysteresis

peripheral devices normally place data on the bus eight bits at a time. Thus, a peripheral device enables eight three-state drivers to drive the bus, all at the same time. Independent enable inputs, as in the '125 and '126, are not necessary.

Thus, to reduce the package size in wide-bus applications, several commonly used MSI parts contain multiple three-state buffers with common enable inputs. For example, Figure 4–44 shows the logic diagram and symbol for a *74LS541* octal noninverting three-state buffer. *Octal* means that the part contains eight individual buffers. Both enable inputs, /G1 and /G2, must be asserted to enable the device's three-state outputs. The little rectangular symbols inside the buffer symbols indicate *hysteresis*, an electrical characteristic of the inputs that improves noise immunity, as we explained in Section 2.11.1. The 74LS541 inputs typically have 0.4 volts of hysteresis.

Figure 4–45 shows part of a microprocessor system with an 8-bit data bus, DB[0–7], and a 74LS541 used as an input port. The microprocessor selects Input

(a)

/G — (19)
DIR — (1)

A1 — (2) ———————————————— (18) B1

A2 — (3) ———————————————— (17) B2

A3 — (4) ———————————————— (16) B3

A4 — (5) ———————————————— (15) B4

A5 — (6) ———————————————— (14) B5

A6 — (7) ———————————————— (13) B6

A7 — (8) ———————————————— (12) B7

A8 — (9) ———————————————— (11) B8

Figure 4–46
The 74LS245 octal three-state transceiver: (a) logic diagram; (b) traditional logic symbol.

(b)

Port 1 by asserting INSEL1 and requests a read operation by asserting READ. The selected 74LS541 responds by driving the microprocessor data bus with user-supplied input data. Other input ports may be selected when a different INSEL line is asserted along with READ.

Many other varieties of octal three-state buffers are commercially available. For example, the *74LS540* is identical to the 74LS541 except that it contains   *74LS540*

Figure 4–47
Bidirectional buses and
transceiver operation.

Table 4–9 Modes of operation for a pair of bidirectional buses.

/ENTFR	ATOB	Operation
0	0	Transfer data from a source on bus B to a destination on bus A.
0	1	Transfer data from a source on bus A to a destination on bus B.
1	x	Transfer data on buses A and B independently.

inverting buffers. The *74LS240* and *74LS241* are similar to the '540 and '541, except that they are split into two 4-bit sections, each with a single enable line.

    A *bus transceiver* contains pairs of three-state buffers connected in opposite directions between each pair of pins, so that data can be transferred in either direction. For example, Figure 4–46 shows the logic diagram and symbol for a *74LS245* octal three-state transceiver. The DIR input determines the direction of transfer, from A to B (DIR=1) or from B to A (DIR=0). The three-state buffer for the selected direction is enabled only if /G is asserted.

    A bus transceiver is typically used between two *bidirectional buses*, as shown in Figure 4–47. Three different modes of operation are possible, depending on the state of /G and DIR, as shown in Table 4–9. As usual, it is the designer's responsibility to ensure that neither bus is ever driven simultaneously by two devices.

74LS240
74LS241
bus transceiver

74LS245

bidirectional bus

* 4.4.2   IEEE Standard Symbols

Three-state-buffer symbols use one more feature of the IEEE standard:

- *Downward-pointing triangle*. This denotes a three-state output.

downward-pointing triangle

Also recall that an enable input, labeled EN, is specifically understood to force all affected outputs a disabled state. For three-state outputs, the disabled state is Hi-Z. Thus, the three-state buffers of Figure 4–40 are drawn as shown in Figure 4–48. Another important feature of the IEEE standard is introduced in the symbols for MSI three-state buffers:

- *Common control block*.   This concept, illustrated in Figure 4–49, may be used with an array of related elements. Inputs to the common control block are understood to affect all the elements of the array.

common control block

    Thus, symbols may be drawn for the 74LS541 and 74LS245 as shown in Figure 4–50. The enable and direction inputs (pins 1 and 19) apply to all elements of the device. The '541 and '245 symbols introduce several other features of the standard:

- *Hysteresis symbol*. Inputs bearing this symbol have hysteresis.

hysteresis symbol

(a)   (b)   (c)   (d)

Figure 4–48   IEEE-standard rectangular-shape symbols for three-state buffers: (a) noninverting, active-high enable; (b) noninverting, active-low enable; (c) inverting, active-high enable; (d) inverting, active-low enable.

Figure 4–49
Common control block in an
IEEE standard symbol.

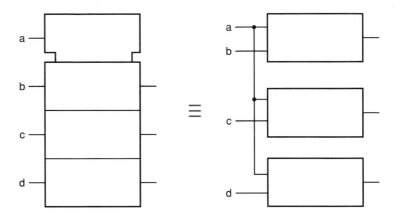

right-pointing
triangle
left-pointing triangle

- *Right-pointing or left-pointing triangle.* These symbols denote "amplifica-tion;" in the case of three-state buffers, they indicate an output circuit that has more fanout and can drive a heavier load than an ordinary output circuit.

arrows

- *Arrows.* These denote the direction of signal flow when it is not strictly left to right, as in the '245 symbol.

identical elements

- *Identical elements.* Only the first of two or more identical elements in an array must be drawn in detail. Thus, the bottom seven elements of the '245 symbol are understood to be identical to the top element (which in this case happens to be divided into two subelements, the three-state buffers for the two directions).

dependency notation

The '245 symbol also introduces *dependency notation*, a means of display-ing some of the more common logical relationships among input and output signals. A few more concepts are needed to make this notation fly:

Figure 4–50
IEEE standard logic symbols for
the 74LS541 and 74LS245.

Figure 4–51
Illustration of AND dependency.

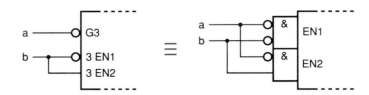

- *Affecting signals.* An input signal (or, occasionally, an output signal) may   *affecting signal*
  affect other inputs or outputs in a way that can be displayed on the symbol.
  Such a signal has a qualifying label Li, where L is a letter that indicates the
  type of relationship or effect, and i is an integer that identifies the affected
  signals. The '245 internal signal labeled G3 is such a signal.

- *Affected signals.* Inputs or outputs that are affected by a signal Li bear the   *affected signal*
  qualifying label i. If an affected signal requires a qualifying label of its
  own, then i is used as a prefix to that label. Thus, the signal labeled 3EN1
  in the '245 symbol affects signals labeled 1, and is itself affected by G3.[6]

- *AND dependency.* This relationship is denoted by Gi, and is a sort of enable   *AND dependency*
  function. Affected signals perform their "normal" functions only if Gi is   *Gi*
  asserted; otherwise, they are negated. In the '245 symbol, inputs 3EN1 and
  3EN2 can "do their thing" only if G3 is asserted. Figure 4–51 shows an
  equivalent notation for the dependent signals. Notice that dependency is
  defined in terms of internal, not external, signal values.

- *Enable dependency.* This dependency is denoted by ENi, and has the same   *enable dependency*
  effect as EN inputs defined earlier. Thus, in the '245 symbol, the internal   *ENi*
  signal 3EN1, which is asserted only if pins 1 and 19 are LOW, enables the
  three-state drivers for pins 2 through 9.

Whewww! You might be thinking that this is a lot of trouble for a few
lousy three-state buffers, but just wait until you see some of the IEEE-standard
counter and shift-register symbols in Chapter 6!

4.5 ENCODERS

A decoder's output code normally has more bits than its input code. If the device's output code has *fewer* bits than the input code, the device is usually called
an *encoder*. For example, consider a device with eight input bits representing an   *encoder*

---

[6]Whoops! I just said "G3" when I should have said "the signal labeled G3." In the standard,
G3 is a qualifying label, and other signals may have the same label. The standard does not specify
unique names for internal and external signals. However, in the rest of this book, we'll take the
liberty of using an internal signal's qualifying label as its name if no ambiguity results.

Figure 4–52
Binary encoder: (a) general
structure; (b) 8-to-3 encoder.

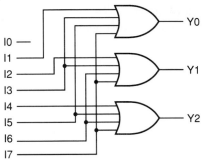

(a)

(b)

unsigned binary number, and two output bits indicating whether the number is
prime or divisible by 7. We might call such a device a lucky/prime encoder.

2^n-to-n encoder
binary encoder

Probably the simplest encoder to build is a 2^n-to-n or *binary encoder*. As
shown in Figure 4–52(a), it has just the opposite function as a binary *decoder*—
its input code is the 1-out-of-2^n code and its output code is n-bit binary. The
equations for an 8-to-3 encoder with inputs I0–I7 and outputs Y0–Y2 are given
below.

$$Y0 \; = \; I1 + I3 + I5 + I7$$
$$Y1 \; = \; I2 + I3 + I6 + I7$$
$$Y2 \; = \; I4 + I5 + I6 + I7$$

Figure 4–52(b) shows the corresponding logic circuit. In general, a 2^n-to-n
encoder can be built from n 2^{n-1}-input OR gates. Bit i of the input code is
connected to OR gate j if bit j in the binary representation of i is 1.

4.5.1 Priority Encoders

The 1-out-of-2^n coded outputs of an n-bit binary decoder are generally used
to control a set of 2^n devices, where at most one device is supposed to be
active at any time. Conversely, consider a system with 2^n *inputs*, each of which
indicates a request for service, as in Figure 4–53. This structure is often found
in microprocessor input/output subsystems, where the inputs might be interrupt
requests.

In this situation, it may seem natural to use a binary encoder of the type
shown in Figure 4–52 to observe the inputs and indicate which one is requesting
service at any time. However, this encoder works properly only if the inputs are
guaranteed to be asserted at most one at a time. If multiple requests can be made
simultaneously, the encoder gives undesirable results. For example, suppose that
inputs I2 and I4 of the 8-to-3 encoder are both 1; then the output is 110, the
binary encoding of 6.

Figure 4–53
A system with 2^n requestors, and a "request encoder" that indicates which request signal is asserted at any time.

[handwritten annotations on figure: "lowest" pointing to REQ1; "highest" pointing to REQN]

Either 2 or 4, not 6, would be a useful output in the preceding example, but how can the encoding device decide which? The solution is to assign *priority* to the input lines, so that when multiple requests are asserted, the encoding device produces the number of the highest-priority requestor. Such a device is called a *priority encoder*.

The logic symbol for an 8-input priority encoder is shown in Figure 4–54(a). Input I7 has the highest priority. Outputs A2–A0 contain the number of the highest-priority asserted input, if any. The IDLE output is asserted if no inputs are asserted.

In order to write logic equations for the priority encoder's outputs, we first define eight intermediate variables H0–H7, such that Hi is 1 if and only if Ii is the highest priority 1 input:

$$H7 = I7$$
$$H6 = I6 \cdot I7'$$
$$H5 = I5 \cdot I6' \cdot I7'$$
$$\ldots$$
$$H0 = I0 \cdot I1' \cdot I2' \cdot I3' \cdot I4' \cdot I5' \cdot I6' \cdot I7'$$

Using these signals, the equations for the A2–A0 outputs are similar to the ones for a simple binary encoder:

$$A2 = H4 + H5 + H6 + H7$$
$$A1 = H2 + H3 + H6 + H7$$
$$A0 = H1 + H3 + H5 + H7$$

The IDLE output is 1 if no inputs are 1:

$$IDLE = (I0 + I1 + I2 + I3 + I4 + I5 + I6 + I7)'$$
$$= I0' \cdot I1' \cdot I2' \cdot I3' \cdot I4' \cdot I5' \cdot I6' \cdot I7'$$

[margin notes:]

priority

priority encoder

convert 1 asserted input to an n-bit binary #. output shows which one was asserted.

If more than 2 inputs are asserted – will cause incorrect output

Priority Encoder – output represents input w/ highest priority.

Figure 4–54
Logic symbol for 8-input
priority encoders: (a) generic;
(b) 74LS148.

4.5.2 The 74LS148 Priority Encoder

74LS148

The *74LS148* is a commercially available, MSI 8-input priority encoder. Its logic symbol is shown in Figure 4–54(b) and its schematic is shown in Figure 4–55. The main difference between this IC and the "generic" priority encoder of Figure 4–54(a) is that its inputs and outputs are active low. Also, it has an enable input, /EI, that must be asserted for any of its outputs to be asserted. The complete truth table is given in Table 4–10.

Instead of an IDLE output, the '148 has a /GS output that is asserted when the device is enabled and one or more of the request inputs is asserted. The manufacturer calls this "Group Select," but it's easier to remember as "Got Something." The /EO signal is an enable *output* designed to be connected to the /EI input of another '148 that handles lower-priority requests. /EO is asserted if /EI is asserted but no request input is asserted; thus, a lower-priority '148 may be enabled. Figure 4–56 shows how four 74LS148s can be connected in this way

Table 4–10 Truth table for a 74LS148 8-input priority encoder.

Inputs									Outputs				
/EI	/I0	/I1	/I2	/I3	/I4	/I5	/I6	/I7	/A2	/A1	/A0	/GS	/EO
1	x	x	x	x	x	x	x	x	1	1	1	1	1
0	x	x	x	x	x	x	x	0	0	0	0	0	1
0	x	x	x	x	x	x	0	1	0	0	1	0	1
0	x	x	x	x	x	0	1	1	0	1	0	0	1
0	x	x	x	x	0	1	1	1	0	1	1	0	1
0	x	x	x	0	1	1	1	1	1	0	0	0	1
0	x	x	0	1	1	1	1	1	1	0	1	0	1
0	x	0	1	1	1	1	1	1	1	1	0	0	1
0	0	1	1	1	1	1	1	1	1	1	1	0	1
0	1	1	1	1	1	1	1	1	1	1	1	1	0

Figure 4–55   Logic diagram for the 74LS148 8-input priority encoder, including pin numbers for a standard 16-pin dual in-line package.

Figure 4–56
Four 74LS148s cascaded to handle 32 requests.

to accept 32 request inputs and produce a 5-bit output, RA4–RA0, indicating the highest-priority requestor. Since the A2–A0 outputs of at most one '148 will be enabled at any time, the outputs of the individual '148s can be OR'ed to produce RA2–RA0. Likewise, the individual /GS outputs can be combined in a 4-to-2 encoder to produce RA4 and RA3. The RGS output is asserted if any GS output is asserted.

* 4.5.3   IEEE Standard Symbols

Figure 4–57 shows the IEEE standard symbol for a 74LS148. Still more features of the standard are used in this symbol:

- *Solidus.* The slash in a label such as 0/Z10, called a *solidus* in the standard, separates multiple functions of a single internal signal. An equivalent notation is shown in Figure 4–58.   *solidus*

- *OR dependency.* This relationship is denoted by Vi; the Vi signal is OR'ed with affected signals. Thus, affected signals perform their "normal" functions only if the Vi signal is negated; otherwise, they are asserted.   *OR dependency*
  Vi

- *Virtual inputs and outputs.* These are internal signals, denoted by a horizontal line going nowhere, such as the ones labeled 10–17 in the '148 symbol. These signals have no external connection but affect or are affected by other signals via dependency notation.   *virtual input*
  virtual output

Figure 4–57
IEEE standard logic symbol for the 74LS148 8-input priority encoder.

Figure 4–58
Equivalent notation for solidi.

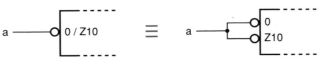

interconnection dependency
Zi

- *Interconnection dependency.* This relationship is denoted by Zi, and indicates an internal connection. Affected signals equal the Zi signal, unless modified by additional dependency notation. Think of the Z as a zig-zag internal wire.

Greek letters

- *Greek letters.* A Greek letter may be used instead of an integer in qualifying labels to avoid ambiguity when the affected signals have a numeric function label, as in the inputs or outputs of a coder.

If you understand these and the previously introduced features of the standard, you can "read" the functional behavior of a '148 right from its symbol. However, most people do the opposite—already knowing how a '148 works, they try to deduce how the standard works from the '148 symbol!

4.6 MULTIPLEXERS

multiplexer

A *multiplexer* is a digital switch—it connects data from one of n sources to its output. Figure 4–59(a) shows the inputs and outputs of an n-input, b-bit multiplexer. There are n sources of data, each of which is b bits wide, and there are b output bits. In typical commercially available multiplexers, $n = 1, 2, 4, 8$ or 16, and $b = 1, 2,$ or 4. There are s inputs that select among the n sources, so $s = \lceil \log_2 n \rceil$. An enable input EN allows the multiplexer to "do its thing;" when EN = 0, all of the outputs are 0. A multiplexer is often called a *mux* for short.

mux

Figure 4–59
Multiplexer structure: (a) inputs and outputs; (b) functional equivalent.

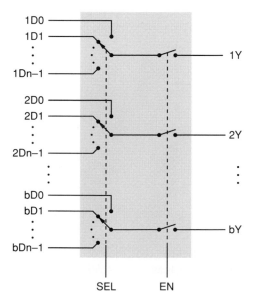

Figure 4–59(b) shows a switch circuit that is roughly equivalent to the multiplexer. However, unlike a mechanical switch, a multiplexer is a unidirectional device: information flows only from inputs (on the left) to outputs (on the right).

We can write a general logic equation for a multiplexer output:

$$iY \;=\; \sum_{j=0}^{n-1} EN \cdot M_j \cdot iDj$$

Here, the summation symbol represents a logical sum of product terms. Variable iY is a particular output bit $(1 \le i \le b)$, and variable iDj is input bit i of source j $(0 \le j \le n - 1)$. M_j represents minterm j of the s select inputs. Thus, when the multiplexer is enabled and the value on the select inputs is j, each output Yi equals the corresponding bit of the selected input, iDj.

Multiplexers are obviously useful devices in any application in which data must be switched from multiple sources to a destination. A common application in computers is the multiplexer between the processor's registers and its arithmetic logic unit (ALU). For example, consider a 16-bit processor in which each instruction has a 3-bit field that specifies one of eight registers to use. This 3-bit field is connected to the select inputs of an 8-input, 16-bit multiplexer. The multiplexer's data inputs are connected to the eight registers, and its data outputs are connected to the ALU to execute the instruction using the selected register.

4.6.1 Standard MSI Multiplexers

The sizes of commercially available MSI multiplexers are limited by the number of pins available in an inexpensive IC package. The most commonly used muxes come in 16-pin packages. At one extreme, we have the *74LS151*, shown in   *74LS151* Figure 4–60, which selects among eight 1-bit inputs. The select inputs are named C, B, and A; as in a decoder, C is most significant numerically. The enable input is active low; both active-high (Y) and active-low (/Y) versions of the output are provided.

The 74LS151's truth table is shown in Table 4–11. Here we have once again extended our notation for truth tables. Up until now, our truth tables have specified an output of 0 or 1 for each input combination. In the 74LS151's table, only a few of the inputs are listed under the "Inputs" heading. Each output is specified as 0, 1, or a simple logic function of the remaining inputs (e.g., D0 or D0'). This notation saves eight columns and eight rows in the table, and presents the logic function more clearly than a larger table would.

At the other extreme of muxes in 16-pin packages, we have the *74LS157*,   *74LS157* shown in Figure 4–61, which selects between two 4-bit inputs. Just to confuse things, the manufacturer has named the select input S and the active-low enable input /G. Also note that the data sources are named A and B instead of D0

Figure 4–60
The 74LS151 8-input, 1-bit multiplexer: (a) logic diagram, including pin numbers for a standard 16-pin dual in-line package; (b) traditional logic symbol.

74LS153

and D1 as in our generic example. Our extended truth-table notation makes the 74LS157's description very compact, as shown in Table 4–12.

Intermediate between the 74LS151 and 74LS157 is the *74LS153*, a 4-input, 2-bit multiplexer. As shown in Figure 4–62, this device has separate enable inputs (1G, 2G) for each bit. Its function is very straightforward.

Some multiplexers have three-state outputs. The enable input of such a multiplexer, instead of forcing the outputs to zero, forces them to the Hi-Z

Table 4–11
Truth table for a 74LS151
8-input, 1-bit multiplexer.

	Inputs			Outputs	
/EN	C	B	A	Y	/Y
1	x	x	x	0	1
0	0	0	0	D0	D0'
0	0	0	1	D1	D1'
0	0	1	0	D2	D2'
0	0	1	1	D3	D3'
0	1	0	0	D4	D4'
0	1	0	1	D5	D5'
0	1	1	0	D6	D6'
0	1	1	1	D7	D7'

Figure 4–61
The 74LS157 2-input, 4-bit
multiplexer: (a) logic diagram,
including pin numbers for a
standard 16-pin dual in-line
package; (b) traditional logic
symbol.

Table 4–12
Truth table for a 74LS157
2-input, 4-bit multiplexer.

Inputs		Outputs			
/G	S	1Y	2Y	3Y	4Y
1	x	0	0	0	0
0	0	1A	2A	3A	4A
0	1	1B	2B	3B	4B

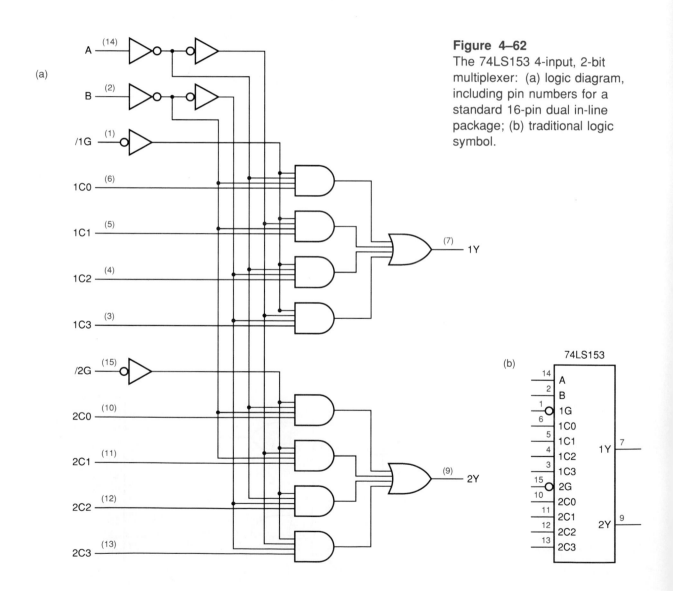

(a)

Figure 4–62
The 74LS153 4-input, 2-bit
multiplexer: (a) logic diagram,
including pin numbers for a
standard 16-pin dual in-line
package; (b) traditional logic
symbol.

(b)

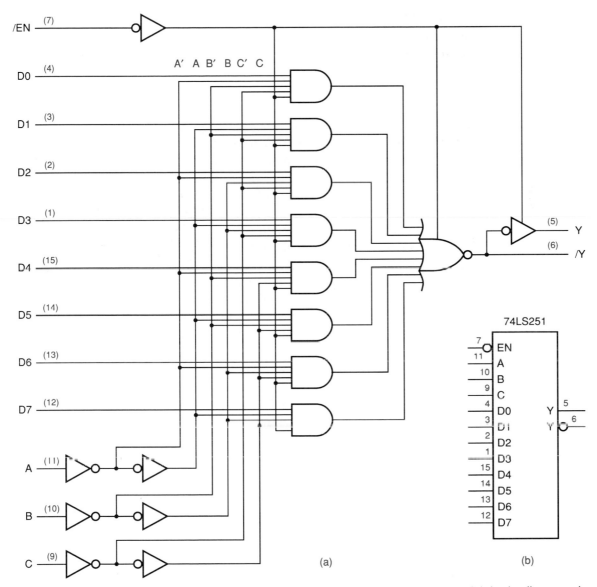

Figure 4–63   The 74LS251 8-input, 1-bit multiplexer with three-state output: (a) logic diagram, including pin numbers; (b) traditional logic symbol.

state. For example, the logic diagram and symbol for the 74LS251 are shown in Figure 4–63. The '251 is identical to the '151, except for the three-state outputs. Similarly, the *74LS253* and *74LS257* are three-state versions of the '153 and '157. The three-state outputs are especially useful when *n*-input muxes are combined to form larger muxes, as shown in the next subsection.

74LS253
74LS257

4.6.2   Expanding Multiplexers

Seldom does the size of an MSI multiplexer match the characteristics of the problem at hand. For example, we suggested earlier that an 8-input, 16-bit multiplexer might be used in the design of a computer processor. This function could be performed by 16 74LS151 8-input, 1-bit multiplexers or equivalent ASIC cells, each handling one bit of all the inputs and the output. The processor's 3-bit register field would be connected to the A, B, and C inputs of all 16 muxes, so they would all select the same register source at any given time.

Of course, the device that produces the 3-bit register field in this example must have enough fanout to drive 16 loads. With 74LS-series ICs this is possible, because typical devices have a fanout of 20 LS-TTL loads. However, we're lucky that the '151 was designed so that each of the A, B, and C inputs presents only one LS-TTL load to the circuit driving it. Theoretically, the '151 could have been designed without the first rank of three inverters shown on the select inputs in Figure 4–60, but then each select input would have presented five LS-TTL loads, and the drivers in our present application would need a fanout of 80.

Another dimension in which multiplexers can be expanded is the number of data sources. For example, suppose we needed a 32-input, 1-bit multiplexer. Figure 4–64 shows one way to build it. Five select bits are required. A 2-to-4 decoder (one-half of a 74LS139) decodes the two high-order select bits to enable one of four 74LS151 8-input multiplexers. Since only one '151 is enabled at a time, the '151 outputs can simply be OR'ed to obtain the final output.

The 32-to-1 multiplexer can also be built using 74LS251s. The circuit is identical to Figure 4–64, except that the output NAND gate is eliminated. Instead, the Y (and, if desired, /Y) outputs of the four '251s are simply tied together, as shown in Figure 4–65. The '139 decoder ensures that at most one of the '251s has its three-state outputs enabled at any time. If the '139 is disabled (/XEN is negated), then all of the '251s are disabled, and the XOUT and /XOUT outputs are undefined. However, if desired, resistors may be connected between these signals and +5 volts to pull the output HIGH in this case.

Notice how bubble-to-bubble logic design helps our understanding of these examples. Since the decoder outputs and the multiplexer enable inputs are all active low, we can hook them up directly. We can ignore the inversion bubbles when thinking about the logic function that is performed—we simply say that when a particular decoder output is asserted, the corresponding multiplexer is enabled.

Bubble-to-bubble design also gives us two options for the final OR function in Figure 4–64. The most intuitive design uses a 4-input OR gate connected to the Y outputs. However, for faster operation, we use an inverting gate, a 4-input NAND connected to the /Y outputs. This eliminates the delay of two inverters—

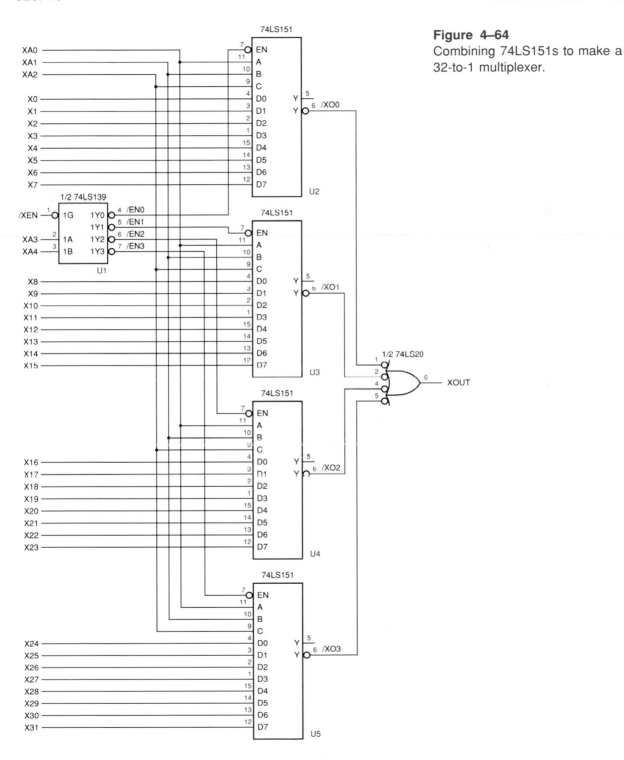

Figure 4–64
Combining 74LS151s to make a
32-to-1 multiplexer.

Figure 4–65
Another design for a 32-to-1 multiplexer.

Table 4-13
General truth table for a
3-variable logic function, F(X,Y,Z).

Row	X	Y	Z	F
0	0	0	0	F_0
1	0	0	1	F_1
2	0	1	0	F_2
3	0	1	1	F_3
4	1	0	0	F_4
5	1	0	1	F_5
6	1	1	0	F_6
7	1	1	1	F_7

the one used inside the '151 to generate Y from /Y, and the extra transistor
inverter that is used to obtain an OR function from a NOR circuit in TTL.

4.6.3 Multiplexers as Function Generators

A 2^s-input multiplexer can perform any logic function of s variables. For example, Table 4–13 shows the truth table for a general logic function of three
variables. The function's value for each input combination can be an arbitrary
value, denoted by F_0, F_1, and so on. Regardless of the values chosen in the truth
table, the logic function can always be performed by connecting logic 0s and 1s to
the corresponding inputs of an 8-input multiplexer, as shown in Figure 4–66(a).
A specific example, a function in which F is 1 if and only if an even number of
inputs are 1, is shown in (b).

Multiplexers can be used even more effectively using a simple form of *functional*
functional decomposition. Consider the general truth table for a logic function of *decomposition*

Figure 4-66
Realizing logic functions with
multiplexers: (a) general
scheme; (b) F = $\Sigma_{X,Y,Z}(0,3,5,6)$.

Table 4–14
Decomposed truth table for a
3-variable logic function, F(X,Y,Z),
after "pulling out" variable Z.

Row	X	Y	Z	F
0,1	0	0	x	$F_{00}(Z)$
2,3	0	1	x	$F_{01}(Z)$
4,5	1	0	x	$F_{10}(Z)$
6,7	1	1	x	$F_{11}(Z)$

three variables, which we showed earlier in Table 4–13. As shown in Table 4–14,
we can "decompose" this table using only half as many rows. Each row displays
a particular combination of variables X and Y, and indicates that the output is
strictly a function of Z for that input combination. To perform a 3-variable logic
function with a 4-input multiplexer, we connect each data input of the multiplexer
to the appropriate function of Z, according to the decomposed truth table. But
there are only four different functions of Z possible: 0, 1, Z, and Z'. Thus, we
need only the multiplexer, constant 0 and 1 logic signals, and if Z' appears in
the decomposed truth table, an inverter for Z.

Figure 4–67 shows the decomposition for the example that we did previously using an 8-input multiplexer. Figure 4–68 shows an example where a
4-variable function is performed by an 8-input multiplexer. In general, an $s + 1$-
variable function can be performed by a 2^s-input multiplexer.

4.6.4 Multiplexers, Demultiplexers, and Buses

demultiplexer

A multiplexer can be used to select one of n sources of data to transmit on a bus.
At the far end of the bus, a *demultiplexer* can be used to route the bus data to one
of m destinations. Such an application, using a 1-bit bus, is depicted in terms of
our switch analogy in Figure 4–69(a). In fact, block diagrams for logic circuits

Figure 4–67
Realizing $F = \Sigma_{X,Y,Z}(0,3,5,6)$ with
a 4-input multiplexer.

Row	X	Y	Z	F
0	0	0	0	1
1	0	0	1	0
2	0	1	0	0
3	0	1	1	1
4	1	0	0	0
5	1	0	1	1
6	1	1	0	1
7	1	1	1	0

(0,1) } Z'
(2,3) } Z
(4,5) } Z
(6,7) } Z'

(a)

(b)

(a)

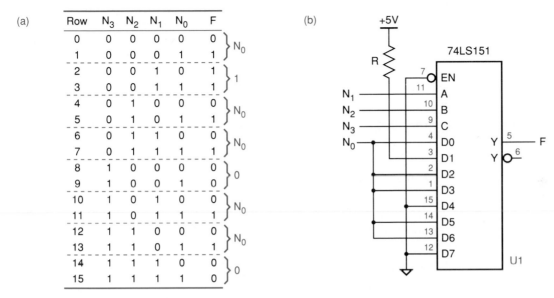

Row	N_3	N_2	N_1	N_0	F	
0	0	0	0	0	0	$\}N_0$
1	0	0	0	1	1	
2	0	0	1	0	1	$\}1$
3	0	0	1	1	1	
4	0	1	0	0	0	$\}N_0$
5	0	1	0	1	1	
6	0	1	1	0	0	$\}N_0$
7	0	1	1	1	1	
8	1	0	0	0	0	$\}0$
9	1	0	0	1	0	
10	1	0	1	0	0	$\}N_0$
11	1	0	1	1	1	
12	1	1	0	0	0	$\}N_0$
13	1	1	0	1	1	
14	1	1	1	0	0	$\}0$
15	1	1	1	1	0	

(b)

Figure 4–68 Realizing $F = \Sigma_{N_3,N_2,N_1,N_0}(1,2,3,5,7,11,13)$ with an 8-input multiplexer.

often depict multiplexers and demultiplexers using the wedge-shaped symbols in (b), to suggest visually how a selected one of multiple bits gets directed onto a bus and routed to a selected one of multiple destinations.

The function of a demultiplexer is just the inverse of a multiplexer's. For

Figure 4–69
A multiplexer driving a bus, with a demultiplexer receiving the bus: (a) switch equivalent; (b) block diagram symbols.

(a)   2-to-4 decoder   (b)   1/2 74LS139

Figure 4–70   Using a 2-to-4 binary decoder as a 1-bit, 4-output demultiplexer: (a) generic decoder; (b) 74LS139.

example, a 1-bit, n-output demultiplexer has one data input and s inputs to select one of $n = 2^s$ data outputs. In normal operation, all outputs except the selected one are 0; the selected output equals the data input. This definition may be generalized for a b-bit, n-output demultiplexer; such a device has b data inputs, and its s select inputs choose one of $n = 2^s$ sets of b data outputs.

A binary decoder with an enable input can be used as a demultiplexer, as shown in Figure 4–70. The decoder's enable input is connected to the data line, and its select inputs determine which of its output lines is driven with the data bit. The remaining output lines are negated. Thus, the 74LS139 can be used as a 2-bit, 4-output demultiplexer with active-low data inputs and outputs, and the 74LS138 can be used as a 1-bit, 8-output demultiplexer. In fact, the manufacturer's catalog typically lists these ICs as "decoders/demultiplexers."

74LS155   Another standard TTL MSI decoder/demultiplexer is the *74LS155*, shown in Figure 4–71. This IC is most useful as a 2-bit, 4-output demultiplexer. It's not as flexible as the 74LS139 when used as a dual 2-to-4 decoder, because both sections use the same select inputs.

* 4.6.5   IEEE Standard Symbols

The IEEE standard provides a special notation for multiplexers and demultiplexers. For example, Figure 4–72 shows the IEEE symbols for multiplexer ICs that we've discussed in this section. The general qualifying symbol MUX identifies a *bit-grouping symbol* multiplexer. The bracket is called a *bit-grouping symbol* and indicates that the *internal value* grouped inputs produce an *internal value* that is a weighted sum. The weights are given by the qualifying labels on the individual inputs; if the weights are all powers of 2, they may be replaced by the corresponding exponents, and all but the first and last exponents may be omitted "if no confusion is likely." Thus, the weights of pins 9, 10, and 11 of the 74LS151 are 2^2, 2^1, and 2^0, respectively; if the input signal on these pins is 110, the internal value is 6.

The internal value produced by bit grouping affects other internal values *range notation* or outputs according to the qualifying label written to the right of the grouping symbol. In the 74LS151 multiplexer symbol, the notation $G\frac{0}{7}$ indicates AND

Figure 4–71 The 74LS155 2-bit, 4-output demultiplexer: (a) logic diagram, including pin numbers for a standard 16-pin dual in-line package; (b) traditional logic symbol.

Figure 4–72 IEEE standard symbols for multiplexers.

Figure 4–73
IEEE standard symbols for demultiplexers.

dependency with signals whose labels are in the range 0–7. In other words, input i is selected (and transferred to the output) if and only if the internal value is i. There are two outputs, normally equal to the selected input and its complement. However, these outputs are asserted only if the enable input EN is asserted.

The 74LS153 symbol also uses bit grouping for the select inputs, and it uses a common control block to indicate that the select inputs affect both sections of the multiplexer. Since the bottom half of the multiplexer is identical in function to the top, its qualifying labels are not repeated. Notice that each half has an independent EN input.

The symbol for the 74LS157 does not use bit grouping because it has only one select input. Instead, the select input is labeled G1, indicating that it has an AND dependency with signals bearing the identifier "1." Thus, pin 3 is selected only if G1 is asserted. Pin 2, on the other hand, bears the identifier "$\overline{1}$;" The overbar indicates that pin 2 is selected only if G1 is negated. All four sections are controlled by G1 in this way.

Figure 4–73 shows the IEEE standard demultiplexer symbols for the TTL decoder/demultiplexer ICs that we've discussed in this section. The general qualifying symbol DMUX identifies a demultiplexer. The notation $G\frac{0}{7}$ indicates AND dependency with outputs whose labels are in the range 0–7. In other words, output i may be asserted only if the internal value is i. In addition, all of the other inputs (pins 4–6) must be asserted.

A similar notation is used in the 74LS139, which contains two independent demultiplexers. Since the second demultiplexer is identical in function to the first, the qualifying labels are not repeated.

The symbol for the 74LS155 uses a common control block to show that the same select inputs (and internal value produced by bit grouping) are used for both sections of the multiplexer. Also, the input labeled "4" has an AND dependency on the input labeled "G4," so the selected output is asserted only if both of these inputs are asserted. Now, is that all clear?

4.7 EXCLUSIVE OR GATES AND PARITY CIRCUITS

4.7.1 EXCLUSIVE OR and EXCLUSIVE NOR Gates

An *EXCLUSIVE OR (XOR)* gate is a 2-input gate whose output is 1 if exactly one of its inputs is 1. Stated another way, an XOR gate produces a 1 output if its inputs are different. An *EXCLUSIVE NOR (XNOR)* or *EQUIVALENCE* gate is just the opposite—it produces a 1 output if its inputs are the same. A truth table for these functions is shown in Table 4–15. The XOR operation is sometimes denoted by the symbol "$\oplus$", that is,

EXCLUSIVE OR (XOR)
EXCLUSIVE NOR (XNOR)
EQUIVALENCE
$\oplus$

$$X \oplus Y = X' \cdot Y + X \cdot Y'$$

Although EXCLUSIVE OR is not one of the basic functions of switching algebra, discrete XOR gates are fairly commonly used in practice. Most switching technologies cannot perform the XOR function directly; instead, they use multigate designs like the ones shown in Figure 4–74.

The logic symbols for XOR and XNOR functions are shown in Figure 4–75. There are four equivalent symbols for each function. All of these alternatives are a consequence of a simple rule:

- Any two signals (inputs or output) of an XOR or XNOR gate may be complemented without changing the resulting logic function.

In bubble-to-bubble logic design, we choose the symbol that is most expressive of the logic function being performed.

Table 4–15
Truth table for XOR and XNOR functions.

X	Y	$X \oplus Y$ (XOR)	$(X \oplus Y)'$ (XNOR)
0	0	0	1
0	1	1	0
1	0	1	0
1	1	0	1

Figure 4–74
Multigate designs for the 2-input
XOR function: (a) AND-OR;
(b) three-level NAND.

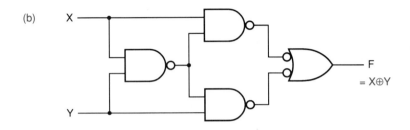

74LS86
74LS266

Four XOR gates are provided in a single 14-pin SSI IC, the *74LS86* shown in Figure 4–76. Likewise, four XNOR gates are provided in the *74LS266*, shown in Figure 4–77. The 74LS266 has open-collector outputs, and therefore requires external pull-up resistors. We'll show how 74LS266s are used to build comparators in Section 4.8.4.

(a)

(b)

Figure 4–75 Equivalent symbols for (a) XOR gates; (b) XNOR gates.

Figure 4–76
Pin diagram for the 74LS86
quadruple 2-input EXCLUSIVE OR
gate.

Figure 4–77
Pin diagram for the 74LS266
quadruple 2-input EXCLUSIVE
NOR gate.

parity = # of 1's in a byte

1.0 10 11 10 1

6 1's – even

even – output 1
odd – output 0.

4.7.2   Parity Circuits

As shown in Figure 4–78(a), n XOR gates may be cascaded to form a circuit
with $n+1$ inputs and a single output. This is called an *odd-parity circuit*, because   *odd-parity circuit*
its output is 1 if an odd number of its inputs are 1. The circuit in Figure 4–78(b)
is also an odd-parity circuit, but it's faster because its gates are arranged in a
tree-like structure. If the output of either circuit is inverted, we get an *even-parity*   *even-parity circuit*
circuit, whose output is 1 if an even number of its inputs are 1.

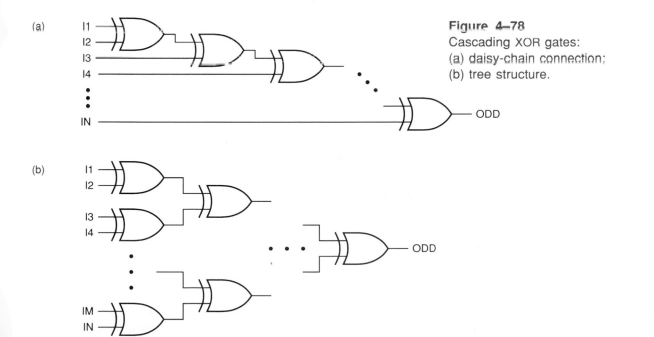

(a)

(b)

Figure 4–78
Cascading XOR gates:
(a) daisy-chain connection;
(b) tree structure.

Figure 4–79
The 74LS280 9-bit odd/even
parity generator: (a) logic
diagram, including pin numbers
for a standard 16-pin dual in-line
package; (b) traditional logic
symbol.

4.7.3   The 74LS280 9-Bit Parity Generator

Rather than build a multibit parity circuit with discrete XOR gates, it is more economical to put all of the XORs in a single MSI package with just the primary inputs and outputs available at the external pins. The *74LS280* 9-bit parity generator, shown in Figure 4–79, is such a device. It has nine inputs and two outputs that indicate whether an even or odd number of inputs are 1.

74LS280

If each XOR gate in Figure 4–79 were built using discrete NAND gates as in Figure 4–74(b), the 74LS280 would be pretty slow, having a propagation delay equivalent to $4 \cdot 3 + 1$ or 13 NAND gates. Instead, the 74LS280 uses a 4-wide AND-OR-INVERT gate to perform the function of each shaded pair of XOR gates with about the same delay as a single NAND gate. The A–I inputs are buffered through two levels of inverters so that each input presents just one unit load to the circuit driving it. Thus, the total propagation delay through the 74LS280 is about the same as five inverting gates, not 13.

4.7.4   Parity Checking Applications

In Section 1.15, we described error-detecting codes that use an extra bit, called a parity bit, to detect errors in the transmission and storage of data. In an even-parity code, the parity bit is chosen so that the total number of 1 bits in a code word is even. Parity circuits like the 74LS280 are used both to generate the

Figure 4–80 Parity generation and checking for an 8-bit-wide memory system.

correct value of the parity bit when a code word is stored or transmitted, and to check the parity bit when a code word is retrieved or received.

Figure 4–80 shows how a parity circuit might be used to detect errors in the memory of a microprocessor system. The memory stores 8-bit bytes, plus a parity bit for each byte. The microprocessor uses a bidirectional bus D[0–7] to transfer data to and from the memory. Two control lines, RD and WR, are used to indicate whether a read or write operation is desired, and an ERROR signal is asserted to indicate parity errors during read operations. Complete details of the memory chips, such as addressing inputs, are not shown; memory chips are described in detail in Section 7.5.

To store a byte into the memory chips, we specify an address (not shown), place the byte on D[0–7], generate its parity bit on PIN, and assert WR. The AND gate on the I input of the 74LS280 ensures that I is 0 except during read operations, so that during writes the 280's output depends only on the parity of the D-bus data. The 280's ODD output is connected to PIN, so that the total number of 1s stored is even.

To retrieve a byte, we specify an address (not shown) and assert RD; the byte value appears on DOUT[0–7] and its parity appears on POUT. A 74LS541

Figure 4-81
Error-correcting circuit for a 7-bit Hamming code.

drives the byte onto the D bus, and the '280 checks its parity. If the parity of the 9-bit word DOUT[0–7],POUT is odd during a read, the ERROR signal is asserted.

Parity circuits are also used with error-correcting codes such as the Hamming codes described in Section 1.15.3. We showed the parity-check matrix for a 7-bit Hamming code in Figure 1–13 on page 40. We can correct errors in this code as shown in Figure 4–81. A 7-bit word, possibly containing an error, is presented on DU[1–7]. Three 74LS280s compute the parity of the three bit-groups defined by the parity-check matrix. The outputs of the '280s form the syndrome, which is the number of the erroneous input bit, if any. A 74LS138 is used to decode the syndrome. If the syndrome is zero, the /NOERROR signal is asserted (this signal also could be named ERROR). Otherwise, the erroneous bit is corrected by complementing it. The corrected code word appears on the /DC bus.[7]

* 4.7.5 IEEE Standard Symbols

The IEEE standard symbols for EXCLUSIVE OR and EXCLUSIVE NOR gates are shown in Figure 4–82. Either of two different notations may be used, depending on how you're thinking about the gate's function. The top symbols, with "=1" inside the symbol outline, assert their outputs when exactly one of the inputs is asserted. The bottom symbols, with "=" inside the symbol outline, assert their outputs when their inputs are equal.

The IEEE standard symbol for 74LS280 9-bit parity generator is shown in Figure 4–83(a). The general qualifying symbol "2k" at the top of the symbol indicates that the outputs are asserted if $2k$ of the inputs are asserted for some integer k. Thus, pin 5, an active-high output, and pin 6, an active-low output, both indicate even parity. The nature of the function, of course, is such that pin 6 could instead be viewed as an active high output that denotes odd parity. The standard symbol for this interpretation of the device function is shown in (b).

4.8 COMPARATORS

Comparing two binary words for equality is a commonly used operation in computer systems and device interfaces. For example, in Figure 1–7(a) on page 32,

---

[7]Note that the active-low outputs of the '138 led us to use an active-low /DC bus. If we required an active-high DC bus, we could put a discrete inverter on each XOR input or output, or use a decoder with active-high outputs, or use XNOR gates.

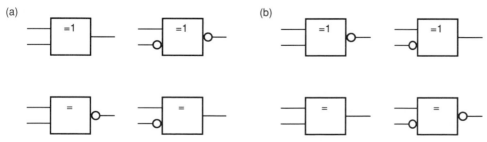

(a)   (b)

Figure 4–82   IEEE standard symbols: (a) EXCLUSIVE OR gates; (b) EXCLUSIVE NOR gates.

Figure 4–83
IEEE standard symbols for the
74LS280 9-bit parity generator:
(a) normal symbol; (b) both
outputs active high.

comparator

we showed a system structure in which devices are enabled by comparing a "device select" word with a predetermined "device ID." A circuit that compares two binary words and indicates whether they are equal is called a *comparator*. Some comparators interpret their input words as signed or unsigned numbers and also indicate an arithmetic relationship (greater or less than) between the words.

4.8.1 Comparator Structure

EXCLUSIVE OR and EXCLUSIVE NOR gates may be viewed as 1-bit comparators. Figure 4–84(a) shows an interpretation of the 74LS86 XOR gate as a 1-bit comparator. The active-high output, DIFF, is asserted if the inputs are different. The outputs of four XOR gates are ORed to create a 4-bit comparator in (b). The DIFF output is asserted if any of the input-bit pairs are different. Given enough XOR gates and wide enough OR gates, comparators with any number of input bits can be built.

Figure 4–84 Comparators using the 74LS86: (a) 1-bit comparator; (b) 4-bit comparator.

Figure 4–85
Comparators using the 74LS266:
(a) 1-bit comparator; (b) 4-bit
comparator.

Figure 4–85(a) shows an interpretation of the 74LS266 XNOR gate as a
1-bit comparator. The active-low output, /DIFF, is asserted if the inputs are
different. Recall that the '266 has an open-collector output, so a pull-up resistor
is required. Four XNOR gates can be combined to create a 4-bit comparator
as shown in (b). Taking advantage of the open-collector outputs, a wired-AND
(NOT-OR-NOT) function is used to create the /DIFF output. Using this structure,
an n-bit comparator can be obtained for any value of n by wire-ANDing the
outputs of n '266s, subject to the electrical limitations[8] of the device.

    The n-bit comparators in this subsection might be called *parallel compara-*   *parallel comparator*
tors because they look at each pair of input bits simultaneously and deliver the
1-bit comparison results in parallel to an n-input OR or AND function. It is also
possible to design an "iterative comparator" that looks at its bits one at a time
and uses a fixed amount of logic per bit. Before looking at this design, we
introduce a general class of circuits with this structure.

* 4.8.2 Iterative Circuits

An *iterative circuit* is a special type of combinational circuit, with the structure   *iterative circuit*
shown in Figure 4–86. The circuit contains n identical modules, each of which
has both *primary inputs and outputs* and *cascading inputs and outputs*. The   *primary inputs and*
leftmost cascading inputs are called *boundary inputs* and are connected to fixed   *outputs*
logic values in most iterative circuits. The rightmost cascading outputs are called   *cascading inputs*
boundary outputs and usually provide important information.   *and outputs*

    Iterative circuits are well suited to problems that can be solved by a simple   *boundary inputs*
iterative algorithm: *boundary outputs*

(1) Set C_0 to its initial value and set i to 0.

---

[8]Mainly HIGH-state output leakage current, I_{OH}; see Exercise 4.47.

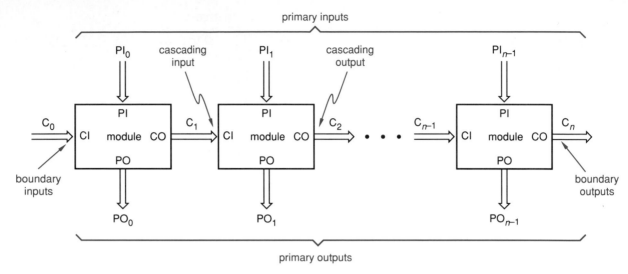

Figure 4–86   General structure of an iterative combinational circuit.

(2) Use C_i and PI_i to determine the values of PO_i and C_{i+1}.

(3) Increment i.

(4) If $i < n$, go to step 2.

In an iterative circuit, the loop of steps 2–4 is "unwound" by providing a separate combinational circuit that performs step 2 for each value of i.

   Examples of iterative circuits are the comparator circuit in the next subsection and the ripple adder in Section 4.9.2. The 74LS85 4-bit comparator and the 74LS283 4-bit adder are examples of MSI circuits that can be used as the individual modules in a larger iterative circuit. In Section 6.5 we'll explore the relationship between iterative circuits and corresponding sequential circuits that execute the 4-step algorithm above in discrete time steps.

* 4.8.3   An Iterative Comparator Circuit

Two n-bit values X and Y can be compared one bit at a time using a single bit EQi at each step to keep track of whether all of the bit-pairs have been equal so far:

(1) Set EQ_0 to 1 and set i to 0.

(2) If EQ_i is 1 and X_i and Y_i are equal, set EQ_{i+1} to 1. Else set EQ_{i+1} to 0.

(3) Increment i.

(4) If $i < n$, go to step 2.

Figure 4–87
An iterative comparator circuit:
(a) module for one bit; (b) complete circuit.

(a)

(b)

Figure 4–87 shows a corresponding iterative circuit. Note that this circuit has no primary outputs; the boundary output is all that interests us. Other iterative circuits, such as the ripple adder of Section 4.9.2, have primary outputs of interest.

Given a choice between the iterative comparator circuit in this subsection and one of the parallel comparators shown previously, we would probably prefer the parallel comparator. The iterative comparator saves little if any cost, and it's very slow because the cascading signals need time to "ripple" from the leftmost to the rightmost module. Iterative circuits that process more than one bit at a time, using modules like the 74LS85 4-bit comparator and 74LS283 4-bit adder, are much more likely to be used in practical designs.

4.8.4   Standard MSI Comparators

Comparator applications are common enough that several MSI comparators have been developed commercially. The *74LS85* is a 4-bit comparator with the logic symbol shown in Figure 4–88(a). It provides a greater-than output (AGTBOUT) and a less-than output (ALTBOUT) as well as an equal output (AEQBOUT). The '85 also has *cascading inputs* (AGTBIN, ALTBIN, AEQBIN) for combining multiple '85s to create comparators for more than four bits. Both the cascading inputs and the outputs are arranged in a 1-out-of-3 code, since in normal operation exactly one input and one output should be asserted.

The cascading inputs are defined so that the outputs of an '85 that compares less-significant bits are connected to the inputs of an '85 that compares more-significant bits, as shown in Figure 4–89 for a 12-bit comparator. This is an

74LS85

cascading inputs

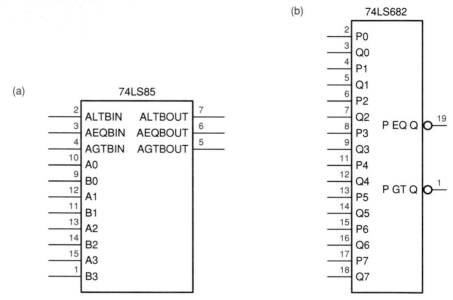

Figure 4–88 Traditional logic symbols for comparators: (a) 74LS85 4-bit comparator in a 16-pin package; (b) 74LS682 8-bit comparator in a 20-pin package.

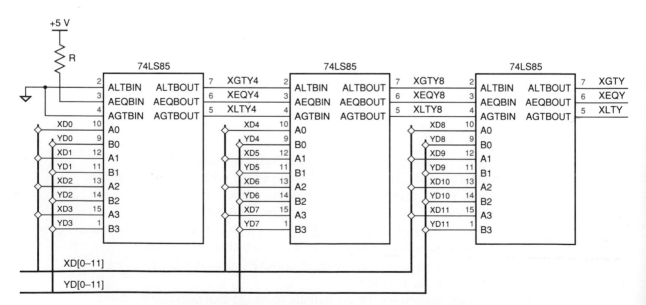

Figure 4–89 A 12-bit comparator using 74LS85s.

iterative circuit according to the definition in Section 4.8.2. Each '85 develops its cascading outputs roughly[9] according to the following equations:

$$AGTBOUT = (A > B) + (A = B) \cdot AGTBIN$$

$$AEQBOUT = (A = B) \cdot AEQBIN$$

$$ALTBOUT = (A < B) + (A = B) \cdot ALTBIN$$

Note that the parenthesized subexpressions above are not normal logic expressions, but indicate an arithmetic comparison that occurs between the A3–A0 and B3–B0 inputs. In other words, AGTBOUT is asserted if $A > B$ or if $A = B$ and AGTBIN is asserted (if the higher-order bits are equal, we have to look at the lower-order bits for the answer). If you're interested, you can develop logic expressions for the arithmetic comparisons, for example,

$$
\begin{aligned}
(A > B) = \ & A3 \cdot B3' + \\
& (A3 \oplus B3)' \cdot A2 \cdot B2' + \\
& (A3 \oplus B3)' \cdot (A2 \oplus B2)' \cdot A1 \cdot B1' + \\
& (A3 \oplus B3)' \cdot (A2 \oplus B2)' \cdot (A1 \oplus B1)' \cdot A0 \cdot B0'
\end{aligned}
$$

Several 8-bit MSI comparators are also available. The simplest of these is the *74LS682*, whose logic symbol was shown in Figure 4–88(b) and whose internal logic diagram is shown in Figure 4–90. The top half of the circuit checks the two 8-bit input words for equality. Each XNOR-gate output is asserted if its inputs are equal, and the /PEQQ output is asserted if all eight input-bit pairs are equal.[10] The bottom half of the circuit compares the input words arithmetically, and asserts /PGTQ if P[7–0] > Q[7–0].

74LS682

Unlike the 74LS85, the 74LS682 does not have cascading inputs. Also unlike the '85, the '682 does not provide a "less than" output. However, any desired condition, including ≤ and ≥, can be formulated as a function of the /PEQQ and /PGTQ outputs, as shown in Figure 4–91.

* 4.8.5 IEEE Standard Symbols

IEEE standard symbols for MSI comparators are shown in Figure 4–92. Like the select inputs of multiplexers, the data inputs have qualifying labels indicating

---

[9]The 74LS85 actually uses (AEQBIN + ALTBIN)' in place of AGTBIN, and (AEQBIN + AGTBIN)' in place of ALTBIN. This produces outputs different from the equations above when all of the cascading inputs are negated, but this input combination is not used in most applications.

[10]The individual 1-bit comparators (XNOR gates) in the '682 are drawn in the opposite sense as the examples of the preceding subsection—outputs are asserted for *equal* inputs and then ANDed, rather than asserted for *different* inputs and then ORed. We can look at a comparator's function either way, as long as we're consistent.

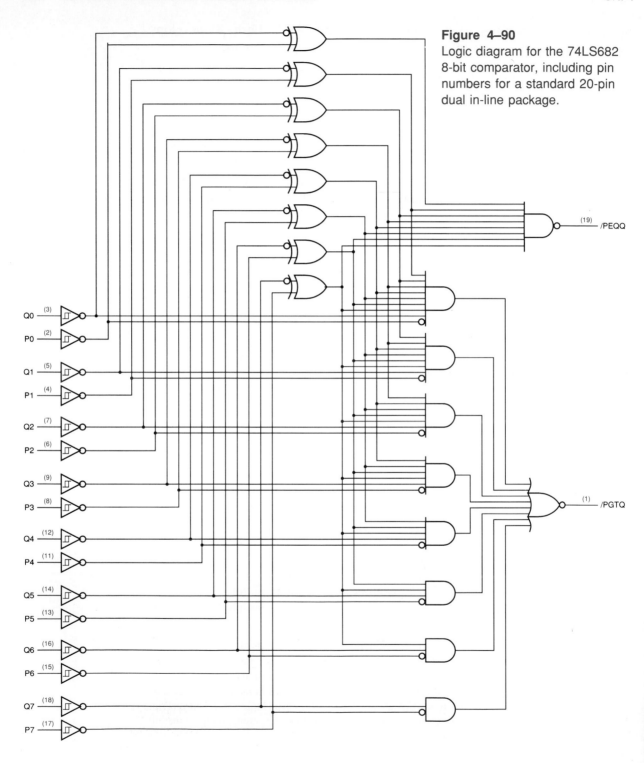

Figure 4–90
Logic diagram for the 74LS682
8-bit comparator, including pin
numbers for a standard 20-pin
dual in-line package.

Figure 4–91
Arithmetic conditions derived
from 74LS682 outputs.

their arithmetic weights in powers of 2 (0–3 corresponding to 2^0–2^3 in the '85). The cascading inputs and outputs are labeled with the appropriate arithmetic function. In the '682 symbol, the right-pointing triangle indicates that the outputs have high fanout capability, and hysteresis is indicated on the data inputs.

Figure 4–92
IEEE standard symbols for MSI
comparators.

* 4.9 ADDERS, SUBTRACTERS, AND ALUS

adder

subtracter

Addition is the most commonly performed arithmetic operation in digital systems. An *adder* combines two arithmetic operands using the addition rules described in Chapter 1. As we showed in Section 1.6, the same addition rules and therefore the same adders are used for both unsigned and two's-complement numbers. An adder can perform subtraction as the addition of the minuend and the complemented (negated) subtrahend, but it is also possible to build *subtracter* circuits that perform subtraction directly. MSI devices called ALUs, described at the end of this section, perform addition, subtraction, or any of several other operations according to an operation code applied to the device's control inputs.

4.9.1 Half Adders and Full Adders

half adder

The simplest adder, called a *half adder*, adds two 1-bit operands X and Y, producing a 2-bit sum. The sum can range from 0 to 2, which requires two bits to express. The low-order bit of the sum may be named HS (half sum), and the high-order bit may be named CO (carry out). We can write the following equations for HS and CO:

$$HS = X \oplus Y$$
$$= X \cdot Y' + X' \cdot Y$$
$$CO = X \cdot Y$$

full adder

To add operands with more than one bit, we must provide for carries between bit positions. The basic building block for this operation is called a *full adder*. Besides the addend-bit inputs X and Y, a full adder has a carry-bit input, CIN. The sum of the three inputs can range from 0 to 3, which can still be expressed with just two output bits, S and COUT, having the following equations:

$$S = X \oplus Y \oplus CIN$$
$$= X \cdot Y' \cdot CIN' + X' \cdot Y \cdot CIN' + X' \cdot Y' \cdot CIN + X \cdot Y \cdot CIN$$
$$COUT = X \cdot Y + X \cdot CIN + Y \cdot CIN$$

Here, S is 1 if an odd number of the inputs are 1, and COUT is 1 if two or more of the inputs are 1. These equations represent the same operation that was specified by the binary addition table in Chapter 1, Table 1–3, on page 8.

One possible circuit that performs the full-adder equations is shown in Figure 4–93(a). The corresponding logic symbol is shown in (b). Sometimes the symbol is drawn as shown in (c), so that cascaded full adders can be drawn more neatly, as in the next subsection.

---

*This section and all of its subsections are optional.

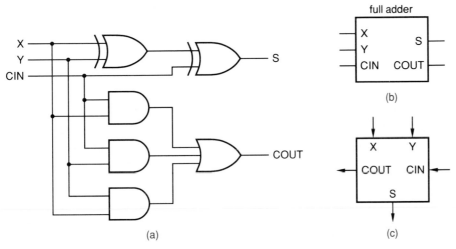

Figure 4–93 Full adder: (a) gate-level circuit diagram; (b) logic symbol; (c) alternate logic symbol suitable for cascading.

4.9.2 Ripple Adders

Two binary words, each with n bits, can be added using a *ripple adder*—a cascade   *ripple adder*
of n full-adder stages, each of which handles one bit. Figure 4–94 shows the
circuit for a 4-bit ripple adder. The carry input to the least signficant bit (c_0)
is normally set to 0, and the carry output of each full adder is connected to the
carry input of the next most significant full adder. The ripple adder is a classic
example of an iterative circuit as defined in Section 4.8.2.

    A ripple adder is slow, since in the worst case a carry must propagate from
the least significant full adder to the most significant one. This occurs if one
addend is $11 \ldots 11$ and the other is $00 \ldots 01$. Assuming that all of the addend
bits are presented simultaneously, the total worst-case delay is

$$t_{ADD} = t_{XYCout} + (n - 2) \cdot t_{CinCout} + t_{CinS}$$

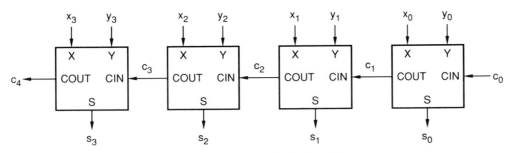

Figure 4–94 A 4-bit ripple adder.

where t_{XYCout} is the delay from X or Y to COUT in the least significant stage, $t_{CinCout}$ is the delay from CIN to COUT in the middle stages, and t_{CinS} is the delay from CIN to S in the most significant stage.

A faster adder can be built by obtaining each sum output s_i with just two levels of logic. This can be accomplished by writing an equation for s_i in terms of x_0–x_i, y_0–y_i, and c_0, "multiplying out" or "adding out" to obtain a sum-of-products or product-of-sums expression, and building the corresponding AND-OR or OR-AND circuit. Unfortunately, beyond s_2, the resulting expressions have too many terms, requiring too many first-level gates and more inputs than typically possible on the second-level gate. For example, even assuming that $c_0 = 0$, a two-level AND-OR circuit for s_2 requires fourteen 4-input ANDs, four 5-input ANDs, and an 18-input OR gate; higher-order sum bits are even worse. Nevertheless, it is possible to build adders with just a few levels of delay using a more reasonable number of gates, as we'll see in Section 4.9.4.

* 4.9.3 Subtracters

full subtracter

A binary subtraction operation analogous to binary addition was also specified in Table 1–3 on page 8. A *full subtracter* handles one bit of the binary subtraction algorithm, having input bits X (minuend), Y (subtrahend), and BIN (borrow in), and output bits D (difference) and BOUT (borrow out). We can write logic equations corresponding to the binary subtraction table as follows:

$$D = X \oplus Y \oplus BIN$$
$$BOUT = X' \cdot Y + X' \cdot BIN + Y \cdot BIN$$

These equations are very similar to equations for a full adder, which should not be surprising. We showed in Section 1.6 that a two's-complement subtraction operation, $X - Y$, can be performed by an addition operation, namely by adding the two's complement of Y to X. The two's complement of Y is $\overline{Y} + 1$, where $\overline{Y}$ is the bit-by-bit complement of Y. We also showed in Exercise 1.23 that a binary adder can be used to perform an unsigned subtraction operation $X - Y$, by performing the operation $X + \overline{Y} + 1$. We can now confirm that these statements are true by manipulating the logic equations above:

$$
\begin{aligned}
BOUT &= X' \cdot Y + X' \cdot BIN + Y \cdot BIN \\
BOUT' &= (X + Y') \cdot (X + BIN') \cdot (Y' + BIN') & \text{(generalized DeMorgan's theorem)} \\
&= X \cdot Y' + X \cdot BIN' + Y' \cdot BIN' & \text{(multiply out)} \\
D &= X \oplus Y \oplus BIN \\
&= X \oplus Y' \oplus BIN' & \text{(complementing XOR inputs)}
\end{aligned}
$$

For the last manipulation, recall that we can complement the two inputs of an XOR gate without changing the function performed.

Figure 4–95   Designing subtracters using adders: (a) full adder; (b) full subtracter; (c) interpreting the device in (a) as a full subtracter; (d) ripple subtracter.

Comparing with the equations for a full adder, the above equations tell us that we can build a full subtractor from a full adder as shown in Figure 4–95. Just to keep things straight, we've given the full adder circuit in (a) a fictitious name, the "74LS999." As shown in (c), we can interpret the function of this same physical circuit to be a full subtracter by giving it a new symbol with active-low carry in, carry out, and subtrahend signals. Thus, to build a ripple subtracter for two n-bit active-high operands, we can use n 74LS999s and inverters, as shown in (d). Note that for the subtraction operation, the borrow input of the least significant bit should be negated (no borrow), which for an active-low input means that the physical pin must be 1 or HIGH. This is just the opposite as in addition, where the same input pin is an active-high carry-in that is 0 or LOW.

By going back to the math in Chapter 1, we can show that this sort of manipulation works for all adder and subtracter circuits, not just ripple adders and subtracters. That is, any n-bit adder circuit can be made to function as a subtracter by complementing the subtrahend and treating the carry-in and carry-out signals as borrows with the opposite active level. The rest of this section

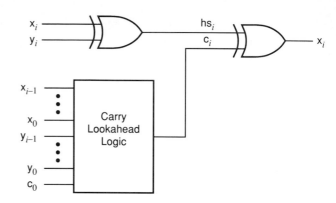

Figure 4–96
Structure of one stage of a carry lookahead adder.

discusses addition circuits only, with the understanding that they can easily be made to perform subtraction.

* 4.9.4   Carry Lookahead Adders

The logic equation for sum bit i of a binary adder can actually be written quite simply:

$$s_i \;=\; x_i \oplus y_i \oplus c_i$$

More complexity is introduced when we expand c_i above in terms of x_0–x_{i-1}, y_0–y_{i-1}, and c_0, and we get a real mess expanding the XORs. However, if we're willing to forego the XOR expansion, we can at least streamline the design of c_i logic using ideas of *carry lookahead* discussed in this subsection.

carry lookahead

Figure 4–96 shows the basic idea. The block labeled "carry lookahead logic" calculates c_i in a fixed, small number of logic levels for any reasonable value of i. Two definitions are the key to carry lookahead logic:

carry generate

- For a particular combination of inputs x_i and y_i, adder stage i is said to *generate* a carry if it produces a carry-out of 1 ($c_{i+1} = 1$) independent of the inputs on x_0–x_{i-1}, y_0–y_{i-1}, and c_0.

carry propagate

- For a particular combination of inputs x_i and y_i, adder stage i is said to *propagate* carries if it produces a carry-out of 1 ($c_{i+1} = 1$) in the presence of an input combination of x_1–x_{i-1}, y_1–y_{i-1}, and c_0 that causes a carry-in of 1 ($c_i = 1$).

Corresponding to these definitions, we can write logic equations for a carry-generate signal, g_i, and a carry-propagate signal, p_i, for each stage of a carry lookahead adder:

$$g_i \;=\; x_i \cdot y_i$$
$$p_i \;=\; x_i + y_i$$

That is, a stage unconditionally generates a carry if both of its addend bits are 1, and it propagates carries if at least one of its addend bits is 1. The carry output of a stage can now be written in terms of the generate and propagate signals:

$$c_{i+1} \;=\; g_i + p_i \cdot c_i$$

To eliminate carry ripple, we recursively expand the c_i term for each stage, and multiply out to obtain a 2-level AND-OR expression. Using this technique, we can obtain the following carry equations for the first four adder stages:

$$
\begin{aligned}
c_1 \;&=\; g_0 + p_0 \cdot c_0 \\
c_2 \;&=\; g_1 + p_1 \cdot c_1 \\
&=\; g_1 + p_1 \cdot (g_0 + p_0 \cdot c_0) \\
&=\; g_1 + p_1 \cdot g_0 + p_1 \cdot p_0 \cdot c_0 \\
c_3 \;&=\; g_2 + p_2 \cdot c_2 \\
&=\; g_2 + p_2 \cdot (g_1 + p_1 \cdot g_0 + p_1 \cdot p_0 \cdot c_0) \\
&=\; g_2 + p_2 \cdot g_1 + p_2 \cdot p_1 \cdot g_0 + p_2 \cdot p_1 \cdot p_0 \cdot c_0 \\
c_4 \;&=\; g_3 + p_3 \cdot c_3 \\
&=\; g_3 + p_3 \cdot (g_2 + p_2 \cdot g_1 + p_2 \cdot p_1 \cdot g_0 + p_2 \cdot p_1 \cdot p_0 \cdot c_0) \\
&=\; g_3 + p_3 \cdot g_2 + p_3 \cdot p_2 \cdot g_1 + p_3 \cdot p_2 \cdot p_1 \cdot g_0 + p_3 \cdot p_2 \cdot p_1 \cdot p_0 \cdot c_0
\end{aligned}
$$

Each equation corresponds to a circuit with just three levels of delay—one for the generate and propagate signals, and two for the sum-of-products shown. A *carry lookahead adder* uses three-level equations such as these in each adder stage for the block labeled "carry lookahead" in Figure 4–96. The sum output for a stage is produced by combining the carry bit with the two addend bits for the stage as we showed in the figure. In the next subsection, we'll study some commercial MSI adders and ALUs that use carry lookahead.

carry lookahead adder

4.9.5 MSI Adders

The *74LS283* is a 4-bit binary adder that forms its sum and carry outputs with just a few levels of logic, using the carry lookahead technique. Figure 4–97 is a logic symbol and Figure 4–98 is a logic diagram for the 74LS283. The older *74LS83* is identical to the '283 except that its pinout is different, including nonstandard locations for power and ground pins.

    As shown in the logic diagram, the '283 design has a few differences from the general carry-lookahead design that we described in the preceding subsection. First of all, its addends are named A and B instead of X and Y; no big deal. Second, it produces active-low versions of the carry-generate (g_i') and carry-propagate (p_i') signals, since inverting gates are faster than noninverting ones.

74LS283

74LS83

Figure 4–97
Traditional logic symbol for the
74LS283 4-bit binary adder.

74LS283

Third, it takes advantage of the fact that we can algebraically manipulate the half-sum equation as follows:

$$
\begin{aligned}
hs_i &= x_i \oplus y_i \\
&= x_i \cdot y_i' + x_i' \cdot y_i \\
&= x_i \cdot y_i' + x_i \cdot x_i' + x_i' \cdot y_i + y_i \cdot y_i' \\
&= (x_i + y_i) \cdot (x_i' + y_i') \\
&= (x_i + y_i) \cdot (x_i \cdot y_i)' \\
&= p_i \cdot g_i'
\end{aligned}
$$

Thus, an AND gate with an inverted input can be used instead of an XOR gate to create each half-sum bit.

Finally, the '283 creates the carry signals using an INVERT-OR-AND structure (the DeMorgan equivalent of an AND-OR-INVERT), which has about the same delay as a single TTL inverting gate. This requires some explaining, since the carry equations that we derived in the preceding subsection are used in a slightly modified form. In particular, the c_{i+1} equation uses the term $p_i \cdot g_i$ instead of g_i. This has no effect on the output, since p_i is always 1 when g_i is 1. However, it allows the equation to be factored as follows:

$$
\begin{aligned}
c_{i+1} &= p_i \cdot g_i + p_i \cdot c_i \\
&= p_i \cdot (g_i + c_i)
\end{aligned}
$$

This leads to the following carry equations, which are used by the circuit:

$$
\begin{aligned}
c_1 &= p_0 \cdot (g_0 + c_0) \\
c_2 &= p_1 \cdot (g_1 + c_1) \\
&= p_1 \cdot (g_1 + p_0 \cdot (g_0 + c_0)) \\
&= p_1 \cdot (g_1 + p_0) \cdot (g_1 + g_0 + c_0)
\end{aligned}
$$

Figure 4–98
Logic diagram for the 74LS283
4-bit binary adder, including pin
numbers for a standard 16-pin
dual in-line package.

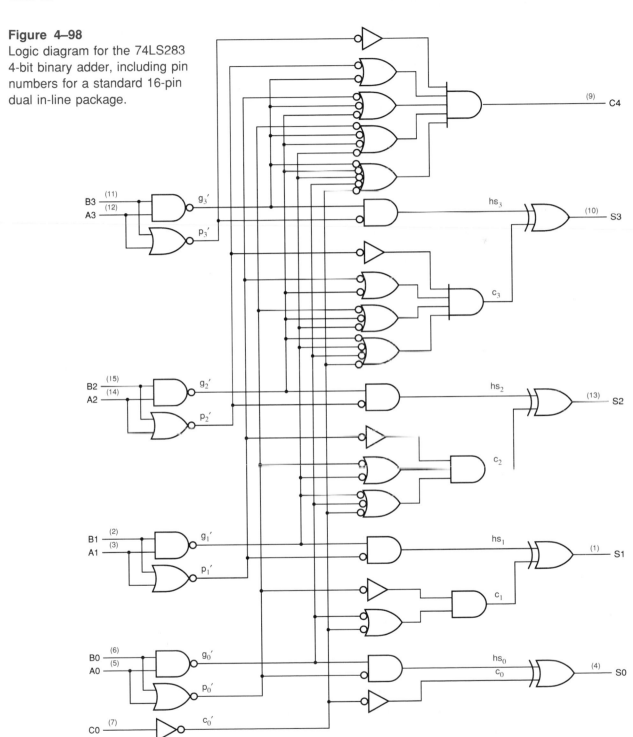

$$c_3 = p_2 \cdot (g_2 + c_2)$$
$$= p_2 \cdot (g_2 + p_1 \cdot (g_1 + p_0) \cdot (g_1 + g_0 + c_0))$$
$$= p_2 \cdot (g_2 + p_1) \cdot (g_2 + g_1 + p_0) \cdot (g_2 + g_1 + g_0 + c_0)$$
$$c_4 = p_3 \cdot (g_3 + p_2 \cdot c_3)$$
$$= p_3 \cdot (g_3 + p_2 \cdot (g_2 + p_1) \cdot (g_2 + g_1 + p_0) \cdot (g_2 + g_1 + g_0 + c_0))$$
$$= p_3 \cdot (g_3 + p_2) \cdot (g_3 + g_2 + p_1) \cdot (g_3 + g_2 + g_1 + p_0) \cdot (g_3 + g_2 + g_1 + g_0 + c_0)$$

If you've followed the derivation of these equations and can obtain the same ones by reading the '283 logic diagram, then, congratulations, you're up to speed on switching algebra! If not, you may want to review Sections 3.1 and 3.2.

group-ripple adder

The propagation delay from the C0 input to the C4 output of the '283 is very short, about the same as two TTL inverting gates. As a result, fairly fast *group-ripple adders* with more than four bits can be made simply by cascading the carry outputs and inputs of '283s, as shown in Figure 4–99 for a 16-bit adder. The total propagation delay from C0 to C16 in this circuit is about the same as that of eight TTL inverting gates.

* 4.9.6   MSI Arithmetic and Logic Units (ALUs)

arithmetic and logic unit (ALU)

An *arithmetic and logic unit (ALU)* is a combinational circuit that can perform any of a number of different arithmetic and logical operations on a pair of *b*-bit operands. The operation to be performed is specified by a set of function-select inputs. Typical MSI ALUs have 4-bit operands and three to five function select inputs, allowing up to 32 different functions to be performed.

74LS181

Figure 4–100 is a logic symbol for the *74LS181* 4-bit ALU. The operation performed by the '181 is selected by the M and S3–S0 inputs, as detailed in Table 4–16. Note that the identifiers A, B, and F in the table refer to the 4-bit words A3–A0, B3–B0, and F3–F0; and the symbols · and + refer to logical AND and OR operations.

The 181's M input selects between arithmetic and logical operations. When M = 1, logical operations are selected, and each output Fi is a function only of the corresponding data inputs, Ai and Bi. No carries propagate between stages, and the CIN input is ignored. The S3–S0 inputs select a particular logical operation; any of the 16 different combinational logic functions on two variables may be selected.

When M = 0, arithmetic operations are selected, carries propagate between the stages, and CIN is used as a carry input to the least significant stage. For operations larger than four bits, multiple '181 ALUs may be cascaded like the group-ripple adder in the Figure 4–99, with the carry-out (COUT) of each ALU

Figure 4–99
A 16-bit group-ripple adder.

Figure 4–100
Logic symbol for the 74LS181
4-bit ALU.

Table 4–16 Functions performed by the 74LS181 4-bit ALU.

Inputs				Function	
S3	S2	S1	S0	M = 0 (arithmetic)	M = 1 (logic)
0	0	0	0	F = A minus 1 plus CIN	F = A′
0	0	0	1	F = A·B minus 1 plus CIN	F = A′ + B′
0	0	1	0	F = A·B′ minus 1 plus CIN	F = A′ + B
0	0	1	1	F = 1111 plus CIN	F = 1111
0	1	0	0	F = A plus (A + B′) plus CIN	F = A′·B′
0	1	0	1	F = A·B plus (A + B′) plus CIN	F = B′
0	1	1	0	F = A minus B minus 1 plus CIN	F = A⊕B′
0	1	1	1	F = A + B′ plus CIN	F = A + B′
1	0	0	0	F = A plus (A + B) plus CIN	F = A′·B
1	0	0	1	F = A plus B plus CIN	F = A⊕B
1	0	1	0	F = A·B′ plus (A + B) plus CIN	F = B
1	0	1	1	F = A + B plus CIN	F = A + B
1	1	0	0	F = A plus A plus CIN	F = 0000
1	1	0	1	F = A·B plus A plus CIN	F = A·B′
1	1	1	0	F = A·B′ plus A plus CIN	F = A·B
1	1	1	1	F = A plus CIN	F = A

connected to the carry-in (CIN) of the next most significant stage. The same function-select signals (M, S3–S0) are applied to all the '181s in the cascade.

To perform two's-complement addition, we use S3–S0 to select the operation "A plus B plus CIN." The CIN input of the least-significant ALU is normally set to 0 during addition operations. To perform two's-complement subtraction, we use S3–S0 to select the operation A minus B minus 1 plus CIN. In this case, the CIN input of the least significant ALU is normally set to 1, since CIN acts as the complement of the borrow during subtraction.

The '181 provides other arithmetic operations, such as "A minus 1 plus CIN," that are useful in some applications (e.g., decrement by 1). It also provides a bunch of weird arithmetic operations, such as "A·B′ plus (A + B) plus CIN," that are almost never used in practice, but that "fall out" of the circuit for free.

Notice that the operand inputs /A3–/A0 and /B3–/B0 and the function outputs /F3–/F0 of the '181 are active low. The '181 can also be used with active-high operand inputs and function outputs. In this case, a different version of the function table must be constructed. When M = 1, logical operations are still performed, but for a given input combination on S3–S0, the function obtained is precisely the dual of the one listed in Table 4–16. When M = 0, arithmetic operations are performed, but the function table is once again different. Refer to a '181 data sheet for more details.

Figure 4–101
Logic symbols for 4-bit ALUs:
(a) 74LS381; (b) 74LS382.

Two other MSI ALUs, the *74LS381* and *74LS382* shown in Figure 4–101, *74LS381* encode their select inputs more compactly, and provide only eight different but *74LS382* useful functions, as detailed in Table 4–17. The only difference between the '381 and '382 is that one provides group-carry lookahead outputs (which we explain next), while the other provides ripple carry and overflow outputs.

* 4.9.7 Group-Carry Lookahead

The '181 and '381 provide *group-carry lookahead* outputs that allow multiple *group-carry* ALUs to be cascaded without rippling carries between 4-bit groups. Like the *lookahead* 74LS283, the ALUs use carry lookahead to produce carries internally. However, they also provide /G and /P outputs that are carry lookahead signals for the entire 4-bit group. The /G output is asserted if the ALU generates a carry, that

Table 4–17
Functions performed by the
74LS381 and 74LS382 4-bit
ALUs.

Inputs			
S2	S1	S0	Function
0	0	0	F = 0000
0	0	1	F = B minus A minus 1 plus CIN
0	1	0	F = A minus B minus 1 plus CIN
0	1	1	F = A plus B plus CIN
1	0	0	F = A⊕B
1	0	1	F = A + B
1	1	0	F = A·B
1	1	1	F = 1111

Figure 4–102
Logic symbol for the 74S182
carry lookahead circuit.

is, if it will produce a carry-out (COUT = 1) independent of whether or not there is a carry-in (CIN = 1):

$$\text{/G} \;=\; (g_3 + p_3 {\cdot} g_2 + p_3 {\cdot} p_2 {\cdot} g_1 + p_3 {\cdot} p_2 {\cdot} p_1 {\cdot} g_0)'$$

The /P output is asserted if the ALU propagates a carry, that is, if it will produce a carry-out if there is a carry-in:

$$\text{/P} \;=\; (p_3 {\cdot} p_2 {\cdot} p_1 {\cdot} p_0)'$$

lookahead carry circuit
74S182

When ALUs are cascaded, the group-carry lookahead outputs may be combined in just two levels of logic to produce the carry input to each ALU. A *lookahead carry circuit*, the *74S182* shown in Figure 4–102, performs this operation. The '182 inputs are C0, the carry input to the least significant ALU ("ALU 0"), and G0–G3 and P0–P3, the generate and propagate outputs of ALUs 0–3. Using these inputs, the '182 produces the carry inputs C1–C3 for ALUs 1–3. Figure 4–103 shows the connections for a 16-bit ALU using four '381s and a '182.

The 182's carry equations are obtained by "adding out" the basic carry lookahead equation of Section 4.9.4:

$$c_{i+1} \;=\; g_i + p_i {\cdot} c_i$$
$$=\; (g_i + p_i) {\cdot} (g_i + c_i)$$

Expanding for the first three values of i, we obtain the following equations:

$$\text{C1} \;=\; (\text{G0} + \text{P0}) {\cdot} (\text{G0} + \text{C0})$$
$$\text{C2} \;=\; (\text{G1} + \text{P1}) {\cdot} (\text{G1} + \text{G0} + \text{P0}) {\cdot} (\text{G1} + \text{G0} + \text{C0})$$
$$\text{C3} \;=\; (\text{G2} + \text{P2}) {\cdot} (\text{G2} + \text{G1} + \text{P1}) {\cdot} (\text{G2} + \text{G1} + \text{G0} + \text{P0}) {\cdot} (\text{G2} + \text{G1} + \text{G0} + \text{C0})$$

The '182 realizes each of these equations with just one level of delay—an INVERT-OR-AND gate.

Figure 4–103
A 16-bit ALU using group-carry lookahead.

When more than four ALUs are cascaded, they may be partitioned into "supergroups," each with its own '182. For example, a 64-bit adder would have four supergroups, each containing four ALUs and a '182. The /G and /P outputs of each '182 can be combined in a next-level '182, since they indicate whether the supergroup generates or propagates carries:

$$\text{/G} = ((G3 + P3)\cdot(G3 + G2 + P2)\cdot(G3 + G2 + G1 + P1)\cdot(G3 + G2 + G1 + G0))'$$
$$\text{/P} = (P0\cdot P1\cdot P2\cdot P3)'$$

* 4.9.8   IEEE Standard Symbols

The IEEE standard uses the general qualifying symbol "Σ" to identify an adder or addition function. Figure 4–104(a) shows the symbol for a 74LS283 4-bit adder. The numbers on the addend inputs and sum output indicate the weight of each pin, as a power of 2.

The IEEE symbol for a 74LS181 4-bit ALU is shown in Figure 4–104(b). The first five inputs of the common control block form a "mode control" word, which the standard designates by the letter M. The weights of the mode control bits are shown as powers of 2, and they designate a mode number in the range 0–31. According to the standard, a separate table accompanies the logic symbol to define the functions performed in each mode. The CP, CG, and CO outputs are enabled in modes 0–15. The four individual ALU blocks are labeled with the weights of the bits they process. The IEEE symbol for the 74S182 carry lookahead circuit is much simpler, and is shown in (c).

Figure 4–104
IEEE standard symbols:
(a) 74LS283; (b) 74LS181;
(c) 74LS182.

(a)   (b)   (c)

Figure 4–105
A 4×4 multiplication element
using a 74284 and a 74285,
each in a standard 16-pin
package.

* 4.10 COMBINATIONAL MULTIPLIERS

In Section 1.8, we outlined an algorithm that uses n shifts and adds to multiply n-bit binary numbers. Although the shift-and-add algorithm emulates the way that we do paper-and-pencil multiplication of decimal numbers, there is nothing inherently "sequential" or "time dependent" about multiplication. That is, given two n-bit input words A and B, it is possible to write a truth table that expresses the $2n$-bit product $P = A \cdot B$ as a *combinational* function of A and B. A *combinational multiplier* is a logic circuit with such a truth table.

The *74284* and *74285* are commercially available MSI 4×4 combinational multipliers. The two chips are used together to obtain the 8-bit product of two 4-bit unsigned inputs. As shown in Figure 4–105, the '284 produces the four high-order bits of the product, and the '285 produces the four low-order bits. Pull-up resistors are shown in the figure because the '284 and '285 have open-collector outputs (Section 2.11.3).

combinational
multiplier
74284
74285

Figure 4–106
Producing an 8×8 product as a
sum of 4×4 products.

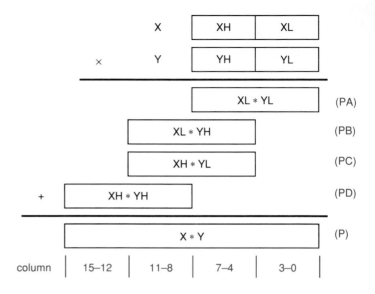

There is no inherent reason for a combinational multiplier to use two separate chips for the low-order and high-order product bits. The '284 and '285 were designed that way because of the pin, power, and gate limitations in TTL technology in 1972. More recently designed combinational multipliers, such as the Cypress Semiconductor 16×16-bit 7C516, use CMOS LSI technology and larger packages to perform larger multiplications.

All by itself, a 4×4 multiplier using the '284 and '285 isn't too useful—you could almost do a 4×4 multiplication in your head in the 60 ns that the '284 and '285 take to do it (well, maybe not quite that fast...). However, most applications require larger multiplications, such as 16×16 or 32×32, that can be performed hierarchically using adders and 4×4 multipliers as building blocks.

Figure 4–106 shows the basic idea for a hierarchical 8×8 multiplier. The 8-bit multiplier and multiplicand X and Y are each broken into two 4-bit components. Each component of X is multiplied by each component of Y, and the partial products are added, with the appropriate shifts. Figure 4–107 is a combinational circuit corresponding to this scheme. This circuit has three levels:

(1) The multipliers (U1–U4)[11] form all of the 4×4 products in parallel.

(2) The second-level adders form a partial product combining the first two operands in columns 11–4 of Figure 4–106.

[11] Each 4×4 multiplier contains a '284 and a '285, designated U1 and U2 in Figure 4–105. When preparing the mechanical documentation for the system, a computer-aided design system must "flatten" the hierarchical schematic, in which case it might give the individual '284s and '285s reference designators like U1.1, U1.2, . . ., U4.1, U4.2.

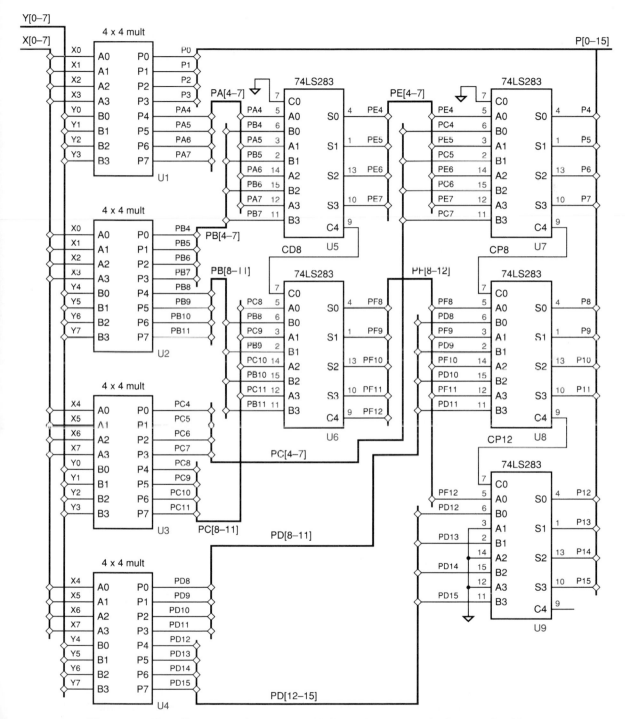

Figure 4–107 Structure of an 8×8 multiplier using 4×4 multipliers and adders.

(3) The third-level adders combine the result of the preceding step with the last operand in columns 15–4 of Figure 4–106. Notice that even though columns 15–12 have only one operand, an adder is required to accommodate the high-order carries from the preceding step.

It is instructive to analyze the total worst-case delay of the 8×8 multiplier circuit. The individual circuit components have the following worst-case delays:

t_{4x4} 60 ns. Input to output delay of a 4×4 multiplier.

t_{c0c4} 16 ns. Delay from the C0 input to the C4 output in the 74LS283.

t_{abc4} 16 ns. Delay from an Ai or Bi input to the C4 output in the 74LS283.

t_{c0s} 21 ns. Delay from the C0 input to any Si output in the 74LS283.

t_{abs} 24 ns. Delay from an Ai or Bi input to any Si output in the 74LS283.

We must determine the worst-case delay from any Xi or Yi input to any Pi output along any path through the circuit. A worst-case delay can be determined by adding the delays of each level in the circuit:

(1) 60 ns. All of the 4×4 multipliers produce their outputs at the same time, t_{4x4} after the inputs change.

(2) 37 ns. The worst-case delay for the second-level (8-bit) adder to have a valid output is t_{abc4} (through U5) plus $\max(t_{c0s}, t_{c0c4})$ (through U6).

(3) 56 ns. The worst-case delay for the third-level (12-bit) adder to have a valid output is t_{abc4} (through U7) plus t_{c0c4} (through U8) plus $\max(t_{c0s}, t_{abs})$ (through U9).

The sum of the three levels' delay is 153 ns. However, this bound is overly pessimistic, since U7 does not need all of the second-level outputs to be valid in order to produce a valid output. Several different paths involving U5–U8 must be analyzed to calculate the actual worst-case delay in Exercise 4.64.

4.11 COMBINATIONAL LOGIC DESIGN WITH MSI FUNCTIONS

We began this chapter by pointing out that practical combinational circuits cannot be designed by brute-force application of theoretical techniques. The key to practical logic design, we said, is structured thinking—decomposing a complex problem into smaller chunks that are easier to swallow. At this point, you should be familiar with the most commonly used combinational MSI chunks, so we can now offer some insights and examples about combinational logic design with MSI.

4.11.1 The Name of the Game

Given the functional and performance requirements for a digital system, the name of the game in practical digital design is to minimize cost. For *board-* *board-level design* *level designs*—systems that are packaged on a single PCB—this usually means minimizing the number of IC packages. If too many ICs are required, they won't all fit on the PCB. "Well, just use a bigger PCB," you say. Unfortunately, PCB sizes are usually constrained by factors such as pre-existing standards (e.g., add-in boards for PCs), packaging constraints (e.g., it has to fit in a toaster), or edicts from above (e.g., in order to get the project approved three months ago, you foolishly told your manager that it would all fit on a 3×5 inch PCB, and now you've got to deliver!). In each of these cases, the cost of using a larger PCB or multiple PCBs may be unacceptable.

Minimizing the number of ICs is usually the rule even though individual IC costs vary. For example, a typical SSI IC may cost 15¢, while an MSI IC may cost 50¢. It may be possible to perform a particular function with two SSI ICs (30¢) or one MSI IC (50¢). In most situations, the more expensive MSI solution is used, not because the designer owns stock in the IC company, but because the MSI solution uses less PCB area.

We'll see this idea repeated when we design with structured logic devices like ROMs and PLDs in Chapter 7. One of these LSI ICs, costing perhaps $2, may replace half a dozen 15–50¢ SSI and MSI ICs. The LSI IC may cost a little more or a little less than the SSI/MSI total, but in any case it saves substantial design time and PCB space.

In *ASIC design*, the name of the game is a little different, but the importance *ASIC design* of MSI (that is, structured, functional) design techniques is the same. Although it's easy to burn hours and weeks creating custom macrocells and minimizing the total gate count of an ASIC, only rarely is this advisable. The per-unit cost reduction achieved by having a 10% smaller chip is negligible except in high-volume applications. In applications with low-to-medium volume (the majority), two other factors are more important: design time and NRE (non-recurring engineering) cost. A shorter design time allows a product to reach the market sooner, increasing revenues over the lifetime of the product. A lower NRE cost also flows right to the "bottom line," and in small companies may be the only way the project can be completed before the company runs out of money (believe me, I've been there!). If the product is successful, it's always possible and profitable to "tweak" the design later to reduce per-unit costs. The need to minimize design time and NRE cost argues in favor of a structured, as opposed to highly optimized, approach to ASIC design, using standard MSI-type functions provided in the ASIC manufacturer's library.

4.11.2   Using MSI Functions for "Random Logic"

random logic

In Section 3.3 we gave many examples of combinational logic functions and showed how to build them using discrete gates. Most designers would classify these functions as *random logic*—functions that do not correspond to a common, structured operation such as decoding, multiplexing, comparing, or adding. Yet, in Sections 4.3.6 and 4.6.3 we showed two structured methods using MSI decoders and multiplexers that can economically perform "random" combinational logic functions with a small number of inputs. For example, compare two designs for the pair of functions $F = \Sigma_{X,Y,Z}(3,6,7)$ and $G = \Sigma_{X,Y,Z}(0,1,3)$:

- The minimal multiple-output sum-of-products (NAND-NAND) circuit based on Figure 3–43(b) on page 190 requires two inverters (2/6 74LS04), four 2-input NANDs (4/4 74LS00), and one 3-input NAND (1/3 74LS10), a total of 1.67 IC packages.

- A structured solution based on Figure 4–33 uses a 74LS138 3-to-8 decoder and two 3-input NANDs (2/3 74LS10), also a total of 1.67 IC packages. At the board level, the MSI solution costs about the same as the SSI solution, but it's a "no-brainer" to design.

Next, compare two designs for a prime-number detector:

- A minimal sum-of-products (NAND-NAND) circuit based on Figure 3–30 on page 180 requires three inverters (3/6 74LS04), one 2-input NAND (1/4 74LS00), three 3-input NANDs (3/3 74LS10), and one 4-input NAND (1/2 74LS20), a total of 2.25 IC packages.

- The structured solution given in Figure 4–68 uses a single 74LS151 8-input multiplexer, just one IC package. Not only is it easier to design, but also it costs substantially less at the PCB level.

Of course, the decoder and multiplexer approaches are uneconomical for circuits with more than a few inputs. However, in Chapter 7 we'll describe structured LSI devices that can handle random-logic functions with lots of inputs.

4.11.3   Almost-MSI Functions

almost-binary decoder

Sometimes a required function almost matches what is provided by a standard MSI IC, but not quite. A common example of such a function is an *almost-binary decoder*. Suppose you needed to generate a set of enable signals according to Table 4–18. Each output is asserted for one row of the table in which /CS and /RD are asserted, and negated for all other input combinations. This almost matches the function of a 3-to-8 decoder with two enable inputs, except that /BILL and /MARY are each asserted for two combinations of the A2–A0 inputs.

Table 4–18
Function table for an almost-binary decoder.

/CS	/RD	A2	A1	A0	Output to assert
1	x	x	x	x	none
x	1	x	x	x	none
0	0	0	0	x	/BILL
0	0	0	x	0	/MARY
0	0	0	1	0	/JOAN
0	0	0	1	1	/PAUL
0	0	1	0	0	/ANNA
0	0	1	0	1	/FRED
0	0	1	1	0	/DAVE
0	0	1	1	1	/KATE

Rather than design a custom decoder from scratch, in a board-level design it is more economical to adapt an MSI decoder as shown in Figure 4–108.

Another example is an *almost-binary-select multiplexer*. Suppose that you needed a circuit that selects one of four 16-bit input buses, A, B, C, or D, to drive a 16-bit output bus F, as specified in Table 4–19 by three control bits (ignore the last two columns for now). One way to design this circuit is to use 16 74LS151 8-input multiplexers, one for each bit connected as shown in Figure 4–109(a). However, this solution is wasteful because it uses 8-input multiplexers to select among only four distinct sources. As shown in (b), a more economical solution uses only eight 74LS153 4-input multiplexers, each of which handles two bits, and a "code converter" circuit. The code converter produces the select inputs for the '153s according to the last two columns of Table 4–19. Only one copy of the code converter is needed to drive all of the '153s, so the net savings with this approach is eight multiplexer packages minus whatever it takes to build the code converter (just one package, according to Exercise 4.67).

almost-binary-select multiplexer

Figure 4–108
Almost-binary decoder circuit.

Table 4–19
Function table for an almost-binary-select multiplexer.

S2	S1	S0	Input to select	MS1	MS0
0	0	0	A	0	0
0	0	1	B	0	1
0	1	0	A	0	0
0	1	1	C	1	0
1	0	0	A	0	0
1	0	1	D	1	1
1	1	0	A	0	0
1	1	1	B	0	1

mode-dependent comparator

A final example is a *mode-dependent comparator*. This example is taken from a system that compares two n-bit words under normal circumstances, but that must sometimes ignore one or two low-order bits of the input words, as specified by two mode-control bits. One way to obtain this function is to use three comparators—an n-bit, an $(n-1)$-bit, and an $(n-2)$-bit one—and combine their outputs with a 3-input multiplexer controlled by the mode bits. However, a more economical solution is shown in Figure 4–110 for an 8-bit mode-dependent comparator. Instead of combining multiple comparator outputs, this design uses gates on a single comparator's low-order input bits, forcing low-order bits to be equal in the appropriate modes. The design can be further simplified using a different encoding for the mode-control bits (see Exercise 4.70).

Figure 4–109   Design of the almost-binary-select multiplexer: (a) duplicating the data inputs; (b) re-encoding the select inputs.

Figure 4–110
Design of an 8-bit mode-
dependent comparator:
(a) mode-control encoding;
(b) logic diagram.

M1	M0	Comparison
0	0	8-bit
0	1	7-bit
1	0	6-bit
1	1	not used

(a)

(b)

4.11.4   Finding the MSI Functions in Larger Problems

Quite often, the requirements for a digital-circuit application are specified in a
way that makes an MSI solution obvious. Consider the following problem.

- Design a combinational circuit whose inputs are two 8-bit unsigned binary
  integers, X and Y, and a control signal MIN/MAX. The output of the circuit is
  an 8-bit unsigned binary integer Z such that $Z = min(X, Y)$ if MIN/MAX = 1,
  and $Z = max(X, Y)$ if MIN/MAX = 0.

This circuit is fairly easy to visualize in terms of MSI functions. Clearly, we
can use a comparator to determine whether $X > Y$. We can use the comparator's
output to control multiplexers that produce $min(X, Y)$ and $max(X, Y)$, and we can
use another multiplexer to select one of these results depending on MIN/MAX.
Figure 4–111(a) is the block diagram of a circuit corresponding to this approach.

Figure 4–111   Min/max circuit: (a) block diagram of a "first cut" solution; (b) block diagram of a more cost-effective solution; (c) logic diagram for the second solution.

Our first solution approach works, but it's more expensive than it needs to be. Although it has three two-input multiplexers, there are only two input words, X and Y, that may ultimately be selected and produced at the output of the circuit. Therefore, we should be able to use just a single two-input mux, and use some other logic to figure out which input to tell it to select. This approach is shown in Figure 4–111(b) and (c). The "other logic" is very simple indeed, just a single XOR gate. The wastefulness of our original design approach may have been obvious to you from the beginning, but it demonstrates an important rule when designing with MSI:

- Use standard MSI parts to handle data, and look for ways that a single MSI part can perform different functions at different times or in different modes. Design control circuits, possibly using SSI, to select the appropriate functions as needed, to reduce the total parts count of the design.

Now let's look at a design problem whose MSI solution is not quite so obvious, a "fixed-point to floating-point encoder." An unsigned binary integer B in the range $0 \leq B < 2^{11}$ can be represented by 11 bits in "fixed-point" format, $B = b_{10}b_9 \ldots b_1b_0$. We can represent numbers in the same range with less precision using only 7 bits in a floating-point notation, $F = M \cdot 2^E$, where M is a 4-bit mantissa $m_3m_2m_1m_0$ and E is a 3-bit exponent $e_2e_1e_0$. The smallest integer in this format is $0 \cdot 2^0$ and the largest is $(2^4 - 1) \cdot 2^7$.

Given an 11-bit fixed-point integer B, we can convert it to our 7-bit floating-point notation by "picking off" four high-order bits beginning with the most significant 1, for example,

$$
\begin{aligned}
11010110100 &= 1101 \cdot 2^7 + 0110100 \\
00100101111 &= 1001 \cdot 2^5 + 01111 \\
00000111110 &= 1111 \cdot 2^2 + 10 \\
00000001011 &= 1011 \cdot 2^0 + 0 \\
00000000010 &= 0010 \cdot 2^0 + 0
\end{aligned}
$$

The last term in each equation is a truncation error that results from the loss of precision in the conversion. Corresponding to this conversion operation, we can write the specification for a fixed-point to floating-point encoder circuit:

- A combinational circuit is to convert an 11-bit unsigned binary integer B into a 7-bit floating point number M, E, where M and E have 4 and 3 bits, respectively. The numbers have the relationship $B = M \cdot 2^E + T$, where T is the truncation error, $0 \leq T < 2^E$.

Starting with a problem statement like the one above, it takes some creativity to come up with an efficient circuit design—the specification gives no clue. However, we can get some ideas by looking at how we converted numbers by hand earlier. We basically scanned each input number from left to right to find the first position containing a 1, stopping at the b_3 position if no 1 was found. We picked off four bits starting at that position to use as the mantissa, and the starting position number determined the exponent. These operations are beginning to sound like MSI functions.

"Scanning for the first 1" is what a generic priority encoder does. The output of the priority encoder is a number that tells us the position of the first 1. The position number determines the exponent; first-1 positions of b_{10}–b_3 imply exponents of 7–0, and positions of b_2–b_0 or no-1-found imply an exponent of 0. Therefore, we can scan for the first 1 with an 8-input priority encoder with inputs I7 (highest priority) through I0 connected to b_{10}–b_3. We can use the priority encoder's A2–A0 outputs directly as the exponent, as long as the no-1-found case produces A2–A0=000.

"Picking off four bits" sounds like a selecting or multiplexing operation. The 3-bit exponent determines which four bits of B we pick off, so we can use the exponent bits to control an 8-input, 4-bit multiplexer that selects the appropriate four bits of B to form M.

A circuit that results from these ideas is shown in Figure 4–112. It contains several optimizations:

- Since the available MSI priority encoder, the 74LS148, has active-low inputs, the input number B is assumed to be available on an active-low bus /B[0–10]. If only an active-high version of B is available, then eight inverters can be used to obtain the active-low version.

- If you think about the conversion operation a while, you'll realize that the most significant bit of the mantissa, m_3, is always 1, except in the no-1-found case. The '148 has a /GS output that indicates this case, allowing us to eliminate the multiplexer for m_3.

- The '148 has active-low outputs, so the exponent bits (/E0–/E2) are produced in active-low form. Naturally, three inverters could be used to produce an active-high version.

- Since everything else is active-low, active-low mantissa bits are used too. Active-high bits are also readily available on the '148 /EO and the '151 /Y outputs.

Strictly speaking, the multiplexers in Figure 4–112 are drawn incorrectly. The 74LS151 has an alternate symbol shown in Figure 4–113. In words, if the multiplexer's data inputs are active low, then the data outputs have an active level opposite that shown in the original symbol. The "active-low-data" symbol should be preferred in Figure 4–112, since the active levels of the '151 inputs

Figure 4–112
A combinational fixed-point to
floating-point encoder.

Figure 4–113
Alternate logic symbol for the
74LS151 8-input multiplexer.

and outputs would then match their signal names. However, in data transfer and storage applications, designers (and the book) don't always "go by the book." It is usually clear from the context that a multiplexer (or a multibit register, in Section 6.2.3) does not alter the active level of its data.

REFERENCES

Digital designers who want to write better should start by reading the classic *Elements of Style* by William Strunk, Jr. and E. B. White (Macmillan, 1979, 3rd ed.). Another book on writing style, especially directed at nerds, is *Effective Writing for Engineers, Managers, and Scientists* by H. J. Tichy (Wiley, 1988, 2nd ed.).

The ANSI/IEEE standard for logic symbols, Std 91-1984, *IEEE Standard Graphic Symbols for Logic Functions*, is available from the IEEE (New York, NY 10017). Another standard of interest to logic designers is ANSI/IEEE 991-1986, *Logic Circuit Diagrams*. These two standards and others, including standard symbols for 10-inch gongs and maid's-signal plugs, can be found in one handy, five-pound reference published by the IEEE, *Electrical and Electronics Graphics Symbols and Reference Designations* (Wiley, 1987, 2nd ed.).

Texas Instruments publishes a series of data books, listed in Table 4–20, that are standard references for digital design with SSI and MSI parts. Other manufacturers that publish fairly comprehensive data books include Motorola, National Semiconductor, and Fairchild Semiconductor. New editions of these books are published every few years. They contain revisions to the specifications for existing parts, along with updated part numbers (e.g., 74LS125A vs. 74LS125), as well as brand new functions in each logic family.

Table 4–20 Texas Instruments logic data books.

Order number	Topics	Title	Year
SDLD001A	74, 74S, 74LS TTL	*TTL Logic Data Book*	1988
SDAD001B	74AS, 74ALS TTL	*ALS/AS Logic Data Book*	1988
SDFD001	74F TTL	*F Logic Data Book*	1987
SCLD001B	74HC, 74HCT CMOS	*High-Speed CMOS Logic Data Book*	1988
SCAD001A	74AC, 74ACT CMOS	*Advanced CMOS Logic Data Book*	1987

On the technical side of digital design, lots of textbooks cover digital design principles, but only a few cover practical aspects of design. A classic is *Digital Design with Standard MSI and LSI* by Thomas R. Blakeslee (Wiley, 1979, 2nd ed.), which includes chapters on everything from structured combinational design with MSI and LSI to "the social consequences of engineering." A more recent, excellent short book focusing on digital design practices is *The Well-Tempered Digital Design*, by Robert B. Seidensticker (Addison-Wesley, 1986). It contains hundreds of readily accessible digital-design "proverbs" in areas ranging from high-level design philosophy to manufacturability.

E X E R C I S E S

4.1 Give three examples of combinational logic circuits that require *billions and billions* of rows to describe in a truth table. For each circuit, describe the circuit's inputs and output(s), and indicate exactly how many rows the truth table contains; you need not write out the truth table. (*Hint:* You can find several such circuits right in this chapter.)

4.2 Determine the exact maximum propagation delay from IN to OUT of the circuit in Figure X4.2 for both LOW-to-HIGH and HIGH-to-LOW transitions, using the timing information given in Table 4–2. Repeat, using a single worst-case delay number for each gate and compare and comment on your results.

Figure X4.2

4.3 Repeat Exercise 4.2, substituting 74LS08s for the 74S00s.

4.4 Determine the exact maximum propagation delay from IN to OUT of the circuit in Figure X4.4 for both LOW-to-HIGH and HIGH-to-LOW transitions, using the timing information given in Table 4–2. Repeat, using a single worst-case delay number for each gate and compare and comment on your results.

Figure X4.4

4.5 A possible definition of a BUT gate (Exercise 3.27) is "Y1 is 1 if A1 and B1 are 1 *but* either A2 or B2 is 0; Y2 is defined symmetrically." Write the truth table and find minimal sum-of-products expressions for the BUT-gate outputs. Draw the logic diagram for a NAND-NAND circuit for the expressions, assuming that only uncomplemented inputs are available. You may use 74LS00, '04, '10, '20, and '30 gates.

4.6 Find a minimum-SSI-package design for the BUT gate defined in Exercise 4.5. You are restricted to using 74LS00, '02, '04, '10, '20, and '30 gates and no more than two packages total. Write the output expressions (which need not be two-level sums-of-products), and draw the logic diagram.

4.7 For each circuit in the two preceding exercises, compute the worst-case delay from input to output, using the delay numbers in Table 4–2. Compare the cost (package count), speed, and input loading of the two designs. Which is better?

butification 4.8 Butify the function $F = \Sigma_{W,X,Y,Z}(3,7,11,12,13,14)$. That is, show how to perform F with a single BUT gate as defined in Exercise 4.5 and a single 2-input OR gate.

4.9 Suppose that a 74LS138 decoder is connected so that all enable inputs are asserted and C B A = 101. Using the information in Table 4–3 and the '138 internal logic diagram, determine the propagation delay from input to all relevant outputs for each possible single-input change. (*Hint:* There are a total of nine delay numbers, since a change on A, B, or C affects two outputs, and a change on any of the three enable inputs affects one output.)

4.10 Suppose that you are asked to design a new MSI part, a decimal decoder that is optimized for applications in which only decimal input combinations are expected to occur. How can the cost of such a decoder be minimized compared to one that is simply a 4-to-16 decoder with six outputs removed? Write the logic equations for all ten outputs of the minimized decoder, assuming active-high inputs and outputs and no enable inputs.

4.11 How many Karnaugh maps would be required to work Exercise 4.10 using the formal multiple-output minimization procedure described in Section 3.3.8?

4.12 Suppose that a system requires a 5-to-32 binary decoder with a single active-low enable input, a design similar to Figure 4–31. With the EN1 input pulled HIGH, either the /EN2 or the /EN3 input in the figure could be used as the enable, with the other input grounded. Discuss the pros and cons of using /EN2 versus /EN3.

4.13 Using the information in Table 4–3, determine the maximum propagation delay from any input to any output in the 5-to-32 decoder circuit of Figure 4–31. You may use the "worst-case" analysis method.

4.14 Repeat Exercise 4.13, performing a detailed analysis for each transition direction, and compare your results.

4.15 Show how to build each of the following single- or multiple-output logic functions using one or more 74LS138 or 74LS139 binary decoders and NAND gates:

(a) $F = \Sigma_{X,Y,Z}(2,4,7)$ (b) $F = \Pi_{A,B,C}(3,4,5,6,7)$

(c) $F = \Sigma_{A,B,C,D}(2,4,6,14)$ (d) $F = \Sigma_{W,X,Y,Z}(0,1,2,3,5,7,11,13)$

(e) $F = \Sigma_{W,X,Y}(1,3,5,6)$ (f) $F = \Sigma_{A,B,C}(0,4,6)$
       $G = \Sigma_{W,X,Y}(2,3,4,7)$ $G = \Sigma_{C,D,E}(1,2)$

4.16 A logic designer built a circuit to perform the function shown in Figure 4–32, only to find out during debugging that a different function was required—minterm 6 should have been included instead of minterm 2. Compare the designer's difficulty in changing the circuit depending on whether Figure 4–32(b) or (c) was built.

4.17 Determine whether the circuits driving the a, b, and c outputs of the 74LS49 seven-segment decoder correspond to minimal product-of-sums expressions for these segments, assuming that the nondecimal input combinations are "don't cares" and /BI = 1.

4.18 Draw the digits created by a 74LS49 seven-segment decoder for the nondecimal inputs 1010 through 1111.

4.19 Redesign the 74LS49 seven-segment decoder so that the digits 6 and 9 have tails as shown in Figure X4.19. Are any of the digit patterns for nondecimal inputs 1010 through 1111 affected by your redesign?

Figure X4.19

4.20 Draw and explain the IEEE standard symbol for a 74LS49 seven-segment decoder.

4.21 A famous logic designer decided to quit teaching and make a fortune by fabricating huge quantities of the MSI circuit shown in Figure X4.21.

(a) Label the inputs and outputs of the circuit with appropriate signal names, including active-level indications.

(b) What does the circuit do? Be specific and account for all inputs and outputs.

(c) Draw the MSI logic symbol that would go on the data sheet of this wonderful device.

(d) With what standard MSI parts does the new part compete? Do you think it would be successful in the marketplace?

Figure X4.21

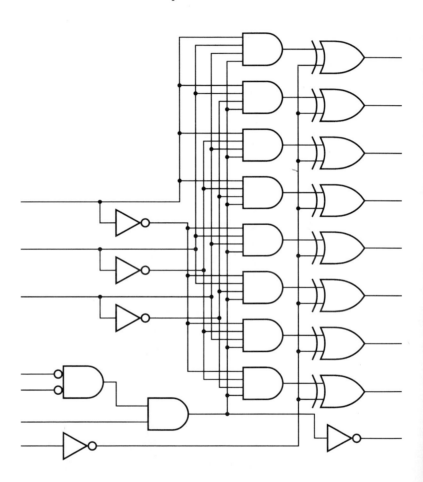

4.22 Draw the IEEE standard symbol for the device in Exercise 4.21. (You may need to use some of the standard's notation that is introduced in Chapter 6.)

4.23 What's terribly wrong with the circuit in Figure X4.23? Suggest a change that eliminates the terrible problem.

4.24 Design a 10-to-4 encoder with inputs in the 1-out-of-10 code and outputs in BCD.

4.25 Draw the logic diagram for a 16-to-4 encoder using just four 8-input NAND gates. What are the active levels of the inputs and outputs in your design?

Figure X4.23

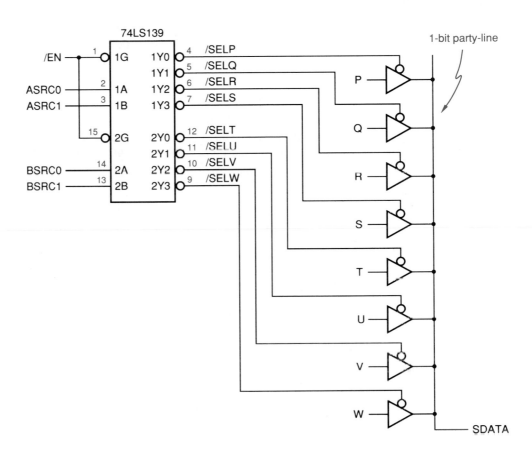

4.26 Starting with the logic diagram for the 74LS148 priority encoder, write logic equations for its /A2, /A1, and /A0 outputs. How do they differ from the "generic" equations given in Section 4.5.1?

4.27 Draw the logic diagram for a circuit that uses the 74LS148 to resolve priority among eight active-high inputs, I0–I7, where I7 has the highest priority. The circuit should produce active-high address outputs A2–A0 to indicate the number of the highest-priority asserted input. If no input is asserted, then A2–A0 should be 111 and an IDLE output should be asserted. You may use discrete SSI gates in addition to the '148. Be sure to name all signals with the proper active levels.

4.28 Draw the logic diagram for a circuit that resolves priority among eight active-low inputs, /I0–/I7, where /I0 has the highest priority. The circuit should produce active-high address outputs A2–A0 to indicate the number of the highest-priority asserted input. If at least one input is asserted, then an AVALID output should be asserted. Be sure to name all signals with the proper active levels. This circuit can be built with a single 74LS148 and no other gates.

4.29 A purpose of Exercise 4.28 was to demonstrate that it is not always possible to maintain consistency in active-level notation unless you are willing to define alternate logic symbols for MSI parts that can be used in different ways. Define an alternate symbol for the 74LS148 that provides this consistency in Exercise 4.28.

4.30 Design a combinational circuit with eight active-low request inputs, /R0–/R7, and eight outputs, A2–A0, AVALID, B2–B0, and BVALID. The /R0–/R7 inputs and A2–A0 and AVALID outputs are defined as in Exercise 4.28. The B2–B0 and BVALID outputs identify the second-highest priority request input that is asserted. You should be able to design this circuit with no more than six SSI and MSI packages, but don't use more than 10 in any case.

4.31 Design a 3-input, 5-bit multiplexer that fits in a 24-pin IC package. Write the truth table and draw a logic diagram and logic symbol for your multiplexer.

4.32 Using the information in Tables 4–2 and 4–3, determine the maximum propagation delay from any input to any output in the 32-to-1 multiplexer circuit of Figure 4–64. You may use the "worst-case" analysis method.

4.33 Write a truth table for the 74LS153 4-input, 2-bit multiplexer.

4.34 Show how, using external connections, the 74LS155 2-bit, 4-output demultiplexer can be made to function as a 1-bit, 8-output demultiplexer.

4.35 Draw the IEEE standard symbol for the 74LS155 used as a dual 2-to-4 decoder.

4.36 Draw the IEEE standard symbol for a 74LS251 multiplexer.

4.37 Write the truth table and a logic diagram for the logic function performed by the CMOS circuit in Figure X4.37. (The circuit contains transmission gates, which were introduced in Exercise 2.39.)

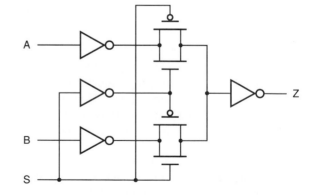

Figure X4.37

4.38 Design a circuit for the BUT gate defined in Exercise 4.5, using a minimum number of MSI packages. You may use any of the 74LS MSI parts described in this chapter, and 74LS04 inverters if necessary. Draw the logic diagram and compare the package count of your circuit with the solutions to Exercises 4.5 and 4.6.

4.39 A famous logic designer decided to quit teaching and make a fortune by fabricating huge quantities of the MSI circuit shown in Figure X4.39.

   (a) Label the inputs and outputs of the circuit with appropriate signal names, including active-level indications.

   (b) What does the circuit do? Be specific and account for all inputs and outputs.

   (c) Draw the MSI logic symbol that would go on the data sheet of this wonderful device.

   (d) With what standard MSI parts does the new part compete? Do you think it would be successful in the marketplace?

Figure X4.39

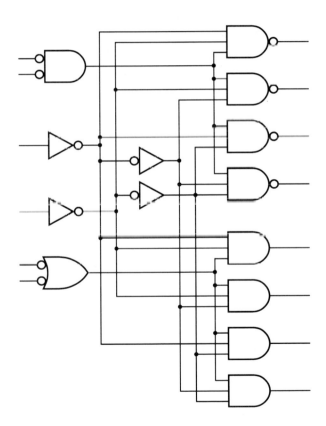

4.40 Draw the IEEE standard symbol for the circuit in Exercise 4.39.

4.41 A 16-bit *barrel shifter* is a combinational logic circuit with 16 data inputs, 16 data outputs, and 4 control inputs. The output word equals the input word, rotated by a number of bit positions specified by the control inputs. For example, if the input   *barrel shifter* word equals ABCDEFGHIJKLMNOP (each letter represents one bit), and the control inputs are 0101 (5), then the output word is FGHIJKLMNOPABCDE. Design a 16-bit barrel shifter using combinational MSI parts discussed in this chapter. Your design should contain 20 or fewer ICs. Do not draw a complete schematic, but sketch and describe your design in general terms and indicate the types and total number of ICs required.

4.42 What logic function is performed by the CMOS circuit shown in Figure X4.42?

Figure X4.42

4.43 Write the truth table and a logic diagram for the logic function performed by the CMOS circuit in Figure X4.43.

4.44 An odd-parity circuit with 2^n inputs can be built with $2^n - 1$ XOR gates. Describe two different structures for this circuit, one of which gives a minimum worst-case input to output propagation delay and the other of which gives a maximum. For each structure, state the worst-case number of XOR-gate delays, and describe a situation where that structure might be preferred over the other.

4.45 A digital designer who built the circuit in Figure 4–80 accidentally used 74LS00s instead of '08s in the circuit, and found that the circuit still worked, except for a change in the active level of the ERROR signal. How was this possible?

4.46 Using the information in Tables 4–2 and 4–3, determine the maximum propagation delay from the DU bus to the DC bus in the error-correction circuit of Figure 4–81. You may use the "worst-case" analysis method.

4.47 Using an n-bit comparator structure similar to Figure 4–85(b), what is the maximum value of n that can be obtained, given the electrical limitations of the 74LS266? Assume that the /DIFF output drives a single 74LS04 input and that normal 74LS TTL noise margins are to be maintained.

4.48 Write a 4-step iterative algorithm corresponding to the iterative comparator circuit of Figure 4–87.

Figure X4.43

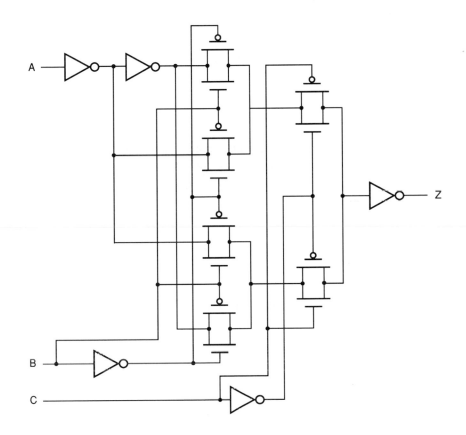

4.49 Design a 16-bit comparator using five 74LS85s in a tree-like structure, such that the maximum delay for a comparison equals twice the delay of one 74LS85.

4.50 Starting with the equations given in Section 4.8.4, write a complete logic expression for the ALTBOUT output of the 74LS85.

4.51 Starting with a manufacturer's logic diagram for the 74LS85, write a logic expression for the ALTBOUT output, and prove that it algebraically equals the expression derived in Exercise 4.50.

4.52 Design a comparator similar to the 74LS85 that uses the opposite cascading order. That is, to perform a 12-bit comparison, the cascading outputs of the high-order comparator would drive the cascading inputs of the mid-order comparator, and the mid-order outputs would drive the low-order inputs. You needn't do a complete logic design and schematic; a truth table and an application note showing the interconnection for a 12-bit comparison are sufficient.

4.53 Starting with the logic diagram for the 74LS682, write a logic expression for the /PGTQ output in terms of the inputs.

4.54 Design a 24-bit comparator using three 74LS682s and additional SSI/MSI components as required. Your circuit should compare three 24-bit unsigned numbers P and Q and produce two output bits that indicate whether P = Q or P > Q.

4.55 Write an algebraic expression for s_2, the third sum bit of a binary adder, as a function of inputs x_0, x_1, x_2, y_0, y_1, and y_2. Assume that $c_0 = 0$, and do not attempt to "multiply out" or minimize the expression.

4.56 Draw a 6-variable Karnaugh map for the s_2 function of Exercise 4.55, and find all of its prime implicants. Using the 6-variable map format of Exercise 3.54, label the variables in the order x_0, y_0, x_2, y_2, x_1, y_1 instead of U, V, W, X, Y, Z. You need not write out the algebraic product corresponding to each prime implicant; simply identify each one with a number (1, 2, 3, . . .) on the map. Then make a list that shows for each prime implicant whether or not it is essential and how many inputs are needed on the corresponding AND gate.

4.57 Starting with the logic diagram for the 74LS283 in Figure 4–98, write a logic expression for the S2 output in terms of the inputs, and prove that it does indeed equal the third sum bit in a binary addition as advertised. You may assume that $c_0 = 0$ (i.e., ignore c_0).

4.58 Using the information in Table 4–3, determine the maximum propagation delay from any input to any output of the 16-bit group ripple adder of Figure 4–99. You may use the "worst-case" analysis method.

4.59 Using the information in Table 4–3, determine the maximum propagation delay from any A or B bus input to any F bus output of the 16-bit carry lookahead adder of Figure 4–103. You may use the "worst-case" analysis method.

4.60 Estimate the number of product terms in a minimal sum-of-products expression for the c_{32} output of a 32-bit binary adder. Be more specific than "billions and billions," and justify your answer.

4.61 Referring to the data sheet of a 74S182 carry lookahead circuit, determine whether or not its outputs match the equations given in Section 4.9.7.

4.62 Draw the logic diagram for a 64-bit ALU using sixteen 74LS181s and five 74S182s for full carry lookahead (two levels of '182s). For the '181s, you need only show the CIN inputs and /G and /P outputs.

4.63 Sketch the structure of a 16×16 multiplication using addition and 4×4 multiplication in the format of Figure 4–106. How many 4-bit adders and 4×4 multipliers are needed to build a 16×16 combinational multiplier?

4.64 Calculate the true worst-case delay of the 8×8 multiplier in Figure 4–107 using the delay numbers given in the text. Describe the signal path over which the worst-case delay occurs.

4.65 Show how to build all four of the following functions using no more than one SSI package and one of the following MSI parts: 74LS138, 74LS139, 74LS151, 74LS153, 74LS157.

$$F1 = X' \cdot Y' \cdot Z' + X \cdot Y \cdot Z \qquad F2 = X' \cdot Y' \cdot Z + X \cdot Y \cdot Z'$$
$$F3 = X' \cdot Y \cdot Z' + X \cdot Y' \cdot Z \qquad F4 = X \cdot Y' \cdot Z' + X' \cdot Y + Z$$

4.66 Design an almost-binary decoder with the function table in Table X4.66. Minimize the number of MSI and SSI IC packages in your design.

Table X4.66

/CS	A2	A1	A0	Output to assert
1	x	x	x	none
0	0	0	x	/BILL
0	0	x	0	/MARY
0	0	1	x	/JOAN
0	0	x	1	/PAUL
0	1	0	x	/ANNA
0	1	x	0	/FRED
0	1	1	x	/DAVE
0	1	x	1	/KATE

4.67 Design a 3-input, 2-output combinational circuit that performs the code conversion specified by Table 4–19, using a single MSI IC.

4.68 Design an almost binary-select multiplexer with four 4-bit input buses P, Q, R, and S, selecting one of the buses to drive a 4-bit output bus T according to Table X4.68. Use the basic structure of Figure 4–109(b), but select a new ordering for data buses on the '153 data inputs that minimizes the size and propagation delay of the code converter.

Table X4.68

S2	S1	S0	Input to select
0	0	0	P
0	0	1	P
0	1	0	P
0	1	1	Q
1	0	0	P
1	0	1	P
1	1	0	R
1	1	1	S

4.69 Design an almost binary-select multiplexer with five 4-bit input buses A, B, C, D, and E, selecting one of the buses to drive a 4-bit output bus T according to Table X4.69. You may use no more than three ICs.

Table X4.69

S2	S1	S0	Input to select
0	0	0	A
0	0	1	B
0	1	0	A
0	1	1	C
1	0	0	A
1	0	1	D
1	1	0	A
1	1	1	E

4.70 Redesign the mode-dependent comparator of Figure 4–110 to minimize the size and delay of the SSI mode-control logic. You may use a different 2-bit code for mode control.

4.71 Replace the SSI mode-control logic of Figure 4–110 with two units of an MSI chip from this chapter, and show that the resulting design can be used with *any* mode-control encoding.

4.72 Repeat Exercise 4.71 using a single MSI chip. *Hint:* Think!

4.73 Design a combinational circuit whose inputs are two 8-bit unsigned binary integers, X and Y, and a control signal MIN/MAX. The output of the circuit is an 8-bit unsigned binary integer Z such that Z = 0 if X = Y; otherwise, Z = min(X,Y) if MIN/MAX = 1, and Z = max(X,Y) if MIN/MAX = 0.

4.74 Design a combinational circuit whose inputs are two 8-bit unsigned binary integers, X and Y, and whose output is an 8-bit unsigned binary integer Z = max(X,Y). Oh, by the way, a U.S. law limiting the import of foreign-made ICs has just raised the price of MSI comparators to over $10, but all of the other SSI and MSI parts in this chapter are still available for about $1. Do a minimum-cost design, of course.

4.75 Repeat Exercise 4.74, assuming that tinkering by Congress has also raised the price of MSI multiplexers to over $10.

4.76 Most MSI parts are designed so that each of their inputs presents only one unit-load to the circuit that is driving it. However, the internal logic diagrams of some MSI parts, including the 74LS148, 74LS157, and 74LS283, show multiple loads on certain inputs. Referring to the electrical specifications of these parts in a TTL data book, list the inputs, if any, that present multiple unit loads, and determine whether or not the load specifications match the requirements implied by the internal logic diagrams.

5

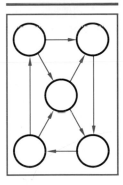

SEQUENTIAL LOGIC DESIGN PRINCIPLES

Recall that logic circuits are classified into two types, "combinational" and "sequential." A *combinational* logic circuit is one whose outputs depend only on its current inputs. The rotary channel selector knob on an inexpensive TV is like a combinational circuit—its "output" selects a channel based only on the current position of the knob ("input").

combinational circuit

A *sequential* logic circuit is one whose outputs depend not only on its current inputs, but also on the past sequence of inputs, possibly arbitrarily far back in time. The circuit controlled by the channel-up and channel-down pushbuttons on a fancy TV or VCR is a sequential circuit—the channel selection depends on the past sequence of up/down pushes, at least since when you started viewing 10 hours before, and perhaps as far back as when you first powered-up the device.

sequential circuit

So it is inconvenient, often impossible, to describe the behavior of a sequential circuit by means of a table that lists outputs as a function of the input sequence that has been received up until the current time. With the fancy TV channel selector, it is impossible to determine what channel is currently selected by looking only at the preceding sequence of presses on the up and down pushbuttons, whether we look at the preceding 10 presses or the preceding 1,000. More information, the current "state" of the channel selector, is needed.

state
state variable

The *state* of a sequential circuit is a collection of *state variables* whose values at any one time contain all the information about the past necessary to account for the circuit's future behavior.[1]

In the channel-selector example, the current channel number is the current state. Inside the TV, this state might be stored as seven binary state variables representing a decimal number between 0 and 127. Given the current state (channel number), we can always predict the next state as a function of the inputs (presses of the up/down pushbuttons). In this example, one highly visible output of the sequential circuit is an encoding of the state itself—the channel-number display. Other outputs, internal to the TV, may be combinational functions of the state alone (e.g., VHF/UHF tuner selection) or of both state and input (e.g., turning off the TV if the current state is 0 and the "down" button is pressed).

State variables need not have direct physical significance, and there are often many ways to choose them to describe a given sequential circuit. For example, in the TV channel selector, the state might be stored as three BCD digits or 12 bits, with many of the bit combinations (4096 possible) being unused.

In a digital logic circuit, state variables are binary values, corresponding to certain logic signals in the circuit, as we'll see in later sections. A circuit with n binary state variables has 2^n possible states. As large as it might be, 2^n is always finite, never infinite, so sequential circuits are sometimes called

finite-state machine

finite-state machines.[2]

clock

The state changes of most sequential circuits occur at times specified by a free-running *clock* signal. Figure 5–1 gives timing diagrams and nomenclature for typical clock signals. By convention, a clock signal is active high if state changes occur at the clock's rising edge or when the clock is HIGH, and active

clock period
clock frequency

low in the complementary case. The *clock period* is the time between successive transitions in the same direction, and the *clock frequency* is the reciprocal of the period. The first edge or pulse in a clock period or sometimes the period itself

clock tick
duty cycle

is called a *clock tick*. The *duty cycle* is the percentage of time that the clock signal is at its asserted level. Typical digital systems, from digital watches to supercomputers, use a quartz-crystal oscillator to generate a free-running clock signal. Clock frequencies range from 32.768 kHz (for a watch) to 400 MHz (for

---

[1] Herbert Hellerman, *Digital Computer Systems Principles*, McGraw-Hill, 1967, p. 237.

[2] A group of mathematicians recently proposed a non-finite-state machine, but they're still working on the state table.... Sorry, that's just a joke. There *are* mathematical models for infinite-state machines, such as Turing machines. They typically contain a small finite-state-machine control unit, and an infinite amount of auxiliary memory, such as an endless tape.

Figure 5–1
Clock signals: (a) active high;
(b) active low.

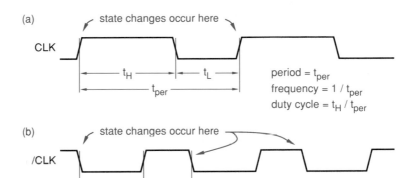

(a)

CLK

state changes occur here

t_H

t_L

t_{per}

period = t_{per}
frequency = $1 / t_{per}$
duty cycle = t_H / t_{per}

(b)

/CLK

state changes occur here

t_L

t_H

t_{per}

duty cycle = t_L / t_{per}

clock duty cycle

$$\frac{t_H}{t\,period}$$

an ECL supercomputer with a cycle time of 2.5 ns); "typical" systems using TTL and CMOS parts have clock frequencies in the 1–25 MHz range.

In this chapter we'll discuss two types of sequential circuits that account for the majority of practical discrete designs. A *feedback sequential circuit* uses ordinary gates and feedback loops to obtain memory in a logic circuit, thereby creating sequential-circuit building blocks such as flip-flops and latches that are used in higher-level designs. A *clocked synchronous state machine* uses these building blocks, in particular edge-triggered D flip-flops, to create circuits whose inputs are examined and whose outputs change with respect to a controlling clock signal. Other sequential circuit types, such as general fundamental mode, multiple pulse mode, and multiphase, are sometimes useful in high-performance systems and VLSI, and are discussed in advanced texts.

feedback sequential circuit

clocked synchronous state machine

5.1 BISTABLE ELEMENTS

The simplest sequential circuit consists of a pair of inverters forming a feedback loop, as shown in Figure 5–2. It has *no* inputs and two outputs, Q and /Q.

5.1.1 Digital Analysis

The circuit of Figure 5–2 is often called a *bistable*, since a strictly digital analysis shows that it has two stable states. If Q is HIGH, then the bottom inverter has

bistable

Figure 5–2
A pair of inverters forming a
bistable element.

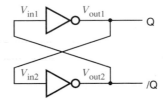

V_{in1} V_{out1} Q

V_{in2} V_{out2} /Q

a HIGH input and a LOW output, which forces the top inverter's output HIGH as we assumed in the first place. But if Q is LOW, then the bottom inverter has a LOW input and a HIGH output, which forces Q LOW, another stable situation. We could use a single state variable, the state of signal Q, to describe the state of the circuit; there are two possible states, Q = 0 and Q = 1.

The bistable element is so simple that it has no inputs and therefore no way of controlling or changing its state. When power is first applied to the circuit, it randomly comes up in one state or the other and stays there forever. Nevertheless, it serves our illustrative purposes very well, and we *will* actually conjure up an application for it in Section 6.2.2.

5.1.2 Analog Analysis

The analysis of the bistable has more to reveal if we consider its operation from an analog point of view. The dark line in Figure 5–3 shows the steady-state (DC) transfer function T for a single inverter; the output voltage is a function of input voltage, $V_{out} = T(V_{in})$. With two inverters connected in a feedback loop as in Figure 5–2, we know that $V_{in1} = V_{out2}$ and $V_{in2} = V_{out1}$; therefore we can plot the transfer functions for both inverters on the same graph with an appropriate labeling of the axes. Thus the dark line is the transfer function for the top inverter in Figure 5–2, and the colored line is the transfer function for the bottom one.

Considering only the steady-state behavior of the bistable's feedback loop, and not dynamic effects, the loop is in equilibrium if the input and output voltages of both inverters are constant DC values consistent with the loop connection and the inverters' DC transfer function. That is, we must have

$$
\begin{aligned}
V_{in1} &= V_{out2} \\
&= T(V_{in2}) \\
&= T(V_{out1}) \\
&= T(T(V_{in1}))
\end{aligned}
$$

Figure 5–3
Transfer functions for inverters in a bistable feedback loop.

Likewise, we must have

$$V_{in2} = T(T(V_{in2}))$$

We can find these equilibrium points graphically from Figure 5–3; they are the points at which the two transfer curves meet. Surprisingly, we find that there are not two but *three* equilibrium points. Two of them, labeled *stable*, correspond to the two states that our "strictly digital" analysis identified earlier, with Q either 0 (LOW) or 1 (HIGH).

stable

    The third equilibrium point, labeled *metastable*, occurs with V_{out1} and V_{out2} about halfway between a valid logic 1 voltage and a valid logic 0 voltage; so Q and /Q are not valid logic signals at this point. Yet the loop equations arc satisfied; if we can get the circuit to operate at the metastable point, it could theoretically stay there indefinitely.

metastable

5.1.3 Metastable Behavior

Closer analysis of the situation at the metastable point shows that it is aptly named. It is not truly stable because random noise will tend to drive a circuit that is operating at the metastable point towards one of the stable operating points as we'll now demonstrate. Suppose the bistable is operating precisely at the metastable point in Figure 5–3. Now let us assume that a small amount of circuit noise reduces V_{in1} by a tiny amount. This tiny change causes V_{out1} to *increase* by a small amount. But since V_{out1} produces V_{in2}, we can follow the first horizontal arrow from near the metastable point to the second transfer characteristic, which now demands a lower voltage for V_{out2}, which is V_{in1}. Now we're back where we started, except we have a much larger change in voltage at V_{in1} than the original noise produced, and the operating point is still changing. This "regenerative" process continues until we reach the stable operating point at the upper left-hand corner of Figure 5–3. However, if we perform a "noise" analysis for either of the stable operating points, we find that feedback brings the circuit back towards the stable operating point, rather than away from it.

    Metastable behavior of a bistable can be compared to the behavior of a ball dropped onto a hill, as shown in Figure 5–4. If we drop a ball from overhead, it will probably immediately roll down to one side of the hill or the other. But if it lands right at the top, it may precariously sit there for a while before random forces (wind, rodents, earthquakes) start it rolling down the hill. Like the ball at the top of the hill, the bistable may stay in the metastable state for an unpredictable length of time before nondeterministically settling into one stable state or the other.

    If the *simplest* sequential circuit is susceptible to metastable behavior, you can be sure that *all* sequential circuits are susceptible. And this behavior is not something that only occurs at power-up.

Figure 5–4
Ball and hill analogy for
metastable behavior.

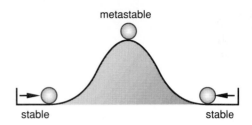

Returning to the ball-and-hill analogy, consider what happens if we try to kick the ball from one side of the hill to the other. Apply a strong force (Pélé), and the ball goes right over the top and lands in a stable resting place on the other side. Apply a weak force (Pee Wee Herman), and the ball falls back to its original starting place. But apply a wishy-washy force (Charlie Brown), and the ball goes to the top of the hill, teeters, and eventually falls back to one side or the other.

This behavior is completely analogous to what happens to flip-flops under marginal triggering conditions. For example, we'll soon study S-R flip-flops, where a pulse on the S input forces the flip-flop from the 0 state to the 1 state. A minimum pulse width is specified for the S input. Apply a pulse of this width or longer, and the flip-flop immediately goes to the 1 state. Apply a very short pulse, and the flip-flop stays in the 0 state. Apply a pulse just under the minimum width, and the flip-flop may go into the metastable state. Once the flip-flop is in the metastable state, its operation depends on "the shape of its hill." Flip-flops built from high-gain, fast technologies tend to come out of metastability faster than ones built from low-performance technologies.

We'll say more about metastability in the next section in connection with specific flip-flop types, and in Section 6.10 with respect to synchronous design methodology and synchronizer failure.

5.2 LATCHES AND FLIP-FLOPS

Latches and flip-flops are the basic building blocks of most sequential circuits. Typical digital systems use latches and flip-flops that are pre-packaged, functionally specified devices in a standard integrated circuit. In ASIC design environments, latches and flip-flops are typically predefined cells specified by the ASIC vendor. However, within a standard IC or an ASIC, each latch or flip-flop is typically designed as a feedback sequential circuit using individual logic gates and feedback loops. We'll study these discrete designs for two reasons—to better understand the behavior of the pre-packaged elements, and to gain the capability of building a latch or flip-flop "from scratch" as is required occasionally in digital-design practice and often in digital-design exams.

Figure 5–5
S-R latch: (a) circuit design using
NOR gates; (b) function table.

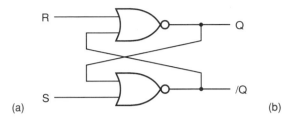

S	R	Q	/Q
0	0	last Q	last /Q
0	1	0	1
1	0	1	0
1	1	0	0

(a)   (b)

All digital designers use the name *flip-flop* for a sequential device that   *flip-flop*
normally samples its inputs and changes its outputs only at times determined by
a clocking signal. On the other hand, most digital designers use the name *latch*   *latch*
for a sequential device that watches all of its inputs continuously and changes its
outputs at any time, independent of a clocking signal. We follow this convention
in this text. However, some textbooks and digital designers may use the name
"flip-flop" for a device that we call a "latch."

In any case, because the functional behaviors of latches and flip-flops are
quite different, it is important for the logic designer to know which type is
being used in a design, either from the device's part number (e.g., 74LS374 vs.
74LS373) or other contextual information. We discuss the most commonly used
types of latches and flip-flops in the following subsections.

5.2.1 S-R Latch

An *S-R (set-reset) latch* based on NOR gates is shown in Figure 5–5(a). The   *S-R latch*
circuit has two inputs, S and R, and two outputs, Q and /Q, where /Q ("not Q")
is normally the complement of Q.

If S and R are both 0, the circuit behaves like the bistable element—we
have a feedback loop that retains one of two logic states, Q=0 or Q=1. As shown
in Figure 5–5(b), either S or R may be asserted to force the feedback loop to a
desired state. S *sets* or *presets* the Q output to 1; R *resets* or *clears* the Q output   *set*
to 0. After the S or R input is negated, the latch remains in the state that it   *preset*
was forced into. Figure 5–6(a) shows the functional behavior of an S-R latch for   *reset*
a typical sequence of inputs. Colored arrows indicate causality, that is, which   *clear*
input transitions cause which output transitions.

In all of the cases discussed so far, the /Q output is the complement of the
Q output. However, /Q is not quite correctly named, because there is one case
where /Q is *not* the complement of Q. If both S and R are 1, as they are in several   *Q vs. /Q*
places in Figure 5–6(b), then both outputs are forced to 0. Once we negate either
input, the outputs return to complementary operation as usual. However, if we
negate both inputs simultaneously, the latch goes to an unpredictable next state,
and it may in fact oscillate or enter the metastable state. Metastability may also
occur if a 1 pulse that is too short is applied to S or R.

Figure 5–6 Typical operation of an S-R latch: (a) "normal" inputs; (b) S and R asserted simultaneously.

Three different logic symbols for the same S-R latch circuit are shown in Figure 5–7. The symbols differ in the treatment of the active-low /Q output. Historically, the first symbol was used, showing the active-low or complemented signal name inside the function rectangle. However, in bubble-to-bubble logic design the second form of the symbol is preferred, showing an inversion bubble outside the function rectangle. The last form of the symbol is obviously wrong.

propagation delay

Figure 5–8 defines timing parameters for an S-R latch. The *propagation delay* is the time it takes for a transition on an input signal to produce a transition on an output signal. A given latch or flip-flop may have several different propagation delay specifications, one for each pair of input and output signals. Also, the propagation delay may be different depending on whether the output makes a LOW-to-HIGH or HIGH-to-LOW transition. With an S-R latch, a LOW-to-HIGH

Figure 5–7
Symbols for an S-R latch: (a) old
style; (b) preferred; (c) incorrect
because of double negation.

Figure 5–8 Timing parameters for an S-R latch.

transition on S can cause a LOW-to-HIGH transition on Q, so a propagation delay $t_{pLH(SQ)}$ occurs as shown in transition (1) in Figure 5–8. Similarly, a LOW-to-HIGH transition on R can cause a HIGH-to-LOW transition on Q, with propagation delay $t_{pHL(RQ)}$ as shown in transition (2). Not shown in the figure are the corresponding transitions on /Q, which would have propagation delays $t_{pHL(S/Q)}$ and $t_{pLH(R/Q)}$.

minimum pulse width

Minimum pulse width specifications are usually given for the S and R inputs. As shown in Figure 5–8, the latch may go into the metastable state and remain there for a random length of time if a pulse shorter than the minimum width $t_{pw(min)}$ is applied to S or R. The latch can be deterministically brought out of the metastable state only by applying a pulse to S or R that meets or exceeds the minimum pulse width requirement.

As we showed previously, the S-R latch may also go into the metastable state if S and R are negated simultaneously. Often, but not always, a commercial latch's specifications define "simultaneously" (e.g., S and R negated within 20 ns of each other). In any case, the minimum delay between negating S and R for them to be considered non-simultaneous is closely related to the mimimum pulse width specification. Both specifications are measures of how long it takes for the latch's feedback loop to stabilize during a change of state.

5.2.2   /S-/R Latch

A */S-/R latch* with active-low set and reset inputs may be built from NAND gates as shown in Figure 5–9(a). In TTL logic families, /S-/R latches are used much more often than S-R latches because NAND gates are preferred over NOR gates.

/S-/R latch

As shown by the function table, Figure 5–9(b), operation of the /S-/R latch is similar to that of the S-R, with two major differences. First, /S and /R are active-low, so the latch remembers its previous state when /S = /R = 1; the active-low inputs are clearly indicated in the symbols in Figure 5–9(c). Second, when both /S and /R are asserted simultaneously, both latch outputs go to 1, not 0 as in the S-R latch. Except for these differences, operation of the /S-/R is the same as the S-R, including timing and metastability considerations.

(a)

(b)

/S /R	Q	/Q
0 0	1	1
0 1	1	0
1 0	0	1
1 1	last Q	last /Q

(c)

Figure 5–9   /S-/R latch: (a) circuit design using NAND gates; (b) function table; (c) logic symbols.

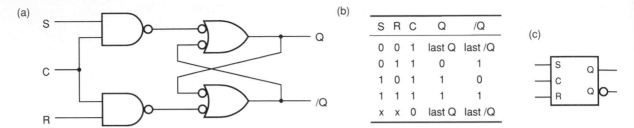

Figure 5–10 presents the circuit, with parts (a), (b), and (c):

(a)

(b)

S	R	C	Q	/Q
0	0	1	last Q	last /Q
0	1	1	0	1
1	0	1	1	0
1	1	1	1	1
x	x	0	last Q	last /Q

(c)

Figure 5–10   S-R latch with enable: (a) circuit using NAND gates; (b) function table; (c) logic symbol.

5.2.3   S-R Latch with Enable

S-R latch with enable

An S-R or /S-/R latch is sensitive to its S and R inputs at all times. However, it may easily be modified to create a device that is sensitive to these inputs only when an enabling input C is asserted. Such an *S-R latch with enable* is shown in Figure 5–10. As shown by the function table, the circuit behaves like an S-R latch when C is 1, and retains its previous state when C is 0. The latch's behavior for a typical set of inputs is shown in Figure 5–11. If both S and R are 1 when C changes from 1 to 0, the circuit behaves like an S-R flip-flop in which S and R are negated simultaneously—the next state is unpredictable and the output may become metastable.

5.2.4   D Latch $Q^* = D$

D latch

S-R latches are useful in control applications, where we often think in terms of setting a flip-flop in response to some condition, and resetting it when conditions change; so we control the set and reset inputs somewhat independently. However, we often need latches simply to store bits of information—each bit of information is presented on a signal line, and we'd like to store it somewhere. A *D latch* may be used in such an application.

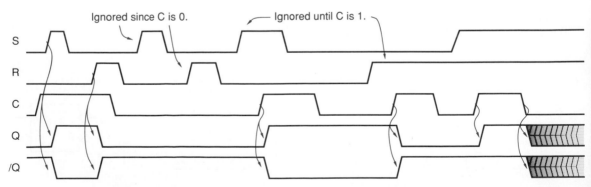

Figure 5–11   Typical operation of an S-R latch with enable.

Figure 5–12   D latch: (a) circuit design using NAND gates; (b) function table; (c) logic symbol.

Figure 5–12 shows a D latch. Its logic diagram is recognizable as that of an S-R latch with enable, with an inverter added to generate S and R inputs from the single D (data) input. This eliminates the troublesome situation in S-R latches, where S and R may be asserted simultaneously. The control input of a D latch, labeled C in Figure 5 12(c), is sometimes named ENABLE, CLK, or G, and is active low in some D-latch designs.

An example of a D latch's functional behavior is given in Figure 5–13. When the C input is asserted, the Q output follows the D input. In this situation, the latch is said to be "open" and the path from D input to Q output is "transparent"; the circuit is often called a *transparent latch* for this reason. When the C input is negated, the latch "closes"; the Q output retains its last value and no longer changes in response to D, as long as C remains negated.

transparent latch

More detailed timing behavior of the D latch is shown in Figure 5–14. Four different delay parameters are shown for signals that propagate from the C or D input to the Q output. For example, at transitions (1) and (4) the latch is initially "closed" and the D input is the opposite of Q output, so that when C goes HIGH the latch "opens up" and the Q output changes after delay $t_{pLH(CQ)}$ or $t_{pHL(CQ)}$. At transitions (2) and (3) the C input is already HIGH and the latch is already open, so that Q transparently follows the transitions on D with delay $t_{pHL(DQ)}$ and $t_{pLH(DQ)}$. Another four parameters exist for signals propagating to the /Q output, not shown.

Although the D latch eliminates the S = R = 1 problem of the S-R latch, it does not eliminate the metastability problem. As shown in Figure 5–14, there is

Figure 5–13   Functional behavior of a D latch for various inputs.

Figure 5–14   Timing parameters for a D latch.

a (shaded) window of time around the falling edge of C when the D input must not change. This window begins at time t_{setup} before the falling (latching) edge of C; t_{setup} is called the *setup time*. The window ends at time t_{hold} afterwards; t_{hold} is called the *hold time*. If D changes at any time during the setup- and hold-time window, the output of the latch is unpredictable and may become metastable, as shown for the last latching edge in the figure.

5.2.5   Edge-Triggered D Flip-Flop $Q^* = D$

positive-edge-
triggered
D flip-flop

master

slave

A *positive-edge-triggered D flip-flop* combines a pair of D latches, as shown in Figure 5–15, to create a circuit that samples its D input and changes its Q and /Q outputs only at the rising edge of a controlling CLK signal. The first latch is called the *master*; it is open and follows the input when CLK is LOW. When CLK goes HIGH, the master latch is closed and its output is transferred to the second latch, called the *slave*. The slave latch is open all the while that CLK is HIGH but changes only at the beginning of this interval, because the master is closed and unchanging during the rest of the interval.

dynamic-input
indicator

The triangle on the D flip-flop's CLK input indicates edge-triggered behavior, and is called a *dynamic-input indicator*. Examples of the flip-flop's functional

Figure 5–15   Positive-edge-triggered D flip-flop: (a) circuit design using D latches; (b) function table; (c) logic symbol.

Figure 5–16 Functional behavior of a positive-edge-triggered D flip-flop.

behavior for several input transitions are presented in Figure 5–16. The QM signal shown is the output of the master latch. Notice that QM changes only when CLK is LOW. When CLK goes HIGH the current value of QM is transferred to Q, and QM is prevented from changing until CLK goes LOW again.

Figure 5–17 shows more detailed timing behavior for the D flip-flop. All propagation delays are measured from the rising edge of CLK, since that's the only event that causes an output change. Different delays may be specified for LOW-to-HIGH and HIGH-to-LOW output changes.

Like a D latch, the edge-triggered D flip-flop has a setup and hold time window during which the D inputs must not change. This window occurs around the triggering edge of CLK, and is indicated by shaded color in Figure 5–17. If the setup and hold times are not met, the flip-flop output will usually go to a stable, though unpredictable, 0 or 1 state. In some cases, however, the output will oscillate or go to a metastable state halfway between 0 and 1, as shown at the second-to-last clock tick in the figure. If the flip-flop goes into the metastable state, it will return to a stable state on its own only after a probabilistic delay, as explained in Section 6.10. It can also be forced into a stable state by applying

Figure 5–17 Timing behavior of a positive-edge-triggered D flip-flop.

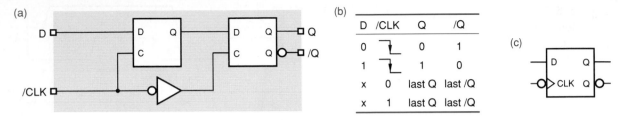

D	/CLK	Q	/Q
0	⌐‾L	0	1
1	⌐‾L	1	0
x	0	last Q	last /Q
x	1	last Q	last /Q

Figure 5–18 Negative-edge triggered D flip-flop: (a) circuit design using D latches; (b) function table; (c) logic symbol.

another triggering clock edge with a D input that meets the setup and hold time requirements, as shown at the last clock tick in the figure.

negative-edge-triggered D flip-flop

A *negative-edge-triggered D flip-flop* simply inverts the clock input, so that all the action takes place on the falling edge of /CLK; its function table and logic symbol are shown in Figure 5–18.

asynchronous inputs

Some D flip-flops have *asynchronous inputs* that may be used to force the flip-flop to a particular state independent of the CLK and D inputs. These inputs, typically labeled PR (preset) and CLR (clear), behave like the set and reset inputs on an S-R flip-flop. The logic symbol and NAND circuit for an edge-triggered D flip-flop with these inputs is shown in Figure 5–19. Although asynchronous inputs are used by some logic designers to perform tricky sequential functions, they are best reserved for initialization and testing purposes, to force a sequential circuit into a known starting state; more on this when we discuss synchronous design methodology in Section 6.9.

preset
clear

Commercial TTL positive-edge-triggered D flip-flops do not use the master-slave latch design of Figure 5–15 or 5–19. Instead, flip-flops like the 74LS74

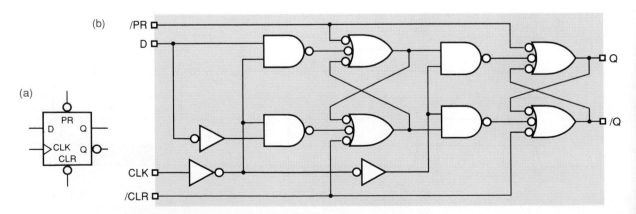

Figure 5–19 Positive-edge-triggered D flip-flop with preset and clear: (a) logic symbol; (b) circuit design using NAND gates.

Figure 5-20
Commercial circuit for a positive-edge-triggered D flip-flop such as 74LS74.

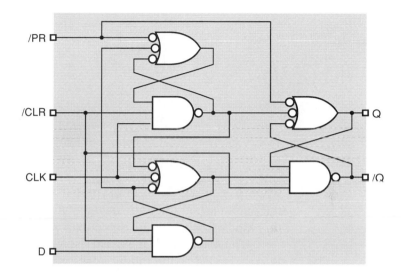

use the six-gate design of Figure 5–20, which is smaller and faster. We'll show how to formally analyze the next-state behavior of both designs in Section 5.5.

* 5.2.6 Master/Slave S-R Flip-Flop

We indicated earlier that S-R latches are useful in "control" applications, where we may have independent conditions for setting and resetting a control bit. If the control bit is supposed to be changed only at certain times with respect to a clock signal, then we need an S-R flip-flop that, like a D flip-flop, changes its outputs only on a certain edge of the clock signal. This subsection and the next two describe flip-flops that are useful for such applications.

If we substitute S-R latches for the D latches in the negative-edge-triggered D flip-flop of Figure 5–18(a), we get a *master/slave S-R flip-flop*, shown in Figure 5–21. Like a D flip-flop, the S-R flip-flop changes its outputs only at the falling edge of a control signal C. However, the new output value depends on input values not just at the falling edge, but during the entire interval in which C is HIGH prior to the falling edge. As shown in Figure 5–22, a short pulse on S any time during this interval can set the master latch; likewise a pulse on R can reset it. The value transferred to the flip-flop output on the falling edge of C depends on whether the master latch was last set or cleared while C was HIGH.

Shown in Figure 5–21(c), the logic symbol for the master/slave S-R flip-flop does not use a dynamic-input indicator, because the flip-flop is not truly edge triggered. It is more like a latch that follows its input during the entire interval that C is HIGH, but that changes its output to reflect the final latched value only

master/slave S-R flip-flop

---

*Throughout this book, optional sections are marked with an asterisk.

Figure 5–21   Master/slave S-R flip-flop: (a) circuit using S-R latches; (b) function table; (c) logic symbol.

postponed-output indicator

pulse-triggered flip-flop

when C goes LOW. In the symbol, a *postponed-output indicator* indicates that the output signal does not change until enable input C is negated. Flip-flops with this kind of behavior are sometimes called *pulse-triggered flip-flops*.

The operation of the master/slave S-R flip-flop is unpredictable if both S and R are asserted at the falling edge of C. In this case, just before the rising edge, both the Q and /Q outputs of the master latch are HIGH. When C goes LOW, the master latch's outputs change unpredictably and may even become metastable. At the same time, the slave latch opens up and propagates this garbage to the flip-flop output.

* 5.2.7 Master/Slave J-K Flip-Flop $Q^* = J \cdot Q' + K' \cdot Q$

master/slave J-K flip-flop

The problem of what to do when S and R are asserted simultaneously is solved in a *master/slave J-K flip-flop*. The J and K inputs are analogous to S and R. However, as shown in Figure 5–23, asserting J asserts the master's S input only if the flip-flop's /Q output is currently 1 (i.e., Q is 0), and asserting K asserts

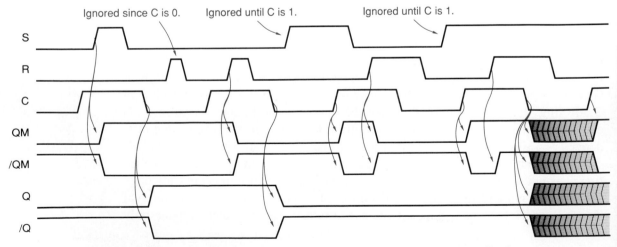

Figure 5–22   Internal and functional behavior of a master/slave S-R flip-flop.

J	K	C	Q	/Q
x	x	0	last Q	last /Q
0	0	⎍	last Q	last /Q
0	1	⎍	0	1
1	0	⎍	1	0
1	1	⎍	last /Q	last Q

(a)   (b)   (c)

Figure 5–23   Master/slave J-K flip-flop: (a) circuit design using S-R latches; (b) function table; (c) logic symbol.

the master's R input only if Q is currently 1. Thus, if J and K are asserted simultaneously, the flip-flop goes to the opposite of its current state.

Figure 5–24 shows the functional behavior of a J-K master/slave flip-flop for a typical set of inputs. Note that the J and K inputs need not be asserted at the end of the triggering pulse for the flip-flop output to change at that time. In fact, because of the gating on the master latch's S and R inputs, it is possible for the flip-flop output to change to 1 even though K and not J is asserted at the end of the triggering pulse. This behavior, known as *1s catching*, is illustrated in the   *1s catching* second-to-last triggering pulse in the figure. An analogous behavior known as *0s catching* is illustrated in the last triggering pulse. Because of this behavior,   *0s catching* the J and K inputs of a J-K master/slave flip-flop must be held valid during the entire interval that C is 1.

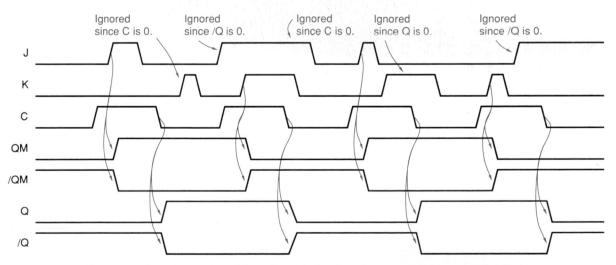

Figure 5–24   Internal and functional behavior of a master/slave J-K flip-flop.

J	K	CLK	Q	/Q
x	x	0	last Q	last /Q
x	x	1	last Q	last /Q
0	0	⌐_	last Q	last /Q
0	1	⌐_	0	1
1	0	⌐_	1	0
1	1	⌐_	last /Q	last Q

(a)   (b)   (c)

Figure 5–25 Edge-triggered J-K flip-flop: (a) equivalent function use an edge-triggered D flip-flop; (b) function table; (c) logic symbol.

* 5.2.8 Edge-Triggered J-K Flip-Flop $\quad Q^* = J \cdot Q' + K' \cdot Q$

edge-triggered J-K flip-flop

The problem of 1s and 0s catching is solved in an *edge-triggered J-K flip-flop*, whose functional equivalent is shown in Figure 5–25. Using an edge-triggered D flip-flop internally, the edge-triggered J-K flip-flop samples its inputs at the rising edge of the clock and produces its next output according to the "characteristic equation" $Q^* = J \cdot Q' + K' \cdot Q$ (see Section 5.2.10).

Typical functional behavior of an edge-triggered J-K flip-flop is shown in Figure 5–26. Like the D input of an edge-triggered D flip-flop, the J and K inputs of a J-K flip-flop must meet published setup and hold time specifications with respect to the triggering clock edge for proper operation.

Because they eliminate the problems of 1s and 0s catching and of simultaneously asserting both control inputs, edge-triggered J-K flip-flops are used much more commonly nowadays than are the older pulse-triggered types of the preceding two subsections. The *74LS109* is a TTL positive-edge-triggered J-/K flip-flop with an active-low K input (named /K or $\overline{K}$). The internal design of the '109 is very similar to that of the 74LS74, which we showed in Figure 5–20. As shown in Figure 5–27, the '109 simply replaces the bottom-left gate of the '74,

74LS109

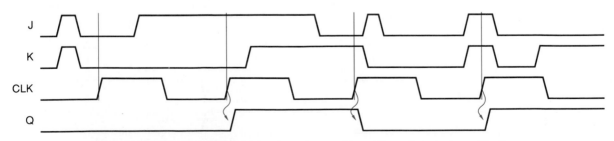

Figure 5–26 Functional behavior of a positive-edge-triggered J-K flip-flop.

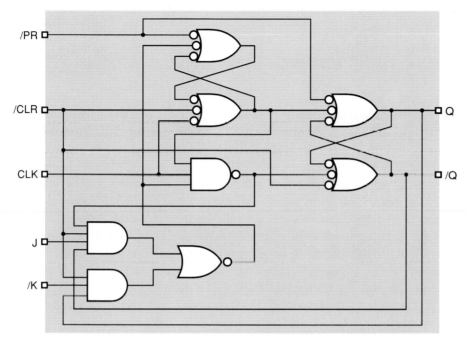

Figure 5–27   Internal logic diagram for a 74LS109 positive-edge-triggered J-/K
flip-flop.

which realizes the characteristic equation $Q^* = D$, with an AND-OR structure that
realizes the J-/K characteristic equation, $Q^* = J \cdot Q' + /K \cdot Q$.

The most common application of J-K flip-flops is in clocked synchronous
state machines. As we'll explain in Section 5.4.5, the next-state logic for J-K flip-
flops is sometimes simpler than for D flip-flops. However, most state machines
are still designed using D flip-flops, because the design methodology is a bit
simpler and most sequential programmable logic devices contain D, not J-K,
flip-flops. Therefore, we'll give most of our attention to D flip-flops.

5.2.9   T Flip-Flop

A *T (toggle) flip-flop* changes state on every tick of the clock. Figure 5–28 shows   *T flip-flop*
the symbol and illustrates the behavior of a positive-edge-triggered T flip-flop.

Figure 5–28   Positive-edge-triggered T flip-flop: (a) logic symbol; (b) functional behavior.

Figure 5–29
Possible circuit designs for a
T flip-flop: (a) using a D flip-flop;
(b) using a J-K flip-flop.

Notice that the signal on the flip-flop's Q output has precisely half the frequency of the T input. Figure 5–29 shows how to obtain a T flip-flop from a D or J-K flip-flop. T flip-flops are most often used in counters and frequency dividers, as we'll show in Section 6.3.

T flip-flop with enable

   In many applications of T flip-flops, the flip-flop need not be toggled on every clock tick. Such applications can use a *T flip-flop with enable*. As shown in Figure 5–30, the flip-flop changes state at the triggering edge of the clock only if the enable signal E is asserted. Like the D, J, and K inputs on other edge-triggered flip-flops, the E input must meet specified setup and hold times with respect to the triggering clock edge. The circuits of Figure 5–29 are easily modified to provide an E input, as shown in Figure 5–31.

5.2.10   Characteristic Equations

characteristic equation

* suffix

The functional behavior of a latch or flip-flop can be described by a *characteristic equation* that specifies the flip-flop's next state as a function of its current state and inputs. The characteristic equations of the flip-flops in this section are listed in Table 5–1. By convention, the * suffix in Q* means "the next value of Q." Notice that the characteristic equation does not describe detailed timing behavior of the device (latching vs. edge-triggered, etc.), only the functional response

(a)   (b)

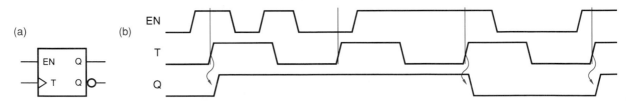

Figure 5–30   Positive-edge-triggered T flip-flop with enable: (a) logic symbol; (b) functional behavior.

Figure 5–31   Possible circuit designs for a T flip-flop with enable: (a) using a D flip-flop; (b) using a J-K flip-flop.

Table 5–1
Latch and flip-flop characteristic equations.

Device type	Characteristic equation
S-R latch	$Q* = S + R' \cdot Q$
D latch	$Q* = D$
Edge-triggered D flip-flop	$Q* = D$
Master/slave S-R flip-flop	$Q* = S + R' \cdot Q$
Master/slave J-K flip-flop	$Q* = J \cdot Q' + K' \cdot Q$
Edge-triggered J-K flip-flop	$Q* = J \cdot Q' + K' \cdot Q$
T flip-flop	$Q* = Q'$
T flip-flop with enable	$Q* = EN \cdot Q' + EN' \cdot Q$

to the control inputs. This simplified description is useful in the analysis and synthesis of state machines, as we'll show in the next two sections.

5.3 CLOCKED SYNCHRONOUS STATE-MACHINE ANALYSIS

Although latches and flip-flops, the basic building blocks of sequential circuits, are themselves feedback sequential circuits that can be formally analyzed, we'll first study the operation of *clocked synchronous state machines*, since they are the easiest to understand. "State machine" is a generic name given to these sequential circuits; "clocked" refers to the fact that their storage elements (flip-flops) employ a clock input; and "synchronous" means that all of the flip-flops use the same clock signal. Such a state machine changes state only when a triggering edge or "tick" occurs on the clock signal.

clocked synchronous state machine

5.3.1 State-Machine Structure

Figure 5–32 shows the general structure of a clocked synchronous state machine. The *state memory* is a set of n flip-flops that store the current state of the machine, and has 2^n distinct states. The flip-flops are all connected to a common clock signal that causes the flip-flops to change state at each *tick* of the clock. What constitutes a tick depends on the flip-flop type (edge triggered, pulse triggered, etc.).

state memory

tick

 The next state of the state machine in Figure 5–32 is determined by the *next-state logic F* as a function of the current state and input. The *output logic G* determines the output as a function of the current state and input. Both F and G are strictly combinational logic circuits. We can write

next-state logic
output logic

$$\text{Next state} \ = \ F(\text{current state}, \text{input})$$
$$\text{Output} \ = \ G(\text{current state}, \text{input})$$

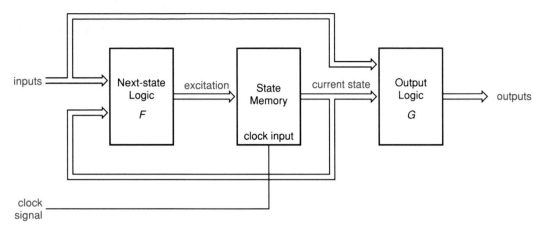

Figure 5–32   Clocked synchronous state-machine structure (Mealy machine).

Mealy machine

A sequential circuit whose output depends on both state and input as stated above is called a *Mealy machine*. In some sequential circuits, the output depends on the state alone:

$$\text{Output} \;=\; G(\text{current state})$$

Moore machine

Such a circuit is called a *Moore machine*, and its general structure is shown in Figure 5–33.

Obviously, the only difference between the two state-machine models is in how outputs are generated. In practice, most state machines must be categorized as Mealy machines, because they have one or more *Mealy-type outputs* that depend on input as well as state. However, many of these same machines also have one or more *Moore-type outputs* that depend only on state. We'll find the distinction between output types to be important when we study the design of PLD-based state machines in Section 7.4.

Mealy-type output

Moore-type output

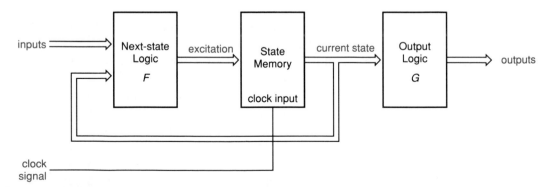

Figure 5–33   Clocked synchronous state-machine structure (Moore machine).

State machines may use positive-edge-triggered D flip-flops for their state memory, in which case a tick occurs at each rising edge of the clock signal. It is also possible for the state memory to use negative-edge-triggered D flip-flops, D latches, or J-K flip-flops. However, inasmuch as most state machines are designed nowadays using programmable logic devices with positive-edge-triggered D flip-flops, that's what we'll concentrate on.

5.3.2   Analysis of State Machines with D Flip-Flops

Consider the formal definition of a Mealy machine:

$$\text{Next state} \;=\; F(\text{current state, input})$$
$$\text{Output} \;=\; G(\text{current state, input})$$

Recalling our notion that "state" embodies all we need to know about the past history of the circuit, the first equation tells us that what we next need to know can be determined from what we currently know and the current input. The second equation tells us that the current output can be determined from the same information. The goal of sequential circuit analysis is to determine the next-state and output functions so that the behavior of a circuit can be predicted.

The analysis of a clocked synchronous state machine has three basic steps:

(1) Determine the next-state and output functions F and G.

(2) Use F and G to construct a *state/output table* that completely specifies the next state and output of the circuit for every possible combination of current state and input.   *state/output table*

(3) (optional) Draw a *state diagram* that presents the information from the previous step in graphical form.   *state diagram*

Figure 5–34 shows a simple state machine with two positive-edge-triggered D flip-flops. To determine the next-state function F, we must first consider the behavior of the state memory. At the rising edge of the clock signal, each D flip-flop samples its D input and transfers this value to its Q output; the characteristic equation of a D flip-flop is Q* = D. Therefore, in order to determine the next value of Q (i.e., Q*), we must first determine the current value of D.

In Figure 5–34 there are two D flip-flops, and we have named the signals on their outputs Q0 and Q1. These two outputs are the state variables; their value is the current state of the machine. We have named the signals on the corresponding D inputs D0 and D1. These signals provide the *excitation* for the D flip-flops at each clock tick. Logic equations that express the excitation signals   *excitation*

Figure 5–34 Clocked synchronous state machine using positive-edge-triggered D flip-flops.

excitation equation

as functions of the current state and input are called *excitation equations*, and can be derived from the circuit diagram:

$$D0 \;=\; Q0 \cdot EN' + Q0' \cdot EN$$
$$D1 \;=\; Q1 \cdot EN' + Q1' \cdot Q0 \cdot EN + Q1 \cdot Q0' \cdot EN$$

* suffix

By convention, the next value of a state variable after a clock tick is denoted by appending a star to the state-variable name, for example, $Q0*$ or $Q1*$. Using the characteristic equation of D flip-flops, $Q* = D$, we can describe the next-state function of the example machine with equations for the next value of the state variables:

$$Q0* \;=\; D0$$
$$Q1* \;=\; D1$$

Substituting the excitation equations for D0 and D1, we can write

$$Q0* \;=\; Q0 \cdot EN' + Q0' \cdot EN$$
$$Q1* \;=\; Q1 \cdot EN' + Q1' \cdot Q0 \cdot EN + Q1 \cdot Q0' \cdot EN$$

Table 5–2
Transition table for the state machine in Figure 5–34.

Q1 Q0	EN	
	0	1
00	00	01
01	01	10
10	10	11
11	11	00
	Q1* Q0*	

These equations, which express the next value of the state variables as a function of current state and input, are called *transition equations*.   *transition equation*

For each combination of current state and input value, the transition equations predict the next state. Each state is described by two bits, the current values of Q0 and Q1: (Q1 Q0) = 00, 01, 10, or 11. (The reason for "arbitrarily" picking the order (Q1 Q0) instead of (Q0 Q1) will become apparent shortly.) For each state, our example machine has only two possible input values, EN = 0 or EN = 1, so there are a total of 8 state/input combinations. (In general, a machine with s state bits and i inputs has 2^{s+i} state/input combinations.)

Table 5–2 shows a *transition table* that is created by evaluating the transition   *transition table* equations for every possible state/input combination. Traditionally, a transition table lists the states along the left and the input combinations along the top of the table, as shown in the example.

The function of our example machine is apparent from its transition table— it is a 2-bit binary counter with an enable input EN. When EN=0 the machine maintains its current count, but when EN=1 the count advances by 1 at each clock tick, rolling over to 00 when it reaches a maximum value of 11.

If we wish, we may assign alphanumeric *state names* to each state. The   *state names* simplest naming is 00 = A, 01 = B, 10 = C, and 11 = D. Substituting the state names for combinations of Q1 and Q0 (and Q1* and Q0*) in Table 5–2 produces the *state table* in Table 5–3. Here "S" denotes the current state, and "S*" denotes   *state table* the next state of the machine. A state table is usually easier to understand than a

Table 5–3
State table for the state machine in Figure 5–34.

S	EN	
	0	1
A	A	B
B	B	C
C	C	D
D	D	A
	S*	

Table 5–4
State/output table for the
machine in Figure 5–34.

	EN	
S	0	1
A	A, 0	B, 0
B	B, 0	C, 0
C	C, 0	D, 0
D	D, 0	A, 1
	S∗, MAX	

transition table, because in complex machines we can use state names that have meaning. However, a state table contains less information than a transition table, because it does not indicate the binary values assumed by the state variables in each named state.

Once a state table is produced, we have only the output logic of the machine left to analyze. In the example machine, there is only a single output signal, and it is a function of both current state and input (this is a Mealy machine). So we

output equation   can write a single *output equation*:

$$MAX = Q1 \cdot Q0 \cdot EN$$

The output behavior predicted by this equation can be combined with the next-
state/output table   state information to produce a *state/output table* as shown in Table 5–4.

State/output tables for Moore machines are slightly simpler. For example, in the circuit of Figure 5–34 suppose we removed the EN signal from the AND gate that produces the MAX output. Then MAX becomes a function of the state only, and the state/output table can list MAX in a single column, independent of the input values. This is shown in Table 5–5.

state diagram   A *state diagram* presents the information from the state/output table in a
node   graphical format. It has one circle (or *node*) for each state, and an arrow (or
directed arc   *directed arc*) for each transition. Figure 5–35 shows the state diagram for our example state machine. The letter inside each circle is a state name. Each arrow leaving a state points to the next state for a particular input combination.

Table 5–5
State/output table for a Moore
machine.

	EN		
S	0	1	MAX
A	A	B	0
B	B	C	0
C	C	D	0
D	D	A	1
	S∗		

Figure 5–35
State diagram corresponding to
the state machine of Table 5–4.

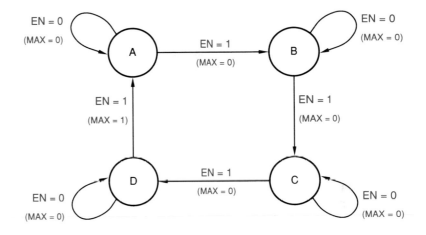

Since there is only one input in our example machine, there are only two possible input combinations, and two arrows leaving each state. In a machine with n inputs, we would have 2^n arrows leaving each state. This is messy if n is large. Later, in Figure 5–39, we'll describe a convention whereby a state needn't have one arrow leaving it for each input combination, only one arrow for each different next state.

The state diagram also shows the output value for each possible state/input combination. However, the notation for output values in Mealy machines is a little misleading. A listed output value is actually produced continuously when the machine is in the indicated state and has the indicated input, not just during the transition to the next state.

The state diagram for a Moore machine can be somewhat simpler. In this case, the output values can be shown inside each state circle since they are functions of state only. The state diagram for a Moore machine using this convention is shown in Figure 5–36.

Figure 5–36
State diagram corresponding to
the state machine of Table 5–5.

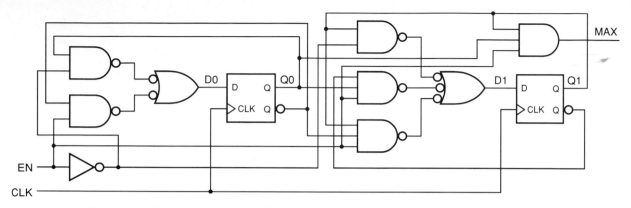

Figure 5-37   Redrawn logic diagram for a clocked synchronous state machine.

The original logic diagram of our example state machine, Figure 5–34, was laid out to match our conceptual model of a Mealy machine. However, nothing requires us to group the next-state logic, state memory, and output logic in this way. Figure 5–37 shows another logic diagram for the same state machine. To analyze this circuit, the designer (or analyzer, in this case) can still extract the required information from the diagram as drawn. The only circuit difference in the new diagram is that we have used the flip-flops' /Q outputs (which are normally the complement of Q) to save a couple of inverters.

In summary, the detailed steps for analyzing a clocked synchronous state machine are as follows:

excitation equations   (1) Determine the excitation equations for the flip-flop control inputs.

transition equations   (2) Substitute the excitation equations into the flip-flop characteristic equations to obtain transition equations.

transition table   (3) Use the transition equations to construct a transition table.

output equations   (4) Determine the output equations.

transition/output table   (5) Add output values to the transition table for each state (Moore) or state/input combination (Mealy) to create a transition/output table.

state names
state/output table   (6) Name the states and substitute state names for state-variable combinations in the transition/output table to obtain a state/output table.

state diagram   (7) (optional) Draw a state diagram corresponding to the state/output table.

We'll go through this complete sequence of steps to analyze another clocked synchronous state machine, shown in Figure 5–38. Reading the logic diagram,

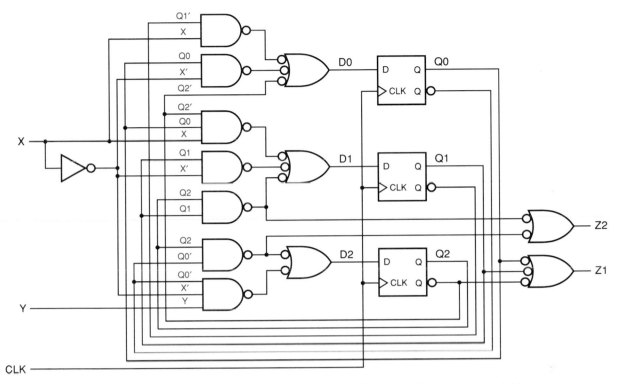

Figure 5–38 A clocked synchronous state machine with three flip-flops and eight states.

we find that the excitation equations are as follows:

$$D0 \;=\; Q1' \cdot X + Q0 \cdot X' + Q2$$
$$D1 \;=\; Q2' \cdot Q0 \cdot X + Q1 \cdot X' + Q2 \cdot Q1$$
$$D2 \;=\; Q2 \cdot Q0' + Q0' \cdot X' \cdot Y$$

Substituting into the characteristic equation for D flip-flops, we obtain the transition equations,

$$Q0* \;=\; Q1' \cdot X + Q0 \cdot X' + Q2$$
$$Q1* \;=\; Q2' \cdot Q0 \cdot X + Q1 \cdot X' + Q2 \cdot Q1$$
$$Q2* \;=\; Q2 \cdot Q0' + Q0' \cdot X' \cdot Y$$

A transition (and output) table based on these equations is shown in Table 5–6. Reading the logic diagram, we can write two output equations,

$$Z1 \;=\; Q2 + Q1' + Q0'$$
$$Z2 \;=\; Q2 \cdot Q1 + Q2 \cdot Q0'$$

Table 5–6
Transition and output table for the state machine in Figure 5–38.

	X Y				
Q2 Q1 Q0	00	01	10	11	Z1 Z2
000	000	100	001	001	10
001	001	001	011	011	10
010	010	110	000	000	10
011	011	011	010	010	00
100	101	101	101	101	11
101	001	001	001	001	10
110	111	111	111	111	11
111	011	011	011	011	11
		Q2* Q1* Q0*			

The resulting output values are shown in the last column of Table 5–6. Assigning state names A–H, we obtain the state/output table shown in Table 5–7.

A state diagram for the example machine is shown in Figure 5–39. Since our example is a Moore machine, the output values are written with each state. Each arc is labeled with a *transition expression*; a transition is taken for input combinations for which the transition expression is 1. Transitions labeled "1" are always taken. Note that the transition expressions on arcs leaving a particular state are mutually exclusive and all inclusive:

mutual exclusion

- No two transition expressions evaluate to 1 for the same input combination, since a machine cannot have two next states for one input combination.

all inclusion

- For every possible input combination, some transition expression is 1, so that all next states are defined.

Table 5–7
State/output table for the state machine in Figure 5–38.

	X Y				
S	00	01	10	11	Z1 Z2
A	A	E	B	B	10
B	B	B	D	D	10
C	C	G	A	A	10
D	D	D	C	C	00
E	F	F	F	F	11
F	B	B	B	B	10
G	H	H	H	H	11
H	D	D	D	D	11
		S*			

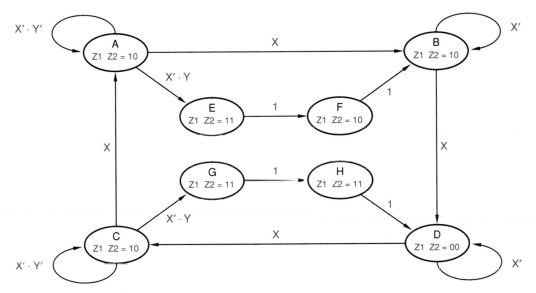

Figure 5–39 State diagram corresponding to Table 5–7.

Starting with the state table, a transition expression for a particular current state and next state can be written as a sum of minterms for the input combinations that cause that transition. If desired, the expression can then be minimized to give the information in a more compact form. Transition expressions are most useful in the *design* of state machines, where the expressions may be developed from the word description of the problem, as we'll show in Section 6.6.2.

*5.3.3 Analysis of State Machines with J K Flip Flops

Clocked synchronous state machines built from J-K flip-flops can also be analyzed by the basic procedure in the preceding subsection. The only difference is that there are two excitation equations for each flip-flop—one for J and the other for K. To obtain the transition equations, both of these must be substituted into the J-K's characteristic equation, $Q* = J \cdot Q' + K' \cdot Q$.

Figure 5–40 is an example state machine using J-K flip-flops. Reading the logic diagram, we can derive the following excitation equations:

$$J0 = X \cdot Y'$$
$$K0 = X \cdot Y' + Y \cdot Q1$$
$$J1 = X \cdot Q0 + Y$$
$$K1 = Y \cdot Q0' + X \cdot Y' \cdot Q0$$

Substituting into the characteristic equation for J-K flip-flops, we obtain the transition equations,

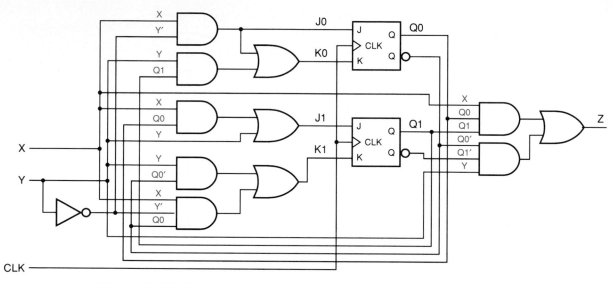

Figure 5–40   A clocked synchronous state machine using J-K flip-flops.

$$Q0* = J0 \cdot Q0' + K0' \cdot Q0$$
$$= X \cdot Y' \cdot Q0' + (X \cdot Y' + Y \cdot Q1)' \cdot Q0$$
$$= X \cdot Y' \cdot Q0' + X' \cdot Y' \cdot Q0 + X' \cdot Q1' \cdot Q0 + Y \cdot Q1' \cdot Q0$$
$$Q1* = J1 \cdot Q1' + K1' \cdot Q1$$
$$= (X \cdot Q0 + Y) \cdot Q1' + (Y \cdot Q0' + X \cdot Y' \cdot Q0)' \cdot Q1$$
$$= X \cdot Q1' \cdot Q0 + Y \cdot Q1' + X' \cdot Y' \cdot Q1 + Y' \cdot Q1 \cdot Q0' + X' \cdot Q1 \cdot Q0 + Y \cdot Q1 \cdot Q0$$

A transition (and output) table based on these equations is shown in Table 5–8. Reading the logic diagram, we can write the output equation,

$$Z = X \cdot Q1 \cdot Q0 + Y \cdot Q1' \cdot Q0'$$

The resulting output values are shown in each column of Table 5–8 along with

Table 5–8
Transition and output table for the state machine in Figure 5–40.

		X Y		
Q1 Q0	00	01	10	11
00	00, 0	10, 1	01, 0	10, 1
01	01, 0	11, 0	10, 0	11, 0
10	10, 0	00, 0	11, 0	00, 0
11	11, 0	10, 0	00, 1	10, 1
	Q1* Q0*, Z			

Table 5–9
State/output table for the state
machine in Figure 5–40.

	X Y			
S	00	01	10	11
A	A, 0	C, 1	B, 0	C, 1
B	B, 0	D, 0	C, 0	D, 0
C	C, 0	A, 0	D, 0	A, 0
D	D, 0	C, 0	A, 1	C, 1
			S*, Z	

the next state. Assigning state names A–D, we obtain the state/output table shown
in Table 5–9. A state diagram is shown in Figure 5–41.[3]

5.4 CLOCKED SYNCHRONOUS STATE-MACHINE DESIGN

The steps for designing a clocked synchronous state machine, starting from a
word description or specification, are just about the reverse of the analysis steps
that we used in the preceding section:

(1) Construct a state/output table corresponding to the word description or spec-   *state/output table*
     ification, using mnemonic names for the states. (It's also possible to start
     with a state diagram; this method is discussed in Section 6.6.2.)

---

[3] An alternative to substituting the excitation equations algebraically into the characteristic equation is to construct an excitation table that shows the excitation values (J0, K0, J1, K1) for each
state/input combination. The characteristic equation may then be applied to each state/excitation
entry to obtain the transition table.

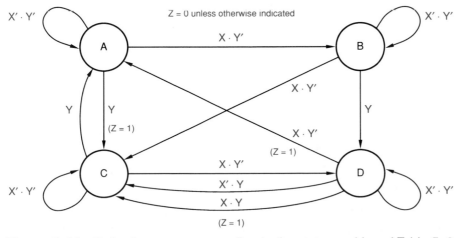

Figure 5–41 State diagram corresponding to the state machine of Table 5–9.

state minimization   (2) (Optional) Minimize the number of states in the state/output table.

state assignment   (3) Choose a set of state variables and assign state-variable combinations to the named states.

transition/output table   (4) Substitute the state-variable combinations into the state/output table to create a transition/output table that shows the desired next state-variable combination and output for each state/input combination.

(5) Choose a flip-flop type (e.g., D or J-K) for the state memory. In most cases, you'll already have a choice in mind at the outset of the design, but this step is your last chance to change your mind.

excitation table   (6) Construct an excitation table that shows the excitation values required to obtain the desired next state for each state/input combination.

excitation equations   (7) Derive excitation equations from the excitation table.

output equations   (8) Derive output equations from the transition/output table.

logic diagram   (9) Draw a logic diagram that shows the state-variable storage elements and realizes the required excitation and output equations.

design   In this section, we'll describe each of these basic steps in state-machine design. The first step is the most important, since it is here that the designer really *designs*, going through the creative process of translating a (perhaps ambiguous) English-language description of the state machine into a formal tabular description. The second step is hardly ever performed by experienced digital designers, but designers bring much of their experience to bear in the third step. Once these steps are completed, all of the remaining steps can be completed by "turning the crank," that is, by following a well-defined synthesis procedure. When state machines are designed using programmable logic devices, all or part of the synthesis procedure is performed by software, as discussed in Section 7.4.

5.4.1 State Table Design

Designing a state table (or equivalently, a state diagram) is a creative process that is like writing a computer program in many ways:

- You start with a fairly precise description of inputs and outputs, but a possibly ambiguous description of the desired relationship between them, and usually no clue about how to actually obtain the desired outputs from the inputs.

- During the design, you may have to identify and choose among different ways of doing things, sometimes using common sense, and sometimes arbitrarily.

- You may have to identify and handle special cases that weren't included in the original description.

- You will probably have to keep track of several ideas in your head during the design process.

- Since the design process is not an algorithm, there's no guarantee that you can complete the state table or program using a finite number of states or lines of code. However, unless you work for the government, you must try to do so.

- When you finally run the state machine or program, it will do exactly what you told it to do—no more, no less.

- There's no guarantee that the thing will work the first time; you may have to debug and iterate on the whole process.

Although state-table design is a challenge, there's no need to be intimidated. If you've made it this far, then you've probably written a few programs that worked, and you can become just as good at designing state tables.

We'll present the state-table design process, as well as the synthesis procedure in later subsections, using the simple design problem below:

> Design a clocked synchronous state machine with two inputs, A and B, and a single output Z. The output should be 1 if A had the same value at each of the two previous clock ticks, or if B has been 1 since the last time the first condition was true. Otherwise, the output should be 0.

The first step is to construct a template for the state table. From the word description, we know that this is a Moore machine—the output depends only on the current state, that is, what happened in previous clock periods. Thus, as shown in Figure 5–42(a), we provide one next-state column for each possible input combination and a single column for the output values. The order in which the input combinations are written doesn't affect this part of the process, but we've written them in Karnaugh-map order to simplify the derivation of excitation equations later. In a Mealy machine we would omit the output column and write the output values along with the next-state values under each input combination. The leftmost column is simply an English-language reminder of the meaning of each state or the "history" associated with it.

The word description isn't specific about what happens when this machine is first started, so we'll just have to improvise. We'll assume that when power

(a)

Meaning	S	00	01	11	10	Z
			A B			
Initial state	INIT					0
. . .						
. . .						
. . .						
			S*			

(b)

Meaning	S	00	01	11	10	Z
			A B			
Initial state	INIT	A0	A0	A1	A1	0
Got a 0 on A	A0					0
Got a 1 on A	A1					0
			S*			

(c)

Meaning	S	00	01	11	10	Z
			A B			
Initial state	INIT	A0	A0	A1	A1	0
Got a 0 on A	A0	OK	OK	A1	A1	0
Got a 1 on A	A1					0
Got two equal A inputs	OK					1
			S*			

(d)

Meaning	S	00	01	11	10	Z
			A B			
Initial state	INIT	A0	A0	A1	A1	0
Got a 0 on A	A0	OK	OK	A1	A1	0
Got a 1 on A	A1	A0	A0	OK	OK	0
Got two equal A inputs	OK					1
			S*			

Figure 5–42   Evolution of a state table.

initial state

is first applied to the system, the machine enters an *initial state*, called INIT in this example.[4] We write the name of the initial state (INIT) in the first row, and leave room for enough rows (states) to complete the design. We can also fill in the value of Z for the INIT state; common sense says it should be 0 because there were *no* inputs beforehand.

Next, we must fill in the next-state entries for the INIT row. The Z output can't be 1 until we've seen at least two inputs on A, so we'll provide two states, A0 and A1, that "remember" the value of A on the previous clock tick, as shown in Figure 5–42(b). In both of these states, Z is 0 since we haven't satisified the conditions for a 1 output yet. The precise meaning of state A0 is "Got A = 0 on the previous tick, A ≠ 0 on the tick before that, and B ≠ 1 at some time since the previous pair of equal A inputs." State A1 is defined similarly.

At this point we know that our state machine has at least three states, and we have created two more blank rows to fill in. Hmmmm, this isn't such a good trend! In order to fill in the next-state entries for *one* state (INIT), we had to

---

[4]For proper system operation, our hardware design should ensure that the state machine enters the INIT state on power-up. Most systems have RESET signals that are asserted during power-up. If the state machine is built using discrete flip-flops with asynchronous preset and clear inputs, the RESET signal can be applied to these inputs to force the machine into the INIT state. If preset and clear inputs are not available, or if reset must be synchronous (as in systems using high-speed microprocessors such as the Intel 80386 and 80486), then the RESET signal may be used as another input to the state machine, with all of the next-state entries going to the INIT state when RESET is asserted.

(a)

Meaning	S	A B				Z
		00	01	11	10	
Initial state	INIT	A0	A0	A1	A1	0
Got a 0 on A	A0	OK	OK	A1	A1	0
Got a 1 on A	A1	A0	A0	OK	OK	0
Got two equal A inputs	OK	?	OK	OK	?	1
		S*				

(b)

Meaning	S	A B				Z
		00	01	11	10	
Initial state	INIT	A0	A0	A1	A1	0
Got a 0 on A	A0	OK0	OK0	A1	A1	0
Got a 1 on A	A1	A0	A0	OK1	OK1	0
Two equal, A=0 last	OK0					1
Two equal, A=1 last	OK1					1
		S*				

(c)

Meaning	S	A B				Z
		00	01	11	10	
Initial state	INIT	A0	A0	A1	A1	0
Got a 0 on A	A0	OK0	OK0	A1	A1	0
Got a 1 on A	A1	A0	A0	OK1	OK1	0
Two equal, A=0 last	OK0	OK0	OK0	OK1	A1	1
Two equal, A=1 last	OK1					1
		S*				

(d)

Meaning	S	A B				Z
		00	01	11	10	
Initial state	INIT	A0	A0	A1	A1	0
Got a 0 on A	A0	OK0	OK0	A1	A1	0
Got a 1 on A	A1	A0	A0	OK1	OK1	0
Two equal, A=0 last	OK0	OK0	OK0	OK1	A1	1
Two equal, A=1 last	OK1	A0	OK0	OK1	OK1	1
		S*				

Figure 5–43 Continued evolution of a state table.

create *two* new states A0 and A1. If we kept going this way, we could end up with 65,535 states by bedtime! Instead, we should be on the lookout for existing states that have the same meaning as new ones that we might otherwise create. Let's see how it goes.

In state A0, we know that input A was 0 at the previous clock tick. Therefore, if A is 0 again, we go to a new state OK with Z = 1, as shown in Figure 5–42(c). If A is 1, then we don't have two equal inputs in a row, so we go to state A1 to remember that we just got a 1. Likewise in state A1, shown in (d), we go to OK if we get a second 1 input in a row, or to A0 if we get a 0.

Once we get into the OK state, the machine description tells us we can stay there as long as B = 1, irrespective of the A input, as shown in Figure 5–43(a). If B = 0, we have to look for two 1s or two 0s in a row on A again. However, we've got a little problem in this case. The current A input may or may not be the second equal input in a row, so we may still be "OK" or we may have to go back to A0 or A1. We defined the OK state too broadly—it doesn't "remember" enough to tell us which way to go.

The problem is solved in Figure 5–43(b) by splitting OK into two states, OK0 and OK1, that "remember" the previous A input. All of the next states for OK0 and OK1 can be selected from existing states, as shown in (c) and (d). For example, if we get A = 0 in OK0, we can just stay in OK0; we don't have to create a new state that "remembers" three 0s in a row, because the machine's description doesn't require us to distinguish that case. Thus, we have achieved "closure" of the state table, which now describes a *finite*-state machine.

(a)

Meaning	S	00	01	11	10	Z
Initial state	INIT	A0	A0	A1	A1	0
Got a 0 on A	A0	OK00	OK00	A1	A1	0
Got a 1 on A	A1	A0	A0	OK11	OK11	0
Got 00 on A	OK00	OK00	OK00	OKA1	A1	1
Got 11 on A	OK11	A0	OKA0	OK11	OK11	1
OK, got a 0 on A	OKA0	OK00	OK00	OKA1	A1	1
OK, got a 1 on A	OKA1	A0	OKA0	OK11	OK11	1
	S*					

(b)

Meaning	S	00	01	11	10	Z
Initial state	INIT	A0	A0	A1	A1	0
Got a 0 on A	A0	OK00	OK00	A1	A1	0
Got a 1 on A	A1	A0	A0	OK11	OK11	0
Got 00 on A	OK00	OK00	OK00	A001	A1	1
Got 11 on A	OK11	A0	A110	OK11	OK11	1
Got 001 on A, B=1	A001	A0	AE10	OK11	OK11	1
Got 110 on A, B=1	A110	OK00	OK00	AE01	A1	1
Got bb...10 on A, B=1	AE10	OK00	OK00	AE01	A1	1
Got bb...01 on A, B=1	AE01	A0	AE10	OK11	OK11	1
	S*					

Figure 5–44   Nonminimal state tables equivalent to Figure 5–43(d).

5.4.2   State Minimization

Figure 5–43(d) is a "minimal" state table for our original word description, in the sense that it contains the fewest possible states. However, Figure 5–44 shows other state tables, with more states, that also do the job. Formal procedures can be used to minimize the number of states in such tables. The basic idea of *equivalent states* formal minimization procedures is to identify *equivalent states*, where two states are equivalent if it is impossible to distinguish the states by observing only the current and future *outputs* of the machine (and *not* the internal state variables). A pair of equivalent states can be replaced by a single state.

It can be proved that two states S1 and S2 are equivalent if two conditions are true. First, S1 and S2 must produce the same values at the state-machine output(s); in a Mealy machine, this must be true for all input combinations. Second, for each input combination, S1 and S2 must have the same next state or equivalent next states.

Thus, a formal state-minimization procedure shows that states OK00 and OKA0 in Figure 5–44(a) are equivalent because they produce the same output and their next-state entries are identical. Since the states are equivalent, state OK00 may be eliminated and its occurrences in the table replaced by OKA0, or vice versa. Likewise, states OK11 and OKA1 are equivalent.

To minimize the state table in Figure 5–44(b), a formal procedure must use a bit of circular reasoning. States OK00, A110, and AE10 all produce the same output and have *almost* identical next-state entries, so they *might* be equivalent. They are equivalent only if A001 and AE01 are equivalent. Similarly, OK11, A001, and AE01 are equivalent only if A110 and AE10 are equivalent. In other words, the states in the first set are equivalent if the states in the second set are, and vice versa. So, let's just go ahead and say they're equivalent.

Details of formal state-minimization procedures are discussed in advanced textbooks, cited in the References. However, these procedures are seldom used

by most digital designers. By carefully matching state meanings to the requirements of the problem, experienced digital designers produce state tables for small
problems with a minimal or near-minimal number of states, without using a formal minimization procedure. Also, there are situations where *increasing* the number of states may simplify the design, so even an automated state-minimization
procedure doesn't necessarily help. A designer can do more to simplify a state
machine during the state-assignment phase of the design, discussed next.

5.4.3   State Assignment

The next step in the design process is to determine how many binary variables
are required to represent the states in the state table, and to assign a specific
combination to each named state. We'll call the binary combination assigned to
a particular state a *coded state*. The *total number of states* in a machine with n *coded state*
flip-flops is 2^n, so the number of flip-flops needed to code s states is $\lceil \log_2 s \rceil$,   *total number of*
the smallest integer greater than or equal to $\log_2 s$.   *states*

    For reference, the state/output table of our example machine is repeated in
Table 5–10. It has five states, so it requires three flip-flops. Of course, three flip-
flops provide a total of eight states, so there will be $8-5 = 3$ *unused states*. We'll   *unused states*
discuss alternatives for handling the unused states at the end of this subsection.
Right now, we have to deal with lots of choices for the five coded states.

    The number of different ways to choose m coded states out of a set of
n possible states is given by a *binomial coefficient*, denoted $\binom{n}{m}$, whose value   *binomial coefficient*
is $\frac{n!}{m!\cdot(n-m)!}$. (We used binomial coefficients previously in Section 1.10, in the
context of decimal coding.) In our example, there are $\binom{8}{5}$ different ways to
choose five coded states out of eight possible states, and 5! ways to assign the
five named states to each different choice. So there are $\frac{8!}{5!\cdot 3!}\cdot 5!$ or 6,720 different
ways to assign the five states of our example machine to combinations of three
binary state variables. We won't have time to look at all of them.

    The simplest assignment of s coded states to 2^n possible states is to use the

Table 5–10
State and output table for
example problem.

	A B				
S	00	01	11	10	Z
INIT	A0	A0	A1	A1	0
A0	OK0	OK0	A1	A1	0
A1	A0	A0	OK1	OK1	0
OK0	OK0	OK0	OK1	A1	1
OK1	A0	OK0	OK1	OK1	1
			S*		

Table 5–11
Possible state assignments for
the state machine in Table 5–10.

State name	Simplest Q1–Q3	Decomposed Q1–Q3	One-hot Q1–Q5	Almost one-hot Q1–Q4
INIT	000	000	00001	0000
A0	001	100	00010	0001
A1	010	101	00100	0010
OK0	011	110	01000	0100
OK1	100	111	10000	1000

first s binary integers in binary counting order, as shown in the first assignment column of Table 5–11. However, the simplest state assignment does not always lead to the simplest excitation equations, output equations, and resulting logic circuit. In fact, the state assignment often has a major effect on circuit cost, and may interact with other factors, such as the choice of storage elements (e.g., D vs. J-K flip-flops) and the realization approach for excitation and output logic (e.g., sum-of-products, product-of-sums, or ad hoc).

So, how do we choose the best state assignment for a given problem? In general, the only formal way to find the *best* assignment is to try *all* the assignments. That's too much work, even for students. Instead, most digital designers rely on experience and several practical guidelines for making reasonable state assignments:

- Choose an initial coded state into which the machine can easily be forced at reset ($00\ldots00$ or $11\ldots11$ in typical circuits).

- Minimize the number of state variables that change on each transition.

- Maximize the number of state variables that don't change in a group of related states (that is, a group of states in which most of transitions stay in the group).

- Exploit symmetries in the problem specification and the corresponding symmetries in the state table. That is, suppose that one state or group of states means almost the same thing as another. Once an assignment has been established for the first, a similar assignment, differing only in one bit, should be used for the second.

- If there are unused states (i.e., if $s < 2^n$ where $n = \lceil \log_2 s \rceil$), then choose the "best" of the available state-variable combinations to achieve the foregoing goals. That is, don't limit the choice of coded states to the first s n-bit integers.

- Decompose the set of state variables into individual bits or fields where each bit or field has a well-defined meaning with respect to the input effects or output behavior of the machine.

- Consider using more than the minimum number of state variables to make a decomposed assignment possible.

Some of these ideas are incorporated in the "decomposed" state assignment in Table 5–11. As before, the initial state is 000, which is easy to force either asynchronously (applying the RESET signal to the flip-flop CLR inputs) or synchronously (by AND'ing RESET′ with all of the D flip-flop inputs). After this point, the assignment takes advantage of the fact that there are only four states in addition to INIT, which is a fairly "special" state that is never re-entered once the machine gets going.[5] Therefore, Q1 can be used to indicate whether or not the machine is in the INIT state, and Q2 and Q3 can be used to distinguish among the four non-INIT states.

The non-INIT states in the "decomposed" column of Table 5–11 appear to have been assigned in binary counting order, but that's just a coincidence. State bits Q2 and Q3 actually have individual meanings in the context of the state machine's inputs and output. Q3 gives the previous value of A, and Q2 indicates that the conditions for a 1 output are satisfied in the current state. By decomposing the state-bit meanings in this way, we can expect the next-state and output logic to be simpler than in a "random" assignment of Q2,Q3 combinations to the non-INIT states. We'll continue the state-machine design based on this assignment in later subsections.

Another useful state assignment, one that can be adapted to any state machine, is the *one-hot assignment* shown in Table 5–11. This assignment uses more than the minimum number of state variables—it uses one bit per state. In addition to being simple, a one-hot assignment has the advantage of usually leading to small excitation equations, since each flip-flop must be set to 1 for transitions into only one state. An obvious disadvantage of a one-hot assignment, especially for machines with many states, is that it requires (many) more than the minimum number of flip-flops. However, the one-hot encoding is ideal for a machine with s states that is required to have a set of 1-out-of-s coded outputs indicating its current state. The one-hot-coded flip-flop outputs can be used directly for this purpose, with no additional combinational output logic.

one-hot assignment

The last column of Table 5–11 is an "almost one-hot assignment" that uses the "no-hot" combination for the initial state. This makes a lot of sense for

---

[5] But don't get the idea that *all* state machines visit their initial state only during reset. Many machines are designed with an "idle" state that is entered both at reset and whenever the machine has nothing in particular to do.

two reasons: it's easy to initialize most storage devices to the all-0s state, and the initial state in this machine is never revisited once the machine gets going. Completing the state-machine design using this state assignment is considered in Exercises 5.24 and 5.27.

unused states   We promised earlier to consider the disposition of *unused states* when the number of states available with n flip-flops, 2^n, is greater than the number of states required, s. There are two approaches that make sense, depending on the application requirements:

- *Minimal risk.* This approach assumes that it is possible for the state machine somehow to get into one of the unused (or "illegal") states, perhaps because of a hardware failure, an unexpected input, or a design error. Therefore, all of the unused state-variable combinations are identified, and explicit next-state entries are made so that, for any input combination, the unused states go to the "initial" state, the "idle" state, or some other "safe" state. This is an automatic consequence of some design methodologies if the initial state is coded $00\ldots00$.

- *Minimal cost.* This approach assumes that the machine will never enter an unused state. Therefore, in the transition and excitation tables, the next-state entries of the unused states can be marked as "don't-cares." In most cases, this simplifies the excitation logic. However, the machine's behavior if it ever does enter an unused state may be totally whacko.

We'll look at both of these approaches as we complete the design of our example state machine.

5.4.4 Synthesis Using D Flip-Flops

Once we've assigned coded states to the named states of a machine, the rest of the design process is pretty much "turning the crank." In fact, in Section 7.4 we'll describe software tools that can turn the crank for you. Just so that you'll appreciate those tools, however, we'll go through the process by hand in this subsection.

transition table   Coded states are substituted for named states in the (possibly minimized) state table to obtain a *transition table*. The transition table shows the next coded state for each combination of current coded state and input. Table 5–12 shows the transition and output table that is obtained from the example state machine of Table 5–10 using the "decomposed" assignment of Table 5–11.

excitation table   The next step is to write an *excitation table* that shows, for each combination of coded state and input, the flip-flop excitation input values needed to make the machine go to the desired next coded state. This structure and content of this table depend on the type of flip-flops that are used (D, J-K, T, etc.). We *usually*

Table 5–12
Transition and output table for example problem.

Q1 Q2 Q3	A B 00	01	11	10	Z
000	100	100	101	101	0
100	110	110	101	101	0
101	100	100	111	111	0
110	110	110	111	101	1
111	100	110	111	111	1
	Q1* Q2* Q3*				

have a particular flip-flop type in mind at the beginning of a design—and we *certainly* do in this subsection, given its title. In fact, most state-machine designs nowadays use D flip-flops, because of their availability in both discrete packages and programmable logic devices, and because of their ease of use (compare with J-K flip-flops in the next subsection).

Of all flip-flop types, a D flip-flop has the simplest characteristic equation, $Q* = D$. Each D flip-flop in a state machine has a single excitation input, D, and the excitation table must show the value required at each flip-flop's D input for each coded-state/input combination. Table 5–13 shows the excitation table for our example problem. Since $D = Q*$, the excitation table is identical to the transition table, except for labeling of its entries. Thus, with D flip-flops, you don't really need to write a separate excitation table; you can just call the first table a *transition/excitation table*.[6]

The excitation table is like a truth table for three combinational logic functions (D1, D2, D3) of five variables (A, B, Q1, Q2, Q3). Accordingly, we can design circuits to realize these functions using any of the combinational design

transition/excitation table

---

[6]If I weren't getting paid by the page, I wouldn't have bothered to write a separate excitation table either.... Sorry, that's just another joke. Writing this for you is so much fun, they don't have to pay me (but they'd better!).

Table 5–13
Excitation and output table for Table 5–12, using D flip-flops.

Q1 Q2 Q3	A B 00	01	11	10	Z
000	100	100	101	101	0
100	110	110	101	101	0
101	100	100	111	111	0
110	110	110	111	101	1
111	100	110	111	111	1
	D1 D2 D3				

Figure 5–45
Excitation maps for D1, D2, and
D3 assuming that unused states
go to state 000.

excitation maps

methods at our disposal. In particular, we can transfer the information in the excitation table to Karnaugh maps, which we may call *excitation maps*, and find a minimal sum-of-products or product-of-sums expression for each function.

Excitation maps for our example state machine are shown in Figure 5–45. Each function, such as D1, has five variables and therefore uses a *5-variable Karnaugh map*. A 5-variable map is drawn as a pair of 4-variable maps, where cells in the same position in the two maps are considered to be adjacent. These maps are a bit unwieldy, but if you want to design by hand any but the most trivial state machines, you're going to get stuck with 5-variable maps and worse. At least we had the foresight to label the input combinations of the original state table in Karnaugh-map order, which makes it easier to transfer information to the maps in this step. However, note that the *states* were not assigned in Karnaugh-map order; in particular, the rows for states 110 and 111 are in the opposite order in the map as in the excitation table.

It is in this step, transferring the excitation table to excitation maps, that we discover why the excitation table is not quite a truth table—it does not specify functional values for *all* input combinations. In particular, the next-state information for the unused states, 001, 010, and 011, is not specified. Here we must make a choice, discussed in the preceding subsection, between a minimal-risk and a minimal-cost strategy for handling the unused states. Figure 5–45 has taken the minimal-risk approach: the next state for each unused state and input combination is 000, the INIT state. The three rows of colored 0s in each Karnaugh map are the result of this choice. With the maps completely filled in, we can now obtain minimal sum-of-products expressions for the flip-flop excitation inputs:

5-variable Karnaugh map

Figure 5–46
Excitation maps for D1, D2, and
D3 assuming that next states of
unused states are "don't-cares."

$$D1 = Q1 + Q2' \cdot Q3'$$
$$D2 = Q1 \cdot Q3' \cdot A' + Q1 \cdot Q3 \cdot A + Q1 \cdot Q2 \cdot B$$
$$D3 = Q1 \cdot A + Q2' \cdot Q3' \cdot A$$

If we choose instead to derive minimal-cost excitation equations, we write "don't-cares" in the next-state entries for the unused states. The colored d's in Figure 5–46 are the result of this choice. The excitation equations obtained from this map are somewhat simpler than before:

$$D1 = 1$$
$$D2 = Q1 \cdot Q3' \cdot A' + Q3 \cdot A + Q2 \cdot B$$
$$D3 = A$$

An output equation can easily be developed directly from the information in Table 5–13. The output equation is simpler than the excitation equations, because the output is a function of state only. We could use a Karnaugh map, but it's easy to find a minimal-risk output function algebraically, by writing it as the sum of the two coded states (110 and 111) in which Z is 1:

$$Z = Q1 \cdot Q2 \cdot Q3' + Q1 \cdot Q2 \cdot Q3$$
$$= Q1 \cdot Q2$$

For a minimal-cost output function, the value of Z is a "don't-care" for the unused states. This leads to an even simpler output function, $Z = Q2$.

Figure 5–47
Logic diagram for example state
machine using D flip-flops and
minimal-cost excitation logic.

At this point, we're just about done with the state-machine design. If the
state machine is going to be built with discrete flip-flops and gates, then the final
step is to draw a logic diagram, such as Figure 5–47. On the other hand, if we are
using a programmable logic device, then we only have to enter the excitation and
output equations into a computer file that specifies how to program the device,
as discussed in Section 7.3. Or, if we're lucky, we specified the machine using
a high-level state-machine description language (Section 7.4.3) in the first place,
and the computer did all the work in this subsection for us!

* 5.4.5 Synthesis Using J-K Flip-Flops

At one time, J-K flip-flops were popular for discrete SSI state-machine designs,
since a J-K flip-flop embeds more functionality than a D flip-flop in the same
size SSI package. By "more functionality" we mean that the combination of J
and K inputs yields more possibilities for controlling the flip-flop than a single D
input does. As a result, a state machine's excitation logic may be simpler using
J-K flip-flops than using D flip-flops, reducing package count when SSI gates are
used for the excitation logic.

However, the name of the game has changed in designs that use PLDs
and ASICs. As you'll realize after studying Section 7.2, the need to provide
separate AND-OR arrays for the J and K inputs of a J-K flip-flop would be a
distinct disadvantage in a PLD. And in ASIC technologies, a J-K flip-flop might
require over 25% more chip area than a D flip-flop. For example, in LSI Logic
Corp.'s LCA10000 series of CMOS gate arrays, an FJK1 J-K flip-flop macrocell

Table 5–14
Application table for J-K flip-flops.

Q	Q*	J	K
0	0	0	d
0	1	1	d
1	0	d	1
1	1	d	0

uses 9 "gate cells," while an FD1 D flip-flop macrocell uses only 7 gate cells. Therefore, a more cost-effective design usually results from sticking with D flip-flops and using the extra chip area for more complex excitation logic in just the cases where it's really needed. Still, we'll go through the J-K synthesis process just for fun.

Up through the state-assignment step, the design procedure with J-K flip-flops is basically the same as with D flip-flops. The only difference is that a designer might select a slightly different state assignment, knowing the sort of behavior that can easily be obtained from J-K flip-flops (for example, "toggling" by setting J and K to 1).

The big difference occurs in the derivation of an excitation table from the the transition table. With D flip-flops, the two tables are identical; using the D's characteristic equation, $Q* = D$, we simply substitute $D = Q*$ for each entry. With J-K flip-flops, each entry in the excitation table has twice as many bits as in the transition table, since there are two excitation inputs per flip-flop.

A J-K flip-flop's characteristic equation, $Q* = J \cdot Q' + K' \cdot Q$, cannot be rearranged to obtain independent equations for J and K. Instead, the required values for J and K are expressed as functions of Q and Q* in a *J-K application table*,   *J-K application table*
Table 5–14. According to the first row, if Q is currently 0, all that is required to obtain 0 as the next value of Q is to set J to 0; the value of K doesn't matter. Similarly, according to the third row, if Q is currently 1, the next value of Q will be 0 if K is 1, regardless of J's value. Each desired transition can be obtained by either of two different combinations on the J and K inputs, so we get a "don't-care" entry in each row of the application table.

To obtain a J-K excitation table, the designer must look at both the current and desired next value of each state bit in the transition table and substitute the corresponding pair of J and K values from the application table. For the transition table in Table 5–12, these substitutions produce the excitation table in Table 5–15. For example, in state 100 under input combination 00, Q1 is 1 and the required Q1* is 1; therefore, "d0" is entered for J1 K1. For the same state/input combination, Q2 is 0 and Q2* is 1, so "1d" is entered for J2 K2. Obviously, it takes quite a bit of patience and care to fill in the entire excitation table (a job best left to a computer).

Table 5–15   Excitation and output table for Table 5–12, using J-K flip-flops.

	A B				
Q1 Q2 Q3	00	01	11	10	Z
000	1d, 0d, 0d	1d, 0d, 0d	1d, 0d, 1d	1d, 0d, 1d	0
100	d0, 1d, 0d	d0, 1d, 0d	d0, 0d, 1d	d0, 0d, 1d	0
101	d0, 0d, d1	d0, 0d, d1	d0, 1d, d0	d0, 1d, d0	0
110	d0, 0d, 0d	d0, 0d, 0d	d0, 0d, 1d	d0, d1, 1d	1
111	d0, d1, d1	d0, 0d, d1	d0, 0d, d0	d0, 0d, d0	1
	J1 K1, J2 K2, J3 K3				

As in the D synthesis example of the preceding subsection, the excitation table is *almost* a truth table for the excitation functions. This information is transferred to Karnaugh maps in Figure 5–48. The excitation table does not specify next states for the unused states, so once again we must choose between the minimal-risk and minimal-cost approaches. The colored entries in the Karnaugh maps result from taking the minimal-risk approach. Even though the "safe" next state for unused states is 000, we can't just put 0s in the corresponding map cells, as we did in the D case. Instead, we still have to work with the application table to determine the proper combination of J and K needed to get $Q* = 0$ for each unused state entry. The following sum-of-products excitation equations can be obtained from the maps:

$$
\begin{aligned}
J1 &= Q2' \cdot Q3' & K1 &= 0 \\
J2 &= Q1 \cdot Q3' \cdot A' + Q1 \cdot Q3 \cdot A & K2 &= Q1' + Q3' \cdot A \cdot B' + Q3 \cdot A' \cdot B \\
J3 &= Q2' \cdot A + Q1 \cdot A & K3 &= Q1' + A'
\end{aligned}
$$

These equations take two more gates to realize than do the preceding subsection's minimal-risk equations using D flip-flops, so J-K flip-flops didn't save us anything in this example, least of all design time.

Excitation maps for the minimal-cost approach would have been somewhat easier to construct, since we could have just put d's in all of the unused state entries. Sum-of-products excitation equations obtained from the minimal-cost maps (not shown) are as follows:

$$
\begin{aligned}
J1 &= 1 & K1 &= 0 \\
J2 &= Q1 \cdot Q3' \cdot A' + Q3 \cdot A & K2 &= Q3' \cdot A \cdot B' + Q3 \cdot A' \cdot B \\
J3 &= A & K3 &= A'
\end{aligned}
$$

The state encoding for the J-K circuit is the same as in the D circuit for the transition table, so the output equation is the same, $Z = Q1 \cdot Q2$ for minimal risk, $Z = Q2$ for minimal cost.

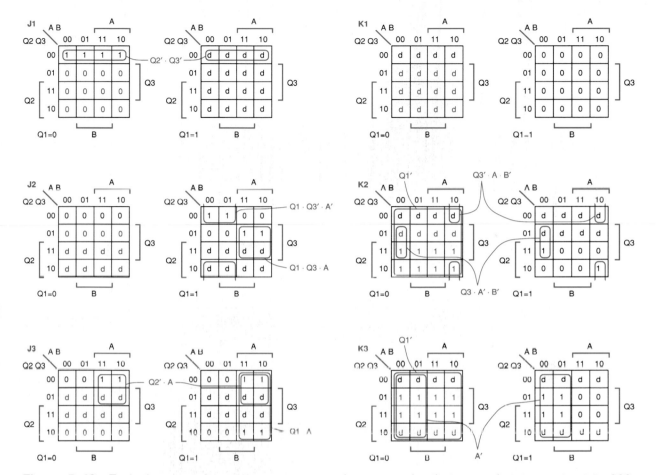

Figure 5–48 Excitation maps for J1, K1, J2, K2, J3, and K3, assuming that unused states go to state 000.

A logic diagram corresponding to the minimal-cost equations is shown in Figure 5–49. This circuit has two more gates than the minimal-cost D circuit in Figure 5–47, so J-K flip-flops still didn't save us anything.

5.4.6 More Design Examples Using D Flip-Flops

We'll conclude this section with two more state-machine design examples using D flip-flops. The first example is a "1s-counting machine":

Design a clocked synchronous state machine with two inputs, X and Y, and a single output, Z. The output should be 1 if the number of 1 inputs on X and Y since the machine started is a multiple of 4, and 0 otherwise.

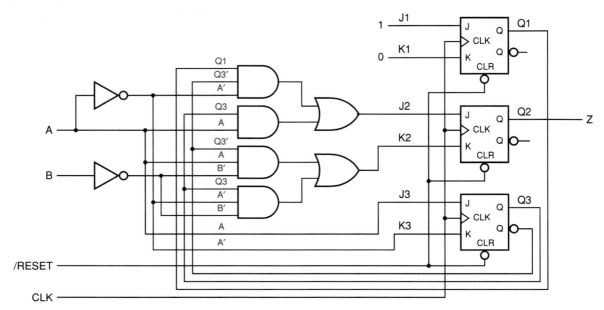

Figure 5–49 Logic diagram for example state machine using J-K flip-flops and minimal-cost excitation logic.

At first glance, you might think the machine needs an infinite number of states, to count the number of 1 inputs that occur over an arbitrarily long time. However, since the output indicates the number of inputs received *modulo 4*, four states are sufficient. We'll name the states S0–S3, where the total number of 1s received in state Si is i modulo 4. Table 5–16 is the resulting state and output table.

The 1s-counting machine can use two state variables to code its four states, with no unused states. In this case, there are only 4! possible assignments of coded states to named states. Still, we'll try only one of them. We'll assign coded states to the named states in Karnaugh-map order (00, 01, 11, 10) for two reasons: in this state table, it minimizes the number of state variables that

Table 5–16
State and output table for 1s-counting machine.

Meaning	S	X Y				Z
		00	01	11	10	
Got zero 1s (modulo 4)	S0	S0	S1	S2	S1	1
Got one 1 (modulo 4)	S1	S1	S2	S3	S2	0
Got two 1s (modulo 4)	S2	S2	S3	S0	S3	0
Got three 1s (modulo 4)	S3	S3	S0	S1	S0	0
				S*		

Table 5–17
Transition/excitation and output
table for 1s-counting machine.

Q1 Q2	X Y				Z
	00	01	11	10	
00	00	01	11	01	1
01	01	11	10	11	0
11	11	10	00	10	0
10	10	00	01	00	0
	Q1* Q2* or D1 D2				

change for most transitions, potentially simplifying the excitation equations; and
it simplifies the mechanical transfer of information to excitation maps.

A transition/excitation table based on our chosen state assignment is shown
in Table 5–17. Since we're using D flip-flops, the transition and excitation tables are the same. Corresponding Karnaugh maps for D1 and D2 are shown in
Figure 5–50. Since there are no unused states, all of the information we need is
in the excitation table; no choice is required between minimal-risk and minimal-
cost approaches. The excitation equations can be read from the maps, and the
output equation can be read directly from the transition/excitation table:

$$D1 \; = \; Q2 \cdot X' \cdot Y + Q1' \cdot X \cdot Y + Q1 \cdot X' \cdot Y' + Q2 \cdot X \cdot Y'$$

$$D2 \; = \; Q1' \cdot X' \cdot Y + Q1' \cdot X \cdot Y' + Q2 \cdot X' \cdot Y' + Q2' \cdot X \cdot Y$$

$$Z \; = \; Q1' \cdot Q2'$$

A logic diagram using D flip-flops and AND-OR or NAND-NAND excitation logic
can be drawn from these equations.

The second example is a "combination lock" state machine that activates
an "unlock" output when a certain binary input sequence is received:

Figure 5–50
Excitation maps for D1 and D2 in
1s-counting machine.

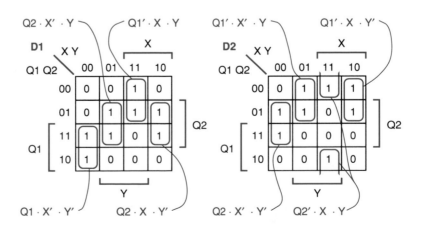

Table 5–18
State and output table for
combination-lock machine.

Meaning	S	X 0	1
Got zip	A	B, 01	A, 00
Got 0	B	B, 00	C, 01
Got 01	C	B, 00	D, 01
Got 011	D	E, 01	A, 00
Got 0110	E	B, 00	F, 01
Got 01101	F	B, 00	G, 01
Got 011011	G	E, 00	H, 01
Got 0110111	H	B, 11	A, 00
		S*, UNLK HINT	

Design a clocked synchronous state machine with one input, X, and
two outputs, UNLK and HINT. The UNLK output should be 1 if and
only if X is 0 and the sequence of inputs received on X at the preceding
seven clock ticks was 0110111. The HINT output should be 1 if and
only if the current value of X is the correct one to move the machine
closer to being in the "unlocked" state (with UNLK = 1).

It should be apparent from word description that this is a Mealy machine. The
UNLK output depends on both the past history of inputs and X's current value,
and HINT depends on both the state and the current X (indeed, if the current X
produces HINT = 0, the user may want to change X before the clock tick).

A state and output table for the combination lock is presented in Table 5–18.
In the initial state, A, we assume that we have received no inputs in the required
sequence; we're looking for the first 0 in the sequence. Therefore, as long as
we get 1 inputs, we stay in state A, and we move to state B when we receive
a 0. In state B, we're looking for a 1. If we get it, we move on to C; if we
don't, we can stay in B, since the 0 we just received might still turn out to be
the first 0 in the required sequence. In each successive state, we move on to
the next state if we get the correct input, and we go back to A or B if we get
the wrong one. An exception occurs in state G; if we get the wrong input (a 0)
there, the previous three inputs might still turn out to be the first three inputs of
the required sequence, so we go back to state E instead of B. In state H, we've
received the required sequence, so we set UNLK to 1 if X is 0. In each state, we
set HINT to 1 for the value of X that moves us closer to state H.

The combination lock's eight states can be coded with three state variables,
leaving no unused states. There are 8! state assignments to choose from. To
keep things simple, we'll use the simplest, and assign the states in binary counting order, yielding the transition/excitation table in Table 5–19. Corresponding

Table 5–19
Transition/excitation table for
combination-lock machine.

Q1 Q2 Q3	X 0	X 1
000	001, 01	000, 00
001	001, 00	010, 01
010	001, 00	011, 01
011	100, 01	000, 00
100	001, 00	101, 01
101	001, 00	110, 01
110	100, 00	111, 01
111	001, 11	000, 00

Q1* Q2* Q3*, UNLK HINT

Karnaugh maps for D1, D2, and D3 are shown in Figure 5–51. The excitation
equations can be read from the maps:

$$D1 = Q1 \cdot Q2' \cdot X + Q1' \cdot Q2 \cdot Q3 \cdot X' + Q1 \cdot Q2 \cdot Q3'$$

$$D2 = Q2' \cdot Q3 \cdot X + Q2 \cdot Q3' \cdot X$$

$$D3 = Q1 \cdot Q2' \cdot Q3' + Q1 \cdot Q3 \cdot X' + Q2' \cdot X' + Q3' \cdot Q1' \cdot X' + Q2 \cdot Q3' \cdot X$$

The output values are transferred from the transition/excitation and output table
to another set of maps in Figure 5–52. The corresponding output equations are:

$$UNLK = Q1 \cdot Q2 \cdot Q3 \cdot X'$$

$$HINT = Q1' \cdot Q2' \cdot Q3' \cdot X' + Q1 \cdot Q2' \cdot X + Q2' \cdot Q3 \cdot X + Q2 \cdot Q3 \cdot X' + Q2 \cdot Q3' \cdot X$$

Note that some product terms are repeated in the excitation and output equations,
yielding a slight savings in the cost of the AND-OR realization. If we went to the

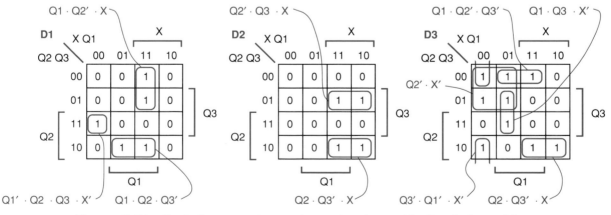

Figure 5–51   Excitation maps for D1, D2, and D3 in combination-lock machine.

Figure 5–52
Karnaugh maps for output functions UNLK and HINT in combination-lock machine.

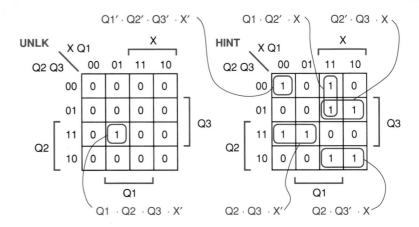

trouble of performing a formal multiple-output minimization of all five excitation and output functions, we could save two more gates (see Exercise 5.40).

* 5.5   FEEDBACK SEQUENTIAL CIRCUITS

The simple bistable and the various latches and flip-flops that we studied earlier in this chapter are all feedback sequential circuits. Each has one or more feedback loops that, ignoring their behavior during state transitions, store a 0 or a 1 at all times. The feedback loops are memory elements, and circuits' behavior depends on both the current inputs and the values stored in the loops.

Only rarely does a logic designer encounter a situation where a feedback sequential circuit must be analyzed or designed. The most commonly used feedback sequential circuits are the flip-flops and latches that are used as the building blocks in larger sequential-circuit designs. Their internal design and operating specifications are supplied by an IC manufacturer. Even an ASIC designer typically does not design gate-level flip-flop or latch circuits, since these elements are supplied in a "library" of commonly used functions in the particular ASIC technology. Still, you may be curious about how off-the-shelf flip-flops and latches "do their thing;" this section will show you how to analyze any such circuit built from conventional gates.

5.5.1   Analysis

fundamental-mode circuit

Feedback sequential circuits are the most common example of *fundamental-mode circuits*. In such circuits, inputs are not normally allowed to change simultaneously. The analysis procedure assumes that inputs change one at a time, allowing

---

*This section and all of its subsections are optional.

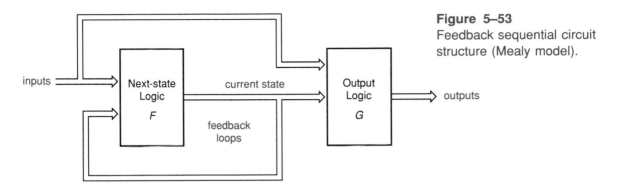

Figure 5–53
Feedback sequential circuit
structure (Mealy model).

enough time between successive changes for the circuit to settle into a stable internal state. This differs from clocked circuits, in which multiple inputs can change at almost arbitrary times without affecting the state, and all input values are sampled and state changes occur with respect to a clock signal.

Like clocked synchronous state machines, feedback sequential circuits may be structured as Mealy or Moore circuits, as shown in Figures 5–53 and 5–54. A circuit with n feedback loops has n binary state variables and 2^n states.

To analyze a feedback sequential circuit, we must break the feedback loops in Figure 5–53 or 5–54 so that the next value stored in each loop can be predicted as a function of the circuit inputs and the current value stored in all loops. Figure 5–55 shows how to do this for the NAND circuit for a D latch, which has only one feedback loop. We conceptually break the loop by inserting a fictional buffer in the loop as shown.[7] The output of the buffer, named Y, is the single state variable for this example.

Let us assume that the propagation delay of the fictional buffer is 10 ns (but any nonzero number will do), and that all of the other circuit components

---

[7]The way the circuit is drawn, it may look like there are two feedback loops. However, once we make one break as shown, there are no more loops. That is, each signal can be written as a combinational function of the other signals, not including itself.

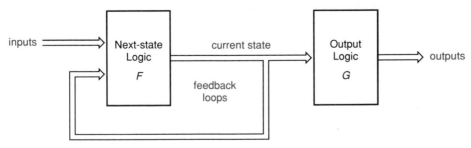

Figure 5–54   Feedback sequential circuit structure (Moore model).

Figure 5–55
Feedback analysis of a D latch.

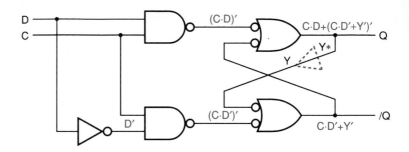

have zero delay. If we know the circuit's current state (Y) and inputs (D and C), then we can predict the value Y will have in 10 ns. The next value of Y, denoted Y*, is a combinational function of the current state and inputs. Thus, reading the

excitation equation   circuit diagram, we can write an *excitation equation* for Y*:

$$Y* = (C \cdot D) + (C \cdot D' + Y')'$$
$$= C \cdot D + C' \cdot Y + D \cdot Y$$

Now the state of the feedback loop (and the circuit) can be written as a function

transition table   of the current state and input, and enumerated by an *transition table* as shown in Table 5–20. Each cell in the transition table shows the fictional-buffer output value that will occur 10 ns (or whatever delay you've assumed) after the corresponding state and input combination occurs.

A transition table has one row for each possible combination of the state variables, so a circuit with n feedback loops has 2^n rows in its transition table. The table has one column for each possible input combination, so a circuit with m inputs has 2^m columns in its transition table.

By definition, a fundamental-mode circuit such as a feedback sequential circuit does not have a clock to tell it when to sample its inputs. Instead, we can imagine that the circuit is evaluating its current state and input *continuously* (or every 10 ns, if you prefer). As the result of each evaluation, it goes to a next state predicted by the transition table. Most of the time, the next state is the same as the current state; this is the essence of fundamental-mode operation. We make some definitions below that will help us study this behavior in more detail.

Table 5–20
Transition table for the D latch in
Figure 5–55.

Y	C D 00	01	11	10
0	0	0	1	0
1	1	1	1	0
		Y*		

Table 5-21
State table for the D latch in Figure 5-55, showing stable total states.

	C D			
S	00	01	11	10
S0	(S0)	(S0)	S1	(S0)
S1	(S1)	(S1)	(S1)	S0
			S*	

In a fundamental-mode circuit, a *total state* is a particular combination of *internal state* (the values stored in the feedback loops) and *input state* (the current value of the circuit inputs). A *stable total state* is a combination of internal state and input state such that the next internal state predicted by the transition table is the same as the current internal state. If the next internal state is different, then the combination is an *unstable total state*. We have rewritten the transition table for the D latch in Table 5-21 as a *state table*, giving the names S0 and S1 to the states and drawing a circle around the stable total states.

total state
internal state
input state
stable total state

unstable total state
state table

To complete the analysis of the circuit, we must also determine how the outputs behave as functions of the internal state and inputs. There are two outputs, and hence two *output equations*:

output equation

$$Q = C \cdot D + C' \cdot Y + D \cdot Y$$
$$/Q = C \cdot D' + Y'$$

Note that Q and /Q are *outputs*, not state variables. Even though the circuit has two output variables, which can theoretically take on four combinations, it has only one state variable Y, and hence only two states.

The output values predicted by the Q and /Q equations can be incorporated in a combined state and output table that completely describes the operation of the circuit, as shown in Table 5-22. In spite of their names, it is possible for Q and /Q to have the same value (1) momentarily, during the transition from S0 to S1 under the CD = 11 column of Table 5-22.

We can now predict the behavior of the circuit from the transition and output table. First of all, notice that we have written the column labels in Table 5-22 in

Table 5-22
State and output table for the D latch in Figure 5-55.

	C D			
S	00	01	11	10
S0	(S0), 01	(S0), 01	S1 , 11	(S0), 01
S1	(S1), 10	(S1), 10	(S1), 10	S0 , 01
			S*, Q /Q	

Table 5–23

Analysis of the D latch in
Figure 5–55 for a few transitions.

S	00	01	11	10
		C D		
S0	(S0), 01	(S0), 01	S1 , 11	(S0), 01
S1	(S1), 10	(S1), 10	(S1), 10	S0 , 01

S*, Q /Q

"Karnaugh map" order, so that only a single input bit changes between adjacent columns of the table. This layout helps our analysis, because we assume that only one input changes at a time, and that the circuit always reaches a stable total state before another input changes.

At any time, the circuit is in a particular internal state and a particular input is applied to it; we called this combination the total state of the circuit. Let us start at the stable total state "S0/00" (S = S0, C D = 00), as shown in Table 5–23. Now suppose that we change D to 1. The total state moves to one cell to the right; we have a new stable total state, S0/01. The D input is different but the internal state and output are the same as before. Next, let us change C to 1. The total state moves one cell to the right to S0/11, which is unstable. The next-state entry in this cell sends the circuit to internal state S1, so the total state moves down one cell, to S1/11. Once again we look at the next-state entry in the new cell, and we find that we have reached a stable total state. We can trace the behavior of the circuit for any desired sequence of single input changes in this way.

Now we can revisit the question of simultaneous input changes. Even though "almost simultaneous" input changes may occur in practice, we must assume that nothing happens simultaneously in order to analyze the behavior of sequential circuits. The impossibility of simultaneous events is supported by the varying delays of circuit components themselves, which depend on voltage, temperature, and fabrication parameters. What this tells us is that a set of n inputs that appear to us to change "simultaneously" may actually change in any of $n!$ different orders from the point of view of the circuit operation.

For example, consider the operation of the D latch as shown in Table 5–24. Let us assume that it starts in stable total state S1/11. Now suppose that C and D are both "simultaneously" set to 0. In reality, the circuit behaves as if one or the other input went to 0 first. Suppose that C changes first. Then the sequence of two left-pointing arrows in the table tells us that the circuit goes to stable total state S1/00. However, if D changes first, then the other sequence of arrows tells us that the circuit goes to stable total state S0/00. So the final state of the circuit is unpredictable, a clue that the feedback loop may actually become metastable

Table 5–24
Multiple input changes for a D latch.

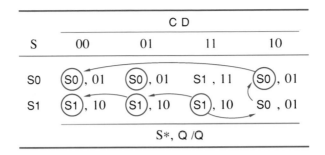

	C D			
S	00	01	11	10
S0	(S0), 01	(S0), 01	S1 , 11	(S0), 01
S1	(S1), 10	(S1), 10	(S1), 10	S0 , 01
			S*, Q /Q	

if we set C and D to 0 simultaneously. The time span over which this view of simultaneity is relevant is the setup and hold time window of the D latch.

Simultaneous input changes don't always cause unpredictable behavior. However, we must analyze the effects of all possible orderings of signal changes to determine this; if all orderings give the same result, then the circuit output is predictable. For example, consider the behavior of the D latch starting in total state S0/00 with C and D simultaneously changing from 0 to 1; it always ends up in total state S1/11.

5.5.2 Analyzing Circuits with Multiple Feedback Loops

In circuits with multiple feedback loops, we must break all of the loops, creating one fictional buffer and state variable for each loop that we break. There are many possible ways, which mathematicians call *cut sets*, to break the loops in a given circuit, so how do we know which one is best? The answer is that any *minimal cut set*—a cut set with a minimum number of cuts—is fine. Mathematicians can give you an algorithm for finding a minimal cut set, but as a digital designer working on small circuits, you can just eyeball the circuit to find one.

cut set
minimal cut set

Different cut sets for a circuit lead to different excitation equations, transition tables, and state/output tables. However, the stable total states derived from one minimal cut set correspond one-to-one to the stable total states derived from any other minimal cut set for the same circuit. That is, state/output tables derived from different minimal cut sets display the same input/output behavior, with only the names and coding of the states changed (to protect the innocent?).

If you use more than the minimal number of cuts to analyze a feedback sequential circuit, the resulting state/output table will still describe the circuit correctly. However, it will use 2^m times as many states as necessary, where m is the number of extra cuts. Formal state-minimization procedures can be used to reduce this larger table to the proper size, but it's a much better idea to select a minimal cut set in the first place.

A good example of a sequential circuit with multiple feedback loops is the commercial circuit design for a positive edge-triggered D flip-flop that we

Figure 5–56
Simplified positive edge-triggered
D flip-flop for analysis.

showed in Figure 5–20. The circuit is redrawn in simplified form in Figure 5–56, assuming that the original circuit's /PR and /CLR inputs are never asserted, and also showing fictional buffers to break the three feedback loops. Interestingly, the three loops give rise to eight states, compared with the minimum of four states used by the two-loop design in Figure 5–19. We'll readdress this curious difference later.

The following excitation and output equations can be derived from the logic diagram in Figure 5–56:

$$Y1* = Y2 \cdot D + Y1 \cdot CLK$$
$$Y2* = Y1 + CLK' + Y2 \cdot D$$
$$Y3* = Y1 \cdot CLK + Y1 \cdot Y3 + Y3 \cdot CLK' + Y2 \cdot Y3 \cdot D$$
$$Q = Y1 \cdot CLK + Y1 \cdot Y3 + Y3 \cdot CLK' + Y2 \cdot Y3 \cdot D$$
$$/Q = Y3' + Y1' \cdot Y2' \cdot CLK + Y1' \cdot CLK \cdot D'$$

The corresponding transition table is shown in Table 5–25, with the stable total states circled. Before we analyze this table further, we must introduce the concept of "races."

5.5.3 Races

race

In a feedback sequential circuit, a *race* is said to occur when multiple internal variables change state as a result of a single input changing state. In the example of Table 5–25, a race occurs in stable total state 011/00 when CLK is changed

Table 5–25
Transition table for the D flip-flop
in Figure 5–56.

	CLK D			
Y1 Y2 Y3	00	01	11	10
000	010	010	(000)	(000)
001	011	011	000	000
010	(010)	110	110	000
011	(011)	111	111	000
100	010	010	111	111
101	011	011	111	111
110	010	(110)	111	111
111	011	(111)	(111)	(111)
		Y1* Y2* Y3*		

from 0 to 1. The table indicates that the next internal state is 000, a 2-variable change from 011.

As we've discussed previously, logic signals never really change "simultaneously." Thus, the internal state may make the change 000→011 as either 000→001→011 or 000→010→011. Table 5–26 indicates that the example circuit, starting in total state 011/00, should go to total state 000/10 when CLK changes from 0 to 1. However, it may temporarily visit total state 001/10 or 010/10 along the way. That's OK, because the next internal state for both of these temporary states is 000; therefore, even in the temporary states, the excitation logic continues to drive the feedback loops toward the same stable total state, 000/10. Since the final state does not depend on the order in which the state variables change, this is called a *noncritical race*.

noncritical race

Table 5–26
Portion of a transition table
showing a noncritical race.

	CLK D			
Y1 Y2 Y3	00	01	11	10
000	010	010	(000)	(000)
001	011	011	000	000
010	(010)	110	110	000
011	(011)	111	111	000
		Y1* Y2* Y3*		

Table 5–27
A transition table containing a critical race.

Y1 Y2 Y3	CLK D			
	00	01	11	10
000	010	010	(000)	(000)
001	011	011	000	000
010	(010)	110	110	110
011	(011)	111	111	000
100	010	010	111	111
101	011	011	111	111
110	010	(110)	111	111
111	011	(111)	(111)	(111)
	Y1* Y2* Y3*			

critical race

Now suppose that the next-state entry for total state 010/10 is changed to 110, as shown in Table 5–27, and consider the case that we just analyzed. Starting in stable total state 011/00 and changing CLK to 1, the circuit may end up in internal state 000 or 111 depending on the order and speed of the internal variable changes. This is called a *critical race*. The designer of a feedback-based sequential circuit must ensure that its transition table does not contain any critical races. Otherwise, the circuit may operate unpredictably, with the next state for racy transitions depending on factors like temperature, voltage, and the phase of the moon.

5.5.4 State Tables and Flow Tables

Analysis of the real transition table for our example circuit, Table 5–25, shows that it does not have any critical races; in fact, it has no races except the non-critical one we identified earlier. Once we've determined this fact, we no longer need to refer to state variables. Instead, we can name the state-variable combinations and determine the output values for each state/input combination to obtain a state/output table such as Table 5–28.

The state table shows that for some single input changes, the circuit takes multiple "hops" to get to a new stable total state. For example, in state S0/11, an input change to 01 sends the circuit first to state S2 and then to stable total *flow table* state S6/01. A *flow table* eliminates multiple hops and shows only the ultimate destination for each transition. The flow table also eliminates the rows for unused internal states—ones that are stable for no input combination—and eliminates

Table 5-28
State and output table for the D
flip-flop in Figure 5-56.

	CLK D			
S	00	01	11	10
S0	S2 , 01	S2 , 01	(S0), 01	(S0), 01
S1	S3 , 10	S3 , 10	S0 , 01	S0 , 01
S2	(S2), 01	S6 , 01	S6 , 01	S0 , 01
S3	(S3), 10	S7 , 10	S7 , 10	S0 , 01
S4	S2 , 01	S2 , 01	S7 , 11	S7 , 11
S5	S3 , 10	S3 , 10	S7 , 10	S7 , 10
S6	S2 , 01	(S6), 01	S7 , 11	S7 , 11
S7	S3 , 10	(S7), 10	(S7), 10	(S7), 10
		S*, Q /Q		

the next-state entries for total states that cannot be reached from a stable total
~ate as the result of a single input change. Using these rules, Table 5–29 shows
the flow table for our positive edge-triggered flip-flop example.

The flip-flop's edge-triggered behavior can be observed in the series of state
transitions shown in Table 5–30. Let us assume that the flip-flop starts in internal
state S0/10. That is, the flip-flop is storing a 0 (since $Q = 0$), CLK is 1, and D is
0. Now suppose that D changes to 1; the flow table shows that we move one cell
to the left, still a stable total state with the same output value. We can change D
between 0 and 1 as much as we want, and just bounce back and forth between
these two cells. However, once we change CLK to 0, we move to internal state

Table 5-29
Flow and output table for the D
flip-flop in Figure 5-56.

	CLK D			
S	00	01	11	10
S0	S2 , 01	S6 , 01	(S0), 01	(S0), 01
S2	(S2), 01	S6 , 01	— , —	S0 , 01
S3	(S3), 10	S7 , 10	— , —	S0 , 01
S6	S2 , 01	(S6), 01	S7 , 11	— , —
S7	S3 , 10	(S7), 10	(S7), 10	(S7), 10
		S*, Q /Q		

Table 5–30

Flow and output table showing D flip-flop's edge-triggered behavior.

S	00	01	11	10
	CLK D			
S0	S2 , 01	S6 , 01	(S0), 01	(S0), 01
S2	(S2), 01	S6 , 01	— , —	S0 , 01
S3	(S3), 10	S7 , 10	— , —	S0 , 01
S6	S2 , 01	(S6), 01	S7 , 11	— , —
S7	S3 , 10	(S7), 10	(S7), 10	(S7), 10

S*, Q /Q

S2 or S6, depending on whether D was 0 or 1 at the time; but still the output is unchanged. Once again, we can change D between 0 and 1 as much as we want, this time bouncing between S2 and S6 without changing the output.

The moment of truth finally comes when CLK changes to 1. Depending on whether we are in S2 or S6, we go back to S0 (leaving Q at 0) or to S7 (setting Q to 1). Similar behavior involving S3 and S7 can be observed on a rising clock edge that causes Q to change from 1 to 0.

5.5.5   Comments

In Figure 5–19, we showed a circuit for a positive edge-triggered D flip-flop with only two feedback loops and hence four states. The circuit that we just analyzed has three loops and eight states. Even after eliminating unused states, the flow table has five states. However, a formal state-minimization procedure can be used to show that states S0 and S2 are "compatible," so that they can be merged into a single state SB that handles the transitions for both, as shown in Table 5–31. Thus, the job really could have been done by a four-state circuit. In fact, in Exercise 5.41 you'll show that the circuit in Figure 5–19 does the job specified by Table 5–31.

The feedback sequential circuits that we've analyzed in this section exhibit quite reasonable behavior since, after all, they are latch and flip-flop circuits that have been used for years. However, if we throw together a "random" collection of gates and feedback loops, we won't necessarily get "reasonable" sequential circuit behavior. In a few rare cases, we may not get a sequential circuit at all (see Exercise 5.42), and in many cases, the circuit may be unstable for some or all input combinations (see Exercise 5.47).

Table 5–31
Reduced flow and output table for a positive edge-triggered D flip-flop.

S	00	01	11	10
			CLK D	
SB	(SB), 01	S6 , 01	(SB), 01	(SB), 01
S3	(S3), 10	S7 , 10	— , —	SB , 01
S6	SB , 01	(S6), 01	S7 , 11	— , —
S7	S3 , 10	(S7), 10	(S7), 10	(S7), 10

$$S*, Q /Q$$

* 5.6 FEEDBACK SEQUENTIAL CIRCUIT DESIGN

It's sometimes useful to design a small feedback sequential circuit, such as a specialized latch or a pulse catcher; this section will show you how. It's even possible that you might go on to be an IC designer, and be responsible for designing high-performance latches and flip-flops from scratch. This section will serve as an introduction to the basic concepts you'll need, but you'll still need considerably more study, experience, and finesse to do it right.

5.6.1 Latches

Although the design of feedback sequential circuits is generally a hard problem, some circuits can be designed pretty easily. Any circuit with one feedback loop is just a variation of an S-R or D latch. It has the general structure shown in Figure 5–57, and an excitation equation with the following format:

$$Q* = (\text{forcing term}) + (\text{holding term}) \cdot Q$$

For example, the excitation equations for S-R and D latches are

$$Q* = S + R' \cdot Q$$
$$Q* = C \cdot D + C' \cdot Q$$

Corresponding circuits are shown in Figure 5–58(a) and (b).

Figure 5–57
General structure of a latch.

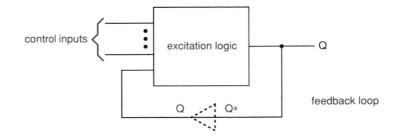

Figure 5–58
Latch circuits: (a) S-R latch;
(b) unreliable D latch; (c) hazard-
free D latch.

(a)

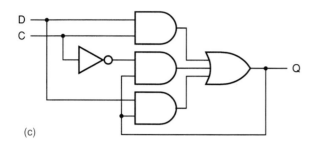

(c)

(b)

hazard-free
excitation logic

In general, *the excitation logic in a feedback sequential circuit must be hazard-free;* we'll demonstrate this fact by way of an example. Figure 5–59(a) is a Karnaugh map for the D-latch excitation circuit of Figure 5–58(b). The map exhibits a static-1 hazard when D and Q are 1 and C is changing. Unfortunately, the latch's feedback loop may not hold its value if a hazard-induced glitch occurs. For example, consider what happens in Figure 5–58(b) when D and Q are 1 and C changes from 1 to 0; the circuit should latch a 1. However, unless the inverter is very fast, the output of the top AND gate goes to 0 before the output of the bottom one goes to 1, the OR-gate output goes to 0, and the feedback loop stores a 0.

Hazards can be eliminated using the methods described in Section 3.5. In the D latch, we simply include the consensus term in the excitation equation:

$$Q* = C \cdot D + C' \cdot Q + D \cdot Q$$

Figure 5–58(c) shows the corresponding hazard-free, correct D-latch circuit.

$$Q* = C \cdot D + C \cdot Q'$$

$$Q* = C \cdot D + C \cdot Q' + D \cdot Q$$

Figure 5–59 Karnaugh maps for D-latch excitation functions: (a) original, containing a static-1 hazard; (b) hazard eliminated.

Now, suppose we need a specialized "D" latch with three data inputs, D1, D2, and D3, that is to store a 1 only if D1–D3 = 010. We can convert this word description into an excitation equation that mimics the equation for the simple D latch:

$$Q* \;=\; C \cdot (D1' \cdot D2 \cdot D3') + C' \cdot Q$$

Eliminating hazards, we get

$$Q* \;=\; C \cdot D1' \cdot D2 \cdot D3' + C' \cdot Q + D1' \cdot D2 \cdot D3' \cdot Q$$

This excitation equation can be realized with discrete gates or in a PLD, as we'll show in Section 7.2.12.

In some cases, the need to cover hazards can cause an explosion in the number of product terms in a two-level realization of the excitation logic. For example, suppose we need a specialized latch two control inputs, C1 and C2, and three data inputs as before. The latch is to be "open" only if both control inputs are 1, and is to store a 1 if any data input is 1. The minimal excitation equation is

$$Q* \;=\; C1 \cdot C2 \cdot (D1 + D2 + D3) + (C1 + C2)' \cdot Q$$
$$\;=\; C1 \cdot C2 \cdot D1 + C1 \; C2 \; D2 + C1 \; C2 \cdot D3 + C1' \cdot Q + C2' \cdot Q$$

However, six consensus terms are needed to eliminate all the hazards (see Exercise 5.49).

5.6.2 Designing Fundamental-Mode Flow Tables

To design feedback sequential circuits more complex than latches, we must first convert the word description into a flow table. Once we have a flow table, we can turn the crank (with some effort) to obtain a circuit.

When we construct the flow table for a feedback sequential circuit, we give each state a meaning in the context of the problem, much like we did in the design of clocked state machines. However, it's easier to get confused when constructing the flow table for a feedback sequential circuit. Therefore, the recommended procedure is to construct a *primitive flow table*—one that has   *primitive flow table* only one stable total state in each row. Since there is only one stable state per row, the output may be shown as a function of state only.

In a primitive flow table, each state has a more precise "meaning" than it might otherwise have, and the table's structure clearly displays the underlying fundamental-mode assumption: inputs change one at a time, with enough time between changes for the circuit to settle into a new stable state. A primitive flow table usually has extra states, but we can "turn the crank" later to minimize the number of states once we have a flow table that we believe to be correct.

Figure 5–60   Typical functional behavior of a pulse-catching circuit.

We'll use the following problem, a "pulse-catching" circuit, to demonstrate flow-table design.

> Design a feedback sequential circuit with two inputs, P (pulse) and R (reset), and a single output Z that is normally 0. The output should be set to 1 whenever a 0-to-1 transition occurs on P, and should be reset to 0 whenever R is 1. Typical functional behavior is shown in Figure 5–60.

Table 5–32 is a primitive flow table for the pulse catcher. Let's walk through how this table was developed.

We assume that the pulse catcher is initially idle, with P and R both 0; this is the IDLE state, with Z = 0. In this state, if reset occurs (R = 1), we could probably stay in the same state, but since this is supposed to be a *primitive* flow table, we create a new state, RES1, so as not to have two stable total states in the same row. On the other hand, if a pulse occurs (P = 1) in the IDLE state, we definitely want to go to a different state, which we've named PLS1, since we've caught a pulse and we must set the output to 1. Input combination 11 is not considered in the IDLE state, because of the fundamental-mode assumption that

Table 5–32   Primitive flow table for pulse-catching circuit.

Meaning	S	00	01	11	10	Z
		P R				
Idle, waiting for pulse	IDLE	(IDLE)	RES1	—	PLS1	0
Reset, no pulse	RES1	IDLE	(RES1)	RES2	—	0
Got pulse, output on	PLS1	PLS2	—	RES2	(PLS1)	1
Reset, got pulse	RES2	—	RES1	(RES2)	PLSN	0
Pulse gone, output on	PLS2	(PLS2)	RES1	—	PLS1	1
Got pulse, but output off	PLSN	IDLE	—	RES2	(PLSN)	0

S*

only one input changes at a time; we assume the the circuit always makes it to another stable state before input combination 11 can occur.

Next, we fill in the next-state entries for the newly created RES1 state. If reset goes away, we can go back to the IDLE state. If a pulse occurs, we must remain in a "reset" state since, according to the timing diagram, a 0-to-1 transition that occurs while R is 1 is ignored. Again, to keep the flow table in primitive form, we must create a new state for this case, RES2.

Now that we have one stable total state in each column, we may be able to go to existing states for more transitions, instead of always defining new states. Sure enough, starting in stable total state PLS1/10, if R = 1 we can go to RES2, which fits the requirement of producing a 0 output. On the other hand, where should we go if P = 0? IDLE is a stable total state in the 00 column, but it produces the wrong output value. In PLS1, we've gotten a pulse and haven't seen a reset, so if the pulse goes away, we should go to a state that still has Z = 1. Thus, we must create a new state PLS2 for this case.

In RES2, we can safely go to RES1 if the pulse goes away. However, we've got to be careful if reset goes away, as shown in the timing diagram. Since we've already passed the pulse's 0-to-1 transition, we can't go to the PLS1 state since that would give us a 1 output. Instead, we create a new state PLS2 with a 0 output.

Finally, we can complete the next-state entries for PLS2 and PLSN without creating any new states. Notice that starting in PLS2, we bounce back and forth between PLS2 and PLS1 and maintain a continuous 1 output if we get a series of pulses without an intervening reset input.

5.6.3 Flow-Table Minimization

As we mentioned earlier, a primitive flow table usually has more states than required. However, there exists a formal procedure, discussed in the References, for minimizing the number of states in a flow table. This procedure is often complicated by the existence of don't-care entries in the flow table.

Table 5–33
Reduced flow table for pulse-catching circuit.

		P R			
S	00	01	11	10	Z
IDLE	(IDLE)	(IDLE)	RES	PLS	0
PLS	(PLS)	IDLE	RES	(PLS)	1
RES	IDLE	IDLE	(RES)	(RES)	0
		S*			

Fortunately, our example flow table is small and simple enough to minimize by inspection. States IDLE and RES1 produce the same output, and they have the same next-state entry for input combinations where they are both specified. Therefore, they are compatible and may be replaced by a single state (IDLE) in a reduced flow table. The same can be said for states PLS1 and PLS2 (replaced by PLS) and RES2 and PLSN (replaced by RES). The resulting reduced flow table, which has only three states, is shown in Table 5–33.

5.6.4 Race-Free State Assignment

The next somewhat creative (read, "difficult") step in feedback sequential circuit design is to find a race-free assignment of coded states to the named states in the reduced flow table. Recall from Section 5.5.3 that a race occurs when multiple internal variables change state as a result of a single input change. A feedback-based sequential circuit must not contain any critical races; otherwise, the circuit may operate unpredictably. As we'll see, eliminating races often necessitates *increasing* the number of states in the circuit.

state adjacency diagram

A circuit's potential for having races in its transition table can be analyzed by means of a *state adjacency diagram* for its flow table. The adjacency diagram is a simplified state diagram that omits self-loops and does not show the direction of other transitions (A→B looks the same as B→A), or the input combinations that cause them. Table 5–34 is an example fundamental-mode flow table and Figure 5–61(a) is the corresponding adjacency diagram.

adjacent states

Two states are *adjacent* if there is an arc between them in the state adjacency diagram. For race-free transitions, adjacent coded states must differ in only one bit. If two states A and B are adjacent, it doesn't matter whether the orginal flow table had transitions from A to B, from B to A, or both. Any one of these transitions is a race if A and B differ in more than one variable. That's why we don't bother to show the direction of transitions in an adjacency diagram.

The problem of finding a race-free assignment of states to n state variables

Table 5–34
Example flow table for state assignment problem.

S	00	01	11	10
		X Y		
A	(A)	B	(A)	B
B	(B)	(B)	D	(B)
C	(C)	B	A	(C)
D	(D)	B	(D)	C
		S*		

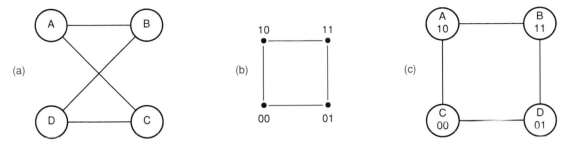

Figure 5–61   State-assignment example: (a) adjacency diagram; (b) a 2-cube; (c) one of eight possible race-free state assignments.

is equivalent to the problem of mapping the nodes and arcs of the adjacency diagram onto the nodes and arcs of an n-cube. In Figure 5–61, the problem is to map the adjacency diagram (a) onto a 2-cube (b). You can visually identify eight ways to do this (four rotations times two flips), one of which produces the state assignment shown in (c).

Figure 5–62(a) is an adjacency diagram for our pulse-catching circuit, based on the reduced flow table in Table 5–34. Clearly, there's no way to map this "triangle" of states onto a 2-cube. At this point, we have to go back and modify the original flow table. In particular, the flow table tells us the destination state that we *eventually* must reach for each transition, but it doesn't prevent us from going through other states on the way. As shown in Table 5–35, we can create a new state RESA and make the transition from PLS to RES by going through RESA. The modified state table has the new adjacency diagram shown in Figure 5–62(b), which has many race-free assignments possible. A transition table based on the assignment in (c) is shown in Table 5–36. Note that the PLS→RESA→RES transition will be slower than the other transitions in the original flow table, because it requires two internal state changes, with two propagation delays through the feedback loops.

Even though we added a state in the previous example, we were still able to

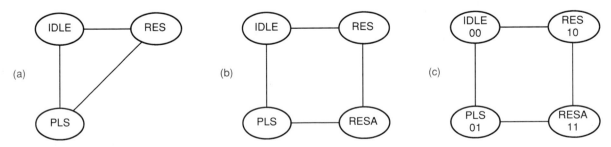

Figure 5–62   Adjacency diagrams for the pulse catcher: (a) using original flow table; (b) after adding a state; (c) showing one of eight possible race-free state assignments.

Table 5–35
State table allowing a race-free assignment for pulse-catching circuit.

S	P R 00	01	11	10	Z
IDLE	(IDLE)	(IDLE)	RES	PLS	0
PLS	(PLS)	IDLE	RESA	(PLS)	1
RESA	—	—	RES	—	–
RES	IDLE	IDLE	(RES)	(RES)	0
			S*		

get by with just two state variables. However, we may sometimes have to add one or more state variables to make a race-free assignment. Figure 5–63(a) shows the worst possible adjacency diagram for four states—every state is adjacent to every other state. Clearly, this adjacency diagram cannot be mapped onto a 2-cube. However, there is a race-free assignment of states to a 3-cube, shown in (b), where each state in the original flow table is represented by two equivalent states in the final state table. Both states in a pair, such as A1 and A2, are equivalent and produce the same output. Each state is adjacent to one of the states in every other pair, so a race-free transition may be selected for each next-state entry.

In the general case of a flow table with 2^n rows, it can be shown that a race-free assignment can be obtained using $2n - 1$ state variables (see the References). However, most designers never design fundamental-mode circuits with more than a few states, so the general case is of little more than academic interest.

5.6.5 Excitation Equations

Once we have a race-free transition table for a circuit, we can just "turn the crank" to obtain excitation equations for the feedback loops. Figure 5–64 shows

Table 5–36
Race-free transition table for pulse-catching circuit.

Y1 Y2	P R 00	01	11	10	Z
00	(00)	(00)	10	01	0
01	(01)	00	11	(01)	1
11	—	—	10	—	–
10	00	00	(10)	(10)	0
			Y1* Y2*		

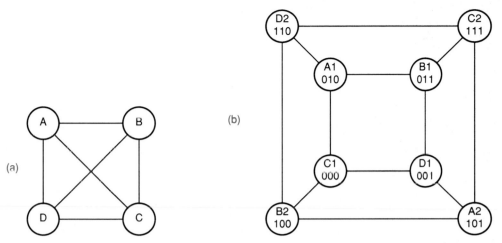

Figure 5–63 A worst-case scenario: (a) 4-state adjacency diagram; (b) assignment
using pairs of equivalent states.

Karnaugh maps derived from Table 5–36, the pulse catcher's transition table.
Notice that "don't-care" next-state and output entries give rise to corresponding
entries in the maps, simplifying the excitation and output logic. The minimal
sum-of-products excitation and output equations derived from the maps are as
follows:

$$Y1* = P \cdot R + P \cdot Y1$$
$$Y2* = Y2 \cdot R' + Y1' \cdot Y2 \cdot P + Y1' \cdot P \cdot R'$$
$$Z = Y2$$

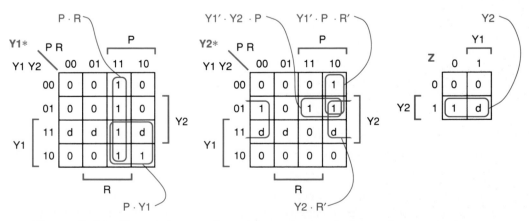

Figure 5–64 Karnaugh maps for pulse-catcher excitation and output logic.

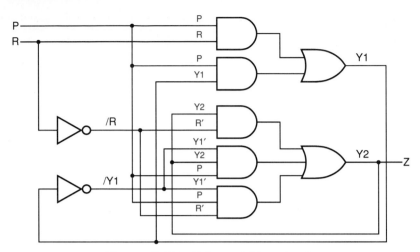

Figure 5–65   Pulse-catching circuit.

Recall that the excitation logic in a feedback sequential circuit must be hazard free. The sum-of-products expressions above happen to be hazard free as well as minimal. The logic diagram of Figure 5–65 uses these expressions to build the pulse-catching circuit.

5.6.6   Essential Hazards

After all this effort, you'd think that we'd have a pulse-catching circuit that would work reliably all of the time. Well, we're not quite there yet. A fundamental-mode circuit must generally satisfy five requirements for proper operation:

(1) Only one input signal changes at a time, with a minimum bound between successive input changes.

(2) There is a maximum propagation delay through excitation logic and feed-back paths; this maximum is less than the time between successive input changes.

(3) The state assignment (transition table) is free of critical races.

(4) The excitation logic is hazard free.

(5) The minimum propagation delay through excitation logic and feedback paths is greater than the maximum timing skew through "input logic."

Without the first requirement, it would be impossible to satisfy the major premise of fundamental-mode operation—that the circuit has time to settle into a stable total state between successive input changes. The second requirement says that the excitation logic is fast enough to do just that. The third requirement ensures

Table 5–37
Transition table for pulse-catching circuit, exhibiting an essential hazard.

Y1 Y2	P R 00	01	11	10	Z
00	00	00	10	01	0
01	01	00	11	01	1
11	—	—	10	—	—
10	00	00	10	10	0

Y1* Y2*

that the proper state changes are made even if the excitation circuits for different state variables have different delays. The fourth requirement ensures that state variables that aren't supposed to change on a particular transition don't.

The last requirement deals with subtle timing-dependent errors that can occur in fundamental-mode circuits, even ones that satisfy the first four requirements. An *essential hazard* is the possibility of a circuit going to an erroneous *essential hazard* next state as the result of a single input change; the error occurs if the input change is not seen by all of the excitation circuits before the resulting state-variable transition(s) propagate back to the inputs of the excitation circuits. In a world where "faster is better" is the usual rule, a designer may sometimes have to *slow down* excitation logic to mask these hazards.

Essential hazards are best explained in terms of an example, our pulse-catching circuit. Suppose we built the our circuit on a PCB or a chip, and we (or, more likely, our CAD system) inadvertently connected input signal P through a long, slow path at the point shown in Figure 5–66. Let's assume that this delay is longer than the propagation delay of the AND-OR excitation logic.

Now consider what can happen if P R = 10, the circuit is in internal state 10, and P changes from 1 to 0. According to the transition table, which is repeated in Table 5–37, the circuit should go to internal state 00, and that's that. However, let's look at the actual operation of the circuit as traced in Figure 5–66:

- (Changes shown in non-italic black.) The first thing that happens after P changes is that Y1 changes from 1 to 0, and /Y1 changes from 0 to 1 shortly thereafter. Now the circuit is in internal state 00.

- (Changes shown in italic black.) The change in /Y1 at AND gate A causes its output to go to 1, which in turn forces Y2 to 1. Whoops, now the circuit is in internal state 01.

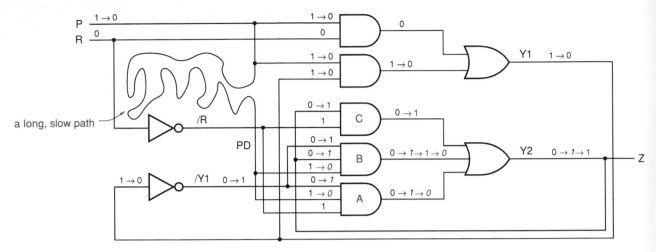

Figure 5–66   Physical conditions in pulse-catching circuit for exhibiting an essential hazard.

- (Changes shown in non-italic color.) The change in Y2 at AND gates B and C causes their outputs to go to 1, reinforcing the 1 output at Y2. All this time, we've been waiting for the 1-to-0 change in P to appear at point PD.

- (Changes shown in italic color.) Finally, PD changes from 1 to 0, forcing the outputs of AND gates A and B to 0. However, AND gate C still has a 1 output, and the circuit remains in state 01—*the wrong state.*

The only way to avoid this erroneous behavior in general is to ensure that changes in P arrive at the inputs of all the excitation circuits before any changes in state variables do. Thus, the inevitable difference in input arrival times, called *timing skew*, must be less than the propagation delay of the excitation circuits and feedback loops. This timing requirement can generally be satisfied only by careful design *at the electrical circuit level.*

timing skew

In the example circuit, it would appear that the hazard is easily masked, even by non-electrical engineers, since the designer need only ensure that a straight wire has shorter propagation delay than an AND-OR structure, easy in most technologies. However, other circuits, such as the edge-triggered D flip-flop in Figure 5–19, have essential hazards in which the input skew paths include inverters. In such cases, the input inverters must be guaranteed to be faster than the excitation logic; that's not so trivial in either board-level or IC design.[8]

---

[8]If the excitation circuit in Figure 5–66 were physically built using AND-OR-INVERT gates, the delay from input changes to /Y1 could be very short indeed, as short as the delay through a single inverter.

Essential hazards are called "essential" because they are inherent in the flow table for a particular sequential function, and will appear in any circuit realization of that function. They can be masked only by controlling the delays in the circuit. Compare with static hazards in combinational logic, where we could eliminate hazards by adding consensus terms to a logic expression.

Essential hazards can be found in most but not all fundamental-mode circuits. There's an easy rule for detecting them; in fact, this is the definition of "essential hazard" in some texts:

- A fundamental-mode flow table contains an essential hazard for a stable total state S and an input variable X if, starting in state S, the stable total state reached after three successive transitions in X is different from the stable total state reached after one transition in X.

Thus, the essential hazard in the pulse catcher is detected by the arrows in Table 5–37, starting in internal state 10 with $P\,R = 10$.

A fundamental-mode circuit must have at least three states to have an essential hazard, so latches don't have them. On the other hand, all flip-flops (circuits that sample inputs on a clock edge) do.

5.6.7 Summary

In summary, the following steps are used to design a feedback sequential circuit:

(1) Construct a primitive flow table from the circuit's word description.

(2) Minimize the number of states in the flow table.

(3) Find a race-free assignment of coded states to named states, adding auxiliary states or splitting states as required.

(4) Construct the transition table.

(5) Construct excitation maps and find a hazard-free realization of the excitation equations.

(6) Draw the logic diagram.

(7) Check for essential hazards. Modify the circuit if necessary to ensure that minimum excitation and feedback delays are greater than maximum inverter or other input-logic delays.

Note that some circuits routinely violate the basic fundamental-mode assumption that inputs change one at a time. For example, in a positive-edge-triggered D flip-flop, the D input may change at the same time that CLK changes from 1 to 0, and the flip-flop still operates properly. The same thing certainly cannot be said at the 0-to-1 transition of CLK. Such situations require analysis of the transition table and circuit on a case-by-case basis if proper operation in "special cases" is to be guaranteed.

That leaves us with one final question. Given the difficulty of designing fundamental-mode circuits that work properly, let alone ones that are fast or compact, how did anyone ever come up with the 6-gate, positive-edge-triggered, commercial D flip-flop design in Figure 5–20? Don't ask me, I don't know!

R E F E R E N C E S

The problem of metastability has been around for a long time. Greek philosophers wrote about the problem of indecision thousands of years ago. A group of modern philosophers named Devo sing about metastability in the title song of their *Freedom of Choice* album.

Most ASICs and MSI-based designs use the sequential-circuit types described in this chapter. However, there are other types that are used in both older discrete designs and in modern, custom VLSI designs.

pulse-mode circuit
pulse input

For example, clocked synchronous state machines are a special case of a more general class of *pulse-mode circuits*. Such circuits have one or more *pulse inputs* such that (a) only one pulse occurs at a time; (b) nonpulse inputs are stable when a pulse occurs; (c) only pulses can cause state changes; and (d) a pulse causes at most one state change. In clocked synchronous state machines, the clock is the single pulse input, and a "pulse" is the triggering edge of the clock. However, it is also possible to build circuits with multiple pulse inputs, and it is possible to use storage elements other than the familiar edge-triggered flip-flops. These possibilities are discussed thoroughly by Edward J. McCluskey in *Logic Design Principles* (Prentice-Hall, 1986).

two-phase latch
machine

A particularly important type of pulse-mode circuit discussed by McCluskey and others is the *two-phase latch machine*. The rationale for a two-phase clocking approach in VLSI circuits is discussed by Carver Mead and Lynn Conway in *Introduction to VLSI Systems* (Addison-Wesley, 1980).

Fundamental-mode circuits need not use feedback loops as the memory elements. For example, McCluskey's *Introduction to the Theory of Switching Circuits* (McGraw-Hill, 1965) gives several examples of fundamental-mode circuits built from S-R flip-flops. His 1986 book shows how transmission gates are used to create memory in CMOS latches.

Methods for reducing both completely and incompletely specified state tables are described in advanced logic design texts, including McCluskey's 1986 book. A more mathematical discussion of these methods and other "theoretical" topics in sequential machine design appears in *Switching and Finite Automata Theory* by Zvi Kohavi (McGraw-Hill, 1978, 2nd ed.).

E X E R C I S E S

5.1 Sketch the outputs of an S-R latch of the type shown in Figure 5–5 for the input waveforms shown in Figure X5.1. Assume that input and output rise and fall times are zero, that the propagation delay of a NOR gate is 10 ns, and that each time division below is 10 ns.

Figure X5.1

5.2 Repeat Exercise 5.1 using the input waveforms shown in Figure X5.2. Although you may find the result unbelievable, this behavior can actually occur in real devices whose transition times are short compared to their propagation delay.

Figure X5.2

5.3 Explain how metastability occurs in a D latch when the setup and hold times are not met, analyzing the behavior of the feedback loop inside the latch.

5.4 Compare the circuit in Figure X5.4 with the D latch in Figure 5–12. Prove that the circuits function identically. In what way is Figure X5.4, which is used in some commercial D latches, better?

Figure X5.4

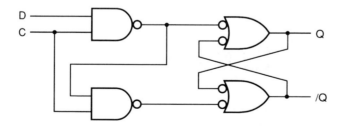

5.5 What is the minimum setup time of a pulse-triggered flip-flop such as a master/slave J-K or S-R flip-flop? (*Hint:* It depends on certain characteristics of the clock.)

5.6 Describe a situation, other than the metastable state, in which the Q and /Q outputs of a 74LS74 edge-triggered D flip-flop may be noncomplementary for an arbitrarily long time.

5.7   Figure 5–31 showed how to build a T flip-flop with enable using a D flip-flop and combinational logic. Show how to build a D flip-flop using a T flip-flop with enable and combinational logic.

5.8   Show how to build a J-K flip-flop using a T flip-flop with enable and combinational logic.

5.9   Show how to build a flip-flop equivalent to the 74LS74 positive-edge-triggered D flip-flop using a 74LS109 positive-edge-triggered J-/K flip-flop and *no* other components.

5.10   Suppose that a clocked synchronous state machine with the structure of Figure 5–32 is designed using D latches with active-high C inputs as storage elements. For proper next-state operation, what relationships must be satisfied among the following timing parameters?

t_{Fmin}, t_{Fmax} Minimum and maximum propagation delay of the next-state logic.

t_{CQmin}, t_{CQmax} Minimum and maximum clock-to-output delay of a D latch.

t_{DQmin}, t_{DQmax} Minimum and maximum data-to-output delay of a D latch.

t_{setup}, t_{hold} Setup and hold times of a D latch.

t_H, t_L Clock HIGH and LOW times.

5.11   Analyze the clocked synchronous state machine in Figure X5.11. Write excitation equations, excitation/transition table, and state/output table (use state names A–D for Q1 Q2 = 00–11).

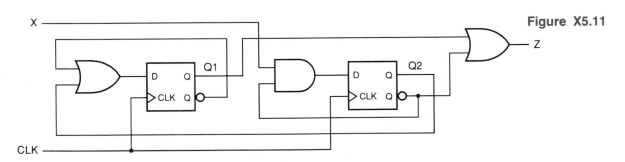

Figure X5.11

5.12   Repeat Exercise 5.11, swapping AND and OR gates in the logic diagram. Is the new state/output table the "dual" of the original one? Explain.

5.13   Redesign the state machine in Figure X5.11 using just three inverting gates—NAND or NOR—and no inverters.

5.14   Analyze the clocked synchronous state machine in Figure X5.14. Write excitation equations, excitation/transition table, and state table (use state names A–H for Q2 Q1 Q0 = 000–111).

Figure X5.14

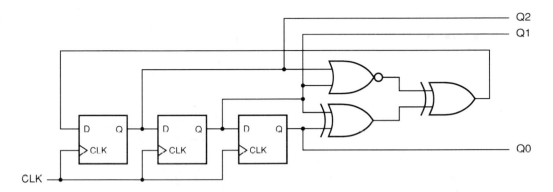

5.15 Analyze the clocked synchronous state machine in Figure X5.15. Write excitation equations, excitation/transition table, and state/output table (use state names A–H for Q1 Q2 Q3 = 000–111).

Figure X5.15

5.16 Analyze the clocked synchronous state machine in Figure X5.16. Write excitation equations, transition equations, transition table, and state/output table (use state names A–D for Q1 Q0 = 00–11). Draw a state diagram, and draw a timing diagram for CLK, EN, Q1, and Q0 for 10 clock ticks, assuming that the machine starts in state 00 and X is continuously 1.

Figure X5.16

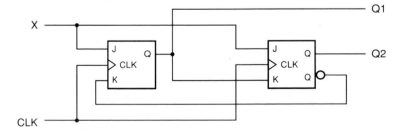

5.17 Analyze the clocked synchronous state machine in Figure X5.17. Write excitation equations, transition equations, transition table, and state/output table (use state names A–D for Q1 Q2 = 00–11). Draw a state diagram, and draw a timing diagram for CLK, X, Q1, and Q2 for 10 clock ticks, assuming that the machine starts in state 00 and X is continuously 1.

Figure X5.17

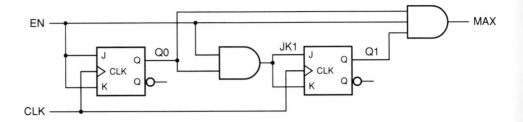

5.18 Analyze the clocked synchronous state machine in Figure X5.18. Write excitation equations, excitation/transition table, and state/output table (use state names A–D for Q1 Q2 = 00–11).

Figure X5.18

5.19 Draw a state diagram for a clocked synchronous state machine with two inputs, INIT and X, and one Moore-type output Z. As long as INIT is asserted, Z is continuously 0. Once INIT is negated, Z should remain 0 until X has been 0 for two successive ticks and 1 for two successive ticks, regardless of the order of occurrence. Then Z should go to 1 and remain 1 until INIT is asserted again. Your state diagram should be neatly drawn and planar (no crossed lines). (*Hint:* No more than ten states are required).

5.20 Design a clocked synchronous state machine that checks a serial data line for even parity. The circuit should have two inputs, SYNC and DATA, in addition to CLOCK, and one Moore-type output, ERROR. Devise a state/output table that does the job using just four states, and include a description of each state's meaning in the table. Choose a 2-bit state assignment, write transition and excitation equations, and draw the logic diagram. Your circuit may use D flip-flops, J-K flip-flops, or one of each.

5.21 Design a clocked synchronous state machine with the state/output table shown in Table X5.21, using D flip-flops. Use two state variables, Q1 Q2, with the state assignment A = 00, B = 01, C = 11, D = 10.

5.22 Repeat Exercise 5.21 using J-K flip-flops.

Table X5.21

S	0	1	Z
A	B	D	0
B	C	B	0
C	B	A	1
D	B	C	0
	S*		

(The columns 0 and 1 are under the heading $\overline{X}$)

5.23 Write a new transition table and derive minimal-cost excitation and output equations for the state table in Table 5–10 using the "simplest" state assignment in Table 5–11 and D flip-flops. Compare the cost of your excitation and output logic (when realized with a two-level AND-OR circuit) with the circuit in Figure 5–47.

5.24 Repeat Exercise 5.23 using the "almost one-hot" state assignment in Table 5–11.

5.25 Suppose that the state machine in Figure 5–47 is to be built using 74LS74 D flip-flops. What signals should be applied to the flip-flop preset and clear inputs?

5.26 Write new transition and excitation tables and derive minimal-cost excitation and output equations for the state table in Table 5–10 using the "simplest" state assignment in Table 5–11 and J-K flip-flops. Compare the cost of your excitation and output logic (when realized with a two-level AND-OR circuit) with the circuit in Figure 5–49.

5.27 Repeat Exercise 5.26 using the "almost one-hot" state assignment in Table 5–11.

5.28 Construct a new excitation table and derive minimal-cost excitation and output equations for the state machine of Table 5–12 using T flip-flops with enable inputs (Figure 5–30). Compare the cost of your excitation and output logic (when realized with a two-level AND-OR circuit) with the circuit in Figure 5–47.

5.29 Determine the full 8-state table of the circuit in Figure 5–47. Use the names U1, U2, and U3 for the unused states (001, 010, and 011). Draw a state diagram and explain the behavior of the unused states.

5.30 Repeat Exercise 5.29 for the circuit of Figure 5–49.

5.31 Write a transition table for the nonminimal state table in Figure 5–44(a) that results from assigning the states in binary counting order, INIT–OKA1 = 000–110. Write corresponding excitation equations for D flip-flops, assuming a minimal-cost disposition of the unused state 111. Compare the cost of your equations with the minimal-cost equations for the minimal state table presented in the text.

5.32 Write the application table for a T flip-flop with enable.

5.33 In many applications, the outputs produced by a state machine during or shortly after reset are irrelevant, as long as the machine begins to behave correctly a short time after the reset signal is removed. If this idea is applied to Table 5–10, the INIT state can be removed and only two state variables are needed to code the remaining four states. Redesign the state machine using this idea. Write a new state table, transition table, excitation table for D flip-flops, minimal-cost excitation and output equations, and logic diagram. Compare the cost of the new circuit with that of Figure 5–47.

5.34 Repeat Exercise 5.33 using J-K flip-flops, and use Figure 5–49 to compare cost.

5.35 Redesign the 1s-counting machine of Table 5–16, assigning the states in binary counting order (S0–S3 = 00, 01, 10, 11). Compare the cost of the resulting sum-of-products excitation equations with the ones derived in the text.

5.36 Repeat Exercise 5.35 using J-K flip-flops.

5.37 Repeat Exercise 5.35 using T flip-flops with enable.

5.38 Redesign the combination-lock machine of Table 5–18, assigning coded states in Gray-code order (A–H = 000, 001, 011, 010, 110, 111, 101, 100). Compare the cost of the resulting sum-of-products excitation equations with the ones derived in the text.

5.39 Find a 3-bit state assignment for the combination-lock machine of Table 5–18 that results in less costly excitation equations than the ones derived in the text. (*Hint:* Use the fact that inputs 1–3 are the same as inputs 4–6 in the required input sequence.)

5.40 What changes would be made to the excitation and output equations for the combination-lock machine in Section 5.4.6 as the result of performing a formal multiple-output minimization procedure (Section 3.3.8) on the five functions? You need not construct 31 product maps and go through the whole procedure; you should be able to "eyeball" the excitation and output maps in Section 5.4.6 to see what savings are possible.

5.41 Analyze the feedback sequential circuit in Figure 5–19, assuming that the /PR and /CLR inputs are always 1. Derive excitation equations, construct a transition table, and analyze the transition table for critical and noncritical races. Name the states, and write a state/output table and a flow/output table. Show that the flow table performs the same function as Table 5–31.

5.42 Draw the logic diagram for a circuit that has one feedback loop, but that is *not* a sequential circuit. That is, the circuit's output should be a function of its current input only. In order to prove your case, break the loop and analyze the circuit as if it were a feedback sequential circuit, and demonstrate that the outputs for each input combination do not depend on the "state."

5.43 A *BUT flop* may be constructed from a single NBUT gate as shown in Figure X5.43. (An *NBUT* gate is simply a BUT gate with inverted outputs; see Exercise 4.5 for the definition of a BUT gate.) Analyze the BUT flop as a feedback sequential circuit and obtain excitation equations, transition table, and flow table. Is this circuit good for anything, or is it a flop?

BUT flop
NBUT

Figure X5.43

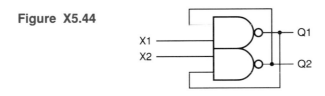

5.44 Repeat Exercise 5.43 for the BUT flop in Figure X5.44.

Figure X5.44

5.45 A clever student designed the circuit in Figure X5.45 as a solution to Exercise 4.38. However, the circuit didn't work correctly 100% of the time. Analyze the circuit and explain why.

Figure X5.45

5.46 Show that a 4-bit ones'-complement adder with end-around carry is a feedback sequential circuit.

5.47 Analyze the feedback sequential circuit in Figure X5.47. Break the feedback loops, write excitation equations, and construct a transition and output table, showing the stable total states. What application might this circuit have?

Figure X5.47

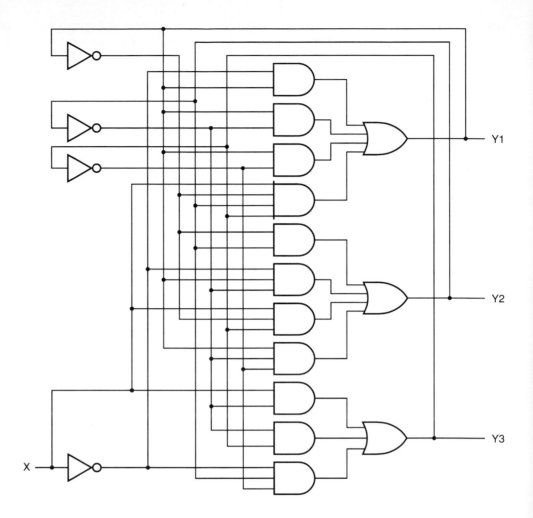

5.48 We claimed in Section 5.6.1 that all single-loop feedback sequential circuits have an excitation equation of the form

$$Q* \;=\; (\text{forcing term}) + (\text{holding term}) \cdot Q$$

Why aren't there any practical circuits whose excitation equation substitutes Q' for Q above?

5.49 Design a latch with two control inputs, C1 and C2, and three data inputs, D1, D2, and D3. The latch is to be "open" only if both control inputs are 1, and it is to store a 1 if any of the data inputs is 1. Use hazard-free two-level sum-of-products circuits for the excitation functions.

5.50 Repeat Exercise 5.49, but minimize the number of gates required; the excitation circuits may have multiple levels of logic.

5.51 Redraw the timing diagram in Figure 5–60, showing the internal state variables of the pulse-catching circuit of Figure 5–65, assuming that it starts in state 00.

5.52 Design a fundamental-mode flow table for a pulse-catching circuit similar to the one described in Section 5.6.2, except that the circuit should detect both 0-to-1 and 1-to-0 transitions on P.

5.53 Design a fundamental-mode flow table for a double-edge-triggered D flip-flop, one that samples its inputs and changes its outputs on both edges of the clock signal.

5.54 Design a fundamental-mode flow table for a circuit with two inputs, EN and CLKIN, and a single output, CLKOUT, with the following behavior. A clock period is defined to be the interval between successive rising edges of CLKIN. If EN is asserted during an entire given clock period, then CLKOUT should be "on" during the next clock period, that is, it should be identical to CLKIN. If EN is negated during an entire given clock period, then CLKOUT should be "off" (constant 1) during the next clock period. If EN is both asserted and negated during a given clock period, then CLKOUT should be on in the next period if it had been off, and off if it had been on. After writing the fundamental-mode flow table, reduce it by combining "compatible" states if possible.

5.55 Design a circuit that meets the specifications of Exercise 5.54 using edge-triggered D flip-flops (74LS74) or J-K flip flops (74LS109) and NAND and NOR gates without feedback loops. Give a complete circuit diagram and word description of how your circuit achieves the desired behavior.

5.56 Which of the circuits of the two preceding exercises is (are) subject to metastability, and under what conditions?

5.57 The general solution for obtaining a race-free state assignment of 2^n states using $2n - 1$ state variables yields the adjacency diagram shown in Figure X5.57 for the $n = 2$ case. Compare this diagram with Figure 5–63. Which is better, and why?

Figure X5.57

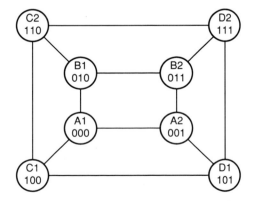

5.58 For the flow table in Table X5.58, find an assignment of state variables that avoids all critical races. Additional states may be added as necessary, but use as few state variables as possible. Assign the all-0s combination to state A. Draw the adjacency diagram for the original flow table, and write the modified flow table and another adjacency diagram to support your final state-variable assignment.

Table X5.58

S	00	01	11	10
		X Y		
A	B	C	—	(A)
B	(B)	E	—	(B)
C	F	(C)	—	E
D	(D)	F	—	B
E	D	(E)	—	(E)
F	(F)	(F)	—	A
		S*		

5.59 Prove that the fundamental-mode flow table of any flip-flop that samples input(s) and change(s) outputs on the rising edge only of a clock signal CLK contains an essential hazard.

5.60 Locate the essential hazard(s) in the flow table for a positive-edge triggered D flip-flop, Table 5–31.

5.61 Identify the essential hazards, if any, in the flow table developed in Exercise 5.60.

5.62 Build a verbal flip-flop — a logical word puzzle that can be answered correctly in either of two ways depending on state. How might such a device be adapted to the political arena?

6

SEQUENTIAL LOGIC DESIGN PRACTICES

The purpose of this chapter is to familiarize you with the most commonly used and dependable sequential-circuit design methods. Therefore, our emphasis is on *synchronous systems*, that is, systems in which all flip-flops are clocked by the same, common clock signal. Although it's true that all the world does not march to the tick of a common clock, within the confines of a digital system or subsystem we can make it so. When we interconnect digital systems or subsystems that use different clocks, we can usually identify a limited number of asynchronous signals that need special treatment, as we'll show later.

    We begin this chapter with a quick summary of sequential circuit documentation standards. Then we introduce the most basic building blocks of sequential-circuit design—flip-flops, counters, and shift registers—and show applications of some of the most common MSI device types. Next, we describe some practical approaches to state-machine design and synthesis, in preparation for our discussion of PLD-based state-machine design in Chapter 7. Finally, we show how all these elements come together in synchronous systems and how the inevitable asynchronous inputs are handled.

synchronous system

437

6.1   SEQUENTIAL CIRCUIT DOCUMENTATION STANDARDS

6.1.1   General Requirements

Basic documentation standards in areas like signal naming, logic symbols, and schematic layout, which we introduced in Chapter 4, apply to digital systems as a whole and therefore to sequential circuits in particular. However, there are several ideas to highlight for system elements that are specifically "sequential":

- *State-machine layout.* Within a logic diagram, a collection of flip-flops and combinational logic that forms a state machine should be drawn together in a logical format on the same page, so the fact that it is a state machine is obvious. (That is, you shouldn't have to flip pages to find the feedback path!)

- *Cascaded elements.* In a similar way, registers, counters, and shift registers that use multiple ICs should have the ICs grouped together in the schematic so that the cascading structure is obvious.

- *Flip-flops.* The symbols for individual sequential-circuit elements, especially flip-flops, should rigorously follow the appropriate drawing standards, so that the type, function, and clocking behavior of the elements are clear.

- *State-machine descriptions.* State machines should be described by state tables, state diagrams, transition lists, ASM charts, or text files in a state-machine description language. The last three types of documentation are described later in this chapter and the next.

- *Timing diagrams.* The documentation package for sequential circuits should include timing diagrams that show the general timing assumptions and timing behavior of the circuit.

- *Timing specifications.* A sequential circuit should be accompanied by a specification of the timing requirements for proper internal operation (e.g., maximum clock frequency), as well as the requirements for any externally supplied inputs (e.g., setup and hold time requirements with respect to the system clock, minimum pulse widths, etc.).

6.1.2   Logic Symbols

We introduced traditional symbols for flip-flops in Chapter 5. Flip-flops are always drawn as rectangular-shaped symbols, and follow the same general guidelines as other rectangular-shaped symbols—inputs on the left, outputs on the right, bubbles for active levels, and so on. In addition, some specific guidelines apply to flip-flop symbols:

- A dynamic indicator is placed on edge-triggered clock inputs.

- A postponed-output indicator is placed on master/slave outputs that change at the end interval during which the clock is asserted.

- Asynchronous preset and clear inputs may be shown at the top and bottom of a flip-flop symbol—preset at the top and clear at the bottom.

The logic symbols for larger-scale sequential elements, such as the counters and shift register described later in this chapter, are generally drawn with all inputs, including presets and clears, on the left, and all outputs on the right. Bidirectional signals may be drawn on the left or the right, whichever is convenient.

Like individual flip-flops, larger-scale sequential elements use a dynamic indicator to indicate edge-triggered clock inputs. In "traditional" symbols, the names of the inputs and outputs give a clue of their function, but they are sometimes ambiguous. For example, the CLR input of a 74LS163 is sampled only at the clock edge, while the CLR input of a 74LS161, which has exactly the same traditional symbol, is asynchronous. IEEE standard symbols, which we introduce for all of the sequential elements in this chapter, provide an unambiguous definition of every signal's function.

6.1.3 State-Machine Descriptions

So far we have dealt with three different representations of state machines—word descriptions, state tables, and state diagrams. A word description is usually imprecise, of course, but a properly constructed state table or state diagram unambiguously defines the state machine's behavior. In this chapter, we will present two more unambiguous ways to describe state machines—transition lists and ASM charts. And in the next chapter, we'll introduce yet another way, state-machine description language.

You might think that having all these different ways to represent state machines is a problem—too much for you to learn! Well, they're *not* all that different or difficult to learn, but there *is* a subtle problem here. Consider a similar problem in assembly-language programming, where high-level "pseudo-code" might be used to document how a program works. The pseudo-code may express the programmer's intentions very well, but errors, misinterpretations, and typos can occur when the pseudo-code is translated into real code. In any creative process, inconsistencies can occur when there are multiple representations of how things work.

The same kind of inconsistencies can occur in state-machine design. A logic designer may document a machine's desired behavior with a 100%-correct hand-drawn state diagram, but there are still plenty of opportunities to mess

*inconsistent
state-machine
representations*

up while "turning the crank" to translate the state diagram into a state table, transition table, excitation equations, and logic diagram.

The solution to this problem is similar to the one adopted by programmers who write self-documenting code using a high-level language. The key is to select a representation that is both expressive of the designer's intentions *and* that can be translated into a physical realization using an error-free, automated process. (You don't hear many programmers screaming "Compiler bug!" when their programs don't work the first time.) The transition-list and ASM-chart notations discussed later in this chapter are a step in the right direction for state-machine designers. However, the ultimate solution (for now, at least) is to write state-machine "programs" in a high-level state-machine description language, allowing automatic translation into a PLD-based design. We'll study this solution in detail in Section 7.4.

6.1.4   Timing Diagrams and Specifications

We showed many examples of timing diagrams in Chapters 4 and 5. In the design of synchronous systems, most timing diagrams show the relationship between the clock and various input, output, and internal signals.

Figure 6–1 shows a fairly typical timing diagram that specifies the require-ments and characteristics of input and output signals in a synchronous circuit. The first line shows the system clock and its nominal timing parameters. The remaining lines show a range of delays for other signals.

For example, the second line shows that flip-flops and other clocked devices change their outputs at some time between the rising edge of CLOCK and time t_{ffpd} afterwards. External circuits that sample these signals should not do so while they are changing. The timing diagram is drawn as if the minimum value

Figure 6–1
A detailed timing diagram
showing propagation delays
and setup and hold times with
respect to the clock.

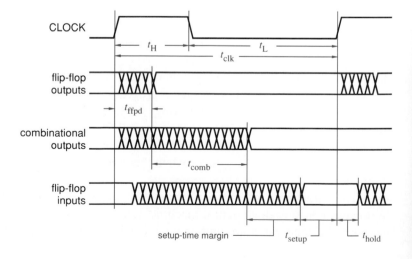

of t_{ffpd} is zero; a complete documentation package would include a timing table indicating the actual minimum, typical, and maximum values of t_{ffpd} and all other timing parameters.

The third line of the timing diagram shows the additional time, t_{comb}, required for the flip-flop output changes to propagate through combinational logic elements, such as flip-flop excitation logic. The excitation inputs of flip-flops and other clocked devices require a setup time of t_{setup}, as shown in the fourth line. For proper circuit operation we must have $t_{clk} - t_{ffpd} - t_{comb} > t_{setup}$.

Timing margins indicate how much "worse than worst-case" the individual *timing margin* components of a circuit can be without causing the circuit to fail. Well-designed systems have positive, nonzero timing margins to allow for unexpected circumstances (marginal components, brown-outs, engineering errors, etc.) and clock skew (Section 6.9.3).

The difference $t_{clk} - t_{ffpd(max)} - t_{comb(max)} - t_{setup(min)}$ is called the *setup-time* *setup-time margin* *margin*; if this number is negative, the circuit won't work. Maximum propagation delays and minimum setup time requirements are used to calculate setup-time margin. Another timing margin involves the hold-time requirement t_{hold}; the sum of the *minimum* values of t_{ffpd} and t_{comb} must be greater than t_{hold}, and the *hold-time margin* is $t_{ffpd(min)} + t_{comb(min)} - t_{hold(min)}$. *hold-time margin*

The timing diagram in Figure 6–1 does not show the timing differences between different flip-flop inputs or combinational-logic signals, even though such differences exist in most circuits. For example, one flip-flop's Q output may be connected directly to another flip-flop's D input, so that t_{comb} for that path is zero, while another's may go the ripple-carry path of a 32-bit adder before reaching a flip-flop input. When proper synchronous design methodology is used, these relative timings are not critical, since none of these signals affect the state of the circuit until a clock edge occurs. You merely have to find the longest delay path in one clock period to determine whether the circuit will work. However, you may have to analyze several different paths to find the worst-case one.

Another, perhaps more common, type of timing diagram shows only functional behavior and is not concerned with actual delay amounts; an example is shown in Figure 6–2. Here, the clock is "perfect." Whether to show signal changes as vertical or slanted lines is strictly a matter of personal taste in this and all other timing diagrams, unless rise and fall times must be explicitly indicated. Clock transitions are shown as vertical lines in this and other figures in keeping with the idea that the clock is a "perfect" reference signal.

The other signals in Figure 6–2 may be flip-flop outputs, combinational outputs, or flip-flop inputs. Shading is used to indicate "don't-care" signal values; cross-hatching as in Figure 6–1 could be used instead. All of the signals are shown to change immediately after the clock edge. In reality, the outputs change

Figure 6–2 Functional timing of a synchronous circuit.

sometime later, and inputs may change just barely before the next clock edge. However, "lining up" everything on the clock edge allows the timing diagram to display more clearly which functions are performed during each clock period. Signals that are lined up with the clock are simply understood to change sometime *after* the clock edge, with timing that meets the setup and hold time requirements of the circuit. Many timing diagrams of this type appear in this chapter.

6.2 LATCHES AND FLIP-FLOPS

6.2.1 SSI Latches and Flip-Flops

Several different types of discrete latches and flip-flops are available as SSI parts. These devices are most commonly used in the design of state machines and "unstructured" sequential circuits that don't fall into the categories of shift registers, counters, and other sequential MSI functions that you'll study later in this chapter. However, SSI latches and flip-flops are being eliminated to a large extent in modern designs as their functions are embedded in larger, structured logic devices such as PLDs, as we'll show in Section 7.2. Nevertheless, a handful of these discrete building blocks still appear in most digital systems, so it's important to be familiar with them.

74LS375

Figure 6–3 shows the pinouts for several SSI sequential devices. The only latch in the figure is the *74LS375*, which contains four D latches, similar in function to the "generic" D latches described in Section 5.2.4. Because of pin limitations, the latches are arranged in pairs with a common C control line for each pair.

74LS74

Among the devices in Figure 6–3, the most important is the *74LS74*, which contains two independent positive-edge-triggered D flip-flops with preset and clear inputs. We described the functional operation, timing, and internal structure of edge-triggered D flip-flops in general, and the 74LS74 in particular, in Section 5.2.5. Besides the 74LS74's use in "random" sequential circuits, faster versions of the part, such as the *74F74*, find application in synchronizers for asynchronous input signals, as discussed in Section 6.10.

74F74

Figure 6–3
Pinouts for SSI latches
and flip-flops.

The *74LS109* is a positive-edge-triggered J-/K flip-flop with an active-low   *74LS109*
K input (named /K or $\overline{K}$). We discussed the internal structure of the '109 in
Section 5.2.8. Another J-K flip-flop is the *74LS112*, which has an active-low   *74LS112*
clock input.

* 6.2.2   Switch Debouncing

A very common application of simple bistables and latches is in debouncing
switches. We're all familiar with electrical switches from experience with lights,
garbage disposals, and other appliances. Switches connected to sources of constant logic 0 and 1 are often used in digital systems to supply "user inputs."
However, in digital logic applications we must consider another aspect of switch
operation, the time dimension. A simple make or break operation, which occurs
instantly as far as we slow-moving humans are concerned, actually has several
phases that are discernable by high-speed digital logic.

Figure 6–4(a) shows how a single-pole, double-throw (SPDT) switch might
be used to generate a single logic input. As shown in Figure 6–4(b), the switch
contacts and wiper have a "break before make" behavior, so the wiper terminal
is "floating" at some time halfway through the switch depression. The figure
assumes that the gate interprets this floating input as a 1, but in a real TTL
circuit we should provide a pull-up resistor on the /SW signal to guarantee a
valid logic signal when the input is floating.

In any case, once the wiper hits the bottom contact, it doesn't stay there
for long; it bounces a few times before finally settling. It bounces far enough to

---

*Throughout this book, *optional sections* are marked with an asterisk.

Figure 6–4
Switch input without debouncing.

(a)

(b)

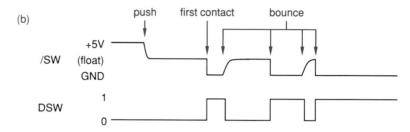

lose its connection to the bottom contact, though not far enough to touch the top
contact again. The result is that several transitions are seen on the /SW and /DSW

contact bounce

logic signals for each single switch depression. This behavior is called *contact bounce*. Typical switches bounce for 10–20 ms, a very long time compared to
the switching speeds of logic gates.

Contact bounce may or may not be a problem, depending on the switch
application. For example, many computers have configuration information spec-

DIP switch

ified by small switches, called *DIP switches* because they are the same size as
a dual in-line package (DIP). Since DIP switches are normally changed only
when the computer is inactive, there's no problem. Contact bounce *is* a problem
if a switch is being used to count or signal some event (e.g., laps in a race).
Then we must provide a circuit (or, in microprocessor-based systems, software)

debounce

to *debounce* the switch—to provide just one signal change or pulse for each
external event.

Switch debouncing is a good application for the simplest sequential circuit, the TTL bistable element of Section 5.1, which can be used as shown in
Figure 6–5. Before the switch is pushed, the top contact holds SW at 0 V, a valid
logic 0, and the top inverter produces a logic 1 on /SW and on the bottom contact. When the button is pushed and contact is broken, feedback in the bistable
continues to hold SW at V_{OL} (≤ 0.5 V for 74LS TTL), still a valid logic 0.

When the wiper hits the bottom contact, the circuit operates unconvention-
ally for a moment. The top inverter in the bistable is trying to maintain a logic
1 on the /SW signal; the top transistor in its totem-pole output is "on" and connecting /SW through a small resistance to +5 V. Suddenly, the switch contact
makes a metallic connection of /SW to ground, 0 V. Not surprisingly, the switch
contact wins.

Figure 6–5
Switch input using a bistable for debouncing.

(a)

(b)

A short time later (30 ns for 74LS TTL), the forced logic 0 on /SW prop-agates through the two inverters of the bistable, so that the top inverter gives up its vain attempt to drive a 1, and instead drives a logic 0 onto /SW. At this point, the top inverter output is no longer shorted to ground, and feedback in the bistable maintains the logic 0 on /SW even if the wiper bounces off the bottom contact, as it does.

Advantages of this circuit compared to other debouncing approaches are that it has a low chip count (one-third of a 74LS04), no pull-up resistors are required, and both polarities of the input signal (active-high and active-low) are produced. In situations where momentarily shorting gate outputs must be avoided, a similar circuit can be designed using a /S-/R latch and pull-up resistors, as suggested in Figure 6–6.

Figure 6–6
Switch input using a /S-/R latch for debouncing.

Figure 6–7   MSI registers: (a) logic diagram for the 74LS175 4-bit register, including pin numbers for a standard 16-pin dual in-line package; (b) traditional logic symbol for the 74LS175; (c) traditional logic symbol for the 74LS174 6-bit register.

6.2.3   Multibit Registers and Latches

register

A collection of two or more D flip-flops with a common clock input is called a *register*. Registers are often used to store a collection of related bits, such as a byte of data in a computer. However, a single register can also be used to store unrelated bits of data or control information; the only real constraint is that all of the bits are stored using the same clock signal.

74LS175

74LS174

Figure 6–7(a) and (b) shows the logic diagram and logic symbol for a commonly used MSI register, the *74LS175*. The 74LS175 contains four edge-triggered D flip-flops and provides both active-high and active-low outputs at the external pins of the device. The *74LS174*, shown in (c), is similar, except that it eliminates the active-low outputs and provides two more flip-flops instead.

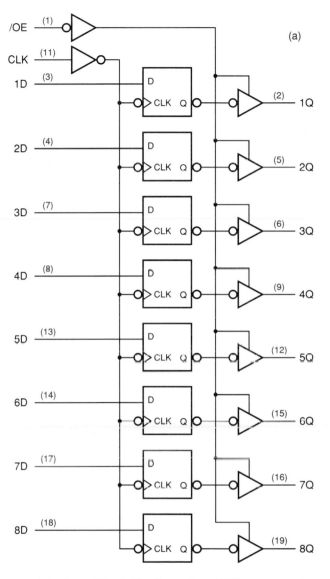

Figure 6–8
The 74LS374 8-bit register:
(a) logic diagram, including
pin numbers for a standard
20-pin dual in-line package;
(b) traditional logic symbol.

The individual flip-flops in a '175 are negative-edge triggered, as indicated by the inversion bubbles on their CLK inputs. However, the circuit also contains an inverter that makes the flip-flops positive-edge triggered with respect to the device's external CLK input pin. A common, active-low, clear signal (/CLR) is connected to the asynchronous clear inputs of all four flip-flops. Both CLK and /CLR are buffered before fanning out to the four flip-flops, so that a device driving one of these inputs sees only one unit load instead of four. This is especially important if a common clock or clear signal must drive many such registers.

Figure 6–9

Traditional logic symbols for
more 8-bit storage elements:
(a) the 74LS373 8-bit latch;
(b) the 74LS377 8-bit register
with gated clock.

Many digital systems, including computers, telecommunications devices, and stereo equipment, process information 8 or 16 bits at a time; as a result, ICs *74LS374* that handle eight bits are very popular. One such MSI IC is the *74LS374* octal edge-triggered D flip-flop, also known simply as an 8-bit register. As shown in Figure 6–8, the 74LS374 contains eight edge-triggered D flip-flops that all sample their inputs and change their outputs on the rising edge of a common CLK input. Each flip-flop output drives a three-state buffer that in turn drives an active-high output. All of the three-state buffers are enabled by a common /OE (output enable) input. As in the other registers that we've studied, the control inputs (CLK and /OE) are buffered so that they present only one unit load to a device that drives them.

Logic symbols for two variations of the 74LS374 are shown in Figure 6–9. *74LS373* The *74LS373* uses D latches instead of edge-triggered flip-flops. Thus, the 373's outputs follow the corresponding inputs whenever C is asserted, and latch the last input values when C is negated.

74LS377 The *74LS377* is an edge-triggered register like the '374, but it does not have three-state outputs. Instead, pin 1 is used as an active-low clock enable input /G. If /G is asserted (LOW) at the rising edge of the clock, then the flip-flops are loaded from the data inputs; otherwise, they retain their present values. The logical equivalent of this function is shown in Figure 6–10.

* 6.2.4 IEEE Standard Symbols

There are a few differences between traditional and IEEE symbols for latches and flip-flops. Figure 6–11 shows IEEE standard symbols for SSI flip-flops. A major difference is that the asynchronous preset and clear inputs are drawn on *S (set)* the left, not on the top and bottom. The names for these inputs and are different, *R (reset)* too: S *(set)* and R *(reset)*. Also, a clock input is simply named C_i, where i is an

Figure 6–10
Logical equivalent of one bit of a
74LS377.

integer and all other inputs labeled with i (e.g., iD) are controlled by Ci; this is
just another instance of dependency notation:

- *Control dependency.* This relationship is denoted by Ci and, like enable   *control dependency*
  dependency, is a sort of enable function. It is intended to be used only for   *Ci*
  clock or timing inputs. Affected signals are enabled when the Ci input is in
  the internal 1-state or, if the Ci input has a dynamic indicator, on the 0-to-1
  change in internal state.

The symbols for the '74, '109, and '112 follow the usual IEEE convention
that only the first of two or more identical elements must be drawn in detail when
they are abutted in an array. Alternatively, the two independent sections of each

Figure 6–11
IEEE standard symbols for
SSI latches and flip-flops.

Figure 6–12 IEEE standard symbols for MSI latches and registers.

SSI package in Figure 6–11 may be drawn separately, as are other traditional and IEEE symbols for devices with independent sections, such as the 74LS139. This is especially useful if the logic functions performed by the two sections in a particular application are unrelated.

Figure 6–12 shows the IEEE standard symbols for MSI latches and registers. All of these symbols use make good use of the IEEE notation for common control blocks. In the '373 and '374 symbols, the label C1 on the clock input indicates that it controls all inputs that bear the label 1, that is, all of the 1D inputs. The downward-pointing triangles indicate three-state outputs; by convention, an input labeled EN enables such outputs.

In the '377 symbol, the input labeled G1 is an enable for inputs bearing the label 1, that is, for the clock input 1C2. The clock input in turn controls all of the inputs bearing the label 2, that is, data inputs 2D. As usual, only the first of the eight identical flip-flop elements is drawn in detail.

6.3 COUNTERS

counter
modulus
modulo-m counter
divide-by-m counter
n-bit binary counter

The name *counter* is generally used for any clocked sequential circuit whose state diagram contains a single cycle, as in Figure 6–13. The *modulus* of a counter is the number of states in the cycle. A counter with m states is called a *modulo-m counter* or, sometimes, a *divide-by-m counter*.

Probably the most commonly used counter type is an *n-bit binary counter*. Such a counter has n flip-flops and has 2^n states, which are visited in the sequence $0, 1, 2, \ldots, 2^n - 1, 0, 1, \ldots$. Each of the foregoing states is encoded as the corresponding n-bit binary integer.

Figure 6–13
General structure of a counter's
state diagram—a single cycle.

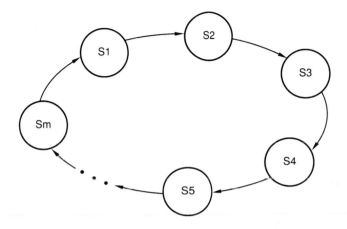

6.3.1 Ripple Counters

An n-bit binary counter can be constructed with just n flip-flops and no other
components, for any value of n. Figure 6–14 shows such a counter for $n = 4$.
Recall that a T flip-flop changes state (toggles) on every rising edge of its clock
input. Thus, each bit of the counter toggles if and only if the immediately
preceding bit changes from 1 to 0. This corresponds to a normal binary counting
sequence—when a particular bit changes from 1 to 0, it generates a carry to the
next most significant bit. The counter in Figure 6–14 is called a *ripple counter*   *ripple counter*
because the carry information ripples from the less significant bits to the more
significant bits one bit at a time.

Figure 6–14
A 4-bit binary ripple counter.

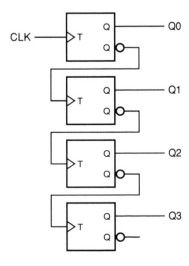

Figure 6–15
A synchronous 4-bit binary
counter with serial enable logic.

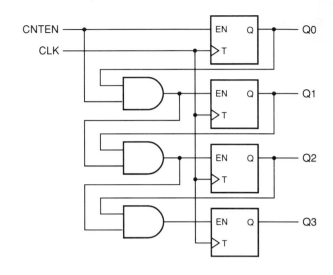

6.3.2 Synchronous Counters

Although a ripple counter requires fewer components than any other type of
binary counter, it does so at a price—it is slower than any other type of binary
counter. In the worst case, when the most significant bit must change, the output
is not valid until time $n \cdot t_{\mathrm{pTQ}}$ after the rising edge of CLK, where t_{pTQ} is the
propagation delay from input to output of a T flip-flop.

synchronous counter   A *synchronous counter* connects all of its flip-flop clock inputs to the same
common CLK signal, so that all of the flip-flop outputs change at the same time,
after only t_{pTQ} ns of delay. As shown in Figure 6–15, this requires the use of
T flip-flops with enable inputs; the output toggles on the rising edge of T if and
only if EN is asserted. Combinational logic on the EN inputs determines which,
if any, flip-flops toggle on each rising edge of T. As shown in the figure, it is
also possible to provide a master count-enable signal CNTEN. Each T flip-flop
toggles if and only if CNTEN is asserted and all of the lower-order counter bits
are 1. Like the binary ripple counter, a synchronous n-bit binary counter can be
built with a fixed amount of logic per bit—in this case, a T flip-flop with enable
and a 2-input AND gate.

synchronous serial   The counter structure in Figure 6–15 is sometimes called a *synchronous*
counter   *serial counter* because the combinational enable signals propagate serially from
the least significant to the most significant bits. If the clock period is too short,
there may not be enough time for a change in the counter's LSB to propagate to
the MSB. This problem is eliminated in Figure 6–16 by driving each EN input
synchronous   with a dedicated AND gate, just a single level of logic. Called a *synchronous*
parallel counter   *parallel counter*, this is the fastest binary counter structure.

Figure 6–16
A synchronous 4-bit binary
counter with parallel enable logic.

6.3.3 MSI Counters and Applications

The most popular MSI counter is the *74LS163*, a synchronous 4-bit binary   *74LS163*
counter with load and clear inputs, whose internal logic diagram is shown in
Figure 6–17.[1] The '163 uses D flip-flops rather than T flip-flops internally to
facilitate the load and clear functions. Each D input is driven by a 2-input multiplexer consisting of an OR gate and two AND gates. The multiplexer output
is 0 if the /CLR input is asserted. Otherwise, the top AND gate passes the data
input (A, B, C, or D) to the output if /LD is asserted. If neither /CLR nor /LD
is asserted, then the bottom AND gate passes the output of an XNOR gate to the
multiplexer output.

The XNOR gates perform the counting function in the '163. One input of
each XNOR is the corresponding count bit (QA, QB, QC, or QD); the other input
is 1, which complements the count bit, if and only if both enables ENP and ENT
are asserted and all of the lower-order count bits are 1. The RCO ("ripple carry
out") signal indicates a carry from the most significant bit position, and is 1 when
all of the count bits are 1 and ENT is asserted. Figure 6–18 shows a traditional
logic symbol for the 74LS163, and Table 6–1 summarizes its function.

Even though most MSI counters have enable inputs, they are often used in
a *free-running* mode in which they are enabled continuously. Figure 6–19 shows   *free-running counter*
the connections to make a '163 operate in this way, and Figure 6–20 shows the

[1]Actually, this is the logic diagram for the Texas Instruments 74S163, which achieves the
same functional behavior with a more straightforward circuit than the TI 'LS163. We've further
improved the TI 'S163 logic diagram to display its functions more clearly by using bubble-to-
bubble drawing standards. (Compare with the 'S163 logic diagram in the TI *TTL Logic Data
Book*.)

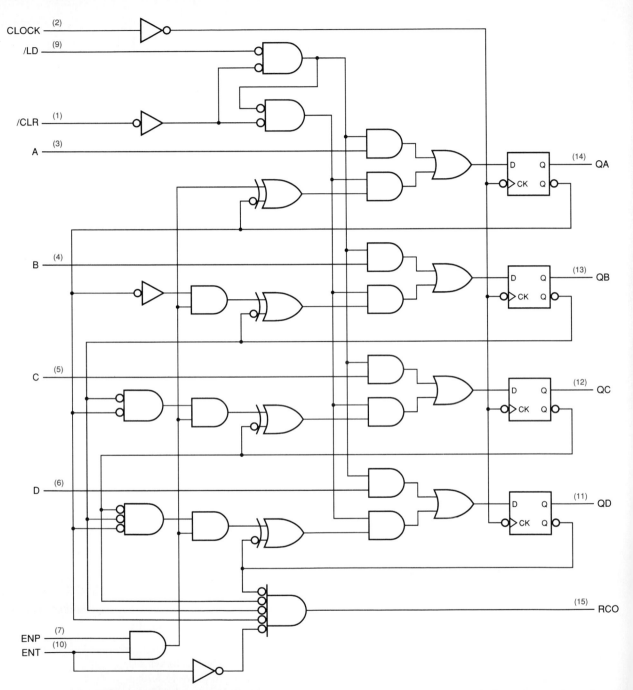

Figure 6–17 Logic diagram for the 74S163 synchronous 4-bit binary counter, including pin numbers for a standard 16-pin dual in-line package.

Figure 6–18
Traditional logic symbol for the 74LS163.

resulting output waveforms. Notice that starting with QA, each signal has half the frequency of the preceding one. Thus, a free-running '163 can be used as a divide-by-2, -4, -8, or -16 counter, ignoring any unnecessary high-order output bits.

Table 6–1   State table for a 74LS163 4-bit binary counter.

Inputs				Current state				Next state			
/CLR	/LD	ENT	ENP	QD	QC	QB	QA	QD*	QC*	QB*	QA*
0	x	x	x	x	x	x	x	0	0	0	0
1	0	x	x	x	x	x	x	D	C	B	A
1	1	0	x	x	x	x	x	QD	QC	QB	QA
1	1	x	0	x	x	x	x	QD	QC	QB	QA
1	1	1	1	0	0	0	0	0	0	0	1
1	1	1	1	0	0	0	1	0	0	1	0
1	1	1	1	0	0	1	0	0	0	1	1
1	1	1	1	0	0	1	1	0	1	0	0
1	1	1	1	0	1	0	0	0	1	0	1
1	1	1	1	0	1	0	1	0	1	1	0
1	1	1	1	0	1	1	0	0	1	1	1
1	1	1	1	0	1	1	1	1	0	0	0
1	1	1	1	1	0	0	0	1	0	0	1
1	1	1	1	1	0	0	1	1	0	1	0
1	1	1	1	1	0	1	0	1	0	1	1
1	1	1	1	1	0	1	1	1	1	0	0
1	1	1	1	1	1	0	0	1	1	0	1
1	1	1	1	1	1	0	1	1	1	1	0
1	1	1	1	1	1	1	0	1	1	1	1
1	1	1	1	1	1	1	1	0	0	0	0

Figure 6–19
Connections for the 74LS163 to operate in a free-running mode.

Note that the '163 is fully synchronous, that is, its outputs change only on the rising edge of CLK. A few applications require an asynchronous clear function, which is provided in the *74LS161*. The '161 has the same pinout as the '163, but its /CLR input is connected to the asynchronous clear inputs of the internal flip-flops.

74LS161

The *74LS160* and *74LS162* are more variations with the same pinouts and general functions as the '161 and '163, except that the counting sequence is modified to go to state 0 after state 9. In other words, these are modulo-10 counters, sometimes called *decade counters*. Figure 6–21 shows the output waveforms for a free-running '160 or '162. Notice that although the QD and QC outputs have one-tenth of the CLK frequency, they do not have a 50% duty cycle, and the QC output, with one-fifth of the input frequency, does not have a constant duty cycle. We'll show the design of a divide-by-10 counter with a 50% duty-cycle output later in this subsection.

74LS160
74LS162

decade counter

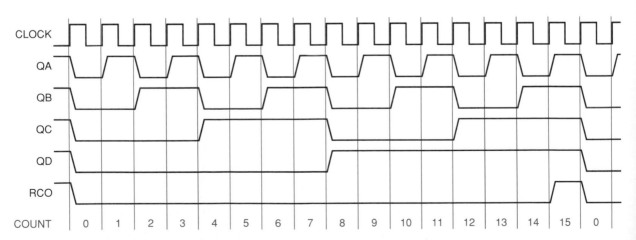

Figure 6–20   Clock and output waveforms for a free-running divide-by-16 counter.

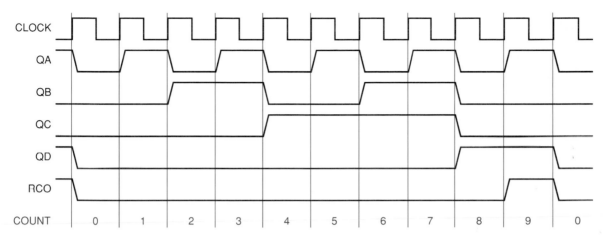

Figure 6–21   Clock and output waveforms for a free-running divide-by-10 counter.

Although the '163 is a modulo-16 counter, it can be made to count in a modulus less than 16 by using the /CLR or /LD input to shorten the normal counting sequence. For example, Figure 6–22 shows one way of using the '163 as a modulo-11 counter. The RCO output, which detects state 15, is used to force the next state to 5, so that the circuit will count from 5 to 15 and then start at 5 again, for a total of 11 states per counting cycle.

A different approach for modulo-11 counting with the '163 is shown in Figure 6–23. This circuit uses a NAND gate to detect state 10 and force the next state to 0. Notice that only a 2-input gate is used to detect state 10 (binary 1010). Although a 4-input gate would normally be used to detect the condition

Figure 6–22
Using the 74LS163 as a modulo-11 counter with the counting sequence 5, 6, ..., 15, 5, 6, ...

Figure 6–23
Using the 74LS163 as a modulo-11 counter with the counting sequence 0, 1, 2, ..., 10, 0, 1, 2, ...

CNT10 = Q3·Q2′·Q1·Q0′, the 2-input gate takes advantage of the fact that no other state in the normal counting sequence of 0–10 has Q3 = 1 and Q1 = 1. In general, to detect state N in a binary counter that counts from 0 to N, we need to AND only the state bits that are 1 in the binary encoding of N.

There are many other ways to make a modulo-11 counter using a '163. The choice of approach—one of the preceding or a combination of them (as in Exercise 6.10)—depends on the application. As another example, in Section 1.10 we promised to show you how to build a circuit that counts in the excess-3 decimal code, shown in Table 1–9 on page 25. Figure 6–24 shows the connections for a '163 to count in the excess-3 sequence. A NAND gate detects state 1100

Figure 6–24
A 74LS163 used as an excess-3 decimal counter.

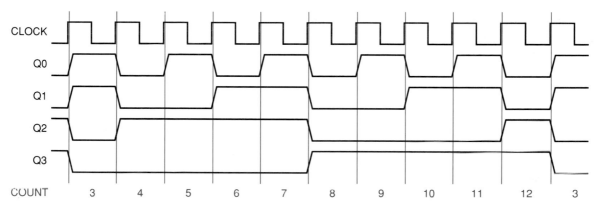

Figure 6–25   Timing waveforms for the '163 used as an excess-3 decimal counter.

and forces 0011 to be loaded as the next state. Figure 6–25 shows the resulting timing waveforms. Notice that the Q3 output has a 50% duty cycle, which may be desirable for some applications.

A binary counter with a modulus greater than 16 can be built by cascading 74LS163s. Figure 6–26 shows the general connections for such a counter. The CLK, /CLR, and /LD inputs of all the '163s are connected in parallel, so that all of them count or are cleared or loaded at the same time. A master count-enable (CNTEN) signal is connected to the low-order '163. The RCO4 output is asserted if and only if the low-order '163 is in state 15 *and* CNTEN is asserted; RCO4 is connected to the enable inputs of the high-order '163. Thus, both the carry information and the master count-enable ripple from the output of one 4-bit counter stage to the next. Like the synchronous serial counter of Figure 6–15, this scheme can be extended to build a counter with any desired number of bits, although the maximum counting speed is limited by the propagation delay of the ripple carry signal through all of the stages (but see Exercise 6.13).

Even experienced digital designers are sometimes confused about the difference between the ENP and ENT enable inputs of the '163 and similar counters, since both must be asserted for the counter to count. However, a glance at the 163's internal logic diagram, Figure 6–17, shows the difference quite clearly— ENT goes *To* the ripple carry output as well. In many applications, this distinction is important.

For example, Figure 6–27 shows an application that uses two '163s as a modulo-193 counter that counts from 63 to 255. The MAXCNT output detects state 255 and stops the counter until /GO is asserted. When /GO is asserted, the counter is reloaded with 63 and counts up to 255 again. (Note that the value of /GO is relevant only when the counter is in state 255.) To keep the counter stopped, MAXCNT must be asserted in state 255 even while the counter is stopped. Therefore, the low-order counter's ENT input is always asserted, its RCO output

Figure 6–26
General cascading connections
for 74LS163-based counters.

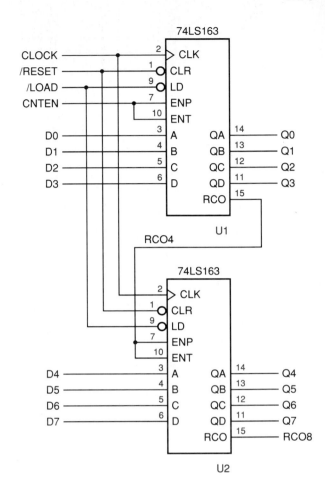

is connected to the high-order ENT input, and MAXCNT detects state 255 even
if CNTEN is not asserted (compare with the behavior of RCO8 in Figure 6–26).
To enable counting, CNTEN is connected to the ENP inputs in *Parallel*. A NAND
gate asserts /RELOAD to go back to state 63 only if /GO is asserted and the
counter is in state 255.

74LS169   Another counter with functions similar to 74LS163's is the *74LS169*, whose
logic symbol is shown in Figure 6–28. One difference in the '169 is that its
carry output and enable inputs are active low. More importantly, the '169 is an
up/down counter *up/down counter*; it counts in ascending or descending binary order depending on
the value of an input signal, UP/DN. With a signal name like UP/DN, figuring
out which value is UP and which is DN is easy if you realize that, using our
active-level conventions, the signal could have been named UP or /DN. Thus,
the '169 counts up when UP/DN is 1 and down when UP/DN is 0. Just to be
sure, some people use the name UP/$\overline{\text{DN}}$ for this signal.

Figure 6–27
Using 74LS163s as a modulo-193 counter with the counting sequence 63, 64, ..., 255, 63, 64, ...

Figure 6–28
Traditional logic symbol for the 74LS169 synchronous 4-bit binary up/down counter.

Figure 6–29
A modulo-8 binary counter and decoder.

6.3.4 Decoding Binary-Counter States

A binary counter may be combined with a decoder to obtain a set of 1-out-of-m-coded signals, where one signal is asserted in each counter state. This is useful when counters are used to control a set of devices where a different device is enabled in each counter state. In this approach, each output of the decoder enables a different device.

Figure 6–29 shows how a 74LS163 wired as a modulo-8 counter can be combined with a 74LS138 3-to-8 decoder to provide eight signals, each one representing a counter state. Typical timing for this circuit is shown in Figure 6–30. Each decoder output is asserted during a corresponding clock period.

decoding glitches

Notice that the decoder outputs may contain "glitches" on state transitions where two or more counter bits change, even though the '163 outputs are glitch free and the '138 does not have any static hazards. In a synchronous counter like the '163, the outputs don't change at exactly the same time. More important, multiple signal paths in a decoder like the '138 have different delays; for example, the path from B to /Y1 is faster than the path from A to /Y1. Therefore, even if the input changes simultaneously from 011 to 100, the decoder may behave as if the input were temporarily 001, and the /Y1 output may have a glitch. In the present example, it can be shown that the glitches can occur in *any* realization

function hazard

of the binary decoder function; this problem is called a *function hazard*.

In most applications, the decoder output signals portrayed in Figure 6–30 would be used as control inputs to registers, counters, and other edge-triggered devices (e.g., /G in a 74LS377, /LD in a 74LS163, or /ENP in a 74LS169). In such a case, the decoding glitches in Figure 6–30 are not a problem, since they occur *after* the clock tick. They are long gone before the next tick comes along, when the decoder outputs are sampled by other edge-triggered devices. However, the glitches *would* be a problem if they were applied to something like the /S

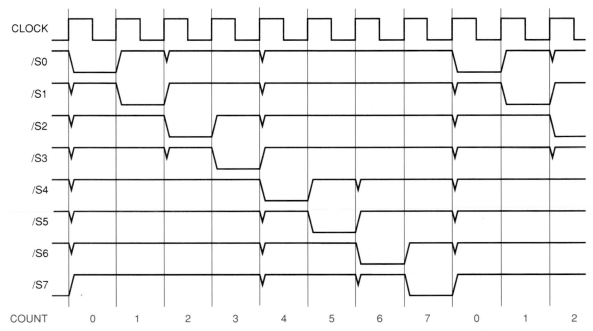

Figure 6–30 Timing diagram for a modulo-8 binary counter and decoder, showing possible decoding glitches.

or /R inputs of a /S-/R latch. Likewise, using such potentially glitchy signals as clocks for edge-triggered devices is a definite no-no.

If necessary, one way to "clean up" the glitches in Figure 6–30 is to connect the '138 outputs to another register that samples the stable decoded outputs on

Figure 6–31 A modulo-8 binary counter and decoder with glitch-free outputs.

the next clock tick, as shown in Figure 6–31. Notice that the decoded outputs have been renamed to account for the 1-tick delay through the register. However, once you decide to pay for an 8-bit register, a less costly solution is to use an 8-bit "ring counter," which provides glitch-free decoded outputs directly, as we'll show in Section 6.4.5.

* 6.3.5   IEEE Standard Symbols

IEEE standard symbols for the counters in this section are shown in Figure 6–32. Like the IEEE symbols for other MSI devices, the counter symbols make good use of the common-control-block and array features of the standard. The general qualifying symbol CTRDIV16 indicates that the device is a divide-by-16 counter, and labels [1]–[8] indicate the arithmetic weight of each counter bit. However, the additional notation used within the common control blocks to describe the devices' functions, though precise, is hardly intuitive. We'll have to describe a few more features of the standard to understand these symbols:

content input
CT=m input

- *Content input.* When an input signal bearing the label CT=m is asserted, the value m is loaded into the device. In the counter symbols, you might read "CT" as "count," but in general it means "content."

The only difference between the '161 and '163 symbols in Figure 6–32 is that the 163's 5CT=0 has a control dependency on the clock input (C5), and is therefore a synchronous clear; the '161 has an asynchronous clear.

content output
CT=m output

- *Content output.* An output bearing the label CT=m is asserted when the content of the device is m.

Figure 6–32   IEEE standard symbols for counters.

In the '161 and '163 symbols, the output 3CT=15 is asserted when the counter is in state 15 and G3 is asserted.

- *Mode dependency.* This dependency is denoted by Mi and, like enable dependency, is a sort of enable function. Affected signals perform their "normal" functions only if Mi is asserted; otherwise, the affected signals have no effect on the device's function and are ignored.

  mode dependency
  Mi

- *Multiple dependencies.* A signal may be affected by several other signals. The identifiers of the affecting signals are listed, separated by commas, in the qualifying label of the affected signal. The dependencies are applied in the order that they are written, from left to right. If all of the dependencies are of the same type (e.g., AND dependency), then the order is irrelevant.

  multiple
    dependencies

- *Multifunction outputs.* A single external output signal, such as pin 15 in the '169 symbol, may have several sets of qualifying labels corresponding to multiple modes of operation. Such a signal may be represented by multiple outputs that are connected together externally. Such outputs normally have a functional OR relationship — the external signal is asserted if any internal signal is asserted.

  multifunction output

In the '161 and '163 symbols, the label 1,5D indicates that the data will be stored when M1 and C5 are asserted (but C5 is "asserted" only on an edge, because of its dynamic indicator). In the '169 symbol, the output 3,5CT=15 is asserted if M3 and G5 are asserted and the counter is in state 15; and 4,5CT−0 is asserted if M4 and G5 are asserted and the counter is in state 0; the external signal (pin 15) is asserted (LOW) if either of these internal signals is asserted.

- *Counting inputs.* When asserted, an input labeled with a + causes the device to count up once. An input labeled with a − causes the device to count down once.

  counting input
  +
  −

According to the '161 and '163 symbols, the device counts up on the rising edge of the signal on pin 2 if M2, G3, and G4 are asserted. The '169 counts up if M2, M3, G5, and G6 are asserted, and down if M2, M4, G5, and G6 are asserted. In each device, the qualifying labels for two separate functions of pin 2—load and count up—are drawn on the same input line, separated by a solidus; they could also have been drawn on two separate input lines, as we showed in Figure 4–58 on page 277.

- *Nonstandardized information.* Descriptive function names and other non-standardized (i.e., helpful) information may be written in brackets next to the qualifying labels in a symbol.

  nonstandardized
    information

The '169 symbol has four "nonstandard" labels to describe the traditional /LD and /ENP inputs. Theoretically, you don't need such labels if you understand the standard. Conversely, if signals are given meaningful names (as in traditional logic symbols), then you don't need the standard!

6.4   SHIFT REGISTERS

6.4.1   Shift-Register Structure

shift register

A *shift register* is an n-bit register with a provision for shifting its stored data by one bit position at each tick of the clock. Figure 6–33 shows the structure of a serial-in, serial-out shift register. The *serial input*, SERIN, specifies a new bit to be shifted into one end at each clock tick. This bit appears at the *serial output*, SEROUT, after n clock ticks, and is lost one tick later. Thus, an n-bit serial-in, serial-out shift register can be used to delay a signal by n clock ticks.

serial input
serial output

A *serial-in, parallel-out shift register*, shown in Figure 6–34, has outputs for all of its stored bits, making them available to other circuits. Such a shift register can be used to perform *serial-to-parallel conversion*, as explained later in this section.

serial-in, parallel-out shift register
serial-to-parallel conversion
parallel-in, serial-out shift register

Conversely, it is possible to build a *parallel-in, serial-out shift register*. Figure 6–35 shows the general structure of such a device. At each clock tick, the register either loads new data from inputs 1D–ND, or it shifts its current contents, depending on the value of the LOAD/SHIFT control input. Internally, the device uses a 2-input multiplexer on each flip-flop's D input to select between the two cases. A parallel-in, serial-out shift register can be used to perform *parallel-to-serial conversion*, as explained later in this section.

parallel-to-serial conversion

Figure 6–33
Structure of a serial-in, serial-out shift register.

Figure 6–34
Structure of a serial-in, parallel-
out shift register.

By providing outputs for all of the stored bits in a parallel-in shift register,
we obtain the *parallel-in, parallel-out shift register* shown in Figure 6–36. Such
a device is general enough to be used in any of the applications of the previous
shift registers.

*parallel-in,
parallel-out shift
register*

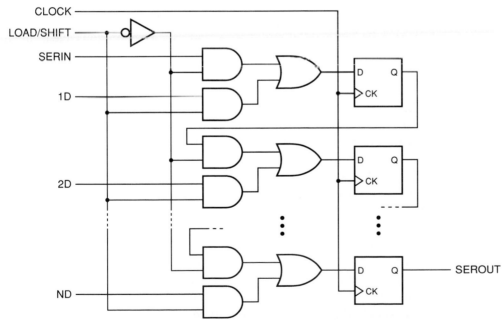

Figure 6–35   Structure of a parallel-in, serial-out shift register.

Figure 6–36 Structure of a parallel-in, parallel-out shift register.

6.4.2 MSI Shift Registers

74LS164

Figure 6–37 shows logic symbols for two popular MSI 8-bit shift registers. The *74LS164* is a serial-in, parallel-out device with an asynchronous clear input (/CLR). It has two serial inputs that are ANDed internally. That is, both SERA and SERB must be 1 for a 1 to be shifted into the first bit of the register.

74LS166

The *74LS166* is a parallel-in, serial-out shift register, also with an asynchronous clear input. This device shifts when SH/LD is 1, and loads new data otherwise. Its clocking is a little unusual, since it has two clock inputs that are connected to the internal flip-flops as shown in Figure 6–37(c). The designers of this device intended for CLK to be connected to a free-running system clock, and for CLKINH to be asserted to inhibit CLK, so that neither shifting nor loading occurs on the next clock tick, and the current register contents are held. However, for this to work, CLKINH must be changed only when CLK is 1; otherwise, undesired clock edges occur on the internal flip-flops. A much better way of obtaining a "hold" function is employed in the next device that we discuss.

74LS194

unidirectional shift register

bidirectional shift register

The *74LS194* is an MSI 4-bit bidirectional, parallel-in, parallel-out shift register whose logic symbol and logic diagram are shown in Figures 6–38 and 6–39. The shift registers that we've studied previously are called *unidirectional shift registers* because they shift in only one direction. The '194 is a *bidirectional shift register* because its contents may be shifted in either of two directions,

Figure 6–37 Traditional logic symbols for MSI shift registers: (a) the 74LS164 8-bit serial-in, parallel-out shift register; (b) the 74LS166 8-bit parallel-in, serial-out shift register; (c) equivalent circuit for 74LS166 clock inputs.

depending on a control input. The two directions are called "left" and "right," even though the logic diagram and the logic symbol aren't necessarily drawn that way. In the '194, *left* means "in the direction from QD to QA," and *right* means "in the direction from QA to QD." Our logic diagram and symbol for the '194 are consistent with these names if you rotate them 90° clockwise.

left

right

    Table 6–2 is the state table for the 74LS194. This state table is highly compressed, since it does not contain columns for most of the inputs (A–D, RIN, LIN) or the current state QA–QD. Still, by expressing each next-state value as a function of these implicit variables, it completely defines the operation of the

Figure 6–38
Traditional logic symbol for the
74LS194.

Figure 6–39
Logic diagram for the 74LS194
4-bit universal shift register,
including pin numbers for a
standard 16-pin dual in-line
package.

Table 6–2
State table for a 74LS194 4-bit
universal shift register.

Function	Inputs		Next state			
	S1	S0	QA*	QB*	QC*	QD*
Hold	0	0	QA	QB	QC	QD
Shift right	0	1	RIN	QA	QB	QC
Shift left	1	0	QB	QC	QD	LIN
Load	1	1	A	B	C	D

'194 for all 2^{12} possible combinations of current state and input, and it sure beats a 4096-row table!

Note that the '194's LIN (left-in) input is conceptually located on the "right-hand" side of the chip, but it is the serial input for *left* shifts. Likewise, RIN is the serial input for right shifts.

The '194 is sometimes called a *universal* shift register because it can be made to function like any of the less general shift register types that we've discussed (e.g., unidirectional; serial-in, parallel-out; parallel-in, serial-out). In fact, many of our design examples in the next few subsections contain '194s configured to use just a subset of their available functions.

universal shift register

The *74LS299* is an 8-bit universal shift register in a 20-pin package; its logic diagram and symbol are given in Figures 6–40 and 6–41. The '299's functions and state table are similar to the '194's, as shown in Table 6–3. To save pins, the '299 uses bidirectional three-state lines for input and output, as shown in the logic diagram. During load operations (S1 S0 = 11), the three-state drivers are disabled and data is loaded through the AQA–HQH pins. At other times, the stored bits are driven onto these same pins if /G1 and /G2 are asserted. The leftmost and rightmost stored bits are available at all times on separate output-only pins, QA and QH.

74LS299

Table 6–3 Function table for a 74LS299 8-bit universal shift register.

Function	Inputs		Next state							
	S1	S0	QA*	QB*	QC*	QD*	QE*	QF*	QG*	QH*
Hold	0	0	QA	QB	QC	QD	QE	QF	QG	QH
Shift right	0	1	RIN	QA	QB	QC	QD	QE	QF	QG
Shift left	1	0	QB	QC	QD	QE	QF	QG	QH	LIN
Load	1	1	AQA	BQB	CQC	DQD	EQE	FQF	GQG	HQH

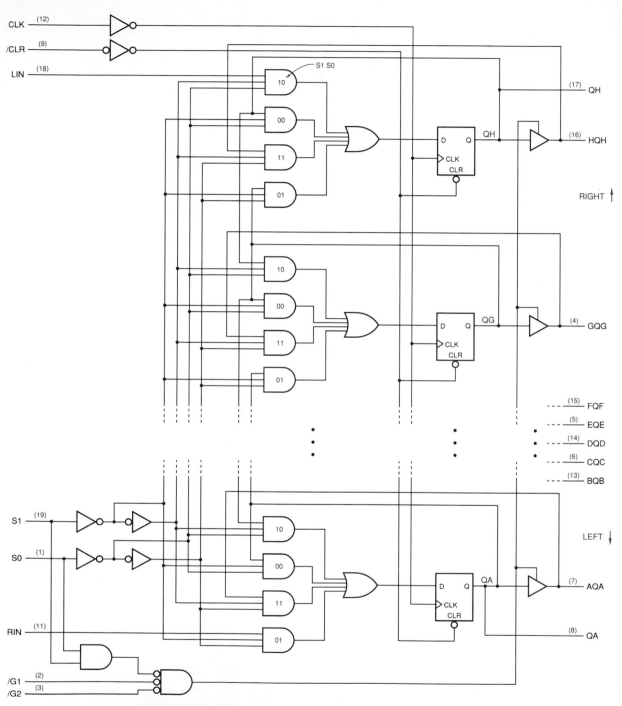

Figure 6–40 Logic diagram for the 74LS299 8-bit universal shift register, including pin numbers for a standard 20-pin dual in-line package.

Figure 6–41
Traditional logic symbol for the
74LS299.

6.4.3   The World's Biggest Shift-Register Application

The most common application of shift registers is to convert parallel data into
serial format for transmission or storage, and to convert serial data back to parallel
format for processing or display (see Section 1.16.1). The most common example
of serial data transmission, one that you almost certainly take part in every day,
is in *digital telephony*.

    For years, TPC (The Phone Company) has been installing digital switching
equipment in its central offices (*COs*). Most home phones have a two-wire analog
connection to the central office. However, modern CO equipment samples the
analog voice signal 8,000 times per second (once every 125 µs) when it enters the
CO, and produces a corresponding sequence of 8,000 8-bit bytes representing the
sign and amplitude of the analog signal at each sampling point. Subsequently,
your voice may be transmitted digitally on 64 kbps *serial channels* throughout the
phone network, until it is converted back to an analog signal at the far-end CO.
Recent advances, such as ISDN (Integrated Services Digital Network), promise
to extend a serial digital channel all the way to your home phone.[2]

    The 64 kbps bandwidth required by a single digital voice signal is far
less than can be obtained on a single digital signal line or switched by digital
ICs. Therefore most digital telephone equipment *multiplexes* multiple 64-kbps
channels onto a single wire, saving both wires and digital ICs for switching. In
the example in the next subsection, 32 channels are processed by a handful of
chips. This is a classic example of a *space/time trade-off* in digital design—
by running the chips faster, you can accomplish a given task with fewer chips.
Indeed, this is the main reason that the telephone network has "gone digital."

digital telephony

CO

serial channel

multiplex

space/time trade-off

---

[2]However, delays in ISDN deployment have led some people in the industry to rename it
"Imaginary Services Delivered Nowhere"!

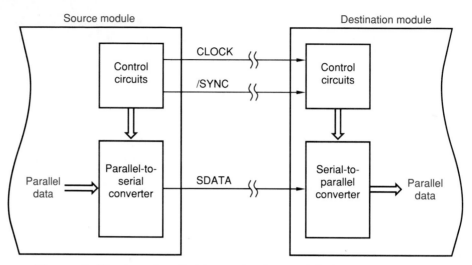

Figure 6–42   A system that transmits data serially between modules.

6.4.4   Serial/Parallel Conversion

A typical application of serial data transfer between two modules (possibly in a piece of CO switching equipment) is shown in Figure 6–42. Three signals are normally connected between a source module and a destination module to accomplish the transfer:

- *Clock.* The clock signal provides the timing reference for transfers, defining the time to transfer one bit. In systems with just two modules, the clock may be part of control circuits located on the source module as shown. In larger systems, the clock may be generated at a common point and distributed to all of the modules.[3]

synchronization pulse
sync pulse

- *Synchronization.* A *synchronization pulse* (or *sync pulse*) provides a reference point for defining the data format, such as the beginning of a byte or word in the serial data stream.

- *Serial data.* The data itself is transmitted on a single line.

The general timing characteristics of these signals in a typical digital telephony application are shown in Figure 6–43(a). The CLOCK signal has a frequency of 2.048 MHz to allow the transmission of $32 \times 8,000$ 8-bit bytes per second. The 1-bit-wide pulse on the /SYNC signal identifies the beginning of

---

[3]Believe it or not, in the phone network, a 2.048 MHz clock is generated in St. Louis and distributed nationwide! The CLOCK signal that is distributed in a particular piece of local CO equipment is normally derived from the national clock.

Figure 6–43 Timing diagram for parallel-to-serial conversion: (a) a complete frame; (b) one byte at the beginning of the frame.

a 125-μs interval called a *frame*. A total of 256 bits are transmitted on SDATA during this interval, which is divided into 32 *timeslots* containing eight bits each. Each timeslot carries one digitally encoded voice signal. Both timeslot numbers and bit positions within a timeslot are located relative to the SYNC pulse.

frame
timeslot

Figure 6–44 shows a circuit that converts parallel data to the serial format of Figure 6–43(a), with detailed timing shown in (b). Two 74LS163 counters are wired as a free-running modulo-256 counter to define the frame. The five high-order and three low-order counter bits are the timeslot number and bit number, respectively.

A 74LS166 parallel-in shift register performs the parallel-to-serial conversion. Bit 0 of the parallel data (D0–D7) is connected to the '166 input closest to the SDATA output, so bits are transmitted serially in the order 0 through 7.[4]

---

[4]Most real serial links for digitized voice actually transmit bit 7 first, because this is the first bit generated by the analog-to-digital converter that digitizes the voice signal. However, to simplify our examples, we transmit bit 0 first so that counter state equals bit number.

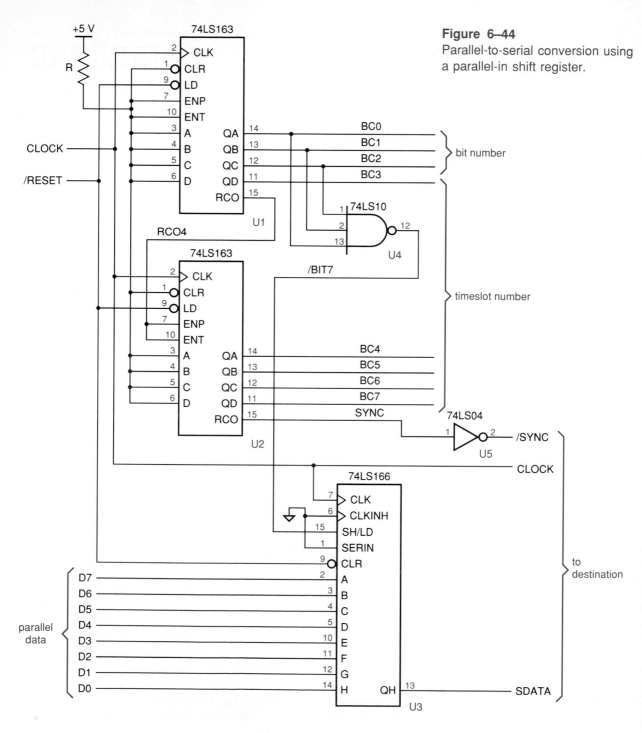

Figure 6–44
Parallel-to-serial conversion using a parallel-in shift register.

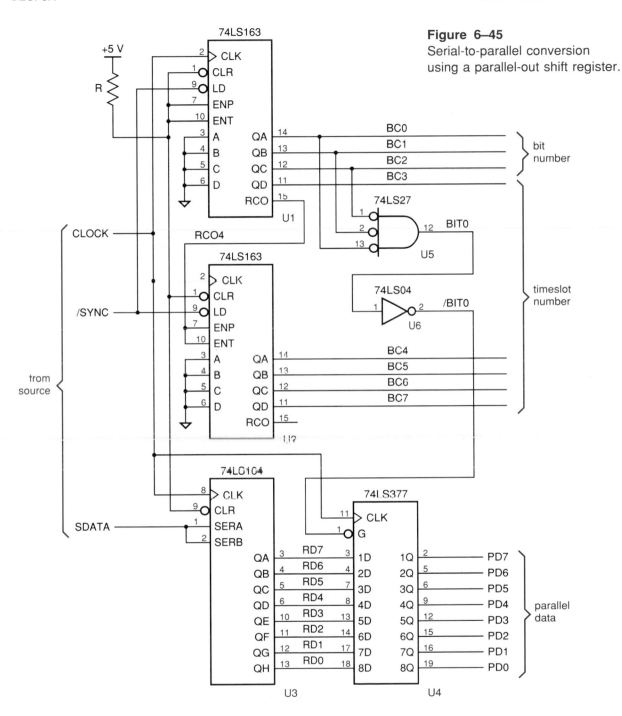

Figure 6–45
Serial-to-parallel conversion
using a parallel-out shift register.

Figure 6–46 Timing diagram for serial-to-parallel conversion.

During bit 7 of each timeslot, the /BIT7 signal is asserted, which causes the '166 to be loaded with D0–D7. The value of D0–D7 is irrelevant except during the setup and hold time window around the clock edge on which the '166 is loaded, as shown by shading in the timing diagram. This leaves open the possibility that the parallel data bus could be used for other things at other times (see Exercise 6.17).

A destination module can convert the serial data back into parallel format using the circuit of Figure 6–45. A modulo-256 counter built from a pair of '163s is used to reconstruct the timeslot and bit numbers. Although /SYNC is asserted during state 255 of the counter on the source module, /SYNC loads the destination module's counter with 0 so that both counters go to 0 on the same clock edge. The counter's high-order bits (timeslot number) are not used in the figure, but they may be used by other circuits in the destination module to identify the byte from a particular timeslot on the parallel data bus (PD0–PD7).

double-buffered data

Figure 6–46 shows detailed timing for the serial-to-parallel conversion circuit. A complete received byte is available at parallel output of the 74LS164 shift register during the clock period following the reception of the last bit (7) of the byte. The parallel data in this example is *double-buffered*—once it is fully received, it is transferred into a 74LS377 register, where it is available on PD0–PD7 for eight full clock periods until the next byte is fully received. The /BIT0 signal enables the '377 to be loaded at the proper time. Additional registers and decoding could be provided to load the byte from each timeslot into a different register, making each byte available for 125 μs (see Exercise 6.19).

Once the received data is in parallel format, it can easily be stored or modified by other digital circuits; we'll give examples in Section 7.1.6. In digital telephony, the received parallel data is converted back into an analog voltage that is filtered and transmitted to an earpiece or speaker for 125 μs, until the next voice sample arrives.

6.4.5   Shift-Register Counters

Serial/parallel conversion is a "data" application, but shift registers have "non-data" applications as well. A shift register can be combined with combinational logic to form a state machine whose state diagram is cyclic. Such a circuit is called a *shift-register counter*. Unlike a binary counter, a shift-register counter does not count in an ascending or descending binary sequence, but it is useful in many "control" applications nonetheless.

shift-register counter

6.4.6   Ring Counters

The simplest shift-register counter uses an n-bit shift register to obtain a counter with n states, and is called a *ring counter*. Figure 6–47(a) is the logic diagram for a 4-bit ring counter. The 74LS194 universal shift register is wired so that it normally performs a left shift. However, when RESET is asserted, it loads 0001 (refer to the '194's function table, Table 6–2). Once RESET is negated, the '194 shifts left on each clock tick. The LIN serial input is connected to the "leftmost" output, so the next states are 0010, 0100, 1000, 0001, 0010, ... Thus, the counter visits four unique states before repeating. A timing diagram is shown in Figure 6–48. In general, an n-bit ring counter visits n states in a cycle.

ring counter

The ring counter in Figure 6–47(a) has one major problem—it is not robust. If its single 1 output is lost due to a temporary hardware problem (e.g., noise), the counter goes to state 0000 and stays there forever. Likewise, if an extra 1 output is set (i.e., state 0101 is created), the counter will go through an incorrect cycle of states and stay in that cycle forever. These problems are quite evident if we draw the *complete* state diagram for the counter circuit, which has 16 states. As shown in Figure 6–49, there are 12 states that are not part of the normal counting cycle. If the counter somehow gets off the normal cycle, it stays off.

A *self-correcting counter* is designed so that all abnormal states have transitions leading to normal states. Self-correcting counters are desirable for the same reason that we use a minimal-risk approach to state assignment in Section 5.4.3: if something unexpected happens, a counter or state machine should go to a "safe" state.

self-correcting counter

A *self-correcting ring counter* circuit is shown in Figure 6–47(b). The circuit uses a NOR gate to shift a 1 into LIN only when the three least significant bits are 0. This results in the state diagram in Figure 6–50; all abnormal states

self-correcting ring counter

Figure 6–47
Four-bit, four-state ring counters:
(a) simplest design, with a single circulating 1; (b) self-correcting, with a single circulating 1;
(c) self-correcting, with a single circulating 0.

lead back into the normal cycle. Notice that, in this circuit, an explicit RESET signal is not necessarily required. Regardless of the initial state of the shift register on power-up, it reaches state 0001 within four clock ticks. An n-bit self-correcting ring counter uses an $n - 1$-input NOR gate, and corrects an abnormal state within $n - 1$ clock ticks.

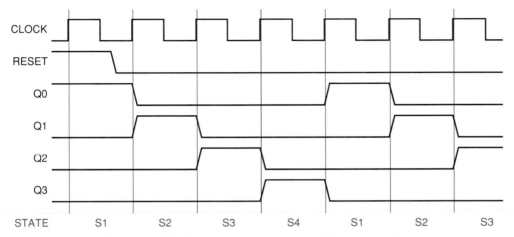

Figure 6–48   Timing diagram for a 4-bit ring counter.

In TTL logic families, NAND gates are easier to come by than NORs, so it may be more convenient to design a self-correcting ring counter as shown in Figure 6–47(c). States in this counter's normal cycle have a single circulating 0.

The major appeal of a ring counter for control applications is that its states appear in 1-out-of-n decoded form directly on the flip-flop outputs. That is, exactly one flip-flop output is asserted in each state. Furthermore, these outputs are "glitch free"; compare with the binary counter and decoder approach of Figure 6–29.

Figure 6–49
State diagram for a simple ring counter.

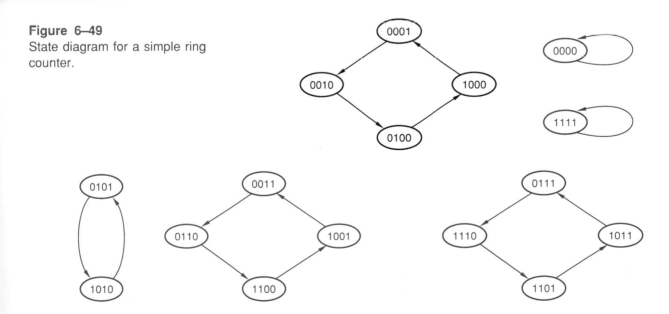

Figure 6–50
State diagram for a self-correcting ring counter.

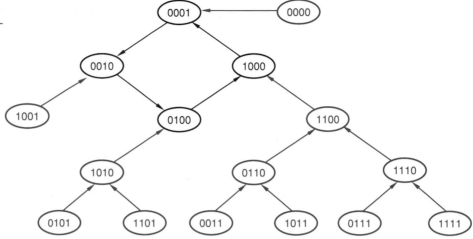

* 6.4.7 Johnson Counters

twisted-ring counter
Moebius counter
Johnson counter

An *n*-bit shift register with the complement of the serial output fed back into the serial input is a counter with $2n$ states and is called a *twisted-ring*, *Moebius*, or *Johnson counter*. Figure 6–51(a) is the basic circuit for a Johnson counter, and Figure 6–52 is its timing diagram. The normal states of this counter are listed in Table 6–4. If both the true and complemented outputs of each flip-flop are available, each normal state of the counter can be decoded with a 2-input AND or NAND gate, as shown in the table. The decoded outputs are glitch free.

self-correcting
  Johnson counter

An *n*-bit Johnson counter has $2^n - 2n$ abnormal states, and is therefore subject to the same robustness problems as a ring counter. A 4-bit *self-correcting Johnson counter* can be designed as shown in Figure 6–51(b). This circuit loads 0001 as the next state whenever the current state is 0xx0. A similar circuit using a single 2-input NOR gate can perform correction for a Johnson counter with any number of bits. The correction circuit must load 00...01 as the next state whenever the current state is 0x...x0. We can prove that the circuit corrects any

Table 6–4
States of a 4-bit Johnson counter.

State name	Q3	Q2	Q1	Q0	Decoding
S1	0	0	0	0	Q3'·Q0'
S2	0	0	0	1	Q1'·Q0
S3	0	0	1	1	Q2'·Q1
S4	0	1	1	1	Q3'·Q2
S5	1	1	1	1	Q3·Q0
S6	1	1	1	0	Q1·Q0'
S7	1	1	0	0	Q2·Q1'
S8	1	0	0	0	Q3·Q2'

Figure 6–51

Four-bit, eight-state Johnson counters: (a) basic; (b) self-correcting.

Figure 6–52   Timing diagram for a 4-bit Johnson counter.

abnormal state as follows. An abnormal state has the coding x...x10x...x, since the only codings that can't be written in this form are normal states (00...00, 11...11, and 01...11). Therefore, within $n-2$ clock ticks, the shift register will contain 10x...x. One tick later it will contain 0x...x0, and one tick after that the normal state 00...01 will be loaded.

* 6.4.8   Linear Feedback Shift Register Counters

linear feedback shift-register (LFSR) counter

maximum-length sequence generator

finite fields

The n-bit shift register counters that we've shown so far have far less than the maximum of 2^n normal states. An n-bit *linear feedback shift-register (LFSR) counter* can have 2^n-1 states, almost the maximum. Such a counter is sometimes called a *maximum-length sequence generator*.

The design of LFSR counters is based on the theory of *finite fields*, which was developed by French mathematician Évariste Galois (1811–1832) shortly before he was killed in a duel with a political opponent. The operation of an LFSR counter corresponds to operations in a finite field with 2^n elements.

Figure 6–53 shows the general structure of an n-bit LFSR counter. The shift-register's serial input is connected to the sum modulo 2 of a certain set of output bits. These feedback connections determine the state sequence of the counter.

maximum-length sequence

Using finite field theory, it can be shown that for any value of n, there exists at least one feedback equation such that the counter cycles through all 2^n-1 nonzero states before repeating. This is called a *maximum-length sequence*. (Regardless of the connection pattern, the next state for the all-0s state is all 0s.) Table 6–5 lists feedback equations that result in maximum-length sequences for

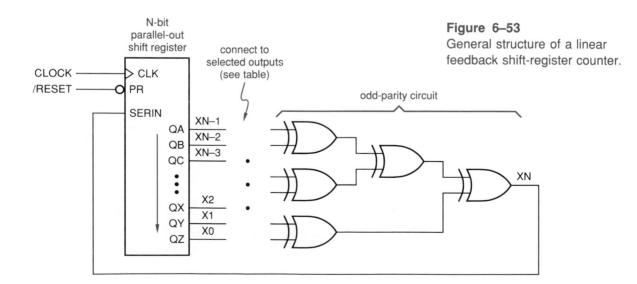

Figure 6–53
General structure of a linear feedback shift-register counter.

Table 6–5
Feedback equations for linear
feedback shift-register counters.

n	Feedback equation
2	$X2 = X1 \oplus X0$
3	$X3 = X1 \oplus X0$
4	$X4 = X1 \oplus X0$
5	$X5 = X2 \oplus X0$
6	$X6 = X1 \oplus X0$
7	$X7 = X3 \oplus X0$
8	$X8 = X4 \oplus X3 \oplus X2 \oplus X0$
12	$X12 = X6 \oplus X4 \oplus X1 \oplus X0$
16	$X16 = X5 \oplus X4 \oplus X3 \oplus X0$
20	$X20 = X3 \oplus X0$
24	$X24 = X7 \oplus X2 \oplus X1 \oplus X0$
28	$X28 = X3 \oplus X0$
32	$X32 = X22 \oplus X2 \oplus X1 \oplus X0$

selected values of n. For each value of n greater than 3, there are many other
feedback equations that result in maximum-length sequences, all different.

The logic diagram for a 3-bit LFSR counter is shown in Figure 6–54. The
state sequence for this counter is shown in the first three columns of Table 6–6.
Starting in any nonzero state, 100 in the table, the counter visits seven states
before returning to the starting state.

An LFSR counter can be modified to have 2^n states, including the all-0s

Figure 6–54
A 3-bit LFSR counter; modifications to include the all-0s state
are shown in color.

Table 6–6
State sequences for the 3-bit LFSR counter in Figure 6–54.

Original sequence			Modified sequence		
X2	X1	X0	X2	X1	X0
1	0	0	1	0	0
0	1	0	0	1	0
1	0	1	1	0	1
1	1	0	1	1	0
1	1	1	1	1	1
0	1	1	0	1	1
0	0	1	0	0	1
1	0	0	0	0	0
.	.	.	1	0	0
.	.	.	.	.	.

state, as shown in color for the 3-bit counter in Figure 6–54. The resulting state sequence is given in the last three columns of Table 6–6. In an n-bit LFSR counter, an extra XOR gate and an $n - 1$ input NOR gate connected to all shift-register outputs except X0 accomplishes the same thing.

The states of an LFSR counter are not visited in binary counting order. However, LFSR counters are typically used in applications where this characteristic is an advantage. A major application of LFSR counters is in generating test inputs for logic circuits. In most cases, the "pseudo-random" counting sequence of an LFSR counter is more likely than a binary counting sequence to detect errors. LFSRs can also be used in the encoding and decoding circuits for certain error-detecting and error-correcting codes, and are often used to "scramble" the data patterns transmitted by high-speed modems.

* 6.4.9 IEEE Standard Symbols

Figure 6–55 shows the IEEE standard symbols for the shift registers in this section. The general qualifying symbol SRGn denotes an n-bit shift register. We've used most of the other notation in these symbols previously; there's just one new thing:

shifting input

$\rightarrow$

$\leftarrow$

- *Shifting inputs.* When asserted, an input labeled with a $\rightarrow$ causes the device to shift its information one position in the direction from left to right or from top to bottom. An input labeled with a $\leftarrow$ causes a shift in the opposite direction.

pros and cons of IEEE standard symbols

These are the last details of the IEEE standard that we cover in this book. At this point, it might be appropriate to consider some of the pros and cons of IEEE standard symbols. The standard has several strengths:

Figure 6–55
IEEE standard symbols for
shift registers.

- It's a standard.

- It's consistent.

- It supports, indeed, defines, the symbology used in bubble-to-bubble design.

- It is very complete. In addition to the basic devices covered in this book, the full IEEE standard covers many other less commonly used devices and structures.

- In most cases, the standard allows the function of a logic device to be defined unambiguously by the device symbol. For example, the '299 symbol in Figure 6–55 conveys as much information as the function table in Table 6–3, and more.

That's the good news; now here's the bad, if you hadn't already noticed:

- Although the standard allows you to figure out, with moderate effort, the precise function of an unfamiliar symbol, it does not provide any standard way to remind you quickly of the function of a familiar symbol. That is, it does not provide any standard, descriptive names for internal signals. Such items are relegated to the status of "nonstandard information," and their use is not encouraged.

- Many signals in IEEE symbols have no qualifying labels, while others have duplicates. This makes it awkward to talk about signals during design and debugging (e.g., "Connect X to the third bit up from the bottom...").

- The standard encourages the omission of qualifying labels "if no confusion will result." That's like saying "Don't use your turn signal if nobody's nearby to see it." It's precisely when someone is driving in your blind spot that the habit of *always* using your turn signal can avert an accident. IEEE symbols have lots of "blind" inputs and outputs that may be hooked up incorrectly while drawing a schematic, or misapplied during debugging.

- Many possible symbols exist for moderately complex devices. For example, in the '194 symbol in Figure 6–55, pin 11 could be handled as shown, or could be drawn as a single input line with the label C4/1→/2← or as three lines with the labels C4, 1→, and 2←. If used, these variations make it even more difficult to "eyeball" the symbol for a familiar device and recognize its functions.

What logic designers really need is a standard set of descriptive, functional, naming conventions for the inputs and outputs of MSI and LSI devices, one that is consistently followed by all data books and CAD systems. For example, it's ridiculous that in the present environment, a single manufacturer can produce 2-, 4-, and 8-input multiplexers with descriptive input names ranging from A–B to C0–C3 to D0–D7! Unfortunately, the industry has too much invested in documentation and CAD software for both inconsistent traditional symbols and unhelpful IEEE standard symbols, for a consistent, helpful naming standard to be deployed anytime soon. At least, today's ASIC designers can work to ensure a modicum of naming consistency and helpfulness in the logic "macrocells" that they define and use in their own designs.

* 6.5   ITERATIVE COMBINATIONAL VS. SEQUENTIAL CIRCUITS

We introduced iterative circuits in Section 4.8.2. The function of an n-module iterative circuit can be performed by a sequential circuit that uses just one copy of the module but requires n steps (clock ticks) to obtain the result. This is an excellent example of a space/time tradeoff in digital design.

As shown in Figure 6–56, flip-flops are used in the sequential-circuit version to store the cascading outputs at the end of each step; the flip-flop outputs are used as the cascading inputs at the beginning of the next step. The flip-flops must be initialized to the boundary-input values before the first clock tick, and they contain the boundary-output values after the nth tick.

Figure 6–56
General structure of the
sequential-circuit version of
an iterative circuit.

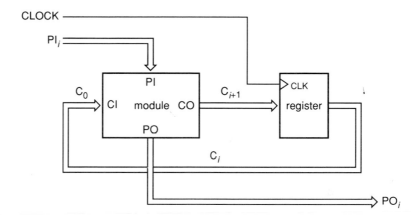

Since an iterative circuit is a combinational circuit, all of its primary and boundary inputs may be applied simultaneously, and its primary and boundary outputs are all available after a combinational delay. In the sequential-circuit version, the primary inputs must be delivered sequentially, one per clock tick, and the primary outputs are produced with similar timing. Therefore, serial-out shift registers are often used to provide the inputs, and serial-in shift registers are used to collect the outputs. For this reason, the sequential-circuit version of an "iterative widget" is often called a "serial widget."

For example, Figure 6–57(a) shows the basic design for a *serial compara-* *serial comparator*
tor circuit. The shaded block is identical to the module used in the iterative comparator of Figure 4–87 on page 303. The circuit is drawn in more detail using TTL chips in (b). In addition, we have provided a synchronous reset input that, when asserted, forces the initial value of the cascading flip-flop to 1 at the next clock tick. (The initial value of the cascading flip-flop corresponds to the boundary input in the iterative comparator.) An n-bit comparison requires $n + 1$ clock ticks. /RESET is asserted at the first clock tick. /RESET is negated and data bits are applied at the next n ticks. The EQI output gives the comparison result during the clock period following the last tick. Timing for two successive 4-bit comparisons is shown in (c).

A *serial binary adder* circuit for addends of any length can be constructed *serial binary adder*
from a full adder and a D flip-flop, as shown in Figure 6–58. The flip-flop, which stores the carry between successive bits of the addition, is cleared to 0 at reset. Addend bits are presented serially on the A and B inputs, starting with the LSB, and sum bits appear on S in the same order.

Because of the large size and high cost of digital logic circuits in the early days, many computers and calculators used serial adders and other serial versions of iterative circuits to perform arithmetic operations. Even though these arithmetic circuits aren't used much today, they provide a powerful reminder of the space/time trade-offs that are possible in digital design.

Figure 6–57
Serial comparator circuit:
(a) simplified; (b) detailed;
(c) timing diagram.

(a)

(b)

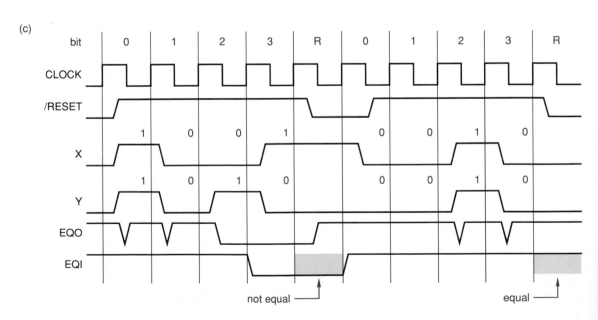

(c)

Figure 6–58
Serial binary adder circuit.

6.6 PRACTICAL STATE-MACHINE DESIGN

Aside from planning the overall architecture of a digital system, designing state machines is probably the most creative task that is performed by a digital designer. In Section 5.4.1 we showed how to design a state machine by designing a state table, and then "turning the crank" to synthesize a circuit. In this section we'll describe several other approaches to state-machine design and synthesis.

6.6.1   Ad Hoc Design

Small state machines can often be designed by directly translating the word description into a circuit. The circuit cost and design effort resulting from this approach may be better or worse than what is obtained with the formal state-table design and synthesis approach described in Section 5.4. As an example, let's reconsider a design problem from that section:

> Design a clocked synchronous state machine with two inputs, A and B, and a single output Z. The output should be 1 if A had the same value at each of the two previous clock ticks, *or if B has been 1 since the last time the first condition was true.* Otherwise, the output should be 0.

If the italicized clause is ignored, the machine is very easy to design. Two flip-flops can be used to store the value of the A input for the previous two clock ticks. An XOR gate can then be connected to the flip-flop outputs to indicate whether A has had the same value for two clock ticks. Figure 6–59(a) is a circuit resulting from these ideas.

The next step in the ad hoc design is to provide for the behavior required by the italicized clause in the problem specification. This can be accomplished by adding another flip-flop, one that gets set to 1 if the first condition is satisfied, and remains 1 as long as B is 1. The Z output is then the OR of this flip-flop's output and the first signal, as shown in Figure 6–59(b). The figure also shows connections for an active-low /RESET input. Comparing with Figure 5–47, the

Figure 6–59   Ad hoc design of the example state machine: (a) ignoring the italicized clause in the design specification; (b) complete design.

ad hoc design has the same number of flip-flops and one more gate. Depending on how your brain works, you may or may not be more confident that the circuit obtained with the ad hoc approach satisfies the design specification (see Exercise 6.32).

finite-memory
machine

An entire class of machines, called *finite-memory machines*, are quite appropriate for ad-hoc design. Such a machine's outputs are completely determined by its current input and its inputs and outputs during the previous n clock ticks, where n is a finite, bounded integer. Any machine that can be realized as shown

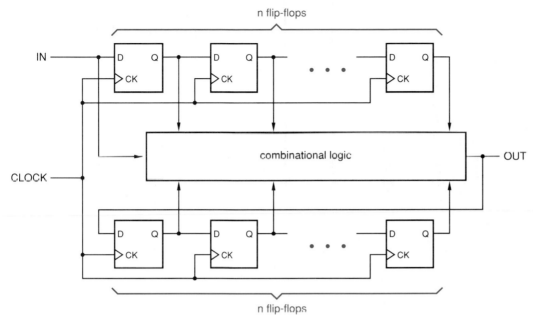

Figure 6–60   Structure of a finite-memory machine with one input and one output.

in Figure 6–60 is a finite-memory machine. Note that a finite-*state* machine need not be a finite-*memory* machine; for example, a modulo-n counter with an enable input and a "MAX" output has only n states, but its output may depend on the value of the enable input at every clock tick since initialization.

The "combination-lock" state machine of Section 5.4.6 is a finite-memory machine; its design specification is repeated below:

> Design a clocked synchronous state machine with one input, X, and two outputs, UNLK and HINT. The UNLK output should be 1 if and only if X is 0 and the sequence of inputs received on X at the preceding seven clock ticks was 0110111. The HINT output should be 1 if and only if the current value of X is the correct one to move the machine closer to being in the "unlocked" state (with UNLK = 1).

If we ignore the HINT output, the combination lock is quite obviously a finite-memory machine: the UNLK output can be obtained by looking at the last seven inputs and determining whether they were 0110111. A shift register can be used to store the last seven inputs, and two gates decode the required pattern, as shown in Figure 6–61.

Compared to the standard state-table design in Section 5.4.6, the circuit in Figure 6–61 is a "no-brainer" to design. It might be preferred for that reason, even though it has more flip-flops. On the other hand, providing the HINT output

Figure 6–61
Design of a combination-lock machine using the finite-memory model.

is a bit tougher. The machine is still a finite-memory machine, but the HINT output is a real mess to design (think about it, or try working Exercise 6.34). For that reason, we may elect to go back to the state-table approach.

In summary, an ad hoc design approach may be appropriate for small machines whose word descriptions translate directly into flip-flops that "remember things" and combinational logic that decodes the things remembered. However, we need better structured approaches to solve more complex problems.

Since the word description or design specification for a state machine is often fuzzy and sometimes nonexistent, a useful design approach must make the designer confident that the machine will really solve the problem at hand. That is, the design approach should highlight, rather than hide, any ambiguities in the design specification, allowing the ambiguities to be resolved. The rest of this section describes two approaches to state-machine design that can provide a high degree of confidence.

6.6.2 Designing State Machines Using State Diagrams

Most people like to take a graphical approach to design—you've probably solved many problems just by doodling. For that reason, state diagrams are often used to design small-to-medium-sized state machines. In this section, we'll give examples of state-diagram design, and describe a simple procedure for synthesizing circuits from the state diagrams. This method is easily adapted to design state machines using programmable logic devices, as we'll show in Section 7.4.

Designing a state diagram is much like designing a state table, which, as we showed in Section 5.4.1, is much like writing a program. However, there is one fundamental difference between a state diagram and a state table, a difference that makes state-diagram design simpler but also more error prone:

- A state table is an exhaustive listing of the next states for each state/input combination. No ambiguity is possible.

- A state diagram contains a set of arcs labeled with transition expressions. Even when there are many inputs, only one transition expression is required per arc. However, when a state diagram is constructed, there is no guarantee that the transition expressions written on the arcs leaving a particular state cover all the input combinations exactly once.

In an improperly constructed (*ambiguous*) state diagram, the next state for some   *ambiguous state*
input combinations may be unspecified, which is generally undesirable, while   *diagram*
multiple next states may be specified for others, which is just plain wrong.
Thus, considerable care must be taken in the design of state diagrams; we'll give
several examples.

Our first example is a state machine that controls the tail lights of a 1965 Ford Thunderbird, shown in Figure 6–62. There are three lights on each side, and for turns they operate in sequence to show the turning direction, as illustrated in Figure 6–63. The state machine has two input signals, LEFT and RIGHT, that indicate the driver's request for a left turn or a right turn. It also has an

Figure 6–62
T-bird tail lights.

LC   LB   LA   RA   RB   RC

Figure 6-63

Flashing sequence for T-bird tail lights: (a) left turn; (b) right turn.

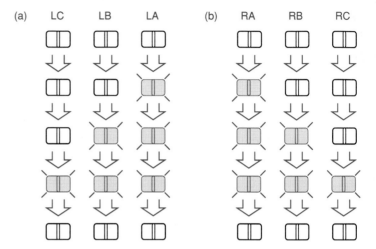

emergency-flasher input, HAZ, that requests the tail lights to be operated in hazard mode—all six lights flashing on and off in unison. We also assume the existence of a free-running clock signal whose frequency equals the desired flashing rate for the lights.

Given the foregoing requirements, we can design a clocked synchronous state machine to control the T-bird tail lights. We will design a Moore machine, so that the state alone determines which lights are on and which are off. For a left turn, the machine should cycle through four states, in which the right-hand lights are off and 0, 1, 2, or 3 of the left-hand lights are on. Likewise, for a right turn, it should cycle through four states in which the left-hand lights are off and 0, 1, 2, or 3 of the right-hand lights are on. In hazard mode, only two states are required—all lights on and all lights off.

Figure 6-64 shows our first cut at a state diagram for the machine. A common IDLE state is defined in which all of the lights are off. When a left turn is requested, the machine goes through three states in which 1, 2, and 3 of the left-hand lights are on, and then back to IDLE; right turns work similarly. In the hazard mode, the machine cycles back and forth between the IDLE state and a state in which all six lights are on. Since there are so many outputs, we've included a separate output table rather than writing output values on the state diagram. Even without assigning coded states to the named states, we can write output equations from the output table, if we let each state name represent a logic expression that is 1 only in that state:

$$LA = L1 + L2 + L3 + LR3 \qquad RA = R1 + R2 + R3 + LR3$$
$$LB = L2 + L3 + LR3 \qquad RB = R2 + R3 + LR3$$
$$LC = L3 + LR3 \qquad RC = R3 + LR3$$

There's one big problem with the state diagram of Figure 6-64—it doesn't

Figure 6–64
Initial state diagram and output
table for T-bird tail lights.

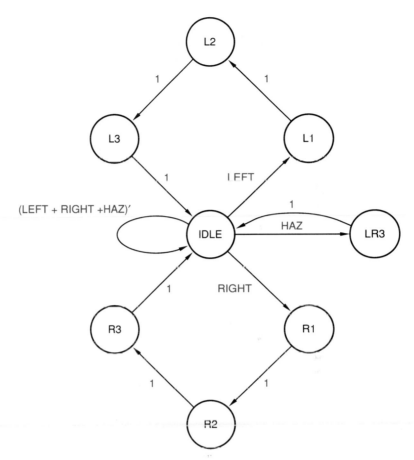

Output Table

State	LC	LB	LA	RA	RB	RC
IDLE	0	0	0	0	0	0
L1	0	0	1	0	0	0
L2	0	1	1	0	0	0
L3	1	1	1	0	0	0
R1	0	0	0	1	0	0
R2	0	0	0	1	1	0
R3	0	0	0	1	1	1
LR3	1	1	1	1	1	1

properly handle the case of multiple inputs asserted simultaneously. For example, what happens in the IDLE state if both LEFT and HAZ are asserted? According to the state diagram, the machine goes to two states, L1 and LR3, which is impossible. In reality, the machine would have only one next state, which could be L1, LR3, or a totally unrelated (and possibly unused) third state, depending on details of the state machine's realization (for example, see Exercise 6.43).

The problem is fixed in Figure 6–65, where we have given the HAZ input priority. Also, we treat LEFT and RIGHT asserted simultaneously as a hazard request, since the driver is clearly confused and needs help.

The new state diagram is unambiguous because the transition expressions on the arcs leaving each state are mutually exclusive and all inclusive. That is, for each state, no two expressions are 1 for the same input combination, and some expression is 1 for every input combination. This can be confirmed algebraically for this or any other state diagram by performing two steps:

Figure 6–65
Corrected state diagram for
T-bird tail lights.

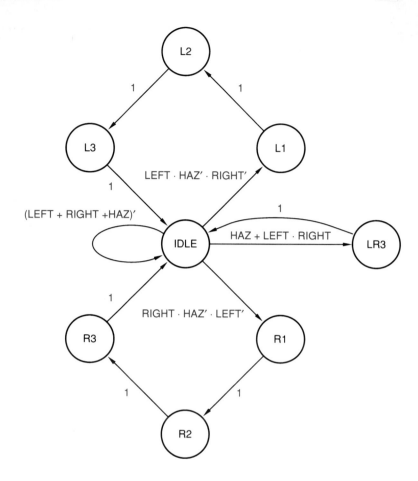

mutual exclusion

(1) *Mutual exclusion.* For each state, show that the logical product of each possible pair of transition expressions on arcs leaving that state is 0. If there are n arcs, then there are $n(n - 1)/2$ logical products to evaluate.

all inclusion

(2) *All inclusion.* For each state, show that the logical sum of the transition expressions on all arcs leaving that state is 1.

If there are many transitions leaving each state, these steps, especially the first one, are very difficult to perform. However, typical state machines, even ones with lots of states and inputs, don't have many transitions leaving each state, since most designers can't dream up such complex machines in the first place. This is where the trade-off between state-table and state-diagram design occurs. In state-table design, the foregoing steps are not required, because the structure of a state table guarantees mutual exclusion and all inclusion. But if there are a lot of inputs, the state table has *lots* of columns.

Table 6-7
State assignment for T-bird tail
lights machine.

State	Q2	Q1	Q0
IDLE	0	0	0
L1	0	0	1
L2	0	1	1
L3	0	1	0
R1	1	0	1
R2	1	1	1
R3	1	1	0
LR3	1	0	0

Verifying that a state diagram is unambiguous may be difficult in principle, but it's not too bad in practice for small state diagrams. In Figure 6–65, most of the states have a single arc with a transition expression of 1, so verification is trivial. Real work is needed only to verify the IDLE state, which has four transitions leaving it. This can be done on a sheet of scratch paper by listing the eight combinations of the three inputs, and checking off the combinations covered by each transition expression. Each combination should have exactly one check. As another example, consider the state diagrams in Figures 5–39 and 5–41 on pages 379 and 381; both can be verified mentally.

Returning to the T-bird tail lights machine, we can now synthesize a circuit from the state diagram if we wish. However, if we want to make any changes in the machine's behavior, now is the time to do it, before we've gone to the trouble of synthesizing a circuit. In particular, notice that once a left- or right-turn cycle has begun, the state diagram in Figure 6–65 allows the cycle to run to completion, even if HAZ is asserted. While this may have a certain aesthetic appeal, it would be safer for the car's occupants to have the machine go into hazard mode as soon as possible. The state diagram is modified to provide this behavior in Figure 6–66.

Now we're finally ready to synthesize a circuit for the T-bird machine. The state diagram has eight states, so we'll need a minimum of three flip-flops to code the states. Obviously, there are many state assignments possible (8! to be exact); we'll use the one in Table 6–7 for the following reasons:

An initial (idle) state of 000 is compatible with most flip-flops and registers, which are easily initialized to the 0 state.

Two state variables, Q1 and Q0, are used to "count" in Gray-code sequence for the left-turn cycle (IDLE→L1→L2→L3→IDLE). This minimizes the number of state-variable changes per state transition, which should simplify the excitation logic.

Figure 6–66
Enhanced state diagram for
T-bird tail lights.

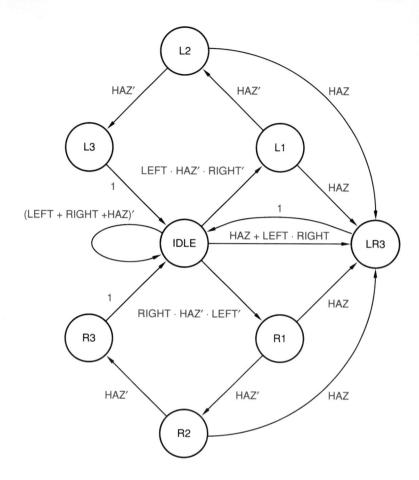

(3) Because of the symmetry in the state diagram, the same sequence on Q1 and Q0 is used to "count" during a right-turn cycle, while Q2 is used to distinguish between left and right.

(4) The remaining state-variable combination is used for the LR3 state.

   The next step is to write a sort of transition table. However, we must use a format different from the transition tables of Section 5.3.2, because the transitions in a state diagram are specified by expressions rather than by an exhaustive tabulation of next states. We'll call the new format a *transition list* because it has one row for each transition or arc in the state diagram.

transition list

   Table 6–8 is the transition list for the state diagram of Figure 6–66 and the state assignment of Table 6–7. Each row contains the current state, next state, and transition expression for one arc in the state diagram. Both the named and coded versions of the current state and next state are shown. The named states

Table 6–8   Transition list for T-bird tail lights machine.

S	Q2	Q1	Q0	Transition expression	S*	Q2*	Q1*	Q0*
IDLE	0	0	0	(LEFT + RIGHT + HAZ)'	IDLE	0	0	0
IDLE	0	0	0	LEFT·HAZ'·RIGHT'	L1	0	0	1
IDLE	0	0	0	HAZ + LEFT·RIGHT	LR3	1	0	0
IDLE	0	0	0	RIGHT·HAZ'·LEFT'	R1	1	0	1
L1	0	0	1	HAZ'	L2	0	1	1
L1	0	0	1	HAZ	LR3	1	0	0
L2	0	1	1	HAZ'	L3	0	1	0
L2	0	1	1	HAZ	LR3	1	0	0
L3	0	1	0	1	IDLE	0	0	0
R1	1	0	1	HAZ'	R2	1	1	1
R1	1	0	1	HAZ	LR3	1	0	0
R2	1	1	1	HAZ'	R3	1	1	0
R2	1	1	1	HAZ	LR3	1	0	0
R3	1	1	0	1	IDLE	0	0	0
LR3	1	0	0	1	IDLE	0	0	0

are useful for reference purposes, while the coded states are used to develop transition equations.

Once we have a transition list, the rest of the synthesis steps are pretty much "turning the crank," especially in gate-level or PLD-based design. If you're interested, synthesis procedures are described in Section 6.7. Although these procedures are sometimes applied manually, more often they are embedded in a CAD software package.

We also encountered one "turn-the-crank" step in this subsection—finding the ambiguities in state diagrams. Even though the procedure we discussed can be easily automated and doesn't require much computer time for most practical-size designs, few CAD programs perform this step. Thus, in most design environments, the designer is responsible for providing a state-machine description that is unambiguous. The algorithmic-state-machine approach described in the next subsection is one way to do this.

6.6.3   Designing State Machines Using ASM Charts

Way back in Section 5.4, we said that designing a state table or diagram is much like writing a computer program; perhaps by now you believe it. The *synthesis* procedures that we'll cover in Section 6.7 may remind you of what compilers do for programs; that is, they convert a high-level description of a state machine into a low-level realization. This subsection is still concerned with ideas in *design*,

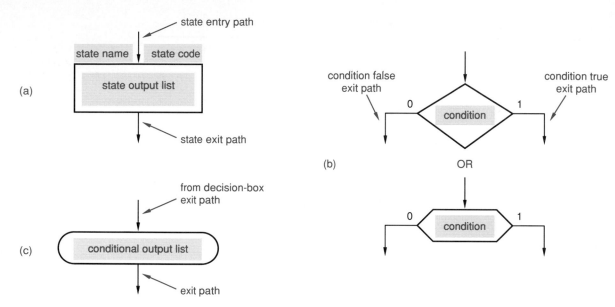

Figure 6–67   Elements of ASM notation: (a) state box; (b) decision box; (c) conditional output box.

and shows another way to create a correct, high-level state-machine description to feed into the CAD software.

algorithmic-state-machine (ASM) chart

An *algorithmic-state-machine (ASM) chart* is a graphical specification of state-machine behavior that is more like a programmer's flowchart than a state diagram. Figure 6–67 shows the basic elements that appear in ASM charts:

state box

- *State box.* An ASM chart has one *state box* per state, showing both the state name and the state code, and containing a list of Moore-type outputs (if any) that are asserted in that state. The main difference between a state box and a state diagram's bubble is that the state box has just a *single exit point* representing the next-state transition, shown by a single transition arrow leaving the box. The arrow leads to a single state box or decision box.

decision box
condition expression

- *Decision box.* A single transition arrow is split into two alternative transitions by a *decision box*, which contains a *condition expression*—a logic expression involving the machine's inputs. For input combinations such that the condition expression is 1, the exit path labeled 1 is taken; otherwise, the exit path labeled 0 is taken. Each exit path leads to a state box or to another decision box. Multiple decision boxes with different condition expressions are placed in series when a state has more than two different next states.

conditional output box

- *Conditional output box.* This element is placed on top of the exit path of a decision box to specify Mealy-type outputs. It lists outputs that are asserted

Figure 6–68
ASM charts: (a) free-running
modulo-4 counter; (b) modulo-4
counter with enable; (c) modulo-4
counter with enable; (d) modulo-
4 counter with a Mealy output.

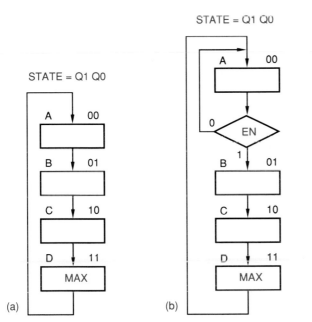

in the current state (as determined by looking back along the path to a state box), given an input combination that would cause that path to be taken at the clock tick. Like the state-diagram notation for Mealy machines, this is a little misleading, since the outputs are normally asserted for the entire clock period when the conditions are satisfied, not just at the clock tick when the transition is taken.

A few simple ASM charts are shown in Figure 6–68. The first example, in (a), is a free-running divide-by-4 counter with a MAX output. A note at the top of the chart indicates the signal names of the state variables, Q1 and Q0 in this example. By looking at the coded states, we can see that the designer has decided to realize the state machine as a binary counter (as opposed to a ring counter, Johnson counter, or some other type).

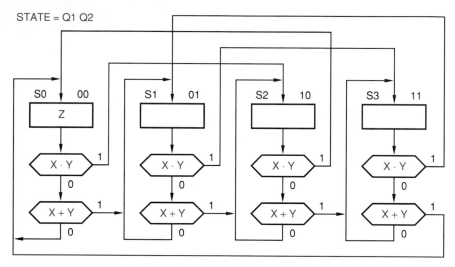

STATE = Q1 Q2

Figure 6–69 ASM chart for the 1s-counting machine.

The ASM chart in Figure 6–68(b) uses a decision box to define an enable input. Note that the enable input is only looked at in state A so that, once started, the machine can only be stopped in state A. A more usual enabling behavior is shown in (c); EN is checked in every state. The machine in (d) uses a different state coding. Also, the MAX output is shown as a Mealy output using a conditional output box; MAX is asserted in state D only if EN = 1.

Another example is a "1s-counting machine," which we previously designed using the state-table approach in Table 5–16 on page 398. The machine's behavior is defined as follows:

> Design a clocked synchronous state machine with two inputs, X and Y, and a single output, Z. The output should be 1 if the number of 1 inputs on X and Y since the machine started is a multiple of 4, and 0 otherwise.

An ASM chart that produces this behavior is shown in Figure 6–69. Note that the conditions X·Y and X + Y in each pair of decision boxes are not mutually exclusive. However, the structure of the ASM chart guarantees that X·Y = 0 before the X + Y test is made, so the chart is not ambiguous. A state diagram would have to use the transition expression X·Y′ + X′·Y instead of X + Y to prevent ambiguity.

A final example, shown in Figure 6–70, is an ASM chart for the T-bird tail-lights machine of Figure 6–66. The IDLE state in the original state diagram has several transitions leaving it, so the ASM chart requires several decision boxes in series to define the transitions. Each state box contains a list of the

Figure 6–70
ASM chart for T-bird tail lights.

outputs that are asserted in that state. The drawing conventions are somewhat relaxed; for example, exit paths leave the most convenient side of a decision box. You are encouraged to study the T-bird tail-lights chart well enough to convince yourself that it does the job.

A big advantage of designing with ASM charts is that a properly constructed chart is guaranteed to provide an unambiguous description of next-state behavior. That is, in each state, every input combination leads to exactly one next state in the chart. This is true because every input combination yields a definite outcome in each decision box and, in a properly constructed chart, all exit paths go to a single next state or to another decision box.[5] Thus, no verification is needed; the mutual-exclusion and all-inclusion properties that we defined for state diagrams are automatic.

Transition lists can be developed directly from an ASM chart, but the procedure is a bit more involved than the one for state diagrams. In a state diagram, each possible transition from a state has its own arc leaving the state bubble. In an ASM chart, each state box has only one exit path; the possible transitions correspond to all possible paths to next states going through zero or more decision boxes. Each path yields one transition expression (and row in the transition

transition lists

---

[5]Some authors allow decision boxes to be placed in parallel; that is, they allow a single exit path to go to multiple decision boxes. In this case, an ambiguous ASM chart can be constructed.

Figure 6-71
Possible paths leaving the IDLE state in T-bird machine.

list). In the T-bird machine, there are four possible paths leaving the IDLE state, as shown in Figure 6–71.

transition
expressions

The transition expression corresponding to an ASM-chart path is the logical product of the current state's minterm and the condition expressions that appear in the decision boxes along the chosen path. Each condition expression is complemented if the chosen path goes through the "false" or "0" exit path of the decision box; otherwise, it appears as-is. Thus, the IDLE state in Figure 6–70 has the transition expressions shown in Table 6–9. The L1, L2, R1, and R2 states have a single decision box and therefore two transition expressions each, and the remaining states have no decision boxes and one transition expression each. Using these expressions, you can construct exactly the same transition list as Table 6–8.

Many digital designers consider ASM charts to be the most convenient way to specify state-machine behavior. However, in much the same way that most programmers have given up flowcharts in favor of block-structured high-level

Table 6–9   Transition expressions for the IDLE state in Figure 6–70.

Condition expressions				
HAZ + LEFT·RIGHT	LEFT	RIGHT	Transition expression	S*
1	–	–	$Q2' \cdot Q1' \cdot Q0' \cdot$(HAZ + LEFT·RIGHT)	LR3
0	1	–	$Q2' \cdot Q1' \cdot Q0' \cdot$(HAZ + LEFT·RIGHT)$' \cdot$(LEFT)	L1
0	0	0	$Q2' \cdot Q1' \cdot Q0' \cdot$(HAZ + LEFT·RIGHT)$' \cdot$(LEFT)$' \cdot$(RIGHT)$'$	IDLE
0	0	1	$Q2' \cdot Q1' \cdot Q0' \cdot$(HAZ + LEFT·RIGHT)$' \cdot$(LEFT)$' \cdot$(RIGHT)	R1

languages and pseudo-code, you can expect to see experienced logic designers give up ASM charts in favor of state-machine description languages. As we'll show in Section 7.4.3, these languages emulate and enhance the features of an ASM chart, and allow a state machine to be described in a text file rather than a hand-drawn picture. A CAD program can then "compile" the text file into a PLD-based state-machine circuit. Therefore, if you're just getting started in this business, don't worry too much about ASM charts; most modern state machines are designed using PLDs and state-machine description languages.

6.7 SYNTHESIZING STATE MACHINES FROM TRANSITION LISTS

Once a machine's state diagram has been designed and a state assignment has been made, the creative part of the design process is pretty much over. The rest of the synthesis procedure can be carried out by CAD programs.

As we showed previously, a transition list can be constructed from a machine's state diagram and state assignment, or from an ASM chart. This section shows how to synthesize a state machine from its transition list. It also delves into some of the options and nuances of state-machine design using transition lists. This material is useful not only for synthesizing machines by hand, but also for understanding the internal operation and the external quirks of CAD programs and languages that deal with state machines.

6.7.1 Transition Equations

The first step in synthesizing a state machine from a transition list is to develop a set of transition equations that define each next-state variable $V*$ in terms of the current state and input. The transition list can be viewed as a sort of hybrid truth table in which current-state state-variable combinations are listed explicitly and input combinations are listed algebraically. Reading down a $V*$ column in a transition list, we find a sequence of 0s and 1s, indicating the value of $V*$ for various (if we've done it right, all) state/input combinations. For reference, the transition list for the T-bird tail-lights machine of Figure 6–66 is repeated in Table 6–10.

A transition equation for a next-state variable $V*$ can be written using a sort of hybrid canonical sum:

$$V* = \sum_{\text{transition-list rows where } V* = 1} (\text{transition p-term})$$

That is, the transition equation has one "transition p-term" for each row of the transition list that contains a 1 in the $V*$ column. A row's *transition p-term*   *transition p-term*
is the product of the current state's minterm and the transition expression. If

Table 6–10   Transition list for T-bird tail-lights machine.

S	Q2	Q1	Q0	Transition expression	S*	Q2*	Q1*	Q0*
IDLE	0	0	0	(LEFT + RIGHT + HAZ)′	IDLE	0	0	0
IDLE	0	0	0	LEFT·HAZ′·RIGHT′	L1	0	0	1
IDLE	0	0	0	HAZ + LEFT·RIGHT	LR3	1	0	0
IDLE	0	0	0	RIGHT·HAZ′·LEFT′	R1	1	0	1
L1	0	0	1	HAZ′	L2	0	1	1
L1	0	0	1	HAZ	LR3	1	0	0
L2	0	1	1	HAZ′	L3	0	1	0
L2	0	1	1	HAZ	LR3	1	0	0
L3	0	1	0	1	IDLE	0	0	0
R1	1	0	1	HAZ′	R2	1	1	1
R1	1	0	1	HAZ	LR3	1	0	0
R2	1	1	1	HAZ′	R3	1	1	0
R2	1	1	1	HAZ	LR3	1	0	0
R3	1	1	0	1	IDLE	0	0	0
LR3	1	0	0	1	IDLE	0	0	0

the transition expression itself happens to be a simple product term, then the transition equation expresses V* in standard sum-of-products form; we'll find this to be important for PLD-based state-machine designs in Section 7.4.

The transition equation for Q2* in the T-bird machine can be written as the sum of eight p-terms:

$$Q2* = Q2'·Q1'·Q0'·(HAZ + LEFT·RIGHT)$$
$$+ Q2'·Q1'·Q0'·(RIGHT·HAZ'·LEFT')$$
$$+ Q2'·Q1'·Q0·(HAZ)$$
$$+ Q2'·Q1·Q0·(HAZ)$$
$$+ Q2·Q1'·Q0·(HAZ')$$
$$+ Q2·Q1'·Q0·(HAZ)$$
$$+ Q2·Q1·Q0·(HAZ')$$
$$+ Q2·Q1·Q0·(HAZ)$$

Some straightforward algebraic manipulations lead to a simplified transition equation that combines the first two, second two, and last four p-terms above:

$$Q2* = Q2'·Q1'·Q0'·(HAZ + RIGHT)$$
$$+ Q2'·Q0·(HAZ)$$
$$+ Q2·Q0$$

Transition equations for Q1* and Q0* may be obtained in a similar manner:

$$
\begin{aligned}
Q1* \; &= \; Q2' \cdot Q1' \cdot Q0 \cdot (HAZ') \\
&\quad + Q2' \cdot Q1 \cdot Q0 \cdot (HAZ') \\
&\quad + Q2 \cdot Q1' \cdot Q0 \cdot (HAZ') \\
&\quad + Q2 \cdot Q1 \cdot Q0 \cdot (HAZ') \\
&= \; Q0 \cdot HAZ' \\
Q0* \; &= \; Q2' \cdot Q1' \cdot Q0' \cdot (LEFT \cdot HAZ' \cdot RIGHT') \\
&\quad + Q2' \cdot Q1' \cdot Q0' \cdot (RIGHT \cdot HAZ' \cdot LEFT') \\
&\quad + Q2' \cdot Q1' \cdot Q0 \cdot (HAZ') \\
&\quad + Q2 \cdot Q1' \cdot Q0 \cdot (HAZ') \\
&= \; Q2' \cdot Q1' \cdot Q0' \cdot HAZ' \cdot (LEFT \oplus RIGHT) + Q1' \cdot Q0 \cdot HAZ'
\end{aligned}
$$

Except for Q1*, there's no guarantee that the transition equations above are in any sense minimal—in fact, the expressions for Q2* and Q0* aren't even in standard sum-of-products or product-of-sums form. The simplified equations, or the original unsimplified ones, merely provide an unambiguous starting point for whatever combinational design method you might choose to synthesize the excitation logic for the state machine—ad hoc, NAND-NAND, MSI-based, or whatever. As you'll see in Section 7.4, in a PLD-based design you could simply plug the equations into a program that calculates minimal sum-of-products expressions for realization in the PLD.

6.7.2 Excitation Equations

While we're on the subject of excitation logic, note that so far we have derived only *transition equations*, not *excitation equations*. However, if we use D flip-flops as the memory elements in our state machines, then the excitation equations are trivial to derive from the transition equations, since the characteristic equation of a D flip-flop is Q* = D. Therefore, if the transition equation for a state variable Qi* is

$$Qi* \; = \; \text{expression}$$

then the excitation equation for the corresponding D flip-flop input is

$$Di \; = \; \text{expression}$$

Efficient excitation equations for other flip-flop types, especially J-K, are not so easy to derive (see Exercise 6.49). For that reason, the vast majority of discrete and PLD-based state-machine designs employ D flip-flops.

* 6.7.3 Variations on the Scheme

There are other ways to obtain transition and excitation equations from a transition list. If the column for a particular next-state variable contains fewer 0s than 1s, it may be advantageous to write that variable's transition equation in terms of the 0s in its column. That is, we write

$$V*' = \sum_{\text{transition-list rows where } V* = 0} (\text{transition p-term})$$

That is, $V*'$ is 1 for all of the p-terms for which $V*$ is 0. Thus, a transition equation for $Q2*'$ may be written as the sum of seven p-terms:

$$
\begin{aligned}
Q2*' \ = \ & Q2' \cdot Q1' \cdot Q0' \cdot ((\text{LEFT} + \text{RIGHT} + \text{HAZ})') \\
& + Q2' \cdot Q1' \cdot Q0' \cdot (\text{LEFT} \cdot \text{HAZ}' \cdot \text{RIGHT}') \\
& + Q2' \cdot Q1' \cdot Q0 \cdot (\text{HAZ}') \\
& + Q2' \cdot Q1 \cdot Q0 \cdot (\text{HAZ}') \\
& + Q2' \cdot Q1 \cdot Q0' \cdot (1) \\
& + Q2 \cdot Q1 \cdot Q0' \cdot (1) \\
& + Q2 \cdot Q1' \cdot Q0' \cdot (1) \\
= \ & Q2' \cdot Q1' \cdot Q0' \cdot \text{HAZ}' \cdot \text{RIGHT}' + Q2' \cdot Q0 \cdot \text{HAZ}' + Q1 \cdot Q0' + Q2 \cdot Q0'
\end{aligned}
$$

Both sides of the simplified equation can be complemented to obtain an equation for $Q2*$.

To obtain an expression for a next-state variable $V*$ directly using the 0s in the transition list, we can complement the right-hand side of the general $V*'$ equation using DeMorgan's theorem, obtaining a sort of hybrid canonical product:

$$V* = \prod_{\text{transition-list rows where } V* = 0} (\text{transition s-term})$$

transition s-term

Here, a row's *transition s-term* is the sum of the maxterm for the current state and the complement of the transition expression. If the transition expression is a simple product term, then its complement is a sum, and the transition equation expresses $V*$ in product-of-sums form.

There are even more things we can do with and to transition lists, as we'll see in examples in the rest of this section. We'll refine our use of transition lists further when we study the design of PLD-based state machines in Section 7.4.

6.7.4 Another Example

Our next state-machine example is a "guessing game" that can be built as an amusing lab project:

Figure 6–72
First try at a state diagram for
the guessing game.

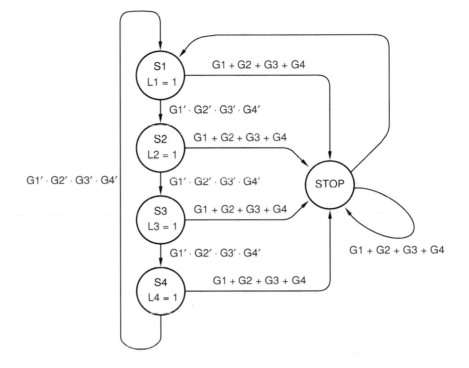

Design a clocked synchronous state machine with four inputs, G1–G4, that are connected to pushbuttons. The machine has four outputs, L1–L4, connected to lamps or LEDs located near the like-numbered pushbuttons. There is also an ERR output connected to a red lamp. In normal operation, the L1–L4 outputs display a 1-out-of-4 pattern. At each clock tick, the pattern is rotated by one position, as in a ring counter; the clock frequency is about 4 Hz.

Guesses are made by pressing a pushbutton, which asserts an input Gi. When any Gi input is asserted, the ERR output is asserted if the "wrong" pushbutton was pressed, that is, if the Gi input detected at the clock tick does not have the same number as the lamp output that was asserted before the clock tick. Once a guess has been made, play stops and the ERR output maintains the same value for one or more clock ticks until the Gi input is negated, then play resumes.

Clearly, we will have to provide four states, one for each position of the rotating pattern, and we'll need at least one state to indicate that play has stopped. A possible state diagram is shown in Figure 6–72. The machine cycles through states S1–S4 as long as no Gi input is asserted, and it goes to the STOP state when a guess is made. Each Li output is asserted in the like-numbered state.

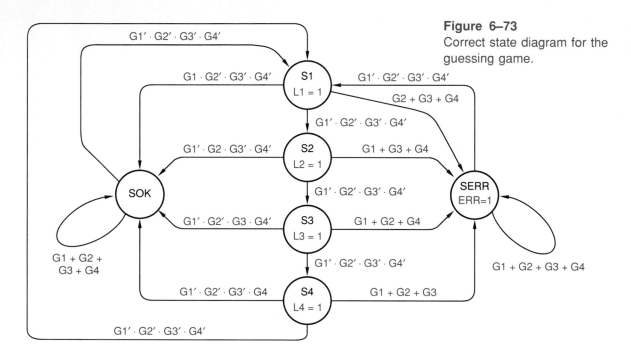

Figure 6–73
Correct state diagram for the guessing game.

Table 6–11 Transition list for guessing-game machine.

Current state				Next state				Output				
S	Q2	Q1	Q0	Transition expression	S*	Q2*	Q1*	Q0*	L1	L2	L3	L4	ERR
S1	0	0	0	$G1' \cdot G2' \cdot G3' \cdot G4'$	S2	0	0	1	1	0	0	0	0
S1	0	0	0	$G1 \cdot G2' \cdot G3' \cdot G4'$	SOK	1	0	0	1	0	0	0	0
S1	0	0	0	$G2 + G3 + G4$	SERR	1	0	1	1	0	0	0	0
S2	0	0	1	$G1' \cdot G2' \cdot G3' \cdot G4'$	S3	0	1	1	0	1	0	0	0
S2	0	0	1	$G1' \cdot G2 \cdot G3' \cdot G4'$	SOK	1	0	0	0	1	0	0	0
S2	0	0	1	$G1 + G3 + G4$	SERR	1	0	1	0	1	0	0	0
S3	0	1	1	$G1' \cdot G2' \cdot G3' \cdot G4'$	S4	0	1	0	0	0	1	0	0
S3	0	1	1	$G1' \cdot G2' \cdot G3 \cdot G4'$	SOK	1	0	0	0	0	1	0	0
S3	0	1	1	$G1 + G2 + G4$	SERR	1	0	1	0	0	1	0	0
S4	0	1	0	$G1' \cdot G2' \cdot G3' \cdot G4'$	S1	0	0	0	0	0	0	1	0
S4	0	1	0	$G1' \cdot G2' \cdot G3' \cdot G4$	SOK	1	0	0	0	0	0	1	0
S4	0	1	0	$G1 + G2 + G3$	SERR	1	0	1	0	0	0	1	0
SOK	1	0	0	$G1 + G2 + G3 + G4$	SOK	1	0	0	0	0	0	0	0
SOK	1	0	0	$G1' \cdot G2' \cdot G3' \cdot G4'$	S1	0	0	0	0	0	0	0	0
SERR	1	0	1	$G1 + G2 + G3 + G4$	SERR	1	0	1	0	0	0	0	1
SERR	1	0	1	$G1' \cdot G2' \cdot G3' \cdot G4'$	S1	0	0	0	0	0	0	0	1

The only problem with the preceding state diagram is that it doesn't "remember" in the STOP state whether the guess was correct, so it has no way to control the ERR output. This problem is fixed in Figure 6–73, which has two "stopped" states, SOK and SERR. On an incorrect guess, the machine goes to SERR, where ERR is asserted; otherwise, it goes to SOK. Although the machine's word description doesn't require it, the state diagram is designed to go to SERR even if the user tries to fool it by pressing two or more pushbuttons simultaneously, or by changing guesses while stopped.

A transition list corresponding to the state diagram is shown in Table 6–11, using a simple 3-bit binary state encoding with Gray-code order for the S1–S4 cycle. Transition equations for Q1* and Q0* can be obtained from the table as follows:

$$
\begin{aligned}
Q1* \;=\; & Q2' \cdot Q1' \cdot Q0 \cdot (G1' \cdot G2' \cdot G3' \cdot G4') \\
& + Q2' \cdot Q1 \cdot Q0 \cdot (G1' \cdot G2' \cdot G3' \cdot G4') \\
=\; & Q2' \cdot Q0 \cdot G1' \cdot G2' \cdot G3' \cdot G4' \\
Q0* \;=\; & Q2' \cdot Q1' \cdot Q0' \cdot (G1' \cdot G2' \cdot G3' \cdot G4') \\
& + Q2' \cdot Q1' \cdot Q0' \cdot (G2 + G3 + G4) \\
& + Q2' \cdot Q1' \cdot Q0 \cdot (G1' \cdot G2' \cdot G3' \cdot G4') \\
& + Q2' \cdot Q1' \cdot Q0 \cdot (G1 + G3 + G4) \\
& + Q2' \cdot Q1 \cdot Q0 \cdot (G1 + G2 + G4) \\
& + Q2' \cdot Q1 \cdot Q0' \cdot (G1 + G2 + G3) \\
& + Q2 \cdot Q1' \cdot Q0 \cdot (G1 + G2 + G3 + G4)
\end{aligned}
$$

Using a logic minimization program, the Q0* expression can be reduced to 11 product terms in two-level sum-of-products form. An expression for Q2* is best formulated in terms of the 0s in the Q2* column of the transition list:

$$
\begin{aligned}
Q2*' \;=\; & Q2' \cdot Q1' \cdot Q0' \cdot (G1' \cdot G2' \cdot G3' \cdot G4') \\
& + Q2' \cdot Q1' \cdot Q0 \cdot (G1' \cdot G2' \cdot G3' \cdot G4') \\
& + Q2' \cdot Q1 \cdot Q0 \cdot (G1' \cdot G2' \cdot G3' \cdot G4') \\
& + Q2' \cdot Q1 \cdot Q0' \cdot (G1' \cdot G2' \cdot G3' \cdot G4') \\
& + Q2 \cdot Q1' \cdot Q0' \cdot (G1' \cdot G2' \cdot G3' \cdot G4') \\
& + Q2 \cdot Q1' \cdot Q0 \cdot (G1' \cdot G2' \cdot G3' \cdot G4') \\
=\; & (Q2' + Q1') \cdot (G1' \cdot G2' \cdot G3' \cdot G4')
\end{aligned}
$$

The last five columns of Table 6–11 show output values. Thus, output equations can be developed in much the same way as transition equations. However, since this example is a Moore machine, outputs are independent of the transition expressions; only one row of the transition list must be considered for each current state. The output equations are

$$L1 = Q2' \cdot Q1' \cdot Q0' \qquad L3 = Q2' \cdot Q1 \cdot Q0 \qquad ERR = Q2 \cdot Q1' \cdot Q0$$
$$L2 = Q2' \cdot Q1' \cdot Q0 \qquad L4 = Q2' \cdot Q1 \cdot Q0'$$

* 6.7.5 Unused States

The state diagram for the guessing game in the preceding subsection has six states, but the actual state machine, built from three flip-flops, has eight. By omitting the unused states from the transition list, we treated them as "don't cares" in a very limited sense:

- When we wrote equations for Q1* and Q0*, we formed a sum of transition p-terms for state/input combinations that had an explicit 1 in the corresponding columns of the transition list. Although we didn't consider the unused states, our procedure implicitly treated them as if they had 0s in the Q1* and Q0* columns.

- Conversely, we wrote the Q2*' equation as a sum of transition p-terms for state/input combinations that had an explicit 0 in the corresponding columns of the transition list. Unused states were implicitly treated as if they had 1s in the Q2* column.

As a consequence of these choices, all of the unused states in the guessing-game machine have a coded next state of 100 for all input combinations. That's acceptable or "safe" behavior should the machine stray into an unused state, since 100 is the coding for one of the normal states (SOK).

To treat the unused states as true "don't-cares," we would have to allow them to go to any next state under any input combination. This is simple in principle but may be difficult in practice.

In Section 5.4.4, we showed how to handle unused states as "don't-cares" in the Karnaugh-map method for developing transition/excitation equations. Unfortunately, for all but the smallest problems, Karnaugh maps are unwieldy. On the other hand, commercially available logic minimization programs can easily handle larger problems, but many of them don't handle "don't-cares." Perhaps over the coming years, as CAD tools are more widely deployed, utilized, and understood, we'll see tools that automatically determine don't-care conditions and synthesize minimum-cost excitation logic directly from a transition list or equivalent state-machine description.

Table 6–12 Transition list for guessing-game machine using outputs as state variables.

	Current state						Next state					
S	L1	L2	L3	L4	ERR	Transition expression	S*	L1*	L2*	L3*	L4*	ERR*
S1	1	0	0	0	0	$G1' \cdot G2' \cdot G3' \cdot G4'$	S2	0	1	0	0	0
S1	1	0	0	0	0	$G1 \cdot G2' \cdot G3' \cdot G4'$	SOK	0	0	0	0	0
S1	1	0	0	0	0	$G2 + G3 + G4$	SERR	0	0	0	0	1
S2	0	1	0	0	0	$G1' \cdot G2' \cdot G3' \cdot G4'$	S3	0	0	1	0	0
S2	0	1	0	0	0	$G1' \cdot G2 \cdot G3' \cdot G4'$	SOK	0	0	0	0	0
S2	0	1	0	0	0	$G1 + G3 + G4$	SERR	0	0	0	0	1
S3	0	0	1	0	0	$G1' \cdot G2' \cdot G3' \cdot G4'$	S4	0	0	0	1	0
S3	0	0	1	0	0	$G1' \cdot G2' \cdot G3 \cdot G4'$	SOK	0	0	0	0	0
S3	0	0	1	0	0	$G1 + G2 + G4$	SERR	0	0	0	0	1
S4	0	0	0	1	0	$G1' \cdot G2' \cdot G3' \cdot G4'$	S1	1	0	0	0	0
S4	0	0	0	1	0	$G1' \cdot G2' \cdot G3' \cdot G4$	SOK	0	0	0	0	0
S4	0	0	0	1	0	$G1 + G2 + G3$	SERR	0	0	0	0	1
SOK	0	0	0	0	0	$G1 + G2 + G3 + G4$	SOK	0	0	0	0	0
SOK	0	0	0	0	0	$G1' \cdot G2' \cdot G3' \cdot G4'$	S1	1	0	0	0	0
SERR	0	0	0	0	1	$G1 + G2 + G3 + G4$	SERR	0	0	0	0	1
SERR	0	0	0	0	1	$G1' \cdot G2' \cdot G3' \cdot G4'$	S1	1	0	0	0	0

* 6.7.6 Output-Coded State Assignment

Let's look at another realization of the guessing-game machine. The machine's outputs are a function of state only; furthermore, a *different* output combination is produced in each named state. Therefore, we can use the outputs as state variables and assign each named state to the required output combination. This sort of *output-coded state assignment* can sometimes result in excitation equations *output-coded state* that are simpler than the set of excitation and output equations obtained with a *assignment* state assignment using a minimum number of state variables.

      Table 6–12 is the transition list resulting from an output-coded state assignment for the guessing-game state diagram. Each transition/excitation equation has very few transition p-terms because there are so few 1s in the next-state columns:

$$
\begin{aligned}
L1* &= L1' \cdot L2' \cdot L3' \cdot L4 \cdot ERR' \cdot (G1' \cdot G2' \cdot G3' \cdot G4') \\
&\quad + L1' \cdot L2' \cdot L3' \cdot L4' \cdot ERR' \cdot (G1' \cdot G2' \cdot G3' \cdot G4') \\
&\quad + L1' \cdot L2' \cdot L3' \cdot L4' \cdot ERR \cdot (G1' \cdot G2' \cdot G3' \cdot G4') \\
L2* &= L1 \cdot L2' \cdot L3' \cdot L4' \cdot ERR' \cdot (G1' \cdot G2' \cdot G3' \cdot G4') \\
L3* &= L1' \cdot L2 \cdot L3' \cdot L4' \cdot ERR' \cdot (G1' \cdot G2' \cdot G3' \cdot G4') \\
L4* &= L1' \cdot L2' \cdot L3 \cdot L4' \cdot ERR' \cdot (G1' \cdot G2' \cdot G3' \cdot G4')
\end{aligned}
$$

$$
\begin{aligned}
\text{ERR*} \;=\; & \text{L1} \cdot \text{L2}' \cdot \text{L3}' \cdot \text{L4}' \cdot \text{ERR}' \cdot (\text{G2} + \text{G3} + \text{G4}) \\
& + \text{L1}' \cdot \text{L2} \cdot \text{L3}' \cdot \text{L4}' \cdot \text{ERR}' \cdot (\text{G1} + \text{G3} + \text{G4}) \\
& + \text{L1}' \cdot \text{L2}' \cdot \text{L3} \cdot \text{L4}' \cdot \text{ERR}' \cdot (\text{G1} + \text{G2} + \text{G4}) \\
& + \text{L1}' \cdot \text{L2}' \cdot \text{L3}' \cdot \text{L4} \cdot \text{ERR}' \cdot (\text{G1} + \text{G2} + \text{G3}) \\
& + \text{L1}' \cdot \text{L2}' \cdot \text{L3}' \cdot \text{L4}' \cdot \text{ERR} \cdot (\text{G1} + \text{G2} + \text{G3} + \text{G4})
\end{aligned}
$$

There are no output equations, of course. The ERR* equation above is the worst in the group, requiring 16 terms to express in minimal sum-of-products or product-of-sums form. As a group, the equations above have about the same complexity as the transition and output equations that we developed from Table 6–11. Even though the output-coded assignment does not produce a simpler set of equations in this example, it can still save cost in a PLD-based design, as we'll show in Section 7.4.

* 6.7.7 "Don't-Care" State Codings

Out of the 32 possible coded states using five variables, only six are used in Table 6–12. The rest of the states are unused and have a next state of 00000 if the machine is built using the equations in the preceding subsection. Another possible disposition for unused states, one that we haven't discussed before, is obtained by careful use of "don't-cares" in the assignment of coded states to current states.

    Table 6–13 shows one such state assignment for the guessing-game machine, derived from the output-coded state assignment of the preceding subsection. In this example, every possible combination of current-state variables corresponds to one of the coded states (e.g., 10111 = S1, 00101 = S3). However, *next states* are coded using the same unique combinations as in the preceding subsection. Table 6–14 shows the resulting transition list.

    In this approach, each unused current state behaves like a nearby "normal" state; Figure 6–74 illustrates this concept. The machine is well-behaved and goes to a "normal state" if it inadvertently enters an unused state. Yet the approach still allows some simplification of the excitation and output logic by introducing

Table 6–13
Current-state assignment for guessing-game machine using don't-cares.

State	L1	L2	L3	L4	ERR
S1	1	x	x	x	x
S2	0	1	x	x	x
S3	0	0	1	x	x
S4	0	0	0	1	x
SOK	0	0	0	0	1
SERR	0	0	0	0	0

Table 6–14 Transition list for guessing-game machine using don't-care state codings.

Current state							Next state					
S	L1	L2	L3	L4	ERR	Transition expression	S*	L1*	L2*	L3*	L4*	ERR*
S1	1	x	x	x	x	$G1'\cdot G2'\cdot G3'\cdot G4'$	S2	0	1	0	0	0
S1	1	x	x	x	x	$G1\cdot G2'\cdot G3'\cdot G4'$	SOK	0	0	0	0	0
S1	1	x	x	x	x	$G2 + G3 + G4$	SERR	0	0	0	0	1
S2	0	1	x	x	x	$G1'\cdot G2'\cdot G3'\cdot G4'$	S3	0	0	1	0	0
S2	0	1	x	x	x	$G1'\cdot G2\cdot G3'\cdot G4'$	SOK	0	0	0	0	0
S2	0	1	x	x	x	$G1 + G3 + G4$	SERR	0	0	0	0	1
S3	0	0	1	x	x	$G1'\cdot G2'\cdot G3'\cdot G4'$	S4	0	0	0	1	0
S3	0	0	1	x	x	$G1'\cdot G2'\cdot G3\cdot G4'$	SOK	0	0	0	0	0
S3	0	0	1	x	x	$G1 + G2 + G4$	SERR	0	0	0	0	1
S4	0	0	0	1	x	$G1'\cdot G2'\cdot G3'\cdot G4'$	S1	1	0	0	0	0
S4	0	0	0	1	x	$G1'\cdot G2'\cdot G3'\cdot G4$	SOK	0	0	0	0	0
S4	0	0	0	1	x	$G1 + G2 + G3$	SERR	0	0	0	0	1
SOK	0	0	0	0	0	$G1 + G2 + G3 + G4$	SOK	0	0	0	0	0
SOK	0	0	0	0	0	$G1'\cdot G2'\cdot G3'\cdot G4'$	S1	1	0	0	0	0
SERR	0	0	0	0	1	$G1 + G2 + G3 + G4$	SERR	0	0	0	0	1
SERR	0	0	0	0	1	$G1'\cdot G2'\cdot G3'\cdot G4'$	S1	1	0	0	0	0

don't-cares in the transition list. When a row's transition p-term is written, current-state variables that are don't-cares in that row are omitted, for example,

$$
\begin{aligned}
\text{ERR*} = \ & \text{L1}\cdot(\text{G2} + \text{G3} + \text{G4}) \\
& + \text{L1}'\cdot\text{L2}\cdot(\text{G1} + \text{G3} + \text{G4}) \\
& + \text{L1}'\cdot\text{L2}'\cdot\text{L3}\cdot(\text{G1} + \text{G2} + \text{G4}) \\
& + \text{L1}'\cdot\text{L2}'\cdot\text{L3}'\cdot\text{L4}\cdot(\text{G1} + \text{G2} + \text{G3}) \\
& + \text{L1}'\cdot\text{L2}'\cdot\text{L3}'\cdot\text{L4}'\cdot\text{ERR}\cdot(\text{G1} + \text{G2} + \text{G3} + \text{G4})
\end{aligned}
$$

Compared with the ERR* equation in the preceding subsection, the one above

Figure 6–74
State assignment using don't-cares for current states.

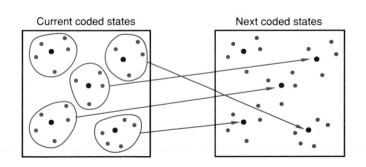

Current coded states   Next coded states

still requires 16 terms to express as a sum of products. However, it requires only five terms in minimal product-of-sums form, which makes its complement more suitable for realization in a PLD.

6.8   OTHER APPROACHES TO STATE-MACHINE DESIGN

6.8.1   MSI Counters and Shift Registers as State Machines

Sometimes a state machine's word description or state diagram almost matches the behavior of an off-the-shelf MSI part. This is especially common for state machines that have an almost cyclic behavior. For these machines, it is often possible to use counters to create the basic cycle and add combinational logic to force the noncyclic transitions.

Figure 6–75 is a straightforward example of a state diagram with an almost cyclic behavior. The main cycle is a modulo-12 counter with an enable input. In certain states, two other inputs are tested, and the counter may be forced into one of two other states. As shown in Figure 6–76, a 74LS163 and some combinational logic can do the job. In addition to counting, a '163 can be synchronously forced into either of two states—0000 and whatever state is defined by the A–D inputs. Thus, the named states of Figure 6–75 may be assigned to the 163's counting states as shown in Table 6–15.

Figure 6–75
State diagram with an almost
cyclic behavior.

Note: No transitions occur
for unspecified
input combinations.

Figure 6–76   Using a 74LS163 binary counter and combinational logic to realize an almost cyclic
state diagram.

The state machine's EN input enables the '163 in Figure 6–76 to count.
State F is coded 0000, and the circuit uses the 163's /CLR input to force state F
as required by the HOP input. Similarly, state M is coded 0110, and is forced
using /LD as required by SKIP. Note that the excitation circuit for the 163's /LD
input uses one of the unused counter states, 1111, as a don't-care.

Sometimes a designer might be willing to bend a design specification a bit
for convenience in an MSI design. An example is the guessing-game machine
of Figure 6–73. It can be built as a Mealy machine using a 74LS194 shift

Table 6–15
Counter states corresponding to
state-machine named states.

Named state	Count	QD	QC	QB	QA
A	7	0	1	1	1
B	8	1	0	0	0
C	9	1	0	0	1
D	10	1	0	1	0
E	11	1	0	1	1
F	0	0	0	0	0
G	1	0	0	0	1
H	2	0	0	1	0
J	3	0	0	1	1
K	4	0	1	0	0
L	5	0	1	0	1
M	6	0	1	1	0

Figure 6–77 MSI realization of the guessing-game machine.

register and a few gates, as shown in Figure 6–77. Since the ERR output is a combinational function of the inputs, players can cheat—once the machine is stopped, the guess can be changed to extinguish the ERR light. (I suppose that if the designer can cheat, it's only fair for the players to cheat, too!)

When the name of the game is to minimize the number of SSI and MSI packages, designers often come up with tricky realizations that coerce an off-the-shelf part into the desired state-machine behavior. An especially tricky design example is shown in Figure 6–78, which realizes the T-bird tail-lights state diagram of Figure 6–65 (without output logic) with one 74LS163 and less than one SSI package (3/4 of a 74LS32 plus 1/6 of a 74LS04). The state encoding for this realization is shown in Table 6–16. The basic idea is that the counter is almost always loading; asserting LEFT, RIGHT, or HAZ loads a non-IDLE state and allows the counter to count until QB = 1. The ENP and ENT inputs can be always asserted since loading and clearing override counting in the '163. The design of the output logic is left as an exercise (6.57).

How do you come up with a tricky realization like Figure 6–78? It's not easy; you just stare at the problem for a long time and experiment with different

Figure 6–78
MSI realization of the T-bird
tail-lights machine.

state encodings and excitation circuits. You might occasionally find this sort
of approach absolutely necessary if you're in a tight situation and don't have
access to PLDs, but the main use of such tricky state-machine realizations by
experienced (insecure?) designers seems to be in showing off how clever they
are.

6.8.2 Decomposing State Machines

Just like large procedures or subroutines in a programming language, large state
machines are difficult to conceptualize, design, and debug. Therefore, when faced
with a large state-machine problem, digital designers often look for opportunities
to solve it with a collection of smaller state machines.

There's a well-developed theory of *state-machine decomposition* that you
can use to analyze any given, monolithic state machine to determine whether it

*state-machine
decomposition*

Table 6–16
State assignment for T-bird tail-
lights machine using a 74LS163.

State	QD	QC	QB	QA
IDLE	0	0	0	0
L1	0	1	0	0
L2	0	1	0	1
L3	0	1	1	0
R1	1	0	0	0
R2	1	0	0	1
R3	1	0	1	0
LR3	1	1	1	1

Figure 6–79

A typical, hierarchical state-machine structure.

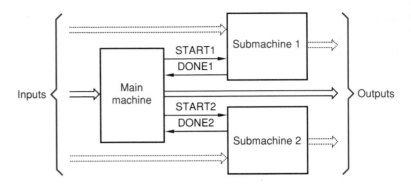

can be realized as a collection of smaller ones. However, decomposition theory is not too useful for designers who want to avoid designing large state machines in the first place. Rather, a practical designer tries to cast the original design problem into a natural, hierarchical structure, so that the uses and functions of submachines are obvious, making it unnecessary ever to write a state table for the equivalent monolithic machine.

main machine
submachine

The simplest and most commonly used type of decomposition is illustrated in Figure 6–79. A *main machine* provides the primary inputs and outputs and executes the top-level control algorithm. *Submachines* perform low-level steps under the control of the main machine, and may optionally handle some of the primary inputs and outputs.

Perhaps the most commonly used submachine is a counter. The main machine starts the counter when it wishes to stay in a particular main state for n clock ticks, and the counter asserts a DONE signal when n ticks have occurred. The main machine is designed to wait in the same state until DONE is asserted. This adds an extra output and input to the main machine (START and DONE), but it saves $n - 1$ states.

A decomposed state machine designed along these lines is based on the guessing game of Section 6.7.4. The original guessing game is easy to win after a minute of practice, because the lamps cycle at a very consistent 4-Hz rate. To make the game more challenging, we can double or triple the clock speed, but allow the lamps to stay in each state for a random length of time. Then the user truly must guess whether a given lamp will stay on long enough for the corresponding pushbutton to be pressed.

A block diagram for the enhanced guessing game is shown in Figure 6–80. The main machine is basically the same as before, except that it only advances from one lamp state to the next if the enable input EN is asserted, as shown by the ASM chart in Figure 6–81.

The low-order bit of an n-bit LFSR counter is used as the enable signal. Thus, the length of time that a particular lamp stays on depends on the pattern in the LFSR. In the best case for the user, the LFSR contains $10\ldots00$, and the lamp

Figure 6–80
Block diagram of guessing game with random delay.

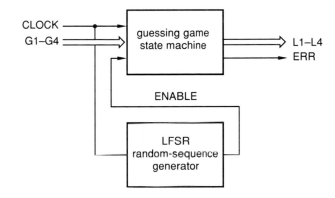

is on for $n - 1$ clock ticks; in the worst case, the LFSR contains $11 \ldots 11$ and shifting occurs for n consecutive clock ticks. At other times, shifting stops for a time determined by the number of consecutive 0s starting with the low-order bit of the LFSR. Of course, all of these cases are unpredictable unless the user has memorized the shifting cycle ($2^n - 1$ states) or is very fast at Galois-field arithmetic. Obviously, a large value of n (≥ 16) provides the most fun.

Another obvious candidate for decomposition is a state machine that per-

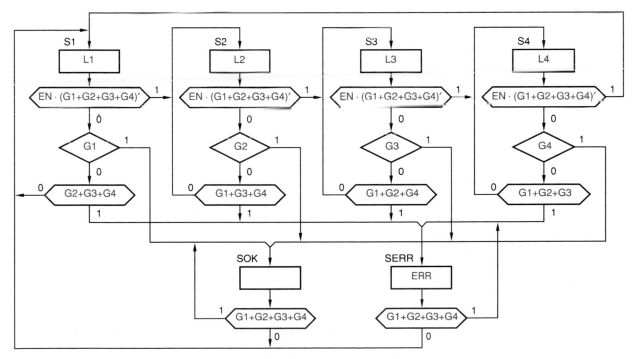

Figure 6–81   ASM chart for guessing machine with enable.

forms binary multiplication using the shift-and-add algorithm, or binary division using the shift-and-subtract algorithm. To perform an n-bit operation, these algorithms require an initialization step, n computation steps, and possible cleanup steps. The main machine for such an algorithm contains states for initialization, generic computation, and cleanup steps, and a modulo-n counter can be used as a submachine to control the number of generic computation steps executed. In Section 6.9.2, we'll show the design of a monolithic state machine for 8-bit shift-and-add multiplication, while the design of a corresponding decomposed machine for n-bit multiplication is left as an exercise (6.59).

6.9 SYNCHRONOUS DESIGN METHODOLOGY

synchronous design methodology

fully synchronous circuit

The basic rule of *synchronous design methodology* is that all of a circuit's flip-flops should be clocked by the same clock, and that preset and clear inputs should not be used, except for system initialization. In such *fully synchronous circuits*, races and hazards are not a problem, for two reasons. First, the only fundamental-mode circuits that might be subject to races or essential hazards are predesigned elements (e.g., discrete flip-flops or ASIC cells), guaranteed by the manufacturer to work properly. Second, even though the combinational circuits that drive flip-flop control inputs may contain static or dynamic or function hazards, these hazards have no effect since the control inputs are sampled only *after* the hazard-induced glitches have had a chance to settle out.

The designer of a synchronous system or subsystem must perform just three well-defined tasks to ensure reliable system operation:

(1) Minimize and determine the amount of clock skew in the system, as discussed in Section 6.9.3.

(2) Ensure that flip-flops have positive setup- and hold-time margins, including an allowance for clock skew, as described in Section 6.1.4.

(3) Identify asynchronous inputs, synchronize them with the clock, and ensure that the synchronizers have an adequately low probability of failure, as described in Sections 6.9.5 and 6.10.

Before discussing some of these timing-related issues, we'll look at a general model for synchronous system structure and an example.

6.9.1 Synchronous System Structure

The sequential-circuit design examples that we've given so far in this chapter are mostly state machines with a small number of states. If a sequential circuit has more than a few flip-flops, then it's not desirable (and often not possible)

Figure 6–82
Synchronous system structure.

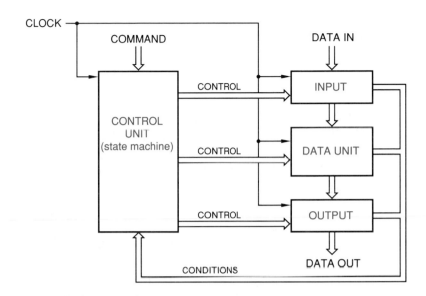

to treat the circuit as a single, monolithic state machine, because the number of states would be too large to handle.

Fortunately, most digital systems or subsystems can be partitioned into two or more parts. Whether the system processes numbers, digitized voice signals, or a stream of spark-plug pulses, a certain part of the system, which we'll call the *data unit*, can be viewed as storing, routing, combining, and generally processing "data." Another part, which we'll call the *control unit*, can be viewed as starting and stopping actions in the data unit, testing conditions, and deciding what to do next according to circumstances. In general, only the control unit is treated as a state machine.   *data unit*   *control unit*

Figure 6–82 is a general block diagram of a system with a control unit and a data unit. We have also included explicit blocks for input and output, but we could have just as easily absorbed these functions into the data unit itself. The control unit is a state machine whose inputs include *command inputs* that indicate how the machine is to function, and *condition inputs* provided by the data unit. The command inputs may be supplied by another subsystem or by a user to set the general operating mode of the control machine (RUN/HALT, NORMAL/TURBO, etc.), while the condition inputs allow the control unit to change its behavior as required by circumstances in the data unit (ZERO DETECT, MEMORY FULL, etc.).   *command input*   *condition input*

A key characteristic of the structure in Figure 6–82 is that the control unit and the data unit both use the same common clock. Figure 6–83 illustrates the operations of both units during a typical clock cycle:

(1) Shortly after the beginning of the clock period, the control-unit state and the data-unit register outputs are valid.

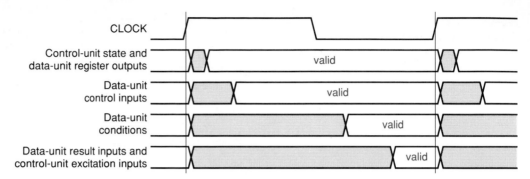

Figure 6–83   Operations during one clock cycle in a synchronous system.

(2) Next, Moore-type outputs of the control-unit state machine become valid. These signals are control *inputs* to the data unit. They determine what data-unit functions are performed in the rest of the clock period, for example, selecting memory addresses, multiplexer paths, and arithmetic operations.

(3) Near the end of the clock period, data-unit condition outputs such as zero- or overflow-detect are valid, and are made available to the control unit.

(4) At the end of the clock period, just before the setup-time window begins, the next-state logic of the control-unit state machine has determined the next state based on the current state and command and condition inputs, and computational results in the data unit are available to be loaded into data-unit registers.

(5) After the clock edge, the whole cycle may repeat.

Control-unit state-machine outputs may be of the Mealy or Moore type; the latter was assumed in Figure 6–83. Moore-type outputs control the data unit's actions strictly according to the current state, which depends on *past* conditions in the data unit. Mealy-type outputs may select different actions in the data unit according to *current* conditions in the data unit. This increases flexibility, but may also increase delay and thereby lengthen the minimum clock period for correct system operation. Also, Mealy-type outputs must not create feedback loops. For example, a signal that adds 1 to an adder's input if the adder output is not 0 causes an oscillation if the adder output is −1.

6.9.2 A Synchronous System Design Example

shift-and-add multiplier   A typical example of a synchronous system with a data unit and a control unit is a *shift-and-add multiplier* for unsigned integers using the algorithm of Section 1.8.

Figure 6–84
Registers and functions used by
the shift-and-add multiplication
algorithm.

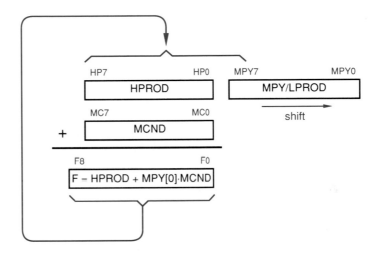

Figure 6–84 illustrates data-unit registers and functions that could be used to
perform an 8-bit multiplication:

MPY/LPROD   A shift register that initially stores the multiplier, and accumulates
            the low-order bits of the product as the algorithm is executed.

   HPROD   Initially cleared, and accumulates the high-order bits of the product
            as the algorithm is executed.

    MCND   Stores the multiplicand throughout the algorithm.

       F   A combinational function equal to the 9-bit sum of HPROD and
            MCND if the low-order bit of MPY/LPROD is 1, and equal to HPROD
            (extended to 9 bits) otherwise.

The MPY/LPROD shift register serves a dual purpose, holding both untested multiplier bits (on the right) and unchanging product bits (on the left) as the algorithm
is executed. At each step it shifts right one bit, discarding the multiplier bit
that was just tested, moving the next multiplier bit to be tested to the rightmost
position, and loading into the leftmost position one more product bit that will
not change for the rest of the algorithm.

   Figure 6–85 is an MSI design for the data unit. An 8-bit bidirectional data
bus, D[0–7] is used to load the multiplier and multiplicand into two registers
before a multiplication begins. When the multiplication is completed, the same
bus is used to read the 16-bit result in two steps. The following signals are used
for these operations:

   /LDMCND When asserted, enables the multiplicand register U1 to be loaded
            at the next clock tick.

Figure 6-85 Data unit of an 8-bit shift-and-add binary multiplier.

MPYS1 MPYS0 When 11, these signals enable the multiplier/low-product register U2 to be loaded at the next clock tick. They are set to 01 during the multiplication operation to enable the register to shift right, and are 00 at other times to preserve the register's contents.

/LPRODOE When asserted, enables the multiplier/low-product register U2 to drive the D[0–7] bus; used to read the low-order result byte.

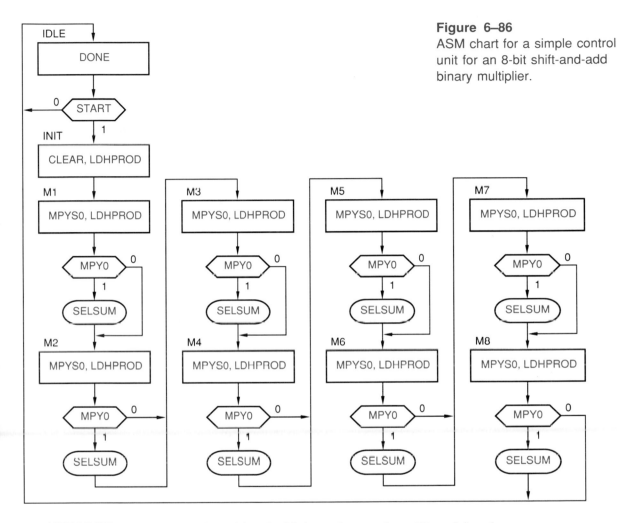

Figure 6–86
ASM chart for a simple control
unit for an 8-bit shift-and-add
binary multiplier.

/HPRODOE When asserted, enables the high-product register U3 to drive the
D[0–7] bus; used to read the high-order result byte.

Among these signals, only MPYS1 is used by the control-unit state machine; the
rest are assumed to be generated elsewhere and should be negated while the
multiplication algorithm is being executed.

An ASM chart for the multiplier's control-unit state machine appears in
Figure 6–86. The state machine has the following inputs and outputs:

START An external command input that tells the machine to perform a multiplication.

MPY0 A condition input from the data unit, the next multiplier bit to be
tested.

MPYS0 A control output that enables the MPY/LPROD register to shift right.

LDHPROD A control output that enables the HPROD register to be loaded.

DONE A status output indicating completion of the multiplication algorithm.

Note that the state machine's MPYS0 output must be combined with an external signal, not shown, that loads the MPY/LPROD register before a multiplication begins; also see Exercise 6.58.

The M1–M8 states in the ASM chart are very similar and repetitive. This is a good indication that the state machine is a good candidate for decomposition using a counter, a possibility explored in Exercise 6.59.

6.9.3 Clock Skew

Synchronous systems using edge-triggered flip-flops work properly only if all flip-flops see the triggering clock edge at the same time. Figure 6–87 shows what can happen otherwise. Here, two flip-flops are theoretically clocked by the same signal, but the clock signal seen by FF2 is delayed by a significant amount relative to FF1's clock. This difference between arrival times of the clock at different devices is called *clock skew*.

clock skew

We've named the delayed clock in Figure 6–87(a) "CLOCKD." If FF1's propagation delay from CLOCK to Q1 is short, and if the physical connection of Q1 to FF2 is short, then the change in Q1 caused by a CLOCK edge may actually reach FF2 *before* the corresponding CLOCKD edge. In this case, FF2 may go to an incorrect next state determined by the *next* state of FF1 instead of the current state, as shown in (b). If the change in Q arrives at FF2 only slightly early relative to CLOCKD, then FF2's hold-time specification may be violated, in which case FF2 may become metastable and produce an unpredictable output.

Figure 6–87
A simple example of clock skew.

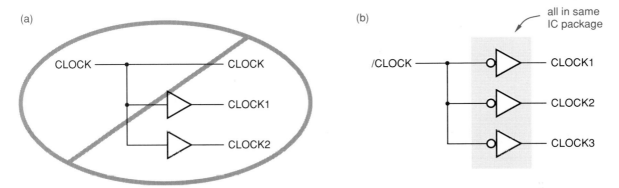

Figure 6–88   Buffering the clock: (a) excessive clock skew; (b) controllable clock skew.

If Figure 6–87 reminds you of our illustration of an essential hazard in Section 5.6.6, you're on to something. The clock-skew problem may be viewed simply as a manifestation of the essential hazards that exist in all edge-triggered devices.

We can determine quantitatively whether clock skew is a problem in a given system by defining t_{skew} to be the amount of clock skew and using the other timing parameters defined in Figure 6–1. For proper operation, we must have

$$t_{ffpd(min)} + t_{comb(min)} - t_{hold(min)} - t_{skew(max)} > 0$$

In other words, clock skew subtracts from the hold-time margin that we defined in Section 6.1.4.

Viewed in isolation, the example in Figure 6–87 may seem a bit extreme. After all, why would a designer provide a short connection path for data and a long one for the clock, when they could just run side-by-side? There are several ways this can happen; some are mistakes, while others are unavoidable.

In a large system, a single clock signal may not have adequate fanout to drive all of the devices with clock inputs, so it may be necessary to provide two or more copies of the clock signal. The buffering method of Figure 6–88(a) obviously produces excessive clock skew, since CLOCK1 and CLOCK2 are delayed through an extra buffer compared to CLOCK.

A recommended buffering method is shown in (b). All of the clock signals go through identical buffers, and have roughly equal delays. Ideally, all the buffers should be part of the same IC package, so that they all have similar delay characteristics and are operating at identical temperature and power-supply voltage. Some manufacturers build special buffers for just this sort of application and specify the worst-case delay variation between buffers in the same package.

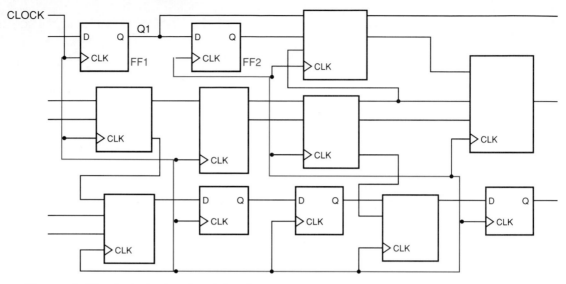

Figure 6–89   A clock-signal path leading to excessive skew in a complex PCB or ASIC.

Even the method in (b) may produce excessive clock skew if one clock signal is loaded much more heavily than the other; transitions on the more heavily loaded clock appear to occur later, because of increases in output-transistor switching delay and signal rise and fall times. Therefore, a careful designer tries to balance the loads on multiple clocks, looking at both DC load (fanout) and AC load (wiring and input capacitance).

Another bad situation can occur when signals on a PCB or in an ASIC are routed automatically by a CAD program. Figure 6–89 shows a PCB or ASIC with many flip-flops and larger-scale elements, all clocked with a common CLOCK signal. The CAD software has laid out CLOCK in a serpentine path that winds its way past all the devices that need it. Most of the other signals are routed point-to-point between an output and a small number of inputs, and so their paths are shorter. To make matters worse, in an ASIC some types of "wire" may be slower than others (polysilicon vs. metal in CMOS technology). As a result, a CLOCK edge may indeed arrive at FF2 quite a bit later than the data change that it produces on Q1.

One way to minimize this sort of problem is to force CLOCK to be distributed in a tree-like structure using the fastest type of wire, as illustrated in Figure 6–90. However, in a complex design with many direct connections from one flip-flop's Q output to another's D input, it still may not be possible to guarantee that clock edges arrive everywhere before the earliest data change. A CAD timing analysis program is typically used to detect these problems, which

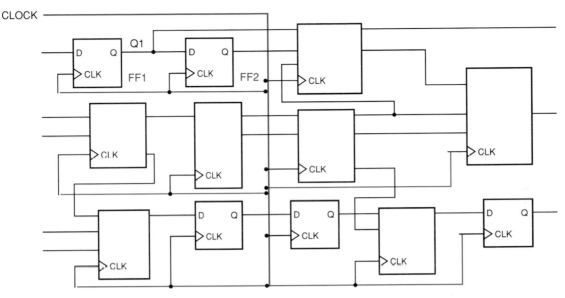

Figure 6–90   Clock-signal routing to minimize skew.

generally can be remedied only by inserting extra delay (e.g., pairs of inverters) in the too-fast data paths.

Although synchronous design methodology simplifies the conceptual operation of large systems, we see that clock skew can be a major problem when edge-triggered flip-flops are used as the storage elements. To control this problem, many high-performance systems and VLSI chips use a *two-phase latch design*, discussed in the References. Such designs effectively split each edge-triggered D flip-flop into its two component latches, and control them with two nonoverlapping clock phases. The nonoverlap between the phases accommodates clock skew.

two-phase latch design

6.9.4   Gating the Clock

Most of the sequential MSI parts that we introduced in this chapter have synchronous function-enable inputs. The first example that we showed was the 74LS377 register with synchronous load-enable input (Figure 6–10); other parts included the 74LS163 counter and 74LS194 shift register with synchronous load-, count-, and shift-enable inputs. Nevertheless, many MSI parts do not have synchronous function-enable inputs; for example, the 74LS374 8-bit register has three-state outputs but no load-enable input.

So, what can a designer do if an application requires an 8-bit register with both a load-enable input *and* three-state outputs? One solution is to use a 74LS377 to get the load-enable, and follow it with a 74LS241 three-state buffer.

Figure 6–91   How not to gate the clock: (a) simple-minded circuit; (b) timing diagram.

gating the clock

However, this increases both cost and delay. Another, somewhat risky alternative is to use a '374, but to suppress its clock input when it's not supposed to be loaded. This is called *gating the clock*.

Figure 6–91 illustrates an obvious but wrong approach to gating the clock. A signal CLKEN is asserted to enable the clock, and is simply ANDed with the clock to produce the gated clock GCLK. There are two problems with this approach:

(1) If CLKEN is a state-machine output or other signal produced by a register clocked by CLOCK, then CLKEN changes some time *after* CLOCK has already gone HIGH. This produces glitches on GCLK, and false clocking of the registers controlled by GCLK.

(2) Even if CLKEN is somehow produced well in advance of CLOCK's rising edge (for example, using a register clocked with the *falling* edge of CLOCK, an especially nasty kludge), the AND-gate delay gives GCLK excessive clock skew, which causes more problems all around.

A method of gating the clock that generates only minimal clock skew is shown in Figure 6–92. Here, both the ungated clock and several gated clocks are

Figure 6–92   An acceptable way to gate the clock: (a) circuit; (b) timing diagram.

generated from the same active-low master clock signal. Gates in the same IC package are used to minimize the possible differences in their delays. The CLKEN signal may change arbitrarily whenever /CLOCK is LOW, which is when CLOCK is HIGH. That's just fine; a CLKEN signal is typically produced by a state machine whose outputs change right after CLOCK goes HIGH.

The approach of Figure 6–92 is acceptable in a particular application only if the clock skew that it generates is acceptable. Furthermore, note that CLKEN must be stable during the entire time that /CLOCK is HIGH (CLOCK is LOW). Thus, the timing margins in this approach are sensitive to the clock's duty cycle, especially if CLKEN suffers a lot of combinational-logic delay (t_{comb}) from the triggering clock edge. A truly synchronous function-enable input, such as the 74LS377's load-enable input in Figure 6–10, can be changed at almost any time during the entire clock period, up until a setup time before the triggering edge.

6.9.5   Asynchronous Inputs

Even though it is theoretically possible to build a computer system that is fully synchronous, you couldn't do much with it, unless you can synchronize your keystrokes with a 16 MHz clock. Digital systems of all types inevitably must deal with *asynchronous input signals* that are not synchronized with the system clock.

asynchronous input signal

Asynchronous inputs are typically requests for service (e.g., interrupts in a computer) or status flags (e.g., a resource has become available). Such inputs normally change slowly compared to the system clock frequency, and need not be recognized at a particular clock tick. If a transition is missed at one clock tick, it can always be detected at the next one. The transition rates of asynchronous signals may range from less than one per second (the keystrokes of a slow typist) to 1 MHz or more (access requests in a multiprocessor system's shared memory).

Ignoring the problem of metastability, it is very easy to build a *synchronizer*, a circuit that samples an asynchronous input and produces an output that meets the setup and hold times required in a synchronous system. As shown in Figure 6–93, a D flip-flop samples the asynchronous input at each tick of the system clock and produces a synchronous output that is valid during the subsequent clock period.

synchronizer

It is essential for asynchronous inputs to be synchronized at only *one place* in a system; Figure 6–94 shows what can happen otherwise. Because of physical delays in the circuit, the two flip-flops will not see the clock and input signals at precisely the same time. Therefore, when asynchronous input transitions occur near the clock edge, there is a small window of time during which one flip-flop may sample the input as 1 and the other may sample it as 0. This inconsistent result may cause improper system operation, as one part of the system responds as if the input were 1, and another part responds as if it were 0.

Figure 6–93
A single, simple synchronizer:
(a) logic diagram; (b) timing.

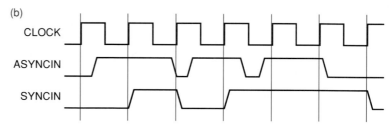

Sometimes combinational logic may hide the fact that there are two synchronizers, as shown in Figure 6–95. Since different paths through the combinational logic typically have different delays, the likelihood of an inconsistent result is even greater. This situation is especially common when asynchronous signals are used as inputs to state machines, since the excitation logic for two or more

Figure 6–94
Two synchronizers for the same
asynchronous input: (a) logic
diagram; (b) possible timing.

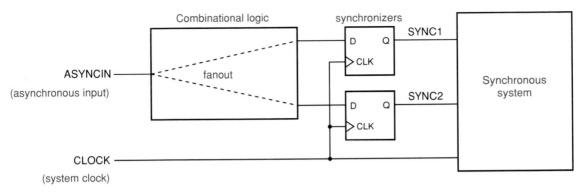

Figure 6–95   An asynchronous input driving two synchronizers through combinational logic.

state variables may depend on the asynchronous input. The proper way to use an asynchronous signal as a state-machine input is shown in Figure 6–96. All of the excitation logic sees the same synchronized input signal, SYNCIN.

As you probably know, even the synchronizers in Figures 6–93 and 6–96 sometimes fail. The reason they fail is that the setup and hold times of the synchronizing flip-flop are sometimes violated, because the asynchronous input can change at any time. "Well, who cares?" you may say. "If the D input changes near the clock edge, then the flip-flop will either see the change this time or miss it and pick it up next time; either way is good enough for me!" The problem is, there is a third possibility, discussed in the next section.

6.10 SYNCHRONIZER FAILURE AND METASTABILITY ESTIMATION

We showed in Section 5.1 that when the setup and hold times of a flip-flop are not met, the flip-flop may go into a third, *metastable* state halfway between 0 and 1. Worse, the length of time it may stay in this state before falling back

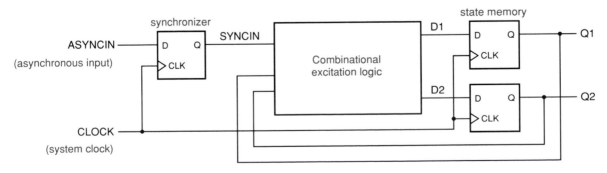

Figure 6–96   An asynchronous state-machine input coupled through a single synchronizer.

into a legitimate 0 or 1 state is theoretically unbounded. When other gates and flip-flops are presented with a metastable input signal, some may interpret it as a 0 and others as a 1, creating the sort of inconsistent behavior that we showed in Figure 6–94. Or, the other gates and flip-flops may produce metastable outputs themselves (after all, they are now operating in the *linear* part of their operating range). Luckily, the probability of a flip-flop output remaining in the metastable state decreases exponentially with time, though never all the way to zero.

6.10.1 Synchronizer Failure

synchronizer failure

Synchronizer failure is said to occur if the system uses a synchronizer output while the output is still in the metastable state. The way to avoid synchronizer failure is to ensure that the system waits "long enough" before using a synchronizer's output, "long enough" so that the mean time between synchronizer failures is several orders of magnitude longer than the designer's expected length of employment.

Metastability is more than an academic problem. More than a few experienced designers of high-speed digital systems have built (and released to production) circuits that suffer from intermittent synchronizer failures. In fact, the initial versions of several commercial ICs are said to have suffered from metastability problems, for example, the AMD 9513 system timing controller, the AMD 9519 interrupt controller, the Zilog Z-80 SIO, the Intel 8048 single-chip microcomputer, and the AMD 29000 RISC microprocessor.

There are two ways to get a flip-flop out of the metastable state:

(1) Force the flip-flop into a valid logic state using input signals that meet the published specifications for minimum pulse width, setup time, and so on.

(2) Wait "long enough," until the flip-flop comes out of metastibility on its own.

Inexperienced designers often attempt to get around metastability in other ways, and they are usually unsuccessful. Figure 6–97 shows an attempt by a designer who thinks that since metastability is an "analog" problem, it must have an

Figure 6–97
A failed attempt to build a metastable-proof /S-/R flip-flop.

"analog" solution. After all, Schmitt trigger inputs and capacitors can normally be used to clean up noisy signals. However, rather than eliminate metastability, this circuit enhances it—with the "right" components, the circuit will oscillate forever once it is excited by negating /S and /R simultaneously. Exercise 6.69 gives an example of a valiant but also failed attempt to eliminate metastability. These examples should give you the sense that synchronizer problems can be very subtle, so you must be careful. The only way to make synchronizers reliable is to wait long enough for metastable outputs to resolve. We answer the question, "How long is 'long enough'?" later in this section.

6.10.2   Metastability Resolution Time

If the setup and hold times of a D flip-flop are met, the flip-flop output settles to a new value within time t_{pd} after the clock edge. If they are violated, the flip-flop output may be metastable for an arbitrary length of time. In a particular system design, we use the parameter t_r, called the *metastability resolution time*, to denote the maximum time that the output can remain metastable without causing synchronizer (and system) failure.

t_r
metastability resolution time

For example, consider the state machine in Figure 6–96. The available metastability resolution time is

$$t_r = t_{clk} - t_{comb} - t_{setup}$$

where t_{clk} is the clock period, t_{comb} is the propagation delay of the combinational excitation logic, and t_{setup} is the setup time of the flip-flops used in the state memory.

t_{clk}
t_{comb}
t_{setup}

6.10.3   Reliable Synchronizer Design

The most reliable synchronizer is one that allows the maximum amount of time for metastability resolution. However, in the design of a digital system, we seldom have the luxury of *slowing down* the clock to make the system work more reliably. Instead, we are usually asked to *speed up* the clock to get higher performance from the system. As a result, we often need synchronizers that work reliably with very short clock periods. We'll show several such designs, and show how to predict their reliability.

We showed previously that a state machine with an asynchronous input, built as shown in Figure 6–96, has $t_r = t_{clk} - t_{comb} - t_{setup}$. In order to maximize t_r for a given clock period, we should minimize t_{comb} and t_{setup}. The value of t_{setup} depends on the type of flip-flops used in the state memory; in general, faster flip-flops have shorter setup times. The minimum value of t_{comb} is zero, and is achieved by the synchronizer design of Figure 6–98, whose operation we explain next.

Figure 6–98 Recommended synchronizer design.

Inputs to flip-flop FF1 are asynchronous with the clock, and may violate the flip-flop's setup and hold times. When this happens, the META output may become metastable and remain in that state for an arbitrary time. However, we assume that the maximum duration of metastability after the clock edge is t_r. (We show how to calculate the probablity that our assumption is correct in the next subsection.) As long as the clock period is greater than t_r plus the FF2's setup time, SYNCIN becomes a synchronized copy of the asynchronous input on the next clock tick without ever becoming metastable itself. The SYNCIN signal is distributed as required to the rest of the system.

6.10.4 Analysis of Metastable Timing

decision window

Figure 6–99 shows the flip-flop timing parameters that are relevant to our analysis of metastability timing. The published setup and hold times of a flip-flop with respect to its clock edge are denoted by t_s and t_h, and bracket an interval called the *decision window*, when the flip-flop samples its input and decides to change its output if necessary. As long as the D input changes outside the decision window, as in (a), the manufacturer guarantees that the output will change and

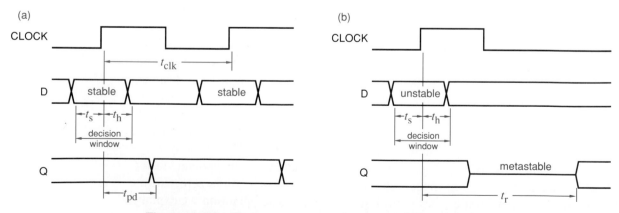

Figure 6–99 Timing parameters for metastability analysis.

settle to a valid logic state before time t_{pd}. If D changes inside the decision window, as in (b), metastability may occur and persist until time t_r.

Theoretical research has suggested and experimental research has confirmed that when asynchronous inputs change during the decision window, the duration of metastable outputs is governed by an exponential formula:

$$\text{MTBF}(t_r) = \frac{\exp(t_r / \tau)}{T_o \cdot f \cdot a}$$

Here $\text{MTBF}(t_r)$ is the mean time between synchronizer failures, where a failure occurs if metastability persists beyond time t_r after a clock edge, where $t_r \geq t_{pd}$. This MTBF depends on f, the frequency of the flip-flop clock; a, the number of asynchronous input changes per second applied to the flip-flop;[6] and T_o and τ, constants that depend on the electrical characteristics of the flip-flop. For a typical 74LS74, $T_o \approx 0.4$ s and $\tau \approx 1.5$ ns.

Now suppose that we build a microprocessor system with a 10-MHz clock and use the circuit of Figure 6–98 to synchronize an asynchronous input. If the asynchronous input changes during the decision window of FF1, its output META may become metastable until time t_r. If META is still metastable at the beginning of the decision window for FF2, then the synchronizer fails because FF2 may have a metastable output; system operation is unpredictable in that case.[7]

Let us assume that the D flip-flops in Figure 6–98 are 74LS74s. The setup time t_s of a 74LS74 is 20 ns, and the clock period in our example microprocessor system is 100 ns, so the t_r required for synchronizer failure is 80 ns. If the asynchronous input changes 100,000 times per second, then the synchronizer MTBF is

$$\text{MTBF}(80\,\text{ns}) = \frac{\exp(80 / 1.5)}{0.4 \cdot 10^7 \cdot 10^5} = 3.6 \cdot 10^{11}\,\text{s}$$

[6]Although a flip-flop output can go metastable *only* if D changes during the decision window, the MTBF formula does not explicitly specify how many such input changes occur. Instead, it specifies the total number of asynchronous input changes per second, a, and assumes that asynchronous input changes are uniformly distributed over the clock period. Therefore, the fraction of input changes that actually occur during the decision window is "built in" to the clock-frequency parameter f— as f increases, the fraction goes up. If the system design is such that input changes might be clustered in the decision window rather than being uniformly distributed (as when synchronizing a slow input with a fixed but unknown phase difference from the system clock), then a frequency equal to the reciprocal of the decision window (based on published setup and hold times), times a safety margin of 10, should be used.

[7]We do not allow metastability, even briefly, on the output of FF2, because we assume that the system has been designed with zero timing margins. If the system can in fact tolerate some increase in the propagation delay of FF2, the MTBF will be somewhat better than predicted here.

That's not bad, about 100 centuries between failures! Of course, if we're lucky enough to sell 10,000 copies of our system, one of them will fail in this way every year. But, no matter, let us consider a more serious problem.

Suppose we upgrade our system to use a brand-new microprocessor chip with a clock speed of 16 MHz. We may have to replace some components in our system to operate at the higher speed, but 74LS74s are still perfectly good at 16 MHz. Or are they? With a clock period of 62.5 ns, the new synchronizer MTBF is

$$\text{MTBF(42.5 ns)} = \frac{\exp(42.5 / 1.5)}{0.4 \cdot 1.6 \cdot 10^7 \cdot 10^5} = 3.1 \text{ s!}$$

The only saving grace of this synchronizer at 16 MHz is that it's so lousy, we're likely to discover the problem in the engineering lab before the product ships! Thank goodness the MTBF wasn't one year.

6.10.5 Better Synchronizers

Given the poor performance of the 74LS74 as a synchronizer at moderate clock speeds, we have a couple of alternatives for building more reliable synchronizers. The simplest solution, which works for most design requirements, is simply to use a flip-flop from a faster technology. Based on published data discussed in the References, Table 6–17 lists the metastability parameters for a number of common flip-flop types and technologies. For reference purposes, note that 10^5 s $\approx$ 1 day, 10^8 s $\approx$ 3 years, and 10^{10} s $\approx$ 3 centuries.

Table 6–17
Metastability parameters for common flip-flops, with f = 25 MHz and a = 100 KHz.

| | | | | MTBF(t_r) (s) | |
Reference	Device	τ (ns)	T_o	t_r=20 ns	t_r=40 ns
TI	74LSxx	1.28	6.7 s	$t_r < t_{pd}$	$2 \cdot 10^0$
TI	74ALSxx	0.87	12.5 ms	$3 \cdot 10^{-1}$	$7 \cdot 10^{10}$
TI	74ASxx	0.34	6.5 μs	$2 \cdot 10^{18}$	$7 \cdot 10^{43}$
Chaney	74S74	1.70	1.0 μs	$5 \cdot 10^{-2}$	$7 \cdot 10^3$
Chaney	74S174	1.20	5.0 μs	$1 \cdot 10^0$	$2 \cdot 10^7$
Chaney	74S373	0.91	60.0 μs	$2 \cdot 10^1$	$8 \cdot 10^{10}$
Chaney	74S374	0.91	0.4 ms	$4 \cdot 10^0$	$1 \cdot 10^{10}$
Chaney	74LS74	1.5	0.4 s	$t_r < t_{pd}$	$4 \cdot 10^{-1}$
Chaney	74F74	0.4	0.2 ms	$1 \cdot 10^{13}$	$5 \cdot 10^{34}$
Chaney	74F373	0.7	4.0 μs	$3 \cdot 10^5$	$7 \cdot 10^{17}$
Chaney	74F374	0.4	0.1 ms	$2 \cdot 10^{13}$	$1 \cdot 10^{35}$

As you can see, the 74LS74 is the worst device in the table. If in the 16 MHz microprocessor system of the preceding section we replace FF1 with a 74ALS74, we get

$$MTBF(42.5\,ns) = \frac{\exp(42.5/0.87)}{12.5 \cdot 10^{-3} \cdot 1.6 \cdot 10^7 \cdot 10^5} = 8.2 \cdot 10^{10}\,s$$

If you're satisfied with a synchronizer MTBF of about 25 centuries per system shipped, you can stop here. However, if FF2 is also replaced with a 74ALS74, the MTBF gets even better, since the 'ALS74 has a shorter setup time than the 'LS74, only 10 ns. With the 'ALS74, the MTBF is about 100,000 times better:

$$MTBF(52.5\,ns) = \frac{\exp(52.5/0.87)}{12.5 \cdot 10^{-3} \cdot 1.6 \cdot 10^7 \cdot 10^5} = 8.1 \cdot 10^{15}\,s$$

Even if we ship a million systems containing this circuit, we (or our heirs) will see a synchronizer failure only once in 240 years. Now that's job security!

Actually, the margins above aren't as large as they might seem. (How large does 240 years seem to *you*?) Most of the numbers given in Table 6–17 are *averages*, and are seldom specified, let alone guaranteed, by the device manufacturer. Furthermore, calculated MTBFs are extremely sensitive to the value of τ, which in turn may depend on temperature, voltage, and the phase of the moon. So the operation of a given flip-flop in an actual system may be much worse (or much better) than predicted by our table.

6.10.6 Other Synchronizer Designs

We promised to describe a couple of ways to build more reliable synchronizers. The first way was to use faster flip-flops, that is, to reduce the value of τ in the MTBF equation. Having said that, the second way is obvious—to *increase* the value of t_r in the MTBF equation.

For a given system clock, the best value we can obtain for t_r using the circuit of Figure 6–98 is t_{clk}, if FF2 has a setup time of 0. However, we can get values of t_r on the order of $n \cdot t_{clk}$ by using the *multiple-cycle synchronizer* circuit of Figure 6–100. Here we divide the system clock by n to obtain a slower synchronizer clock and longer $t_r = (n \cdot t_{clk}) - t_{setup}$. Usually a value of $n = 2$ or $n = 3$ gives adequate synchronizer reliability. If necessary, the output of the synchronizer, which changes on the rising edge of the divide-by-n clock, may be reclocked by the system clock to eliminate skew with the rest of the system.

multiple-cycle synchronizer

The larger the value of n, the longer it takes for an asynchronous input change to be seen by the synchronous system. This is simply a price that must be paid for reliable system operation. In typical microprocessor systems, most asynchronous inputs are for external events—interrupts, DMA requests, and so on—that need not be recognized very quickly, relative to synchronizer delays.

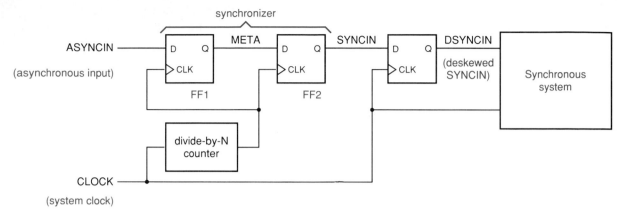

Figure 6–100   Multiple-cycle synchronizer.

In the time-critical area of main memory access, experienced designers use the processor clock to run the memory subsystem too, if possible. This eliminates the need for synchronizers and provides the fastest possible system operation.

At higher frequencies, the feasibility of the multiple-cycle synchronizer design shown in Figure 6–100 tends to be limited by clock skew. For this reason, rather than use a divide-by-n synchronizer clock, some designers use *cascaded synchronizers*. This design approach simply replaces FF1 in the synchronizer of Figure 6–98 with a cascade (shift register) of n flip-flops, all clocked with the high-speed system clock.

cascaded synchronizers

With cascaded synchronizers, the idea is that metastability will be resolved with some probability by the first flip-flop, and failing that, with an equal probability by each successive flip-flop in the cascade. So the overall probability of failure is on the order of the nth power of the failure probability of a single-flip-flop synchronizer at the system clock frequency. While this is true, the MTBF of the cascade is poorer than the MTBF of a multiple-cycle synchronizer with the same delay ($n \cdot t_{\text{clk}}$). With the cascade, the flip-flop setup time t_{setup} must be subtracted from t_r, the available metastability resolution time, n times, but in a multiple-cycle design, it is subtracted only once.

PLDs that contain internal flip-flops can be used in synchronizer designs, where the two flip-flops in Figure 6–98 are simply included in the PLD. This is very convenient in most applications, because it eliminates the need for external, discrete flip-flops. However, a PLD-based synchronizer typically has a poorer MTBF than a corresponding discrete circuit. This happens because each flip-flop in a PLD has combinational logic circuits on its D input that increase its setup time and thereby reduce the amount of time t_r available for metastability resolution during a given system clock period t_{clk}. To maximize t_r, FF2 in Figure 6–98 should be a short-setup-time discrete flip-flop.

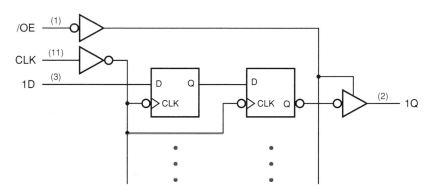

Figure 6–101
Logic diagram for the 74AS4374
octal dual-rank D flip-flop.

6.10.7 Metastable-Hardened Flip-Flops

Going one step further, Texas Instruments has developed a series of SSI and MSI flip-flops that are specifically designed for synchronizer applications. For example, the *74AS4374* is similar to the 74AS374, except that each individual flip-flop is replaced with a pair of flip-flops, as shown in Figure 6–101. Each pair of flip-flops can be used as a synchronizer of the type shown in Figure 6–98, so eight asynchronous inputs can be synchronized with one 74AS4374.

    The 'AS4374 flip-flops are specified with a slightly poorer τ than the average τ that we showed in Table 6–17 for other 74AS flip-flops. The internal design of the 'AS4374 flip-flops actually has been improved to reduce τ compared to other 74AS flip-flops, but TI used this improvement to give better design margins rather than a better spec. The new flip-flops have a *worst-case* spec of $\tau = 0.42$ ns.

    The 'AS4374 flip-flops have a T_o parameter of only 4 ns, compared to 6.5 µs for other 74AS flip-flops. Since T_o appears in the denominator of the MTBF equation, this improves the MTBF by a factor of 1625.

    The biggest improvement in the 'AS4374 is a greatly reduced t_{setup}. Because the entire synchronizer of Figure 6–98 is built on a single chip, there are no input or output buffers between FF1 and FF2, and t_{setup} for FF2 is only 0.5 ns. Compared to a conventional 74AS flip-flop with a 5 ns t_{setup}, this improves the MTBF by a factor of exp(4.5 / .42), or about 45,000.

    Because of the improvements in T_o and t_{setup}, synchronizers using the 'AS4374 have MTBFs that are about $7 \cdot 10^7$ times better than ones built from discrete 74AS flip-flops. Although they cost more than discrete flip-flops, at very high system clock speeds (40 MHz and up) the 'AS4374 and related TI parts are the only 74AS-TTL parts that can reliably synchronize asynchronous inputs using only one clock period of delay, as in Figure 6–98. On the other hand, applications that use slower clocks (20 MHz or less) or that can tolerate the long latency of a multiple-cycle synchronizer can be designed more economically using conventional AS or ALS flip-flops.

74AS4374

REFERENCES

Probably the first comprehensive set of examples of sequential MSI parts and applications appeared in *The TTL Applications Handbook*, edited by Peter Alfke and Ib Larsen (Fairchild Semiconductor, 1973). This highly practical and informative book was invaluable to this instructor and many others who developed digital design curricula in the 1970s.

Another book that emphasizes design with larger-scale combinational and sequential logic functions is Thomas R. Blakeslee's *Digital Design with Standard MSI and LSI* (Wiley, 1979, 2nd ed.). His coverage of the concept of space/time trade-offs is excellent, and he also introduces the microprocessor as "a universal logic circuit."

Even though this is a "practices" chapter, we introduced a few useful theoretical topics. Function hazards are discussed by Edward J. McCluskey in *Logic Design Principles* (Prentice-Hall, 1986). Galois fields and their applications to error-correcting codes, as well as to the LFSR counters of this chapter, are described in introductory books on coding theory, including *Error-Control Techniques for Digital Communication* by A. M. Michelson and A. H. Levesque (Wiley-Interscience, 1985).

Algorithmic-state-machine notation was pioneered at Hewlett-Packard Laboratories by Thomas E. Osborne and was further developed by Osborne's colleague Christopher R. Clare in a book, *Designing Logic Systems Using State Machines* (McGraw-Hill, 1973). It has subsequently found a home in many digital design texts, including *The Art of Digital Design* by F. P. Prosser and D. E. Winkel (Prentice-Hall, 1987, 2nd ed.) and *Digital Design* by M. Morris Mano (Prentice-Hall, 1984). Another notation for describing state machines, an extension of "traditional" state-diagram notation, is the mnemonic documented state (MDS) diagram developed by William I. Fletcher in *An Engineering Approach to Digital Design* (Prentice-Hall, 1980).

The general topics of clock skew and multiphase clocking are discussed in McCluskey's *Logic Design Principles*, while an illuminating discussion of these topics as applied to VLSI circuits can be found in *Introduction to VLSI Systems* by Carver Mead and Lynn Conway (Addison-Wesley, 1980). Mead and *self-timed systems* Conway's book also provides an introduction to the important topic of *self-timed systems* that eliminate the system clock, allowing each logic element to proceed at its own rate. To give credit where credit is due, we note that all of the Mead and Conway material on system timing, including an outstanding discussion of metastability, appears in Chapter 7 written by Charles L. Seitz.

Thomas J. Chaney has spent decades studying and reporting on the metastability problem. One of his more important works, "Measured Flip-Flop Responses to Marginal Triggering" (*IEEE Trans. Comput.*, Vol. C-32, No. 12, December 1983, pp. 1207–1209), reports results that we used in Table 6–17.

Another important reference is "Metastable Characteristics of Texas Instruments Advanced Bipolar Logic Families" by Robert Brueninger and Kevin Frank (TI pub. SDAA004). Besides describing the results for ALS, AS, and LS circuits that we used in Table 6–17, it contains a description of a measurement circuit that you can duplicate (with some care) in your own lab to measure the metastable characteristics of your own devices.

For the mathematically inclined, Lindsay Kleeman and Antonio Cantoni have written "On the Unavoidability of Metastable Behavior in Digital Systems" (*IEEE Trans. Comput.*, Vol. C-36, No. 1, January 1987, pp. 109–112); the title says it all. The same authors posed the question, "Can Redundancy and Masking Improve the Performance of Synchronizers?" (*IEEE Trans. Comput.*, Vol. C-35, No. 7, July 1986, pp. 643–646). Their answer in that paper was "No," but a response from a reviewer caused them to change their minds to "Maybe." Obviously, they've gone metastable themselves! (Having two authors and a reviewer hasn't improved their performance, so the obvious answer to their original question is "No!"!) In any case, Kleeman and Antonio's papers provide a good set of pointers to mainstream scholarly references on metastability.

The most comprehensive set of references on metastability (not including Greek philosophers or Devo) is Martin Bolton's "A Guided Tour of 35 Years of Metastability Research" (*Proc. Wescon 1987*, Session 16, "Everything You Might Be Afraid to Know About Metastability," Wescon Session Records, 8110 Airport Blvd., Los Angeles, CA 90045).

E X E R C I S E S

6.1   What does the TTL Data Book have to say about momentarily shorting the outputs of a gate to ground as we do in the switch debounce circuit of Figure 6–5?

6.2   Investigate the behavior of the switch debounce circuit of Figure 6–5 if 74HCT04 inverters are used; repeat for 74AC04 inverters.

6.3   Suppose you are asked to design a circuit that produces a debounced logic input from an SPST (single-pole, *single*-throw) switch. What inherent problem are you faced with?

6.4   How does the meaning of the label 1D differ between the traditional and the IEEE symbols for registers like the 74LS374? (*Hint:* How might pin 4 of a '374 be labeled in the IEEE standard?)

6.5   Write a formula for the maximum clock frequency of the synchronous serial binary counter circuit in Figure 6–15. In your formula, let t_{pTQ} denote the propagation delay from T to Q in a T flip-flop, t_{setup} the setup time of the EN input to the rising edge of T, and t_{AND} the delay of an AND gate.

6.6   Repeat Exercise 6.5 for the synchronous parallel binary counter circuit shown in Figure 6–16, and compare results.

6.7   Repeat Exercise 6.5 for an n-bit synchronous serial binary counter.

6.8   Repeat Exercise 6.5 for an n-bit synchronous parallel binary counter. Beyond what value of n is your formula no longer valid?

6.9   What is the counting sequence of the circuit shown in Figure X6.9?

Figure X6.9

6.10 Using a 74LS163 4-bit binary counter, design a modulo-11 counter circuit with the counting sequence 3, 4, 5, ..., 12, 13, 3, 4, ...

6.11 What is the behavior of the counter circuit of Figure 6–23 if it is built using a 74LS161 instead of a 74LS163?

6.12 Look up the internal logic diagram for a 74LS162 synchronous decade counter in a data book, and write its state table in the style of Table 6–1, including its counting behavior in the normally unused states 10–15.

6.13 Devise a cascading scheme for 74LS163s, analogous to the synchronous parallel counter structure of Figure 6–16, such that the maximum counting speed is the same for any counter with up to 36 bits (nine '163s). Determine the maximum counting speed using worst-case delays from a manufacturer's data book for the '163s and any SSI components used for cascading.

6.14 Determine the widths of the glitches shown in Figure 6–30 on the /Y2 output of a 74LS138 decoder, assuming that the '138 is internally structured as shown in Figure 4–29(a) on page 253, and that each internal gate has a delay of 10 ns.

6.15 Design a clocked synchronous circuit with a four inputs, N3, N2, N1, and N0, that represent an integer N in the range 0–15. The circuit has a single output Z that is asserted for exactly N clock ticks during any 16-tick interval (assuming that N is held constant during the interval of observation). (*Hints:* Use combinational logic with a 74LS163 set up as a free-running divide-by-16 counter. The ticks in which Z is asserted should be spaced as evenly as possible, that is, every second tick when $N = 8$, every fourth when $N = 4$, and so on.)

6.16 Modify the circuit of Exercise 6.15 so that Z produces N *transitions* in each 16-tick interval. The resulting circuit is called a *binary rate multiplier*, and was once sold as a TTL MSI part, the 7497. (*Hint:* Gate the clock with the level output of the previous circuit.)

binary rate multiplier

6.17 Design a parallel-to-serial conversion circuit with eight 2.048-Mbps, 32-channel serial links and a single 2.048-MHz, 8-bit, parallel data bus that carries 256 bytes per frame. Each serial link should have the frame format defined in Figure 6–43. Each serial data line SDATAi should have its own sync signal /SYNCi; the sync pulses should be staggered so that /SYNCi+1 has a pulse one clock tick after /SYNCi.

Show the timing of the parallel bus and the serial links, and write a table or formula that shows which parallel-bus timeslots are transmitted on which serial links and timeslots. Draw a logic diagram for the circuit using TTL MSI parts from this chapter; you may abbreviate repeated elements (e.g., shift registers), showing only the unique connections to each one.

6.18 Repeat Exercise 6.17, assuming that all serial data lines must reference their data to a single, common SYNC signal. How many more chips does this design require?

6.19 Show how to enhance the serial-to-parallel circuit of Figure 6–45 so that the byte received in each timeslot is stored in its own register for 125 μs, until the next byte from that timeslot is received. Draw the counter and decoding logic for 32 timeslots in detail, as well as the parallel data registers and connections for timeslots 31, 0, and 1. Also draw a timing diagram in the style of Figure 6–46 that shows the decoding and data signals associated with timeslots 31, 0, and 1.

6.20 Suppose you are asked to design a serial computer, one that moves and processes data one bit at a time. One of the first decisions you must make is which bit to transmit and process first, the LSB or the MSB. Which would you choose, and why?

6.21 Design an 8-bit self-correcting ring counter whose states are 11111110, 11111101, ..., 01111111, using only two TTL SSI/MSI packages.

6.22 Design two different 2-bit, 4-state counters, where each design uses just a single 74LS74 package (two edge-triggered D flip-flops) and no other gates.

6.23 Design a 4-bit Johnson counter and decoding for all eight states using just three TTL SSI/MSI packages. Your counter need not be self-correcting.

6.24 Starting with state 0001, write the sequence of states for a 4-bit LFSR counter designed according to Figure 6–53 and Table 6–5.

6.25 Prove that an even number of shift-register outputs must be connected to the odd-parity circuit in an n-bit LFSR counter if it generates a maximum-length sequence. (Note that this is a necessary but not a sufficient requirement. Also, although Table 6–5 is consistent with what you're supposed to prove, simply quoting the table is not a proof!)

6.26 Prove that X0 must appear on the right-hand side of any LFSR feedback equation that generates a maximum-length sequence.

6.27 Suppose that an n-bit LFSR counter is designed according to Figure 6–53 and Table 6–5. Prove that if the odd-parity circuit is changed to an even-parity circuit, the resulting circuit is a counter that visits $2^n - 1$ states, including all of the states except $11\ldots 11$.

6.28 Find a feedback equation for a 3-bit LFSR counter, other than the one given in Table 6–5, that produces a maximum-length sequence.

6.29 Prove that if an n-bit LFSR counter generates a maximum-length sequence ($2^n - 1$ states), then an extra XOR gate and an $n-1$ input NOR gate connected as suggested in Figure 6–54 produces a counter with 2^n states.

6.30 Prove that a sequence of 2^n states is still obtained if a NAND gate is substituted for a NOR above, but that the state sequence is different.

6.31 Design an iterative circuit with one input B_i per stage and two boundary outputs X and Y such that X = 1 if at least two B_i inputs are 1 and Y = 1 if at least two *consecutive* B_i inputs are 1.

6.32 Analyze the state machine in Figure 6–59(b), producing a state and output table. Show that this table is equivalent to Table 5–10 on page 387.

6.33 Demonstrate that the state machine in Figure 6–59(b) can be realized as a finite-memory machine. Draw a logic diagram for a realization of the machine that matches the general structure shown in Figure 6–60.

6.34 Write an equation for a HINT output of the finite-memory machine in Figure 6–61. You need not multiply out or minimize your result. (*Hint:* Consider all of the possible initial sequences of 0110111 that might be stored in the finite memory.)

6.35 Design a circuit with two inputs, CLOCK and /SYNC, that produces the output waveforms shown in Figure X6.35. The /SYNC pulse occurs only once every 256 clock cycles. Your design should use at most one SSI package and two MSI parts that were described in this chapter. The output signals should be glitch-free and must change synchronously on the rising edge of CLOCK.

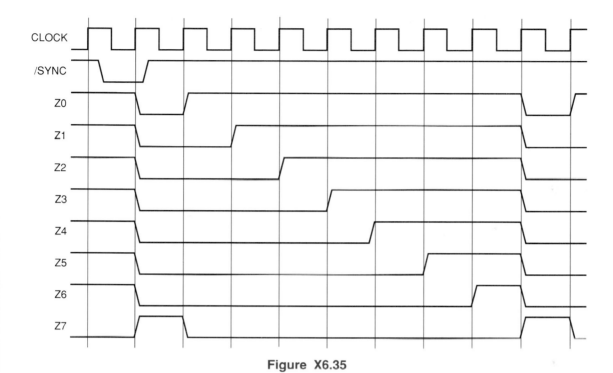

Figure X6.35

6.36 Modify the state diagram of Figure 6–66 so that the machine goes into hazard mode immediately if LEFT and RIGHT are asserted simultaneously during a turn. Write the corresponding transition list.

6.37 Repeat Exercise 6.36 using the ASM chart of Figure 6–70.

6.38 Construct an ASM chart for the combination-lock machine of Section 5.4.6.

6.39 Construct a state diagram for the 1s-counting machine of Table 5–16 and compare with the ASM chart in Figure 6–69.

6.40 Synthesize a circuit from a transition list for the ASM chart of Exercise 6.37. Write a transition equation for each state variable as a sum of p-terms, and write simplified transition/excitation equations for a realization using D flip-flops. Draw a circuit diagram using TTL SSI and MSI components.

6.41 Synthesize a circuit for the state diagram of Figure 6–66 using six variables to encode the state, where the LA–LC and RA–RC outputs equal the state variables themselves. Write a transition list, a transition equation for each state variable as a sum of p-terms, and simplified transition/excitation equations for a realization using D flip-flops. Draw a circuit diagram using TTL SSI and MSI components.

6.42 Synthesize a circuit for the ambiguous state diagram in Figure 6–64. Use the state assignment in Table 6–7. Write a transition list, a transition equation for each state variable as a sum of p-terms, and simplified transition/excitation equations for a realization using D flip-flops. Determine the actual next state of the circuit, starting from the IDLE state, for each of the following input combinations on (LEFT, RIGHT, HAZ): (1,0,1), (0,1,1), (1,1,0), (1,1,1). Comment on the machine's behavior in these cases.

6.43 Suppose that for a state SA and an input combination I, an ambiguous state diagram indicates that there are two next states, SB and SC. The actual next state SD for this transition depends on the state machine's realization. If the state machine is synthesized using the $V* = \Sigma$(p-terms where $V* = 1$) method to obtain transition/excitation equations for D flip-flops, what is the relationship between the coded states for SB, SC, and SD? Explain.

6.44 Repeat Exercise 6.43, assuming that the machine is synthesized using the $V*' = \Sigma$(p-terms where $V* = 0$) method.

6.45 Suppose that for a state SA and an input combination I, an ambiguous state diagram does not define a next state. The actual next state SD for this transition depends on the state machine's realization. Suppose that the state machine is synthesized using the $V* = \Sigma$(p-terms where $V* = 1$) method to obtain transition/excitation equations for D flip-flops. What coded state is SD? Explain.

6.46 Repeat Exercise 6.45, assuming that the machine is synthesized using the $V*' = \Sigma$(p-terms where $V* = 0$) method.

6.47 Synthesize a circuit from the ASM chart of Figure 6–69. Compare your results with the results of Section 5.4.6, and explain why they are the same or different.

6.48 Given the transition equations for a clocked synchronous state machine that is to be built using master/slave S-R flip-flops, how can the excitation equations for the S and R inputs be derived? (*Hint:* Show that any transition equation, $Qi* = $ expr, can be written in the form $Qi* = Qi \cdot expr1 + Qi' \cdot expr2$, and see where that leads.)

6.49 Repeat Exercise 6.48 for J-K flip-flops. How can the "don't-cares" that are possible in a J-K design be specified?

6.50 Starting with the transition list in Table 6–11, find a minimal sum-of-products expression for $Q2*$, assuming that the next states for the unused states are true don't-cares.

6.51 Draw a logic diagram for the output logic of the guessing-game machine in Table 6–11 using a single 74LS139 dual 2-to-4 decoder. (*Hint:* Use active-low outputs.)

6.52 What does the personalized license plate in Figure 6–62 stand for? (*Hint:* It's a computer engineer's version of OTTFFSS.)

6.53 Write a new transition list and transition/excitation equations for the T-bird tail-lights machine of Figure 6–66 using an output-coded state assignment.

6.54 Draw a complete state diagram for the ad hoc realization of the guessing game in Figure 6–77, including the normally unused states.

6.55 Redesign the excitation logic for flip-flop U5 in Figure 6–59 so that it uses only a single AND gate.

6.56 Design a combination-lock machine according to the state table of Table 5–18 on page 400 using a single 74LS163 counter and combinational logic for the /LD, /CLR, and A–D inputs of the '163. Use counter values 0–7 for states A–H.

6.57 Design the output logic for the 74LS163-based T-bird tail-lights machine shown in Figure 6–78. You may use any of the TTL SSI and MSI parts introduced in Chapters 2 and 4, but try to minimize the total number of IC packages.

6.58 Modify the multiplier state machine of Figure 6–86 so that it also controls the loading of operands and the reading of results. During the first two clock ticks after START is asserted, the machine should read the multiplier and multiplicand from the data bus. Then it should perform the multiplication and drive the two result bytes onto the data bus for one clock cycle each before returning to the IDLE state. In addition to an ASM chart or state diagram, draw a timing diagram that shows what operations are performed during each clock cycle.

6.59 Decompose the state machine for the 8-bit shift-and-add multiplier of Figure 6–86, using a 74LS163 to count the addition steps. Draw a block diagram showing the inputs and outputs of the main machine, a state diagram or ASM chart for the main machine, and a logic diagram for the 74LS163-based submachine. Explain how the state-machine design must be changed for an n-bit multiplication.

6.60 Calculate the minimum clock period of the data unit in Figure 6–85. Use the maximum propagation delays given in Table 4–3 for combinational parts. Unless you can get the real numbers from a TTL data book, assume that all registers require a 20-ns minimum setup time on inputs and have a 25-ns propagation delay from clock to outputs. Indicate any assumptions you've made about delays in the control unit.

6.61 Design a data unit and a control-unit state machine for multiplying 8-bit two's-complement numbers using the algorithm discussed in Section 1.8.

6.62 Design a data unit and control-unit state machine for dividing 8-bit unsigned numbers using the shift-and-subtract algorithm discussed in Section 1.9.

6.63 Calculate the MTBF of a synchronizer built according to Figure 6–98 using 74F74s, assuming a clock frequency of 25 MHz and an asynchronous transition rate of 1 MHz. Assume that the setup time of an 'F74 is 5 ns and the hold time is zero.

6.64 Calculate the MTBF of the synchronizer shown in Figure X6.64, assuming a clock frequency of 25 MHz and an asynchronous transition rate of 1 MHz. Assume that the setup time t_{setup} and the propagation delay t_{pd} from clock to Q or /Q in a 74ALS74 are both 10 ns.

Figure X6.64

6.65 Suppose that the SYNCIN signal in Exercise 6.64 is connected to a combinational circuit in the synchronous system, which in turn drives the D inputs of 74ALS74 flip-flops that are clocked by CLOCK. What is the maximum allowable propagation delay of the combinational logic?

6.66 The circuit in Figure X6.66 includes a deskewing flip-flop so that the synchronized output from the multiple-cycle synchronizer is available as soon as possible after the edge of CLOCK. Ignoring metastability considerations, what is the maximum frequency of CLOCK? Assume that for a 74F74, $t_{setup} = 5$ ns and $t_{pd} = 7$ ns.

Figure X6.66

6.67 Using the maximum clock frequency determined in Exercise 6.66, and assuming an asynchronous transition rate of 4 MHz, determine the synchronizer's MTBF.

6.68 Determine the MTBF of the synchronizer in Figure X6.66, assuming an asynchronous transition rate of 4 MHz and a clock frequency of 40 MHz, which is less than the maximum determined in Exercise 6.66. In this situation, "synchronizer failure" really occurs only if DSYNCIN is metastable. In other words, SYNCIN may be allowed to be metastable for a short time, as long as it doesn't affect DSYNCIN. This yields a better MTBF.

6.69 A famous digital designer devised the circuit shown in Figure X6.69(a), which is supposed to eliminate metastability within one period of a system clock. Circuit M is a memoryless analog voltage detector whose output is 1 if Q is in the metastable state, 0 otherwise. The circuit designer's idea was that if line Q is detected to be in the metastable state when CLOCK goes low, the NAND gate will clear the D flip-flop, which in turn eliminates the metastable output, causing a 0 output from circuit M and thus negating the CLR input of the flip-flop. The circuits are all fast enough that this all happens well before CLOCK goes high again; the expected waveforms are shown in Figure X6.69(b).

Unfortunately, the synchronizer still failed occasionally, and the famous digital designer is now designing pockets for blue jeans. Explain, in detail, how it failed, including a timing diagram.

Figure X6.69

7

PROGRAMMABLE LOGIC DEVICES

*programmable logic
device (PLD)*

program a PLD

Discrete SSI and MSI chips are gradually being supplanted by LSI chips that can be programmed to perform the same functions. A *programmable logic device (PLD)* is an LSI chip that contains a "regular" circuit structure, but that allows the designer to customize it for a specific application. When a typical PLD leaves the IC manufacturer's factory, it is not yet customized with a specific function. Instead, it is *programmed* by the purchaser to perform a function required by a particular application.

The "regular" circuit structure in a PLD is usually a rectangular array of identical cells that can be individually programmed. For a given chip area, the potential functionality that can be obtained with such a structure is greater than that of a "random" collection of logic gates. In fact, at just about every point in the development of IC technology, the chips with the greatest functionality per unit area have been memory chips, which contain a rectangular array of memory cells. The first PLDs that we'll study are "read-only memory" chips.

Designs using PLDs are usually smaller and faster than their SSI/MSI equivalents, and they often cost less, too. Just as important, PLDs usually simplify the logic designer's job. A PLD equivalent to a larger SSI/MSI circuit can often be specified with a simple program, and can easily be changed to fix bugs or to improve functionality. In fact, every time you see a Pascal or ABEL program in this chapter, you should think to yourself, "Would I rather write this program or design the corresponding SSI/MSI circuit?"

Figure 7–1
Basic structure of a $2^n \times b$ ROM.

7.1 READ-ONLY MEMORY

A *read-only memory (ROM)* is a combinational circuit with n inputs and b outputs, as shown in Figure 7–1. The inputs are called *address inputs* and are traditionally named A0, A1, ..., An−1. The outputs are called *data outputs*, and are traditionally named D0, D1, ..., Db-1. A ROM "stores" the truth table of an n-input, b-output combinational logic function. For example, Table 7–1 is the truth table of a 3-input, 4-output combinational function; it could be stored in a $2^3 \times 4$ (8×4) ROM. Neglecting propagation delays, a ROM's data outputs at all times equal the output bits in the truth-table row selected by the address inputs.

Since a ROM is a *combinational* circuit, we would be correct to say that it's not really a memory at all. In terms of digital circuit operation, we can treat a ROM like any other combinational logic element. However, we can also think of information as being "stored" in the ROM when it is manufactured or programmed (we'll say more about how this is done in Section 7.1.4).

When we think of ROM as being a type of memory, It has a key difference from most other types of integrated-circuit memory. ROM is *nonvolatile memory*; that is, its contents are preserved even if no power is applied.

read-only memory (ROM)
address input
data output

nonvolatile memory

Table 7–1
Truth table for a 3-input, 4-output combinational logic function.

A2	A1	A0	D3	D2	D1	D0
\multicolumn{3}{Inputs}	\multicolumn{4}{Outputs}

Inputs			Outputs			
A2	A1	A0	D3	D2	D1	D0
0	0	0	1	1	1	0
0	0	1	1	1	0	1
0	1	0	1	0	1	1
0	1	1	0	1	1	1
1	0	0	0	0	0	1
1	0	1	0	0	1	0
1	1	0	0	1	0	0
1	1	1	1	0	0	0

Figure 7–2
A 2-to-4 decoder with output
polarity control.

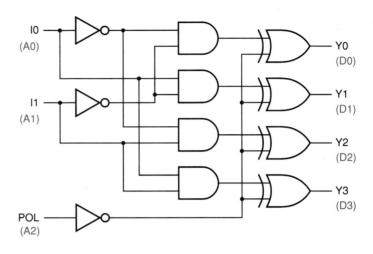

7.1.1 Using ROMs for "Random" Combinational Logic Functions

Table 7–1 is actually the truth table of a 2-to-4 decoder with an output-polarity control, a function that can be built with discrete gates as shown in Figure 7–2. Thus, we have two ways to build the decoder—with discrete gates, or with an 8×4 ROM that contains the truth table, as shown in Figure 7–3.

The assignment pattern of decoder inputs and outputs to ROM inputs and outputs in Figure 7–3 is a consequence of the way that the truth table in Table 7–1 is constructed. Therefore, the physical ROM realization of the decoder is not unique. That is, we could write the rows or columns of the truth table in a different order and use a physically different ROM to perform the same logic function, simply by assigning the decoder signals to different ROM inputs and outputs. Another way to look at this is that we can rename the individual address inputs and data outputs of the ROM.

For example, swapping the bits in the D0 and D3 columns of Table 7–1 would give us the truth table for a physically different ROM. However, the new ROM could still be used to build the 2-to-4 decoder simply by swapping the Y0 and Y3 labels in Figure 7–3. Likewise, if we shuffled the data rows of the truth table as shown in Table 7–2, we would get another different ROM, but it can

Figure 7–3
Connections to build the 2-to-4
decoder using an 8×4 ROM that
stores Table 7–1.

Table 7–2
Truth table with data rows shuffled.

Inputs			Outputs			
A2	A1	A0	D3	D2	D1	D0
0	0	0	1	1	1	0
0	0	1	0	0	0	1
0	1	0	1	1	0	1
0	1	1	0	0	1	0
1	0	0	1	0	1	1
1	0	1	0	1	0	0
1	1	0	0	1	1	1
1	1	1	1	0	0	0

still be used as the 2-to-4 decoder with a rearrangement of the address inputs, A0 = POL, A1 = I0, A2 = I1.

When constructing a ROM to store a given truth table, input and output signals reading from right to left in the truth table are normally assigned to ROM address inputs and data outputs with ascending labels. Each address or data combination may then be read as a corresponding binary integer with the bits numbered in the "natural" way. A data file is typically used to specify the truth table to be stored in the ROM when it is manufactured or programmed. The data file usually gives the address and data values as hexadecimal numbers. For example, a data file may specify Table 7–2 by saying that ROM addresses 0–7 should store the values E, 1, D, 2, B, 4, 7, 8.

An excellent example of a function that can be built with ROM is 4×4 unsigned binary multiplication, which we performed in Section 4.10 with two chips, the 74284 and the 74285. A $2^8 \times 8$ (256×8) ROM can be used instead, with the connections shown in Figure 7–4.

A ROM's contents are normally specified by a file that contains one entry for every address in the ROM. For example, Table 7–3 is a hexadecimal listing

Figure 7–4
Connections to perform a 4×4 unsigned binary multiplication using a 256×8 ROM.

Table 7–3
Hexadecimal text file specifying
the contents of a 4×4 multiplier
ROM.

```
00:  00 00 00 00 00 00 00 00 00 00 00 00 00 00 00 00
10:  00 01 02 03 04 05 06 07 08 09 0A 0B 0C 0D 0E 0F
20:  00 02 04 06 08 0A 0C 0E 10 12 14 16 18 1A 1C 1E
30:  00 03 06 09 0C 0F 12 15 18 1B 1E 21 24 27 2A 2D
40:  00 04 08 0C 10 14 18 1C 20 24 28 2C 30 34 38 3C
50:  00 05 0A 0F 14 19 1E 23 28 2D 32 37 3C 41 46 4B
60:  00 06 0C 12 18 1E 24 2A 30 36 3C 42 48 4E 54 5A
70:  00 07 0E 15 1C 23 2A 31 38 3F 46 4D 54 5B 62 69
80:  00 08 10 18 20 28 30 38 40 48 50 58 60 68 70 78
90:  00 09 12 1B 24 2D 36 3F 48 51 5A 63 6C 75 7E 87
A0:  00 0A 14 1E 28 32 3C 46 50 5A 64 6E 78 82 8C 96
B0:  00 0B 16 21 2C 37 42 4D 58 63 6E 79 84 8F 9A A5
C0:  00 0C 18 24 30 3C 48 54 60 6C 78 84 90 9C A8 B4
D0:  00 0D 1A 27 34 41 4E 5B 68 75 82 8F 9C A9 B6 C3
E0:  00 0E 1C 2A 38 46 54 62 70 7E 8C 9A A8 B6 C4 D2
F0:  00 0F 1E 2D 3C 4B 5A 69 78 87 96 A5 B4 C3 D2 E1
```

of the 4×4 multiplier ROM contents. Each row gives a starting address in the ROM and specifies the 8-bit data values stored at 16 successive addresses.

The nice thing about ROM-based design is that you can usually write a simple program in a high-level language to calculate what should be stored in the ROM. For example, it took me less than five minutes to write a Pascal program, shown in Table 7–4, that generated the contents of Table 7–3.[1]

* 7.1.2   Internal ROM Structure

The mechanism used by ROMs to "store" information varies with different ROM technologies. In most ROMs, the presence or absence of a diode or transistor distinguishes between a 0 and a 1.

Figure 7–5 is the schematic of a primitive 8×4 ROM that you can build yourself using a TTL decoder and a handful of diodes. The address inputs select one of the decoder outputs to be asserted. Each decoder output is called a *word line* because it selects one row or word of the table stored in the ROM. The figure shows the situation with A2–A0 = 101 and /ROW5 asserted.

word line

bit line

Each vertical line in Figure 7–5 is called a *bit line* because it corresponds to one output bit of the ROM. An asserted word line pulls a bit line LOW if a diode is connected between the word line and the bit line. There is only one diode in row 5, so the corresponding bit line (/D1) is pulled LOW. The bit lines

---

[1]Well, not quite. It took me five minutes to write the original version in C, and about 20 minutes to convert the C into Pascal because of Pascal's less capable standard I/O.

*Throughout this book, optional sections are marked with an asterisk.

Table 7–4
Program to generate the text file specifying the contents of a 4×4 multiplier ROM.

```
PROGRAM MultROM(input, output);
TYPE hex = 0..15;
VAR x, y: integer;

{ Procedure to write d as a hex digit.}
PROCEDURE WriteHexDigit(d: hex);
  BEGIN
    IF d<10 THEN write(char(ord('0')+d))
    ELSE write(char(ord('A')+d-10));
  END;

{ Procedure to write i as two hex digits.}
PROCEDURE WriteHex2(i: integer);
  BEGIN
    WriteHexDigit((i DIV 16) MOD 16);
    WriteHexDigit(i MOD 16);
  END;

BEGIN {Main program.}
  FOR x := 0 UNTIL 15 DO {For all values of x.}
    BEGIN
      WriteHex2(x*16); write(':');
      FOR y := 0 UNTIL 15 DO {For all values of y.}
        BEGIN  write(' '); WriteHex2(x*y)  END;
      writeln;
    END;
END.
```

are buffered through inverters to produce the D3–D0 ROM outputs, 0010 for the case shown.[2]

In the ROM circuit of Figure 7–5, each intersection between a word line and a bit line corresponds to one bit of "memory." If a diode is present at the intersection, a 1 is stored; otherwise, a 0 is stored. If you were to build this circuit in the lab, you would "program" the memory by inserting and removing diodes at each intersection. Primitive though it may seem, owners of the DEC PDP-11 minicomputer (circa 1970) made use of similar technology in the M792 32×16 "bootstrap ROM module." The module was shipped with 512 diodes soldered in place, and owners programmed it by clipping out the diode at each location where a 0 was to be stored.

---

[2]Schmitt-trigger inverters are used to improve the circuit's noise margins, because the 0.7-volt drop across the diodes creates a LOW level on the bit lines that is not really so low. Of course, you'd never really build this circuit except in an academic lab—it's easier just to buy and program a commercial ROM chip.

Figure 7–5
Logic diagram of a simple 8×4 ROM.

The diode pattern shown in Figure 7–5 corresponds to the 2-to-4-decoder truth table of Table 7–1. This doesn't seem very efficient—we used a 3-to-8 decoder and a bunch of diodes to build the ROM version of a 2-to-4 decoder. However, we'll show a more efficient ROM structure and more-useful design examples later.

Note that ROM circuit of Figure 7–5 must use diodes, not direct connections, at each location where a 1 is to be stored. Figure 7–6 shows what happens if just a few diodes—such as the ones in row 3—are replaced with direct connections. Suppose the address inputs are 101; then /ROW5 is asserted, and only /D1 is supposed to be pulled LOW to create an output of 0010. However, the *sneak path* direct connections allow current to flow along *sneak paths*, so that bit lines /D2 and /D0 are also pulled LOW, and the (incorrect) output is 0111. When diodes are used, sneak paths are blocked by reverse-biased diodes, and correct outputs are obtained.

MOS ROMs actually use a transistor instead of a diode at each location where a bit is to be stored; Figure 7–7 shows the basic idea. The row decoder

Figure 7–6
Sneak paths in a ROM.

Figure 7–7
MOS transistors as storage
elements in a ROM.

has active-high outputs. When a row line is asserted, the NMOS transistors in that row are turned on, which pulls the corresponding bit lines low. A similar idea can be used in ROMs built with bipolar transistors.

* 7.1.3   Two-Dimensional Decoding

Suppose you wanted to build a 128×1 ROM using the kind of structure described in the preceding subsection. Have you ever thought about what it would take to build a 7-to-128 decoder? Try 128 7-input NAND gates to begin with, and add 14 buffers and inverters with a fanout of 64 each! ROMs with a million bits or more are available commercially; trust me, they do not contain 20-to-1,048,576 *two-dimensional* decoders. Instead, a different structure, called *two-dimensional decoding*, is used *decoding* to reduce the decoding complexity to something on the order of the square root of the number of addresses.

The basic idea in two-dimensional decoding is to arrange the ROM cells in an array that is as close as possible to square. For example, Figure 7–8 shows a possible internal structure for a 128×1 ROM. The three high-order address bits, A6–A4, are used to select a row. Each row stores 16 bits starting at address A6 A5 A4 0000. When an address is applied to the ROM, all 16 bits in the selected row are "read out" in parallel on the bit lines. A 16-input multiplexer selects the desired data bit based on the low-order address bits.

By the way, the diode pattern in Figure 7–8 was not chosen at random. It performs a very useful 7-input combinational logic function that would require 35 4-input AND gates to build as a minimal two-level AND-OR circuit (see Exercise 7.2). The ROM version of this function certainly saves a lot of engineering effort and space in a board-level design.

Two-dimensional decoding allows the 128×1 ROM to be built with a 3-to-8 decoder and a 16-input multiplexer (whose complexity is comparable to that of a 4-to-16 decoder). A 1M×1 ROM could be built with a 10-to-1024 decoder and a 1024-input multiplexer—not easy, but a lot simpler than the one-dimensional alternative.

Besides reducing decoding complexity, two-dimensional decoding has one other benefit—it leads to a chip with physical dimensions that are close to square, important for chip fabrication and packaging. A chip containing a 1M×1 physical array would be *very* long and skinny and could not be built economically.

In ROMs with multiple data outputs, the storage arrays corresponding to each data output may be made narrower in order to achieve an overall chip layout that is closer to square. For example, Figure 7–9 shows the possible layout of a 32K×8 ROM.

Figure 7–8   Internal structure of a 128×1 ROM using two-dimensional decoding.

7.1.4   Commercial ROM Types

Unless you visit computer museums or surplus stores, it's difficult nowadays to find ROM modules built with discrete diodes. A modern ROM is fabricated as a single IC chip; one that stores a million bits can be purchased for under $10. Various methods are used to "program" the information stored in a ROM, as discussed below and summarized in Table 7–5.

Most of the early integrated-circuit ROMs were *mask-programmable ROMs* (or, simply, *mask ROMs*). A mask ROM is programmed by the pattern of connections and no-connections in one of the *masks* used in the IC manufacturing process. To program or write information into the ROM, the customer gives the manufacturer a listing of the ROM contents, using a floppy disk or other medium. The manufacturer uses this information to create one or more customized masks to manufacture ROMs with the required pattern. ROM manufacturers impose a *mask charge* of several thousand dollars for the "customized" aspects of mask-

mask-programmable ROM

mask ROM

mask

mask charge

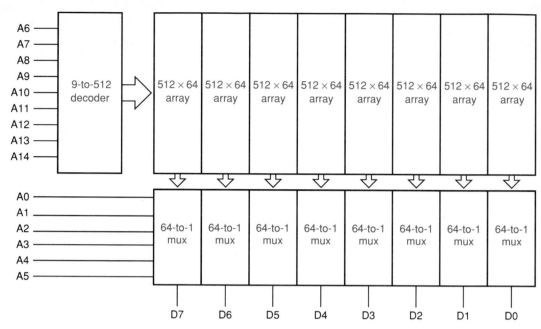

Figure 7–9 Possible layout of a 32K×8 ROM.

ROM production. Because of mask charges and the four-week delay typically required to obtain programmed chips, mask ROMs are normally used today only in high-volume applications. For low-volume applications there are more cost-effective choices, discussed next.

programmable read-only memory (PROM)
PROM programmer

A *programmable read-only memory (PROM)* is similar to a mask ROM, except that the customer may store data values (i.e., "program the PROM") in just a few minutes using a *PROM programmer*. A PROM chip is manufactured with all of its diodes or transistors "connected." This corresponds to having all bits at a particular value, typically 1. The PROM programmer can be used to set desired bits to the opposite value, typically by vaporizing tiny *fusible links* inside the PROM corresponding to each bit. A link is vaporized by selecting

fusible link

Table 7–5 Commercial ROM types.

Type	Technology	Read cycle	Write cycle	Comments
Mask ROM	NMOS, CMOS	25–500 ns	4 weeks	Write once; low power
Mask ROM	Bipolar	<100 ns	4 weeks	Write once; high power; low density
PROM	Bipolar	<100 ns	5 minutes	Write once; high power; no mask charge
EPROM	NMOS, CMOS	25–500 ns	5 minutes	Reusable; low power; no mask charge
EEPROM	NMOS	50–500 ns	i0 ms/byte	10,000 writes/location limit

Figure 7–10
Storage matrix in an EPROM
using floating-gate MOS
transistors.

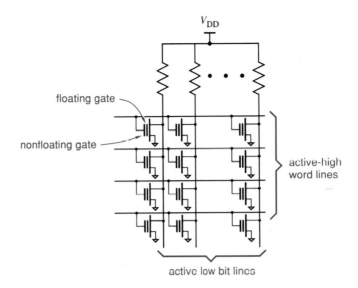

it using the PROM's address and data lines, and then applying a high-voltage pulse (10–30 V) to the device through a special input pin.

Early PROMs had reliability problems. Sometimes the stored bits changed because of incompletely vaporized links that would "grow back," and sometimes intermittent failures occurred because of floating shrapnel inside the IC package. However, these problems have been worked out, and reliable fusible-link technology is used nowadays not only in PROMs, but also in the bipolar PLDs that we introduce in Section 7.2.

An *erasable programmable read-only memory (EPROM)* is programmable like a PROM, but it can also be "erased" to the all-1s state by exposing it to ultraviolet light. No, the light does not cause fuses to grow back! Rather, EPROMs use a different technology, called "floating-gate MOS."

As shown in Figure 7–10, an EPROM has a *floating-gate MOS transistor* at *every* bit location. Each transistor has two gates. The "floating" gate is unconnected and is surrounded by extremely high-impedance insulating material. To program an EPROM, the programmer applies a high voltage to the nonfloating gate at each bit location where a 0 is to be stored. This causes a breakdown in the insulating material and allows a negative charge to accumulate on the floating gate. When the high voltage is removed, the negative charge remains. During subsequent read operations, the negative charge prevents the MOS transistor from turning on when it is selected.

EPROM manufacturers claim that a properly programmed bit will retain 70% of its charge for at least 10 years, even if the part is stored at 125°C, so EPROMs definitely fall into the category of "nonvolatile memory." However, they can also be erased. The insulating material surrounding the floating gate

erasable
programmable
read-only memory
(EPROM)

floating-gate MOS
transistor

EPROM erasing

becomes slightly conductive if it is exposed to ultraviolet light with a certain wavelength. Thus, EPROMs can be erased by exposing the chips to ultraviolet light, typically for 5–20 minutes. An EPROM chip is normally housed in a package with a quartz lid through which the chip may be exposed to the erasing light.

Probably the most common application of EPROMs is to store programs in microprocessor systems. EPROMs are typically used during program development, where the program or other information in the EPROM must be repeatedly changed during debugging. However, ROMs and PROMs usually cost less than EPROMs of similar capacity. Therefore, once a program is finalized, a ROM or PROM may be used in production to save cost. Most of today's PROMs are actually EPROMS housed in inexpensive packages without quartz lids; these are sometimes called *one-time programmable (OTP) ROMs.*

*one-time
programmable
(OTP) ROM
electrically erasable
programmable
read-only memory
(EEPROM)*

An *electrically erasable programmable read-only memory (EEPROM)* is similar to an EPROM, except that individual stored bits may be erased electrically. The floating gates in an EEPROM are surrounded by a much thinner insulating layer, and can be erased by applying a voltage of the opposite polarity as the charging voltage to the nonfloating gate.

As indicated in Table 7–5, programming or "writing" an EEPROM location takes much longer than reading it, so an EEPROM is no substitute for the read/write memories discussed in Section 7.5. Also, because the insulating layer is so thin, it can be worn out by repeated programming operations. As a result, EEPROMs can be reprogrammed only a limited number of times, typically 10,000 times per location. Therefore, EEPROMs are typically used for storing data that must be preserved when the equipment is not powered, but that doesn't change very often, such as the default configuration data for a smart terminal.

*2764
27128
27256
27512*

The most popular ROMs in microprocessors and other moderate-speed ROM applications are the *2764, 27128, 27256,* and *27512* EPROMs, whose logic symbols are shown in Figure 7–11. The figure also indicates pins that must be connected to a constant signal during normal read-only operation. Inputs labeled V_{CC} must be connected to +5 volts, inputs labeled V_{IH} must be connected to a valid HIGH logic signal, and "N.C." means "no connection." Different input configurations are used to program and test the device; the pin labeled VPP is used to apply the programming voltage.

7.1.5 ROM Control Inputs and Timing

*output-enable (OE)
input*

The outputs of a ROM must often be connected to a three-state bus, where different devices may drive the bus at different times. Therefore, most commercial ROM chips have three-state data outputs and an *output-enable (OE) input* that must be asserted to enable the outputs.

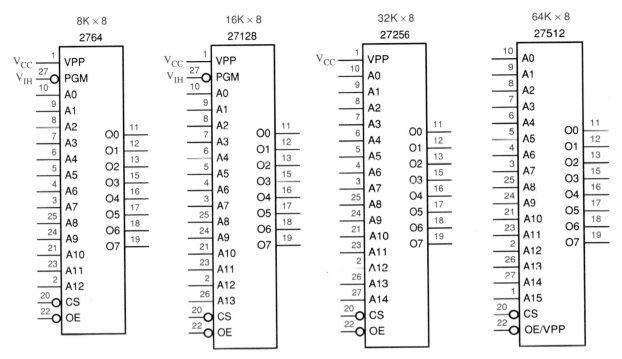

Figure 7–11   Logic symbols for standard EPROMs in 28-pin dual in-line packages.

Many ROM applications, especially program storage, have multiple ROMs connected to a bus, where only one ROM drives the bus at a time. Most ROMs have a *chip-select (CS) input* to simplify the design of such systems. In addition to OE, a ROM's CS input must be asserted to enable the three-state outputs.

Figure 7–12 shows how the OE and CS inputs could be used when connecting four 32K×8 ROMs to a microprocessor system that requires 128 Kbytes of ROM. The microprocessor has an 8-bit data bus and a 20-bit address bus, for a maximum address space of 1 Mbyte (2^{20} bytes). The ROM is supposed to be located in the highest 128K of the address space. Therefore, a NAND gate produces the /HIMEM signal, which is asserted when the address bus contains an address in the highest 128K (A19–A17 = 111). A 74LS139 decoder then selects one of the four 32K×8 ROMs. The selected ROM drives the data bus only when the microprocessor requests a read operation by asserting READ, which is connected to all of the OE inputs.

As we've described it so far, a CS input is no more than a second output-enable input that is ANDed with OE to enable the three-state outputs. However, in many ROMs, CS also serves as a *power-down input*. When CS is negated, power is removed from the ROM's internal decoders, drivers, and multiplexers. In this *standby mode* of operation, a typical ROM consumes less than 10% of

chip-select (CS) input

power-down input

standby mode

Figure 7–12 Address decoding and ROM enabling in a microprocessor system.

active mode

the power it uses in *active mode* with CS asserted. In Figure 7–12, at most one ROM is selected at any time, so the total power consumption of all the ROM chips is much closer to that of one chip than four.

Figure 7–13 shows how CS and OE inputs are used inside a typical ROM. Figure 7–14 shows typical ROM timing, including the following parameters:

t_{AA}
access time from address

t_{AA} *Access time from address.* The access time from address of a ROM is the propagation delay from stable address inputs to valid data outputs. When designers talk about "a 200-ns ROM," they are usually referring to this parameter.

t_{ACS}
access time from chip select

t_{ACS} *Access time from chip-select.* The access time from chip select of a ROM is the propagation delay from the time CS is asserted until the data outputs are valid. In some chips, this is longer than the access time from address, because the chip takes a little while to "power up." In others, this time is shorter, because CS controls only output enabling.

Figure 7–13
Internal ROM structure, showing
use of control inputs.

t_{OE} *Output-enable time.* This parameter is usually much shorter than access t_{OE}
time. The output-enable time of a ROM is the propagation delay from   *output-enable time*
OE and CS both asserted until the three-state output drivers have left the
Hi-Z state. Depending on whether the address inputs have been stable
long enough, the output data may or may not be valid at that point.

t_{OZ} *Output-disable time.* The output-disable time of a ROM is the propagation t_{OZ}
delay from a OE or CS negated until the three-state output drivers have   *output-disable time*
entered the Hi-Z state.

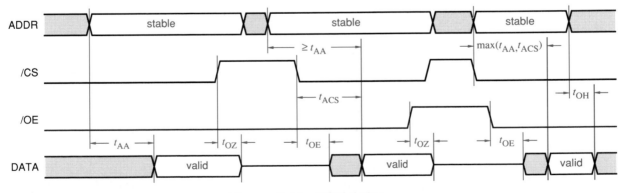

Figure 7–14   ROM timing.

t_{OH}
output-hold time

t_{OH} *Output-hold time.* The output-hold time of a ROM is the length of time that the outputs remain valid after a change in the address inputs, or after /OE or /CS is negated.

As with other components, the manufacturer specifies maximum and, sometimes, typical values for all timing parameters. Usually, minimum values are also specified for t_{OE} and t_{OH}. The minimum value of t_{OH} is usually specified to be 0, that is, the minimum combinational-logic delay through the ROM is 0.[3]

Given the decoding and multiplexing structure in Figure 7–13, you might think that the address access time is shorter from some address inputs than it is from others. In a given application, if some address-input signals were delayed relative to the others, you could recover this delay by connecting the slower signals to the "faster" ROM inputs. After all, any input signal can be connected to any ROM address input if you're willing to rearrange the ROM contents accordingly. However, ROM manufacturers don't specify which, if any, inputs are faster. In fact, the internal electrical characteristics of most ROMs may be such that the difference is not enough to bother with.

7.1.6 ROM Applications

As we mentioned earlier, the most common application of ROMs is for program storage in microprocessor systems. However, there are many applications where ROM provides a low-cost realization of complex or "random" combinational logic functions. In this section, we'll give a couple of examples of ROM-based circuits that are used in the digital telephone network.

When an analog voice signal enters a typical digital telephone system, it is sampled 8000 times per second and converted into a sequence of 8-bit bytes representing the analog signal at each sampling point. In Section 6.4.3, we described how these digital voice samples could be converted between parallel and serial formats, but we did not describe the coding of the 8-bit bytes themselves. We'll do that now, and then show how ROM-based circuits can easily deal with this highly encoded information.

linear encoding
dynamic range

The simplest 8-bit encoding of the sign and amplitude of an analog signal would be an 8-bit integer in the two's-complement or signed-magnitude system; this is called a *linear encoding*. However, an 8-bit linear encoding yields a *dynamic range*—the ratio between range of representable numbers and the smallest representable difference—of only 2^8 or 256. For you audiophiles, this corresponds to a dynamic range in signal power of $20 \log 256$ or about 48 dB. By

---

[3]The actual delay is never 0, of course, but it can easily be just slightly less than what you need to meet a nonzero hold-time requirement elsewhere in a design.

Figure 7–15
Format of a μ-law PCM byte.

7	6	5	4	3	2	1	0
S	E			M			

sign   exponent   mantissa

comparison, compact audio discs use a 16-bit linear encoding with a (theoretical) dynamic range of $20 \log 2^{16}$ or about 96 dB.

Instead of a linear encoding, the North American telephone network uses an 8-bit *companded encoding* called μ-*law PCM* ("mu-law" pulse-code modulation). Figure 7–15 shows the format of an 8-bit coded byte, a sort of floating-point representation containing sign (S), exponent (E), and mantissa (M) fields. The analog value V represented by a byte in this format is given by the formula,

$$V = (1 - 2S) \cdot (2^E) \cdot (2M + 33) - 33$$

companded encoding

μ-law PCM

An analog signal represented in this format can range from $-8159 \cdot k$ to $+8159 \cdot k$, where k is an arbitrary scale factor. The range of signals is $2 \cdot 8159$ and the smallest difference that can be represented is only 2 (when $E = 0$), so the dynamic range is $20 \log 8159$ or about 78 dB, quite an improvement over an 8-bit linear code.[4]

Now let's look at some ROM applications involving μ-law coded voice signals. Believe it or not, in many types of phone connections, your voice is purposely attenuated by a few decibels to make things work better. In an analog phone network, attenuation is performed by a simple, passive analog circuit, but not so in the digital world. Given a μ-law PCM byte, a *digital attenuator* must produce a different PCM byte that represents the original analog signal multiplied by a specified attenuation factor.

digital attenuator

One way to build a digital attenuator is shown in Figure 7–16. The input byte is applied to a μ-law decoder that expands the byte according to the formula given earlier to produce a 14-bit signed-magnitude integer. This 14-bit linear value is then multiplied by a 14-bit binary fraction corresponding to the desired attenuation amount. The fractional bits of the product are discarded, and the result is re-encoded into a new 8-bit μ-law PCM byte. Each block in the figure could be built with perhaps a dozen MSI chips.

Figure 7–17 is the logic diagram for a digital attenuator that uses a single 8K×8 ROM instead. This ROM can apply any of 32 different attenuation factors

---

[4]You don't get something for nothing, of course. In the companded encoding, the difference between successive coded values is greater at high amplitudes than it is at low amplitudes, so the encodings of large analog signals suffer from more quantizing distortion (caused by the difference between the sampled analog value and the nearest available coded representation). Expressed as a percentage of peak signal amplitude, though, quantizing distortion is about constant across the entire range of representable values.

Figure 7–16   Block diagram of a digital attenuator using combinational MSI logic.

to a μ-law input byte—it simply stores 32 different attenuation tables. The high-order address bits select a particular table, and the low-order address bits select an entry. Each entry is a pre-computed μ-law byte corresponding to the given attenuation code and input byte. Table 7–6 is a Pascal program that can be used to generate the ROM contents. The **WriteHex** procedures are not shown, and details of the **UlawToLinear** and **LinearToUlaw** functions are left as an exercise.

By the way, the digital attenuator is another good example of the many cost-effective space/time trade-offs that were made possible when the telephone network "went digital." A single analog voice signal produces an 8-bit μ-law PCM sample once every 125 μs, but the digital attenuator ROM can produce a valid output in a few hundred nanoseconds or less. Thus, in a digital telephone system, a single ROM chip can attenuate hundreds of digital voice streams,

Figure 7–17
A digital attenuator.

Table 7–6 Program to generate the contents of an 8K×8, 32-position attenuator ROM for
μ-law coded bytes.

```
PROGRAM AttenROM(input, output);
TYPE
  hex = 0..15;
  byte = 0..255;
  linear = -8159..8159;
VAR
  i, j, position : integer;
  pcmIN, pcmOUT: byte;
  atten, attenDB: real;
FUNCTION UlawToLinear(in: byte) : linear;
...
FUNCTION LinearToUlaw(x: linear) : byte;
...
BEGIN {Main program.}
  FOR position := 0 UNTIL 31 DO {Make 32 256-byte tables.}
    BEGIN
      readln{attenDB}; {Get amount in dB from designer,
                    negative for attenuation, positive for gain.}
      atten := exp(ln(10)*attenDB/10); {Convert to fraction.}
      FOR i := 0 to 15 DO {Construct output file in rows of 16.}
        BEGIN
          WriteHex4( (position*256) + i*16);  write(':');
          FOR j := 0 UNTIL 15 DO
            BEGIN
              pcmIN := i*16 + j;
              pcmOUT := LinearToUlaw(atten * UlawToLinear(pcmIN));
              write(' '); WriteHex2(pcmOUT);
            END;
          writeln;
        END;
    END;
END.
```

where hundreds of analog attenuator networks would have been required in the
old days.

   Another ROM application in digital telephony is a digital conference circuit.
In the analog telephone network, it's fairly easy to make a conference connection
between three or more parties. Just connect the analog phone wires together, and
you get an analog summing junction, so each person hears everyone else (it's
not quite that easy to do it right, but you get the idea). In the digital network,
of course, chaos would result from just shorting together digital output signals.
Instead, a digital conference circuit must include a digital *adder* that produces
output samples corresponding to sums of the input samples.

Figure 7–18
An adder circuit for μ-law coded bytes.

You know how to build a binary adder for two 8-bit operands, but binary adders cannot process μ-law PCM bytes directly. Instead, the 8-bit PCM bytes must be converted into 14-bit linear format, then added, and then re-encoded. As in the digital attenuator circuit of Figure 7–16, an MSI design for this function would require many chips. Alternatively, the function could be performed by a single 64K×8 ROM, as shown in Figure 7–18. The ROM has 16 address inputs, accommodating two 8-bit μ-law PCM operands. For each pair of operands, the corresponding ROM address contains the precomputed μ-law PCM sum. Table 7–7 is a Pascal program that can be used to generate the ROM contents.

These examples illustrate the many advantages of building complex combinational functions with ROMs. We usually consider a function to be "complex" if we know that an SSI, MSI, or ad hoc circuit for the function will be a real pain to design. However, most "complex" functions, like the examples in this subsection, have fairly straightforward word descriptions. Such a description can generally be translated into a computer program that "figures out" what the function's outputs should be for every possible input combination, so the function can be built simply by dropping its truth table into a ROM. Besides ease and speed of design, a ROM-based circuit has other important advantages:

- For a moderately complex function, a ROM-based circuit is usually faster than a discrete SSI/MSI circuit, and often faster than custom LSI in a comparable technology.

Table 7–7 Program to generate the contents of a 64K×8 adder ROM for μ-law coded bytes.

```
PROGRAM AdderROM(input, output);
CONST
  minLinear = -8159;
  maxLinear = 8159;
TYPE
  hex = 0..15;
  byte = 0..255;
  linear = minLinear..maxLinear;
VAR
  i, j, linearSum : integer;
  pcmINX, pcmINY, pcmOUT: byte;
...
BEGIN {Main program.}
  FOR pcmINY := 0 UNTIL 255 DO {For all Y samples...}
    BEGIN
      FOR i := 0 to 15 DO {Construct output file in rows of 16.}
        BEGIN
          WriteHex4( (pcmINY*256) + i*16); write(':');
          FOR j := 0 UNTIL 15 DO {For all X samples...}
            BEGIN
              pcmINX := i*16 + j;
              linearSum := UlawToLinear(pcmINX) + UlawToLinear(pcmINY);
              {The next two lines perform  "clipping" on overflow.}
              IF linearSum < minLinear THEN linearSum := minLinear;
              IF linearSum > maxLinear THEN linearSum := maxLinear;
              write(' '); WriteHex2(LinearToUlaw(linearSum));
            END;
          writeln;
        END,
    END;
END.
```

- The program that generates the ROM contents can easily be structured to handle unusual or undefined cases that would require additional hardware in any other design. For example, the adder program in Table 7–7 easily handles out-of-range sums. (Also see Exercise 7.15.)

- A ROM's function is easily modified just by changing the stored pattern, usually without changing any external connections to the device. For example, the PCM attenuator and adder ROMs in this subsection can be changed to use 8-bit "A-law" PCM, the standard digital voice coding in Europe.

- The prices of ROMs and other structured logic devices are always dropping, making them more economical, and the densities are always increasing, expanding the scope of problems that can be solved with a single chip.

There are a few disadvantages of ROM-based circuits, too:

- For simple to moderately complex functions, a ROM-based circuit may cost more or consume more power than a discrete SSI/MSI circuit.

- For functions with more than 16–20 inputs, a ROM-based circuit is impractical because of the limit on ROM sizes that are available. For example, you wouldn't want to build a 16-bit adder in ROM—it would require billions and billions of bits.

7.2 COMBINATIONAL PLDS

programmable logic device (PLD)

The generic name *programmable logic device (PLD)* refers to any logic device whose function is specified by the user after the device is manufactured. PLDs are "structured" logic devices, in both their internal circuit design and the software tools that are used to specify their functions. Thus, they are most often used to replace "random" SSI and MSI logic, often saving both device cost and design time.

The ROMs of the preceding section are most often used as "program memories" in microprocessor systems, but in the sense of the definition above they can also be viewed as combinational PLDs. Examples in the preceding section showed how a single ROM could replace a substantial amount of "random" SSI and MSI logic. In this section, we'll introduce other combinational PLD types that can do the same thing.

7.2.1 Programmable Logic Arrays (PLAs)

programmable logic array (PLA)

inputs

outputs

product terms

A *programmable logic array (PLA)* is a combinational, two-level AND-OR device that can be programmed to realize any sum-of-products logic expression, subject to the size limitations of the device. Limitations are the number of inputs (n), the number of outputs (m), and the number of product terms (p). We might describe such a device as "an $n \times m$ PLA with p product terms." In general, p is far less than the number of n-variable minterms. Thus, unlike a $2^n \times m$-bit ROM, a PLA cannot perform arbitrary n-input, m-output logic functions; its usefulness is limited to functions that can be expressed in sum-of-products form using p or fewer product terms.

An $n \times m$ PLA with p product terms contains p $2n$-input AND gates and m p-input OR gates. Figure 7–19 shows a small PLA with four inputs, six AND gates, and three OR gates and outputs. Connections are indicated by X's; the device is programmed by breaking connections that are not needed. Thus, each AND gate can be connected to any subset of the primary input signals and their

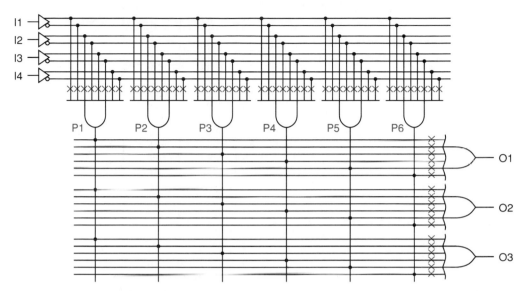

Figure 7–19 A 4×3 PLA with six product terms.

complements. Likewise, each OR gate can be connected to any subset of the
AND-gate outputs. This PLA can perform any three 4-input combinational logic
functions that can be written as sums of products using a total of six or fewer
distinct product terms, for example:

$$O1 \;=\; I1 \cdot I2 + I1' \cdot I2' \cdot I3' \cdot I4'$$
$$O2 \;=\; I1 \cdot I3' + I1' \cdot I3 \cdot I4 + I2$$
$$O3 \;=\; I1 \cdot I2 + I1 \cdot I3' + I1' \cdot I2' \cdot I4'$$

These equations have a total of eight product terms, but the first two terms in
the O3 equation are the same as the first terms in the O1 and O2 equations.

As shown in Figure 7–20, a more compact diagram can be used to repre-
sent a PLA. Moreover, the layout of this diagram more closely resembles the   *PLA diagram*
actual internal layout of a PLA chip. For example, Figure 7–21 shows how the
PLA circuit might be built in a bipolar, TTL-like technology. Each connection
represents a diode, and a diode-AND function is performed. Each diode-AND
line is followed by an inverting buffer, so overall a NAND function is obtained.
The resulting two-level NAND-NAND structure is, of course, functionally equivalent to the original AND-OR structure. The programmed connection pattern in
Figure 7–22 matches the logic equations given previously.

Sometimes a PLA output must be programmed to be a constant 1 or a   *PLA constant*
constant 0. That's no problem, as shown in Figure 7–23. Product term P1 is   *outputs*
always 1 because its product line is connected to no inputs and is therefore
always pulled HIGH; this constant-1 term drives the O1 output. No product term

Figure 7–20
Compact representation of a 4×3
PLA with six product terms.

Figure 7–21
A 4×3 PLA built using TTL-like
open-collector gates and diode
logic.

Figure 7–22
A 4×3 PLA programmed with a
set of three logic equations.

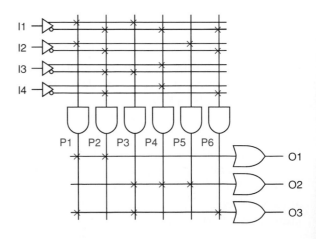

Figure 7-23
A 4×3 PLA programmed to
produce constant 0 and 1
outputs.

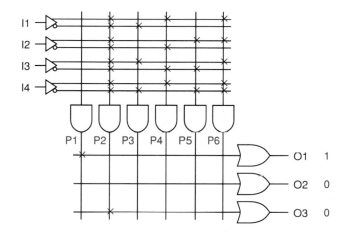

drives the O2 output, which is therefore always 0. Another method of obtaining a constant-0 output is shown for O3. Product term P2 is connected to each input variable and its complement; therefore, it's always 0 ($X \cdot X' = 0$).[5]

Our example PLA has too few inputs, outputs, and AND gates (product terms) to be very useful. An n-input PLA could conceivably use as many as 2^n product terms, to realize all possible n-variable minterms. The actual number of product terms in typical commercial PLAs is far less, on the order of 4 to 12 per output, regardless of the value of n.

The Signetics 82S100 is a typical example of the PLAs introduced in the mid-1970s. It has 16 inputs, 48 AND gates, and 8 outputs. Thus, it has $2 \times 16 \times 48 = 1536$ fusible links for the AND array and $8 \times 48 = 384$ for the OR array.

The locations of diodes in a PLA are specified when the device is programmed. As with ROMs, there are many programming options. Early PLAs used bipolar technology and had fusible links like the ones in bipolar ROMs. CMOS PLDs use floating-gate EPROM technology and are reprogrammable (when packaged with a transparent lid for erasing). Some PLDs have EEPROM-type or read/write memory cells, so that the PLD's logic function can actually be changed while the system is running!

7.2.2 ROMs vs. PLAs

A read-only memory (ROM) can be viewed as a special case of a PLA. An n-input, m-output ROM is equivalent to a PLA with a *fixed* AND array of 2^n *fixed AND array* n-input AND gates that realize all 2^n possible n-variable minterms. The OR array

---

[5]Theoretically, if *all* of the input variables change simultaneously, P2 could have a glitch. This is highly unlikely in typical applications, and is impossible if one input happens to be unused and is connected to a constant logic signal.

is programmable, consisting of m 2^n-input OR gates. Since any combinational logic function can be written as a sum of minterms, the ROM can perform any n-input, m-output combinational logic function. The number of fuses is $m \cdot 2^n$, which is the number of bits "stored" in the ROM. As we showed in Section 7.1, the ROM directly stores the function's truth table.

Of course, you've already learned that a $2^n \times m$-bit ROM isn't really built with 2^n n-input AND gates and m 2^n-input OR gates. The point here is to understand that ROMs are functionally equivalent to a programmable AND-OR structure, and as such they are often useful as truly universal combinational logic building blocks, not just as "storage" devices for microprocessor programs. Thus, when a logic function has too many product terms to be built with a PLA or other PLA-like PLD, a ROM-based design may be a viable alternative. We gave several examples in Section 7.1.6.

7.2.3   Programmable Array Logic (PAL) Devices

programmable array logic (PAL) device

Another special case of a PLA, and today's most commonly used type of PLD, is the *programmable array logic (PAL) device*.[6] Just the opposite of a ROM, a PAL device has a *programmable* AND array and a *fixed* OR array.[7]

PAL16L8

PAL devices were introduced by Monolithic Memories, now part of AMD, in the late 1970s. Key innovations in the first PAL devices, besides the introduction of a catchy acronym, were the use of a fixed OR array and bidirectional input/output pins.[8] These ideas are well illustrated by the *PAL16L8*, shown in Figures 7–24 and 7–25 and probably today's most commonly used combinational PLD. Its programmable AND array has 64 rows and 32 columns, identified for programming purposes by the small numbers in the logic diagram, and $64 \times 32 = 2048$ fusible links. Each of the 64 AND gates in the array has 32 inputs, accommodating 16 variables and their complements (hence, the "16" in "PAL16L8").

Eight AND gates are associated with each output pin of the PAL16L8. Seven of them provide inputs to a fixed 7-input OR gate. The eighth, which we call

---

[6]PAL is a registered trademark of Advanced Micro Devices, Inc. Like other trademarks, it should be used only as an adjective. Use it as a noun or without a trademark notice at your own peril (as I learned in a letter from AMD's lawyers in February, 1989).

[7]To get around AMD's trademark, I suggest that you use a descriptive name that is more indicative of the device's internal structure: a *fixed-OR element (FOE)*. FOE is a trademark of Xanthrax Systems (Sunnyvale, CA).

fixed-OR element FOE

[8]A step *backwards* in MMI's introduction was their popularization of the word "combinatorial" to describe combinational circuits. *Combinational* circuits have no memory—their output at any time depends on the current input *combination*. For well-rounded computer engineers, the word "combinatorial" conjures up images of binomial coefficients, problem-solving complexity, and Donald Knuth.

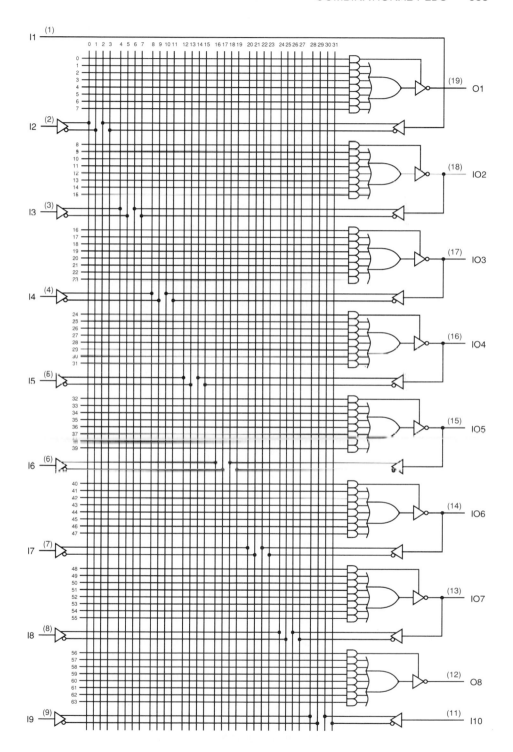

Figure 7–24
Logic diagram
of the PAL16L8.

Figure 7–25
Traditional logic symbol for the
PAL16L8.

output-enable gate

the *output-enable gate*, is connected to the three-state enable input of the output buffer; the buffer is enabled only when the output-enable gate has a 1 output. Thus, an output of the PAL16L8 can perform only logic functions that can be written as sums of seven or fewer product terms. Each product term can be a function of any or all 16 inputs, but only seven such product terms are available.

The worst-case logic function for two-level AND-OR design is an n-input XOR (parity) function, which requires 2^{n-1} product terms. However, less perverse functions with more than seven product terms can often be built by decomposing them into a 4-level structure (AND-OR-AND-OR) that can be realized with two passes through the AND-OR array. Unfortunately, besides using up PLD outputs for the first-pass terms, this doubles the delay, since a first-pass input must pass through the PLD twice to propagate to the output.

Although the PAL16L8 has up to 16 inputs and up to 8 outputs, it is housed in a package with only 20 pins, including two for power and ground (the corner pins, 10 and 20). This magic is achieved through the use of six bidirectional pins (13–18) that may be used as inputs or outputs or both. This and other differences between the PAL16L8 and a PLA structure are summarized below:

- The PAL16L8 has a fixed OR array, with seven AND gates permanently connected to each OR gate. AND-gate outputs cannot be shared; if a product term is needed by two OR gates, it must be generated twice.

- Each output of the PAL16L8 has an individual three-state output enable signal, controlled by a dedicated AND gate (the output-enable gate). Thus, outputs may be programmed as always enabled, always disabled, or enabled by a product term involving the device inputs.

- There is an inverter between the output of each OR gate and the external pin of the device.

- Six of the output pins, called I/O pins, may also be used as inputs. This provides many possibilities for using each I/O pin, depending on how the device is programmed:

  (1) If an I/O pin's output-control gate produces a constant 0, then the output is always disabled and the pin is used strictly as an input.

  (2) If the input signal on an I/O pin is not used by any gates in the AND array, then the pin may be used strictly as an output. Depending on the programming of the output-enable gate, the output may be always enabled, or enabled only for certain input conditions.

  (3) If an I/O pin's output-control gate produces a constant 1, then the output is always enabled, but the pin may still be used as an input too. In this way, outputs can be used to generate first-pass "helper terms" for logic functions that cannot be performed in a single pass with the limited number of AND terms available for a single output. We'll discuss this case in Section 7.2.6.

  (4) In another case with an I/O pin always output-enabled, the output may be used as an input to AND gates that affect the very same output. That is, we can embed a feedback sequential circuit in a PAL16L8. We'll discuss this case in Section 7.2.12.

7.2.4 The ABEL PLD Programming Language

Like a ROM or PROM, a PLD is ultimately programmed by specifying a diode or fuse pattern. However, few designers like to hack away at fuse patterns directly, or even indirectly in a hexadecimal text file. Instead, most designers use a *PLD programming language* to specify logic functions symbolically.

PLD programming language

In this subsection, we'll describe the basic features of an industry-standard PLD programming language called *ABEL*.[9]

ABEL

Other PLD programming languages have similar features, but vary somewhat in their syntax.

A PLD programming language is supported by a *PLD language processor*, which we'll simply call a *compiler*. The compiler's job is to translate a text file written in the language into a fuse pattern for a physical PLD. Even though most PLDs can be physically programmed only with sum-of-products expressions, languages like ABEL allow PLD equations to be written in almost any format. The compiler algebraically manipulates and minimizes the equations to fit, if possible, into the available PLD structure. The most advanced compilers for

PLD language processor
compiler

---

[9]ABEL ('Advanced Boolean Equation Language') is a trademark of Data I/O Corporation (Redmond, WA).

sequential PLDs allow state machines to be defined in a high-level language and automatically select an appropriate PLD, perform state assignments, and develop logic equations.

Table 7–8 shows the typical structure of a PLD program in the ABEL

Table 7–8
Typical structure of an ABEL program.

```
module module name
title string
deviceID device deviceType;
pin declarations
other declarations
equations
equations
end module name
```

Table 7–9   An ABEL program for a memory-decoder PLD.

```
module Memory_Decoder
title 'Memory Decoder PLD
J. Wakerly, Micro Systems Engineering'
MEMDEC device 'P16L8';
@ALTERNATE

" Input pins
LARGE                           pin 1;
A16, A17, A18, A19              pin 2, 3, 4, 5;
A20, A21, A22, A23              pin 6, 7, 8, 9;
/RTEST                          pin 11;
IOSEL                           pin 18;

" Output pins
/ROMCS                          pin 12;
LOCAL                           pin 13;

equations

ROMCS = LARGE * A23 * A22 * A21 * A20 * A19 * A18 * A17 * A16
      + /LARGE * A19 * A18 * A17 * A16
      + RTEST;

LOCAL = LARGE * A23 * A22 * A21 * A20 * A19 * A18 * A17 * A16
      + /LARGE * A19 * A18 * A17 * A16
      + RTEST
      + IOSEL;

end Memory_Decoder
```

language, and Table 7–9 shows an actual program. This example exhibits the following language features:

- A program file begins with a `module` statement, which associates a name (e.g., `Memory_Decoder`) with the program module. Large programs can have multiple modules, each with its own local title, declarations, and equations. The module name can be any valid identifier.

  module

- *Identifiers* must begin with a letter or underscore, may contain up to 31 letters, digits, and underscores, and are case sensitive.

  identifier

- The `title` statement specifies a title string that will be inserted into the documentation files that are created by the compiler.

  title

- A *string* is a series of characters enclosed by single quotes.

  string

- The `device` declaration includes a device identifier (e.g., `MEMDEC`) and a string that denotes the device type (e.g., `'P16L8'` for a PAL16L8). The compiler uses the device identifier in the names of documentation files that it generates, and it uses the device type to determine whether the device can really perform the logic functions requested in the program.

  device

- The `@ALTERNATE` directive tells the compiler to recognize an alternate set of symbols to denote logic operations:

  @ALTERNATE

  * `*` AND.

    *

  + `+` OR.

    +

  / `/` NOT (used as a prefix).

    /

  `:+:` XOR.

    :+:

  `:*:` XNOR.

    :*:

  The default symbols in ABEL for these operations are `&`, `#`, `!`, `$`, and `!$`.

- *Comments* begin with a double quote and end with another double quote or the end of the line, whichever comes first.

  comment

- *Pin declarations* tell the compiler about symbolic names associated with the device's external pins. Signals whose names have the NOT prefix (`/`) are active low at the external pin; others are active high.

  pin declarations

- *Other declarations* allow the designer to define constants and macros to simplify program design and improve program readability.

  other declarations

- The `equations` statement indicates that logic equations defining output signals as functions of input signals will follow.

  equations

Figure 7–26
Logic diagram for memory decoder using a PAL16L8.

equations

- *Equations* are written like assignment statements in a programming language. Each equation is terminated by a semicolon.

end

- The **end** statement marks the end of the module.

unclocked assignment operator, =

/

Equations for combinational PLD outputs use the *unclocked assignment operator, =*. The left-hand side of an equation normally contains a signal name. The right-hand side is a logic expression, not necessarily in sum-of-products form. The signal name on the left-hand side of an equation may be optionally preceded by the NOT operator /; this is equivalent to complementing the right-hand side. The compiler's job is to generate a fuse pattern such that the signal named on the left-hand side realizes the logic expression on the right-hand side.

A diagram showing the PAL16L8 logic symbol and the assignment of input and output signals is given in Figure 7–26; the same information is contained in the program in Table 7–9. The internal fuse configuration for the /ROMCS output that results from the program is shown in Figure 7–27. Several aspects of this figure must be explained.

Figure 7–27 Fuse configuration for the /ROMCS output.

Previously in this book, / has been used as an ordinary character in al-phanumeric signal names, but in ABEL it is used as a prefix operator indicating logical negation. However, this usage is consistent with our signal naming convention, in which / is used as a prefix to indicate an active-low signal. Under our naming convention, if *signal* and */signal* are the names of two different signals, then in normal operation

$$signal = /signal'$$

In ABEL, */signal* is an expression, not a signal name, but the equation above simply expresses a double negation, and is still true.[10]

Let us consider input pin 11 in Figure 7–27, which was declared to be active low (/RTEST). This signal is buffered and driven onto column line 30 (/RTEST), and its complement is driven onto column line 31 (RTEST). If pin 11 was declared to be active high (RTEST), the internal signals would be available on the opposite column lines. Both signals are available to the AND-OR array in either case; the compiler just has to keep track of where to find them.

On the other hand, consider the /ROMCS output signal on pin 12. The output signal is active low, but there is an inverter between the pin and the OR gate that generates the sum-of-products expression. Therefore, the OR gate must produce the signal /ROMCS', or ROMCS. We happened to specify the pin-12 output by writing ROMCS as a sum of three product terms. Therefore, the compiler can use these three terms directly. The effect would be the same if we wrote the equation

```
/ROMCS = /( LARGE * A23 * A22 * A21 * A20
                  * A19 * A18 * A17 * A16
        + /LARGE * A19 * A18 * A17 * A16
        + RTEST );
```

Either way, we get the fuse pattern shown in rows 57–59 of Figure 7–27.

All of the fuses in row 56 are blown so that the output-enable gate produces a constant-1 output and the three-state buffer is always enabled. All of the fuses in rows 60–63 are left intact, so the corresponding, unused AND gates produce constant-0 outputs.

---

[10]Documentation files generated by ABEL use ~ as a *prefix character* in the names of signals on external pins that have been declared to be active low, while continuing to use ! or / as a *negation operator*. Thus, when you see something like "!~ROMCS" in an ABEL DOC file, it means "the complement of the signal on the external pin named ~ROMCS." To be consistent, ABEL should really use the ~ prefix instead of ! or / in a program's pin declarations to denote active-low inputs. In any case, "!~ROMCS" contains a double negation and may be read simply as "ROMCS."

7.2.5　Signal Polarity

Consistent use of active-level notation requires the choice of "true" signal names (without a /) that describe conditions detected and actions performed in a positive sense. In full signal names, a consistent suffix or prefix (/ in this book) is used to denote the active level.

　　When active levels are used consistently in the external signal names in a circuit, PLD equations involving those signals can be written in a quite natural and consistent manner. Equations are written in terms of the "true" signal names, and a complement operator (/, !, or whatever) is used to perform logical negation as required to express the desired logic function. The correspondence between each internal "true" signal and an external signal that has a particular active level is established by the pin declarations or equivalent section in the PLD program. The active levels are easily changed without rewriting any of the logic equations.

　　For input signals, the choice of active level has no effect on whether a particular logic equation can be realized in a PLD. Since both true and complemented versions of each input signal are available at all the AND gates, the compiler can always select the appropriate one.

　　On the other hand, things aren't quite so simple for output signals. In the example in the preceding subsection, we happened to write the ROMCS in just the right form for a sum-of-products realization. Let us now consider the LOCAL equation:

```
LOCAL = LARGE * A23 * A22 * A21 * A20 * A19 * A18 * A17 * A16
      + /LARGE * A19 * A18 * A17 * A16
      + RTEST
      + IOSEL;
```

At first glance, this looks like no problem, since it is a sum of four product terms, and we have seven available. However, since the output pin is declared to be active high (LOCAL), the AND-OR logic must produce an internal signal that is active low (/LOCAL). To obtain an equation for the internal signal, we (or the compiler) can use DeMorgan's theorem to complement both sides of the preceding equation:

```
/LOCAL = (/LARGE + /A23 + /A22 + /A21 + /A20
                 + /A19 + /A18 + /A17 + /A16)
         * (LARGE + /A19 + /A18 + /A17 + /A16)
         * /RTEST
         * /IOSEL;
```

We (or the compiler) can multiply out and simplify this equation to obtain the required form for the PLD structure, a sum of products:

```
/LOCAL = /RTEST * /IOSEL * LARGE * /A23
       + /RTEST * /IOSEL * LARGE * /A22
       + /RTEST * /IOSEL * LARGE * /A21
       + /RTEST * /IOSEL * LARGE * /A20
       + /RTEST * /IOSEL * /A19
       + /RTEST * /IOSEL * /A18
       + /RTEST * /IOSEL * /A17
       + /RTEST * /IOSEL * /A16;
```

This equation illustrates a common problem that occurs when we try to fit expressions into a PLD structure. When a product of n variables is complemented, it may give rise to a sum of n product terms, exceeding the number of product terms available in the PLD. In particular, the equation above is a sum of *eight* product terms, and the PAL16L8 has only seven per output.

   A combinational PLD can handle decoder-like functions very well, since it has many inputs and can easily generate "wide" products of many variables. However, if the output polarity is such that a wide product term must be complemented, we may run out of AND terms in the array. The simplest solution to this problem is to redefine the output to have the opposite polarity, usually requiring an external inverter to generate the polarity that we wanted in the first place. Sometimes this approach is undesirable, or we may be stuck with an expression that has too many product terms in either polarity. Another solution is discussed in the next subsection.

7.2.6 Two Pass Logic

The number of product terms in the /LOCAL equation above can't be reduced, so we can't generate /LOCAL in one pass through a PAL16L8. However, we could split it into two parts by generating a "helper" term, using one of the six output pins that are also available as inputs:

```
/LCLHELP                        pin 14;
...
LCLHELP = /RTEST * /IOSEL * /LARGE * /A23
        + /RTEST * /IOSEL * /LARGE * /A22
        + /RTEST * /IOSEL * /LARGE * /A21
        + /RTEST * /IOSEL * /LARGE * /A20;

/LOCAL = LCLHELP
       + /RTEST * /IOSEL * /A19
       + /RTEST * /IOSEL * /A18
       + /RTEST * /IOSEL * /A17
       + /RTEST * /IOSEL * /A16;
```

Better yet, by observing the close relationship between the ROMCS and LOCAL signals, we could take advantage of the existing logic and simply rewrite the LOCAL equation in terms of ROMCS and IOSEL:

```
LOCAL = ROMCS + IOSEL;
```

Here, we are trying to use an output signal (ROMCS) as an input signal as well. This can be done in a PAL16L8 as long as the signal is assigned to a bidirectional pin. Thus, ROMCS should be reassigned from pin 12 to, say, pin 15.

Applying DeMorgan's theorem to the preceding equation, we find that /LOCAL can be expressed as a single product term:

```
/LOCAL = /ROMCS * /IOSEL;
```

However, both of the solutions presented above require signals to take two passes through the PLD. That is, a transition on a signal such as A23 must propagate through the PLD twice before affecting the LOCAL output.

7.2.7 Combinational PLD Applications

Quite often, combinational PLDs are handy for building special-purpose versions of functions that already exist as MSI parts—multiplexers, decoders, comparators, and the like. For example, the almost-binary-select decoder of Figure 4–108 on page 329 is easily built using a single PAL16L8; Table 7–10 is an ABEL program for the device, and Figure 7–28 is a corresponding logic diagram. Each of the last six equations uses a single AND gate in the PLD. The ABEL compiler can also minimize the BILL and MARY equations to use just one AND gate each. Note that active-high output signals could be obtained just by changing two lines in the declaration section:

```
BILL, MARY, JOAN, PAUL      pin 19, 18, 17, 16;
ANNA, FRED, DAVE, KATE      pin 15, 14, 13, 12;
```

Figure 7–28
Logic diagram for the PAL16L8 used as an almost-binary-select decoder.

Table 7–10   An ABEL program for an almost-binary-select decoder.

```
module Almost_Binary_Decoder
title 'Almost-Binary-Select Decoder PLD
J. Wakerly, Stanford University'
ABINDEC device 'P16L8';
@ALTERNATE

" Input pins
/CS, /RD, A0, A1, A2            pin 1, 2, 3, 4, 5;

" Output pins
/BILL, /MARY, /JOAN, /PAUL   pin 19, 18, 17, 16;
/ANNA, /FRED, /DAVE, /KATE   pin 15, 14, 13, 12;

equations

BILL = CS * RD * (/A2 * /A1 * /A0 + /A2 * /A1 *  A0);
MARY = CS * RD * (/A2 * /A1 * /A0 + /A2 *  A1 * /A0);
JOAN = CS * RD * (/A2 *  A1 * /A0);
PAUL = CS * RD * (/A2 *  A1 *  A0);
ANNA = CS * RD * ( A2 * /A1 * /A0);
FRED = CS * RD * ( A2 * /A1 *  A0);
DAVE = CS * RD * ( A2 *  A1 * /A0);
KATE = CS * RD * ( A2 *  A1 *  A0);

end Almost_Binary_Decoder
```

Since each of the (minimized) equations required a single product of four or five variables, each complemented equation requires a *sum* of four or five variables, less than the seven available in a PAL16L8.

Another example is the almost-binary-select multiplexer of Table 4–19 on page 330. A four-input, three-bit multiplexer with the required behavior can be built in a single PAL16L8 with the logic diagram shown in Figure 7–29. An ABEL program that creates the required behavior is shown in Table 7–11. Since this function uses all of the available pins on the PAL16L8, we had to make the pin assignment carefully. In particular, we were forced to assign two output signals to the two output-only pins (O1 and O8), so as to maximize the number of input pins available.

An 18-bit almost-binary-select multiplexer can be built with six PAL16L8 chips programmed and wired according to Table 7–11 and Figure 7–29. Furthermore, its select coding can be easily changed with a few keystrokes. This compares favorably with the MSI approach of Figure 4–109(b) on page 330, which required nine chips for a 16-bit mux.

Table 7–11   An ABEL program for an almost-binary-select multiplexer.

```
module Almost_Binary_Multiplexer
title 'Almost-Binary-Select Multiplexer PLD
J. Wakerly, Alantec, Inc.'
ABINMUX device 'P16L8';
@ALTERNATE

" Input pins
S0, S1, S2                        pin 16, 17, 18;
A0, A1, A2, B0, B1, B2            pin 1, 2, 3, 4, 5, 6;
C0, C1, C2, D0, D1, D2            pin 7, 8, 9, 11, 13, 14;

" Output pins
F0, F1, F2                        pin 19, 15, 12;

equations

F0 = /S0 * A0
   + (/S2 * /S1 * S0 + S2 * S1 * S0) * B0
   + (/S2 *  S1 * S0) * C0
   + ( S2 * /S1 * S0) * D0;
F1 = /S0 * A1
   + (/S2 * /S1 * S0 + S2 * S1 * S0) * B1
   + (/S2 *  S1 * S0) * C1
   + ( S2 * /S1 * S0) * D1;
F2 = /S0 * A2
   + (/S2 * /S1 * S0 + S2 * S1 * S0) * B2
   + (/S2 *  S1 * S0) * C2
   + ( S2 * /S1 * S0) * D2;

end Almost_Binary_Multiplexer
```

7.2.8 Sets and Relations

We can take this opportunity to introduce another useful capability of PLD programming languages. Most digital systems include buses, registers, and other circuits that handle a group of two or more signals in an identical fashion. In the multiplexer example of Table 7–11, the three output equations were identical except for the bit numbers 0, 1, 2. The ABEL language includes the concept of "sets" for writing such equations more compactly (and therefore with less chance of errors or inconsistencies).

set

A *set* is simply a defined collection of signals that is handled as a unit. When an operation is applied to a set, it is applied to each element of the set.

Figure 7–29
Logic diagram for the PAL16L8
used as an almost-binary-select
multiplexer.

Table 7-12 shows how sets can be used to write the ABEL equations for the almost-binary-select multiplexer. Each set is defined at the beginning of the program by associating a set name (e.g., **ABUS**) with a bracketed list of the set elements (e.g., **[A2,A1,A0]**). The number and order of elements in a set are significant, as we'll see.

Most of the operators in ABEL, including all of the ones in this book, can be used with sets. When an operation is applied to two or more sets, all of the sets must have the same number of elements, and the operation is applied individually to set elements in corresponding positions. For example, the equation

```
FBUS = ABUS + BBUS + (CBUS * DBUS);
```

is equivalent to three equations,

```
F2 = A2 + B2 + (C2 * D2);
F1 = A1 + B1 + (C1 * D1);
F0 = A0 + B0 + (C0 * D0);
```

When an operation includes both set and nonset variables, the nonset variables are combined individually with set elements in each position. Thus, the single **FBUS** equation in Table 7–12 is equivalent to the **F0**, **F1**, and **F2** equations in Table 7–11.

Another interesting feature of ABEL is its ability to convert "relational expressions" into logic expressions. A *relational expression* is a pair of operands combined with one of the *relational operators* listed in Table 7–13. The compiler

relational

expression

relational operator

Table 7–12 Almost-binary-select-multiplexer program using set notation.

```
module Almost_Binary_Multiplexer
title 'Almost-Binary-Select Multiplexer PLD
J. Wakerly, Kalvij Telecom, Inc.'
ABINMXS device 'P16L8';
@ALTERNATE

" Input pins
S0, S1, S2                      pin 16, 17, 18;
A0, A1, A2, B0, B1, B2          pin 1, 2, 3, 4, 5, 6;
C0, C1, C2, D0, D1, D2          pin 7, 8, 9, 11, 13, 14;

" Output pins
F0, F1, F2                      pin 19, 15, 12;

" Set definitions
ABUS = [A2,A1,A0];
BBUS = [B2,B1,B0];
CBUS = [C2,C1,C0];
DBUS = [D2,D1,D0];
FBUS = [F2,F1,F0];

equations

FBUS = /S0 * ABUS
   + (/S2 * /S1 * S0 + S2 * S1 * S0) * BBUS
   + (/S2 *  S1 * S0) * CBUS
   + ( S2 * /S1 * S0) * DBUS;

end Almost_Binary_Multiplexer
```

converts a relational expression into a logic expression that is 1 if and only if the relation is true. Either operand in a relation may be an integer or a set. If the operand is a set, it is treated as an unsigned binary integer with the leftmost variable representing the most significant bit. Thus, we could write the output equations for a 74LS682 comparator in two lines in ABEL:

```
PEQQ = ([P7,P6,P5,P4,P3,P2,P1,P0] == [Q7,Q6,Q5,Q4,Q3,Q2,Q1,Q0]);
PGTQ = ([P7,P6,P5,P4,P3,P2,P1,P0] >  [Q7,Q6,Q5,Q4,Q3,Q2,Q1,Q0]);
```

Of course, the fact we can write the equations easily doesn't mean that they'll fit in a PLD. The **PEQQ** expression above expands into 16 product terms, and **PGTQ** expands into 373 product terms (no, I didn't figure that one out by hand!).

Table 7–13
Relational operators in ABEL.

Symbol	Relation
==	equal
!=	not equal
<	less than
<=	less than or equal
>	greater than
>=	greater than or equal

* 7.2.9   Another Combinational PLD Application

The next application we consider is based on the mode-dependent comparator design of Figure 4–110 on page 331. Comparing, adding, and other "iterative" operations are usually poor candidates for PLD-based design, because an equivalent two-level sum-of-products expression has far too many product terms. Based on the results in the preceding subsection, we certainly wouldn't be able to replace the 74LS682 in Figure 4–110 with a PLD; the 74LS682 8-bit comparator is just about the most efficient possible single chip we can use to perform an 8-bit comparison. However, a PLD-based design is quite reasonable for the "random" mode-control logic.

Taking this idea one step further, suppose we needed to design a 32-bit mode-dependent comparator. Once again, the 74LS682 is the most efficient building block. However, the '682 does not have cascading inputs, so the outputs of four '682s must somehow be combined to perform a 32-bit comparison. The combining logic and the mode-control logic can be put together in a single PAL16L8. Figure 7–30 shows the complete circuit design, and Table 7–14 is the ABEL program for the MODECOMP PLD. The strategy for this design is to use the '682s to compare most of the bits, and to use a PAL16L8 to combine the '682 outputs and to handle the two low-order bits as a function of the mode.

Notice that since the outputs of the MODECOMP PLD are active low, and the equations are written in terms of active-high output signals, the right-hand side of each equation can be multiplied out to obtain directly an expression suitable for programming into the PLD's AND-OR structure. Both outputs use seven product terms, the maximum available in the PAL16L8. (When expanded into a sum of products, the XGTY equation yields eight product terms. However, two terms can be combined; can you see which ones?) If the output pins were declared to be active high, the resulting equations would "blow up," requiring more product terms than are available in the PAL16L8.

* 7.2.10   Other Combinational PLDs and Applications

The *PAL20L8* is another combinational PLD similar to the PAL16L8, except that   *PAL20L8*

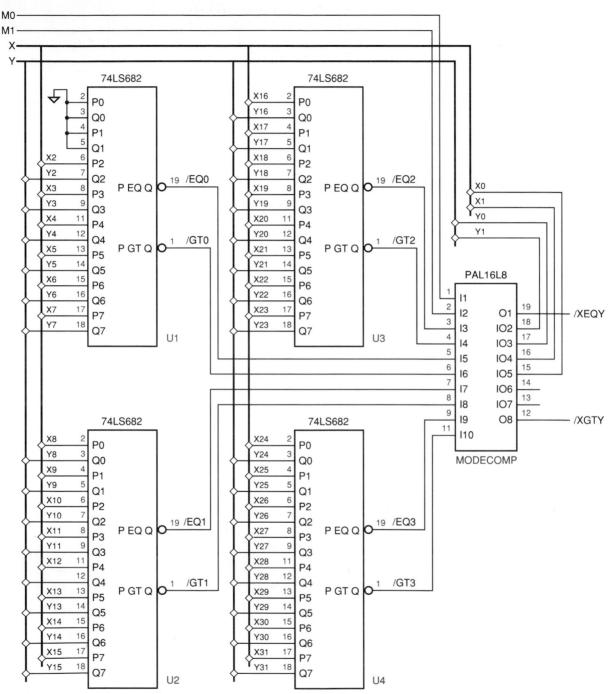

Figure 7–30 A 32-bit mode-dependent comparator.

Table 7-14 An ABEL program for the MODECOMP PLD.

```
module Comparator_Control
title 'Control PLD for Mode-Dependent Comparator
J. Wakerly, Micro Design Resources, Inc.'
MODECOMP device 'P16L8';
@ALTERNATE

" Input pins
M0, M1                        pin 1, 2;
/EQ0, /GT0, /EQ1, /GT1        pin 5, 6, 7, 8;
/EQ2, /GT2, /EQ3, /GT3        pin 3, 4, 9, 11;
X0, X1, Y0, Y1                pin 15, 16, 17, 18;

" Output pins
/XEQY, /XGTY                  pin 19, 12;

equations

XEQY = EQ3 * EQ2 * EQ1 * EQ0 * /M1 * /M0 * (        " Mode 00 (32-bit)
          /X1 * /Y1 * /X0 * /Y0 +
          /X1 * /Y1 *  X0 *  Y0 +
           X1 *  Y1 * /X0 * /Y0 +
           X1 *  Y1 *  X0 *  Y0 )
     + EQ3 * EQ2 * EQ1 * EQ0 * /M1 *  M0 * (        " Mode 01 (31-bit)
         /X1 * /Y1 + X1 * Y1 )
     + EQ3 * EQ2 * EQ1 * EQ0 *  M1 * /M0;           " Mode 10 (30-bit)

XGTY = GT3 + (EQ3 * GT2) + (EQ3 * EQ2 * GT1)        " Easy cases
     + (EQ3 * EQ2 * EQ1 * GT0)                      " Covers all modes
     + EQ3 * EQ2 * EQ1 * EQ0 * /M1 *  M0 * X1 * /Y1 " Mode 01 (31-bit)
     + EQ3 * EQ2 * EQ1 * EQ0 * /M1 * /M0 * (        " Mode 00 (32-bit)
           X1 * /Y1 +
          /X1 * /Y1 *  X0 *  Y0 +                   " No special case
           X1 *  Y1 *  X0 * /Y0 );                  "  needed for 30-bit mode

end Comparator_Control
```

* 7.2.10 Other Combinational PLDs and Applications

The *PAL20L8* is another combinational PLD similar to the PAL16L8, except that   *PAL20L8*
it comes in a 24-pin package and has four more input-only pins. An application
that pushes the PAL20L8 just about to its limit is a 15-input priority encoder,
similar in structure to the MSI 74LS148.

A logic diagram for the priority encoder is given in Figure 7–31. Inputs
P0–P14 are asserted to indicate requests, with P14 having the highest priority. If
EN (enable) is asserted, then the /Y3–/Y0 outputs give the number (active low)

Figure 7–31
Logic diagram for a PLD-based
15-input priority encoder.

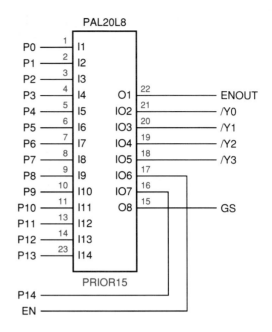

of the highest-priority request, and GS (Got Something) is asserted if any request is present. If EN is negated, then the /Y3–/Y0 outputs are negated and GS is negated. /ENOUT is asserted if EN is asserted and no request is present.

An ABEL program for the priority encoder is given in Table 7–15. ABEL

constant expression   allows *constant expressions* to be defined, which is useful when the same product terms are used in many equations. For the priority encoder, we wrote a series of expressions for constants H0–H14, where Hi is 1 if input Pi is the highest-priority asserted request input. These expressions allow us to write the equations for Y3–Y0 as simple sums, each with seven product terms.

The active levels for the /Y3–/Y0, GS, and /ENOUT outputs were not chosen at random. If the opposite polarity were specified, more product terms would be needed, and the function wouldn't fit in a PAL20L8.

Another priority example uses a single PLD to replace the four MSI parts in the fixed-point to floating-point encoder of Figure 4–112 on page 335. A logic diagram is shown in Figure 7–32 and an ABEL program is given in Table 7–16. The constant terms S0–S7 select exponent values of 0–7 depending on the location of the most significant 1 bit in the fixed-point input number. The exponent E2–E0 is simply a binary encoding of the select terms, and the mantissa bits M3–M0 are generated by four equations that serve as a 4-bit, 8-input multiplexer. Unfortunately, the PAL20L8 has available only seven product terms per output, and these equations require eight product terms each. There's only one spare output, not enough to generate all four required outputs using two-pass logic.

Table 7-15 An ABEL program for a 15-input priority encoder.

```
module Priority_Encoder_15
title '15-Input Priority Encoder
J. Wakerly, DAVID Systems, Inc.'
PRIOR15 device 'P20L8';
@ALTERNATE

" Input pins

P0, P1, P2, P3, P4, P5, P6, P7        pin 1, 2, 3, 4, 5, 6, 7, 8;
P8, P9, P10, P11, P12, P13, P14       pin 9, 10, 11, 13, 14, 23, 16;
EN                                    pin 17;

" Output pins

/Y3, /Y2, /Y1, /Y0                    pin 18, 19, 20, 21;
GS, /ENOUT                            pin 15, 22;

" Constant expressions

H14 = EN*P14;
H13 = EN*/P14*P13;
H12 = EN*/P14*/P13*P12;
H11 = EN*/P14*/P13*/P12*P11;
H10 = EN*/P14*/P13*/P12*/P11*P10;
H9  = EN*/P14*/P13*/P12*/P11*/P10*P9;
H8  = EN*/P14*/P13*/P12*/P11*/P10*/P9*P8;
H7  = EN*/P14*/P13*/P12*/P11*/P10*/P9*/P8*P7;
H6  = EN*/P14*/P13*/P12*/P11*/P10*/P9*/P8*/P7*P6;
H5  = EN*/P14*/P13*/P12*/P11*/P10*/P9*/P8*/P7*/P6*P5;
H4  = EN*/P14*/P13*/P12*/P11*/P10*/P9*/P8*/P7*/P6*/P5*P4;
H3  = EN*/P14*/P13*/P12*/P11*/P10*/P9*/P8*/P7*/P6*/P5*/P4*P3;
H2  = EN*/P14*/P13*/P12*/P11*/P10*/P9*/P8*/P7*/P6*/P5*/P4*/P3*P2;
H1  = EN*/P14*/P13*/P12*/P11*/P10*/P9*/P8*/P7*/P6*/P5*/P4*/P3*/P2*P1;
H0  = EN*/P14*/P13*/P12*/P11*/P10*/P9*/P8*/P7*/P6*/P5*/P4*/P3*/P2*/P1*P0;

equations

Y3 = H8 + H9 + H10 + H11 + H12 + H13 + H14;
Y2 = H4 + H5 + H6 + H7 + H12 + H13 + H14;
Y1 = H2 + H3 + H6 + H7 + H10 + H11 + H14;
Y0 = H1 + H3 + H5 + H7 + H9 + H11 + H13;

GS    = EN*(P14+P13+P12+P11+P10+P9+P8+P7+P6+P5+P4+P3+P2+P1+P0);
ENOUT = EN*/P14*/P13*/P12*/P11*/P10*/P9*/P8*/P7*/P6*/P5*/P4*/P3*/P2*/P1*/P0;

end Priority_Encoder_15
```

Figure 7–32
Fixed-point to floating-point
encoder using a single PLD.

We're stuck—the fixed-point to floating-point encoder cannot be built with a single PAL20L8. However, the equations in Table 7–16 could be used in another PLD with eight or more product terms, such as the PAL20R8 discussed in Section 7.3.

* 7.2.11 Three-State Output Control

The combinational-PLD applications that we've shown so far have used the bidirectional I/O pins (IO2–IO7 on a PAL16L8) statically, that is, always output-enabled or always output-disabled. In such applications, the compiler can take care of programming the output-enable gates appropriately—all fuses blown, or all fuses intact. By default in ABEL, a three-state output pin is programmed to be always enabled if its signal name appears on the left-hand side of an equation, and always disabled otherwise. However, the direction of I/O pins can also be changed dynamically, since the output-enable gate can be programmed to be a nontrivial product of the PLD inputs. An example of such an application is a four-way, 2-bit bus transceiver with the following specifications:

- The transceiver handles four 2-bit bidirectional buses, A[1-2], B[1-2], C[1-2], and D[1-2].

- Each bus has its own output-enable signal, /AOE, /BOE, /COE, or /DOE. There is also a "master" output-enable signal, /MOE. The transceiver drives a particular bus if and only if /MOE and the output-enable signal for that bus are both asserted.

Table 7-16 An ABEL program for the fixed-point to floating-point PLD.

```
module Fixed_to_Floating_Encoder
title 'Fixed-point to Floating-point Encoder
J. Wakerly, Xanthrax Systems'
FPENC device 'P20L8';
@ALTERNATE

" Input pins
B0, B1, B2, B3, B4, B5        pin 1, 2, 3, 4, 5, 6;
B6, B7, B8, B9, B10           pin 7, 8, 9, 10, 11;

" Output pins
/E0, /E1, /E2                 pin 20, 21, 22;
/M0, /M1, /M2, /M3            pin 15, 16, 17, 18;

" Constant expressions
S7 = B10;
S6 = /B10 * B9;
S5 = /B10 * /B9 * B8;
S4 = /B10 * /B9 * /B8 * B7;
S3 = /B10 * /B9 * /B8 * /B7 * B6;
S2 = /B10 * /B9 * /B8 * /B7 * /B6 * B5;
S1 = /B10 * /B9 * /B8 * /B7 * /B6 * /B5 * B4;
S0 = /B10 * /B9 * /B8 * /B7 * /B6 * /B5 * /B4;

equations

E2 = S7 + S6 + S5 + S4;
E1 = S7 + S6 + S3 + S2;
E0 = S7 + S5 + S3 + S1;

M3 = S7*B10 + S6*B9 + S5*B8 + S4*B7 + S3*B6 + S2*B5 + S1*B4 + S0*B3;
M2 = S7*B9  + S6*B8 + S5*B7 + S4*B6 + S3*B5 + S2*B4 + S1*B3 + S0*B2;
M1 = S7*B8  + S6*B7 + S5*B6 + S4*B5 + S3*B4 + S2*B3 + S1*B2 + S0*B1;
M0 = S7*B7  + S6*B6 + S5*B5 + S4*B4 + S3*B3 + S2*B2 + S1*B1 + S0*B0;

end Fixed_to_Floating_Encoder
```

- The source of data to drive the buses is selected by three select inputs, $S[2-0]$, according to Table 7–17. If S2 is 0, the buses are driven with a constant value, otherwise they are driven with one of the other buses. However, when the selected source is a bus, that bus is driven with 00.

Figure 7–33 is a logic diagram for a PAL16L8 with the required inputs and outputs. Since the device has only six bidirectional pins and the specification requires eight, the A bus uses one pair of pins for input and another for output.

Table 7–17
Bus selection codes for a four-way bus transceiver.

S2	S1	S0	Source selected
0	0	0	00
0	0	1	01
0	1	0	10
0	1	1	11
1	0	0	A bus
1	0	1	B bus
1	1	0	C bus
1	1	1	D bus

ENABLE

Table 7–18 is an ABEL program that performs the transceiver function. The **ENABLE** keyword indicates that the following equation applies to the output-enable gate, so each bus is output-enabled if MOE and its own OE are asserted. Each bus is driven with S1 and S0 if S2 is 0, and with the selected bus if a different bus is selected. If the bus itself is selected, the output equation evaluates to 0, and the bus is driven with 00 as required.

* 7.2.12   Latches

We showed in Section 5.2 that basic sequential-circuit building blocks such as latches and flip-flops are normally built using ordinary combinational logic gates and feedback. For example, the excitation equation for an S-R latch is

$$Q* \ = \ S + R' \cdot Q$$

Thus, we could build an S-R latch using one combinational output of a PLD such as the PAL16L8. Furthermore, the S and R signals above could be replaced

Figure 7–33
PLD inputs and outputs for a four-way, 2-bit bus transceiver.

Table 7–18   An ABEL program for four-way, 2-bit bus transceiver.

```
module Four_Way_Transceiver
title 'Four-way 2-bit Bus Transceiver
J. Wakerly, BNR, Inc.'
XCVR4X2 device 'P16L8';
@ALTERNATE

" Input pins
A1I, A2I                        pin 1, 11;
/AOE, /BOE, /COE, /DOE, /MOE    pin 2, 3, 4, 5, 6;
S0, S1, S2                      pin 7, 8, 9;

" Output and bidirectional pins
A1O, A2O                        pin 19, 12;
B1, B2, C1, C2, D1, D2          pin 18, 17, 16, 15, 14, 13;

" Set definitions
ABUSO = [A1O,A2O];
ABUSI = [A1I,A2I];
BBUS  = [B1,B2];
CBUS  = [C1,C2];
DBUS  = [D1,D2];
SEL   = [S2,S1,S0];
CONST = [S1,S0];

" Constants
SELA = [1,0,0];
SELB = [1,0,1];
SELC = [1,1,0];
SELD = [1,1,1];

equations

ENABLE ABUSO = AOE * MOE;
ABUSO = /S2*CONST + (SEL==SELB)*BBUS + (SEL==SELC)*CBUS + (SEL==SELD)*DBUS;

ENABLE BBUS = BOE * MOE;
BBUS = /S2*CONST + (SEL==SELA)*ABUSI + (SEL==SELC)*CBUS + (SEL==SELD)*DBUS;

ENABLE CBUS = COE * MOE;
CBUS = /S2*CONST + (SEL==SELA)*ABUSI + (SEL==SELB)*BBUS + (SEL==SELD)*DBUS;

ENABLE DBUS = DOE * MOE;
DBUS = /S2*CONST + (SEL==SELA)*ABUSI + (SEL==SELB)*BBUS + (SEL==SELC)*CBUS;

end Four_Way_Transceiver
```

with more complex logic functions of the PLD's inputs, limited only by the availability of product terms (seven per output in the PAL16L8) to realize the final excitation equation. Note, however, that the Q output must be continuously enabled; otherwise, the feedback loop would be broken and the latch's state lost. Also, the feedback loop can be created only when Q is assigned to a bidirectional pin, one of IO2–IO7 in the PAL16L8.

Probably the handiest sequential circuit to build out of a combinational PLD is a D latch. The basic excitation equation for a D latch is

$$Q* \; = \; C \cdot D + C' \cdot Q$$

However, we showed in Section 5.6.1 that this equation contains a static hazard, and the corresponding circuit does not latch data reliably. To build a reliable D latch, we must include a consensus term in the excitation equation:

$$Q* \; = \; C \cdot D + C' \cdot Q + D \cdot Q$$

The D input in this equation may be replaced with a more complicated expression, but the equation's structure remains the same:

$$Q* \; = \; C \cdot expression + C' \cdot Q + expression \cdot Q$$

It is also possible to use a more complex expression for the C input, as we showed in Section 5.6.1.

Probably the most common use of a PLD-based latch is to simultaneously decode and latch addresses in microprocessor systems. For example, expanding the memory-decoder example of Table 7–9, suppose we have another input ADDRVALID that indicates that the address is valid and the decoder output should be latched. We can combine the decoding and latching functions in one output with the following equation:

```
ROMCS = ADDRVALID *
        (LARGE * A23 * A22 * A21 * A20 * A19 * A18 * A17 * A16
         + /LARGE * A19 * A18 * A17 * A16 + RTEST)
    + /ADDRVALID * ROMCS
    + ROMCS *
        (LARGE * A23 * A22 * A21 * A20 * A19 * A18 * A17 * A16
         + /LARGE * A19 * A18 * A17 * A16 + RTEST);
```

The equation expands into seven product terms, and thus fits on a PAL16L8 output.

After seeing how easy it is to build S-R and D latches using combinational PLDs, you might be tempted to go further and try to build an edge-triggered D flip-flop. Although this is possible, it is expensive because an edge-triggered flip-flop has four internal states and thus two feedback loops, consuming two PLD outputs. Furthermore, the setup and hold times and propagation delays of such a flip-flop would be quite poor compared to those of a discrete flip-flop in the same technology. Finally, as we discussed in Section 5.6.6, the flow tables of all edge-triggered flip-flops contain essential hazards, which may be difficult to mask in a PLD-based design.

7.3 SEQUENTIAL PLDS

Most digital designs require flip-flops and, as a result, most PLDs contain flip-flops. Such PLDs are often called *registered PLDs*. The word "register" is used because, like the MSI registers in Section 6.2.3, most registered PLDs apply a single, common clock signal to all of their flip-flops.

registered PLDs

    We'll describe a few standard registered PLDs in this section. Registered PLDs are often used to build special-purpose versions of common sequential functions, such as shift registers and counters. They can also be used to reduce the chip count of a circuit by adding storage to a combinational MSI function, for example, by combining a decoder or multiplexer with a register. A few PLD applications of this sort appear later in this section. Registered PLDs can also be used to build "random" sequential circuits, the kind best described as state machines; this is the subject of Section 7.4.

7.3.1 Standard Registered PLDs

Figure 7–34 shows the structure of a *PAL16R6* registered PLD. This device has eight primary inputs, eight outputs, and common clock and output-enable inputs, and fits in a 20-pin package.

PAL16R6

    The PAL16R6's AND-OR array is exactly the same as the one found in the PAL16L8 combinational PLD. However, the PAL16R6 has edge-triggered D flip-flops on six of its outputs, O2–O7. All of the flip-flops are connected to a common clock input, CLK, and change state on the rising edge of the clock. Each flip-flop drives an output pin through a 3-state buffer; the buffers have a common output-enable signal, /OE. Notice that, like the combinational output pins, the registered output pins contain the complement of the signal produced by the AND-OR array. The IO1 and IO8 pins of the PAL16R6 are bidirectional, serving both as inputs and as combinational outputs with separate 3-state enables, just like the PAL16L8's bidirectional pins.

    The possible inputs to the PAL16R6's AND-OR array are the eight primary inputs (I1–I8), the six D flip-flop outputs, and the two bidirectional pins (IO1, IO8). The connection from the D flip-flop outputs into the AND-OR array makes it easy to design shift registers, counters, and general state machines. Unlike the PAL16L8's combinational outputs, the PAL16R6's D flip-flop outputs are

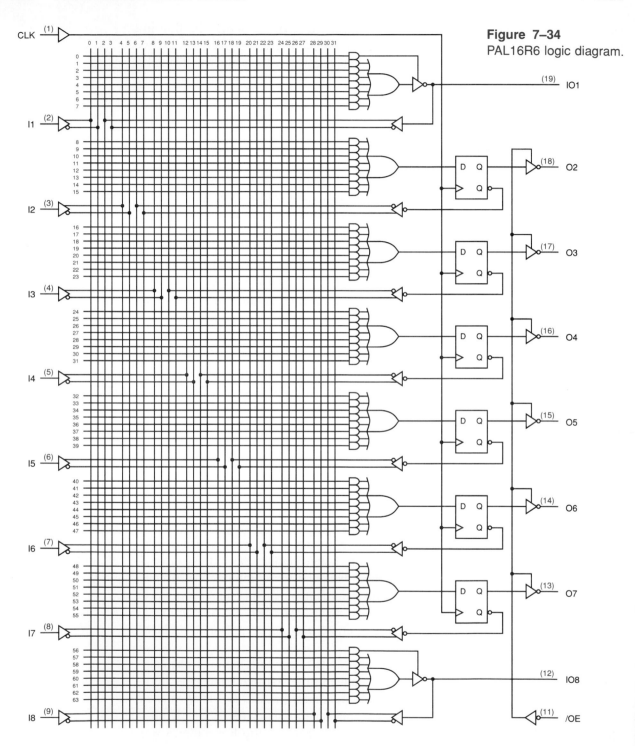

Figure 7–34
PAL16R6 logic diagram.

Table 7–19 Characteristics of common PLDs.

Part number	Package pins	AND-gate inputs	Inputs to AND array			
			Primary inputs	Bidirectional comb. outputs	Registered outputs	Combinational outputs
PAL16L8	20	16	10	6	0	2
PAL16R4	20	16	8	4	4	0
PAL16R6	20	16	8	2	6	0
PAL16R8	20	16	8	0	8	0
PAL20L8	24	20	14	6	0	2
PAL20R4	24	20	12	4	4	0
PAL20R6	24	20	12	2	6	0
PAL20R8	24	20	12	0	8	0

available to the AND-OR array whether or not the O2–O7 three-state drivers are enabled. Thus, the internal flip-flops can go to a next state that is a function of the current state even when the O2–O7 outputs are disabled.

As shown in Table 7–19, there are several PLDs similar to the PAL16R6 with differing numbers and types of inputs and outputs. All of the PAL16xx parts in the table use the same AND-OR array, where each output has eight AND gates, each with 16 variables and their complements as possible inputs. The PAL20xx parts use a similar AND-OR array with 20 variables and their complements as possible inputs. Figure 7–35 gives logic symbols for all of the PLDs in the table.

PAL16R4
PAL16R8
PAL20L8
PAL20R4
PAL20R6
PAL20R8

7.3.2 PLD Timing Specifications

Several timing parameters are associated with registered PLDs, but only three are critical for most applications:

t_{PD} This parameter applies to combinational outputs. It is the propagation delay from a primary, bidirectional, or "feedback" input to the combinational output. A *feedback input* is the output of an internal flip-flop that gets fed back into the AND-OR array.

t_{PD}

feedback input

t_{CLK} This parameter applies to registered outputs. It is the propagation delay from the rising edge of CLK to a primary output or feedback input.

t_{CLK}

t_{SU} This parameter applies to primary, bidirectional, and feedback inputs that affect the D inputs of flip-flops. It is the setup-time requirement for the input signal to be stable before the rising edge of CLK.

t_{SU}

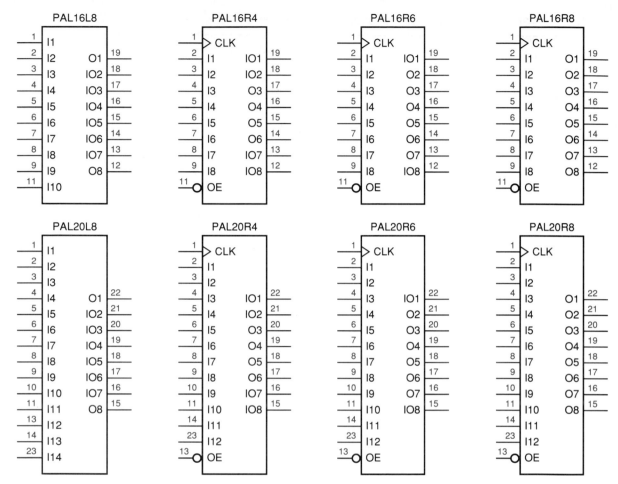

Figure 7–35 Traditional logic symbols for standard combinational and registered PLDs.

Each of the PLDs in Table 7–19 is available in several different speed varieties. However, within a given speed variety, all of the PLDs in the table have the same timing specs. For example, "A-series" PLDs (PAL16L8A, PAL20R4A, etc.) have $t_{PD} = 25$, $t_{CLK} = 15$, and $t_{SU} = 25$ (all worst-case numbers, in nanoseconds). Newer "D-series" PLDs have $t_{PD} = 10$, $t_{CLK} = 8$, and $t_{SU} = 10$.

When registered PLDs are used in applications with critical timing, it's important to remember that they have longer setup times than discrete edge-triggered registers in the same technology, owing to the delay of the AND-OR array on each D input. Conversely, they have shorter hold times, typically *minus* 6 to 15 ns. However, you can't count on having a negative hold time—the worst-case spec is zero.

Figure 7–36
PLD connections for a 4-bit,
2-input multiplexer with storage;
connections for increased
functionality are shown in color.

7.3.3 Registered-PLD Applications

In this subsection we'll describe a few non-state-machine applications of registered PLDs. Instead of an = sign, ABEL equations for registered PLD outputs use the *clocked assignment operator*, :=. Although the compiler already knows from the **device** declaration which outputs are registered and which are not, it makes the designer distinguish between the two types anyway, perhaps as a sanity check.

clocked assignment operator, :=

A simple application of a registered PLD is the combination of a multiplexer and a register. Figure 7–36 shows how a PAL16R4 could be used as a 4-bit wide, 2-input multiplexer with a storage register. At each clock tick, the A or B bus is loaded into the register, as selected by SELB. Table 7–20 is an ABEL program that provides the required behavior. Note that the CLK and /OE pin definitions are not used by the compiler to generate the fuse map; they merely serve to make the documentation complete.

In this example, one PAL16R4 replaces two MSI chips (e.g., a 74LS157 and a 74LS374), not much of a savings. However, once we have a PLD-based design, we can easily modify it or add to its functionality. For example, the PAL16R4 in Figure 7–36 has three spare I/O pins. As shown in color, we could use two inputs pins to provide "hold" and "preset" functions, and one output pin to provide a "zero-detect" function. The modified ABEL program requires only the appropriate pin definitions and the following equations:

```
YBUS := /HOLD * (/SELB * ABUS + SELB * BBUS) + PRESET;
ZERO = (YBUS == 0);
```

Table 7–20   Program for a 4-bit, 2-input multiplexer with storage.

```
module Mux_with_Storage
title '4-Bit, 2-Input Multiplexer with Storage
J. Wakerly, Marquette University'
MUX42R device 'P16R4';
@ALTERNATE

" Input pins
CLK, /OE                              pin 1, 11;
A0, A1, A2, A3, B0, B1, B2, B3        pin 2, 3, 4, 5, 6, 7, 8, 9;
SELB                                  pin 12;

" Output pins
Y0, Y1, Y2, Y3                        pin 17, 16, 15, 14;

" Set definitions
ABUS = [A0,A1,A2,A3];
BBUS = [B0,B1,B2,B3];
YBUS = [Y0,Y1,Y2,Y3];

equations

YBUS := /SELB * ABUS + SELB * BBUS;

end Mux_with_Storage
```

Shift registers are very easy to build with registered PLDs. For example, Table 7–21 is the ABEL program for an 8-bit bidirectional shift register similar to the 74LS194 4-bit part. It illustrates the use of "self-documenting" definitions; the four functions of the device are given names (HOLD–LOAD), the corresponding select-input combinations are listed, and the output equations are written in terms of the names and in a way that displays exactly what's going on.

In Section 6.4.5, we gave many examples of counter circuits based on shift registers. In a PLD design, shift register operation may be combined with the combinational logic required to achieve the desired counting behavior. For example, Table 7–22 is the program for an 8-bit ring counter. We've used the extra capability of the PLD to add two functions not present in our MSI designs: counting occurs only if CNTEN is asserted, and the next state is forced to S0 if RESTART is asserted.

Ring counters are often used to generate multiphase clocks or enable signals in digital systems, and the requirements in different systems are many and varied. The ability to reprogram the counter's behavior easily is a distinct advantage of a PLD-based design. Figure 7–37 shows a set of enable signals that might be required in a digital system with six distinct phases of operation. Each phase lasts

Table 7–21 Program for 8-bit bidirectional shift register.

```
module Eight_Bit_Shift_Register
title '8-bit Bidirectional Shift Register
J. Wakerly, Brother Rice High School'
SREG8 device 'P20R8';
@ALTERNATE

" Input pins
CLK, /OE                          pin 1, 13;
A, B, C, D, E, F, G, H            pin 9, 8, 7, 6, 5, 4, 3, 2;
S0, S1, LIN, RIN                  pin 10, 11, 23, 14;

" Output pins
QA, QB, QC, QD, QE, QF, QG, QH    pin 22, 21, 20, 19, 18, 17, 16, 15;

" Set definitions
DIN    = [A,B,C,D,E,F,G,H];
QOUT   = [QA,QB,QC,QD,QE,QF,QG,QH];
SEL    = [S1,S0];

" Control combinations
HOLD   = (SEL == [0,0]);
RIGHT  = (SEL == [0,1]);
LEFT   = (SEL == [1,0]);
LOAD   = (SEL == [1,1]);

equations

QOUT := HOLD   * QOUT
      + RIGHT  * [RIN,QA,QB,QC,QD,QE,QF,QG]
      + LEFT   * [QB,QC,QD,QE,QF,QG,QH,LIN]
      + LOAD   * DIN;

end Eight_Bit_Shift_Register
```

for two ticks of a master clock signal, MCLK, during which the corresponding active-low phase-enable signal /Pi is asserted. We can obtain this sort of timing from a ring counter if we provide an extra flip-flop to count the two ticks of each phase, so that a shift occurs on the *second* tick of each phase.

The timing generator can be built with a single PAL16R8 with the connections shown in Figure 7–38. The three control inputs, whose active levels can be chosen arbitrarily, are required to have the following behavior:

/RESET When this input is asserted, no outputs are asserted. The counter always starts in phase 1 after /RESET is negated.

Table 7–22 Program for 8-bit ring counter.

```
module Eight_Bit_Ring_Counter
title '8-bit Ring Counter
J. Wakerly, Pizza Roma'
RING8 device 'P16R8';
@ALTERNATE

" Input pins
MCLK, /OE                        pin 1, 11;
CNTEN, RESTART                   pin 2, 3;

" Output pins
S0, S1, S2, S3, S4, S5, S6, S7   pin 19, 18, 17, 16, 15, 14, 13, 12;

equations

S0 := CNTEN * /S0 * /S1 * /S2 * /S3 * /S4 * /S5 * /S6 + /CNTEN * S0 + RESTART;
S1 := CNTEN * /RESTART * S0 + /CNTEN * /RESTART * S1;
S2 := CNTEN * /RESTART * S1 + /CNTEN * /RESTART * S2;
S3 := CNTEN * /RESTART * S2 + /CNTEN * /RESTART * S3;
S4 := CNTEN * /RESTART * S3 + /CNTEN * /RESTART * S4;
S5 := CNTEN * /RESTART * S4 + /CNTEN * /RESTART * S5;
S6 := CNTEN * /RESTART * S5 + /CNTEN * /RESTART * S6;
S7 := CNTEN * /RESTART * S6 + /CNTEN * /RESTART * S7;

end Eight_Bit_Ring_Counter
```

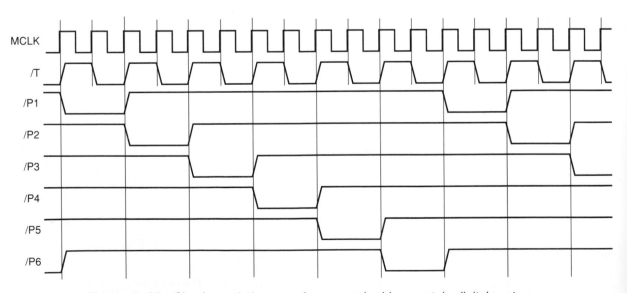

Figure 7–37 Six-phase timing waveforms required in a certain digital system.

Table 7–23 Program for a six-phase timing generator.

```
module Timing_Generator_6
title 'Six-phase Master Timing Generator
J. Wakerly, Unemployed'
TIMEGEN6 device 'P16R8';
@ALTERNATE

" Input pins
CLK, /OE                        pin 1, 11;
/RESET, RUN, /RESTART           pin 2, 3, 4;

" Output pins
/P1, /P2, /P3, /P4, /P5, /P6    pin 19, 18, 17, 16, 15, 14;
/T                              pin 13; " not used externally

equations

T   := /RESET * /RESTART * RUN * /T       " Set only for second tick of a phase
    +  /RESET * /RESTART * /RUN * T;      " Extend second tick when halted

P1  := /RESET * /P1 * /P2 * /P3 * /P4 * /P5 * /P6    " Start-up after reset
    +  /RESET * P1 * /T              " Hold after P1's first tick
    +  /RESET * P1 * T * /RUN        " Hold after second tick if not running
    +  /RESET * P6 * T * RUN         " Wrap-around after phase 6
    +  /RESET * RESTART;             " Assert P1 on restart

P2  := /RESET * /RESTART * P1 * T * RUN   " Assert P2 after P1's second tick
    +  /RESET * /RESTART * P2 * /T         " Hold after P2's first tick
    +  /RESET * /RESTART * P2 * T * /RUN;  " Hold after second tick if not running

P3  := /RESET * /RESTART * P2 * T * RUN   " Assert P3 after P2's second tick
    +  /RESET * /RESTART * P3 * /T         " Hold after P3's first tick
    +  /RESET * /RESTART * P3 * T * /RUN;  " Hold after second tick if not running

P4  := /RESET * /RESTART * P3 * T * RUN   " Assert P4 after P3's second tick
    +  /RESET * /RESTART * P4 * /T         " Hold after P4's first tick
    +  /RESET * /RESTART * P4 * T * /RUN;  " Hold after second tick if not running

P5  := /RESET * /RESTART * P4 * T * RUN   " Assert P5 after P4's second tick
    +  /RESET * /RESTART * P5 * /T         " Hold after P5's first tick
    +  /RESET * /RESTART * P5 * T * /RUN;  " Hold after second tick if not running

P6  := /RESET * /RESTART * P5 * T * RUN   " Assert P6 after P5's second tick
    +  /RESET * /RESTART * P6 * /T         " Hold after P6's first tick
    +  /RESET * /RESTART * P6 * T * /RUN;  " Hold after second tick if not running

end Timing_Generator_6
```

Figure 7–38
Using a PAL16R8 as a six-phase
timing generator.

RUN   This input must be asserted to allow the counter to advance to the
         next phase after the second tick of a phase; otherwise, the current
         phase is extended.

/RESTART   Asserting this input causes the counter to go back to the first tick of
                  phase 1, even if RUN is not asserted.

Table 7–23 is a program that creates the required behavior. The equations are
somewhat complicated by the requirements for the /RESET, RUN, and /RESTART
inputs. Some of the requirements are ambiguous; for example, if RUN is negated
during the first tick of a phase, but asserted during the second, is the phase
extended by one tick? You have to decipher the equations to figure out what
the PLD does in this case. Still, you're better off than with "random logic"—
you can easily change the PLD equations if you don't like their behavior (see
Exercises 7.28 and 7.29). Going a step further, you can use a state-machine
description language (Section 7.4) for an even more understandable design.

* 7.3.4   Designing Binary Counters with Registered PLDs

Binary counters are good candidates for PLD-based design, for many reasons:

- A large state machine can often be decomposed into two or more smaller
  state machines where one of the smaller machines is a binary counter that
  keeps track of how long the other machine should stay in a particular state.
  This may simplify both the conceptual design and the circuit design of the
  machine.

- Many applications require almost-binary-modulus counters with special requirements for initialization, state detection, or state skipping. For example,
  a counter in an elevator controller may skip state 13. Instead of using an

off-the-shelf binary counter and extra logic for the special requirements, a designer can put precisely the required functions in a PLD.

- Most standard MSI counters have only 4 bits, while a single 24-pin PLD can be used to create a binary counter with up to 10 bits.

Like most MSI counters, a PLD-based counter is a *synchronous* counter, that is, all of its output bits change at the rising edge of the clock. The basic logic function for one output bit of a synchronous binary up counter is, in words, "Change state if counting is enabled and all the lower-order bits are 1." (To get a binary *down* counter, just change 1 to 0 in this statement.) A corresponding ABEL equation is written below for bit 4 of a synchronous binary up counter:

```
Q4  := /Q4 * (CNTEN * Q3 * Q2 * Q1 * Q0)     " Complement
     +  Q4 * /(CNTEN * Q3 * Q2 * Q1 * Q0); " Preserve
```

The first line complements Q4 if all of the conditions for changing state are true. Otherwise, the second line preserves the state of Q4. This equation requires six product terms to realize in any of the registered PLDs that we've studied so far, assuming that /Q4 is assigned to a registered output pin. Using ABEL's XOR operator, : + :, we can write the equation in just one line:

```
Q4  := Q4 :+: (CNTEN * Q3 * Q2 * Q1 * Q0);  " Toggle if enabled
```

However, after expansion, the equation still requires a total of six product terms.
Other counter features, like synchronous loading and clearing, can be added quite easily to the preceding equation:

```
Q4  := /CLR * /LD * (Q4 :+: (CNTEN * Q3 * Q2 * Q1 * Q0)) "Toggle
     + /CLR * LD * D4;  " Load
```

The expanded and reduced equation has eight product terms, which just fits on a registered output of a PAL16Rx or PAL20Rx. Higher-order counter bits (Q5, Q6, ...) require even more product terms, and therefore do not fit in the PLDs that we've discussed so far. So, before we look at other modifications to the basic counter equations, we'll introduce a series of PLDs that were specifically designed to reduce the product-term requirements of counter logic.

* 7.3.5 Counter Design with X-Series PLDs

The "X series" of PLDs uses XOR gates to facilitate the design of binary counters. This series includes the *PAL20X8*, whose logic diagram is shown in Figure 7–39, and three similar devices. The other devices have a different allocation of registered and combinational outputs, as shown in Table 7–24 and Figure 7–40. The *PAL20L10* has *no* registered outputs or XOR gates, but is included in the series because it has the same AND-OR array as the other devices.

PAL20X8
PAL20X4
PAL20X10
PAL20L10

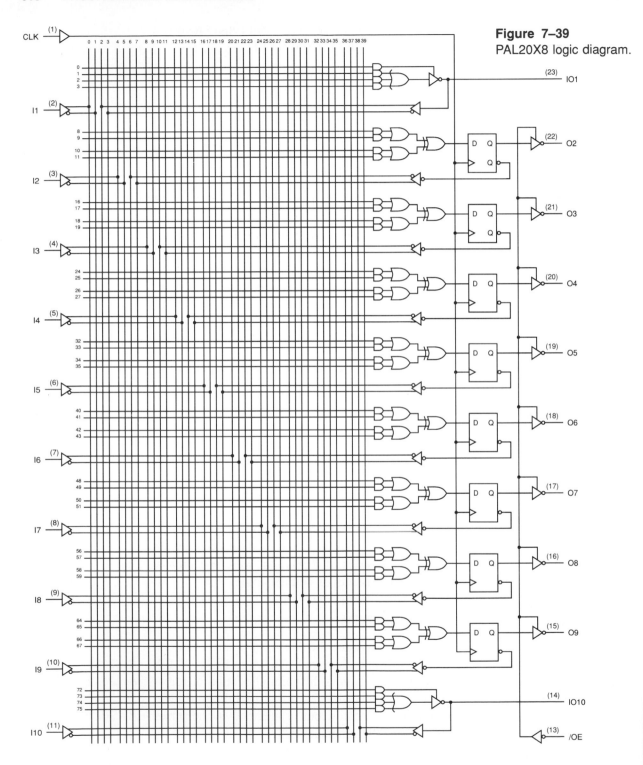

Figure 7–39
PAL20X8 logic diagram.

Table 7–24 Characteristics of "X-series" PLDs.

Part number	Package pins	AND-gate inputs	Inputs to AND array			
			Primary inputs	Bidirectional comb. outputs	Registered outputs	Combinational outputs
PAL20L10	24	20	12	8	0	2
PAL20X4	24	20	10	6	4	0
PAL20X8	24	20	10	2	8	0
PAL20X10	24	20	10	0	10	0

Each X-series output has only four product terms available, which often leaves designers starving for more. The combinational outputs use one of the four terms to control the three-state enable, while the registered outputs perform the XOR of two ORed pairs of product terms. In ABEL, the generic equation for a registered output is

```
OUT := (P1 + P2) :+: (P3 + P4);
```

where P1–P4 are the four product terms. With a little factoring, the equation for output Qi of a binary synchronous up counter maps nicely into this structure:

```
Qi := ( (/CLR * /LD * Qi)                        " Hold
        + 0 )    :+:                             " Unused
      ( (/CLR * /LD * CNTEN * Qi-1 * ... * Q0)   " Toggle
        + (/CLR * LD * Di) );                    " Load
```

Figure 7–40 Traditional logic symbols for X-series PLDs.

The same basic equation works for high-order bits as well as low-order ones, with no increase in the number of product terms. In fact, one product term is not even needed (the explicit 0 on the second line).

The Qi signal in the preceding equation is active high at the flip-flop input and output, and therefore active low at the device pin. To obtain an active-high pin output, the flip-flop output must be active-low, /Qi, so the equations must be modified:

```
/Qi := ( (/PR * /LD * /Qi)                        " Hold
         + 0 )    :+:                              " Unused
       ( (/PR * /LD * CNTEN * Qi-1 * ... * Q0)     " Toggle
         + (/PR * LD * /Di) )                      " Load
```

Note that the functionality of what was the CLR input is now different, since clearing the flip-flop *presets* the active-high output pin to 1. The unused product term can be employed to obtain a true CLR function (see Exercise 7.33). The equations for down counters are identical, except that the toggle term must detect the state /Qi-1*...*/Q0.

If we try to add more features, such as multiple load and clear inputs and nonbinary modulus counting, we quickly run out of product terms in an X-series PLD. Also, PLD compilers aren't smart enough to rearrange an arbitrary expression into the form required by the X-series structure. Let's look at some modifications that *are* possible using at most one extra product term.

In order to obtain a counter that counts from 0 to state N and then returns to state 0 and repeats, we must detect state N and force the counter to return to state 0 on the next count. Suppose that STATEN is a product term that detects state N. Then the expression STATEN·CNTEN is 1 when the counter should be forced into state 0 (unless the /LD input overrides). We can use a combinational output of the PLD to realize this expression:

```
FORCE0 = STATEN * CNTEN;
```

Using FORCE0 as a "helper" term, we can now easily obtain the desired behavior:

```
Qi := ( (/CLR * /LD * Qi * /FORCE0 )              " Hold
        + 0 )    :+:                              " Unused
      ( (/CLR * /LD * CNTEN * Qi-1 *...* Q0 * /FORCE0) " Toggle
        + (/CLR * LD * Di) );                     " Load
```

If we modified the equation above to use STATEN·CNTEN directly instead of FORCE0, we would run out of product terms—STATEN, a product term, blows up into a sum of many terms when complemented.

Another approach to forcing state 0 does not use a combinational "helper" output, but instead uses at most one extra product term for each Qi output, depending on i and N:

(1) If Qi would be 0 in state $N + 1$, do not modify the equation for Qi.

(2) If Qi would normally change from 0 to 1 between states N and $N + 1$, add a product term to cancel this change:

```
Qi := ( (/CLR * /LD * Qi)                    " Hold
        + CNTEN * STATEN )   :+:              " Cancel
      ( (/CLR * /LD * CNTEN * Qi-1 * ... * Q0) " Toggle
        + (/CLR * LD * Di) )                  " Load
```

(3) If Qi would be 1 in both states N and $N + 1$, add a product term to toggle it to 0:

```
Qi := ( (/CLR * /LD * Qi)                    " Hold
        + (/CLR * LD * Di) ) :+:             " Load
      ( (/CLR * /LD * CNTEN * Qi-1 * ... * Q0) " Toggle
        + CNTEN * STATEN )                   " Extra
```

These rules can be extended to force the counter to any desired starting state, M. If Qi has the same value in states M and $N + 1$, do nothing. If the values are different, add a "cancel" term if Qi has different values in states N and $N + 1$, and an "extra" toggle term otherwise.

In general, if you get stuck with an equation that has more than two product terms on either side of the XOR, you can't perform it in a X-series PLD, unless you use an extra combinational output and two-pass logic. However, sometimes our simple rule of thumb about XOR gates can help:

- Any two signals (inputs or output) of an XOR or XNOR gate may be complemented without changing the resulting logic function.

That is, you can complement both inputs or the output and one input without changing the result. For example, suppose an equation has the form

```
OUT := S1 :+: ((P1) + (P2));
```

where S1 is a sum term and P1 and P2 are product terms. If S1 has more than two variables, the equation can't be realized in one level. However, if you can live with a complemented output signal, you can write

```
/OUT := /S1 :+: ((P1) + (P2));
```

Under DeMorgan's theorem, /S1 is a single product term, so there's no problem.

Another trick is to remember that the XOR gate is equivalent to an OR gate *if* the logic functions on the two inputs are never both 1 simultaneously. So, if you need to perform a logic function that is a sum of four product terms, having nothing to do with XOR, you can still use the X-series output structure, if you can just split the terms into two nonoverlapping pairs.

* 7.3.6 Cascading and Carry Outputs

PLDs can be cascaded to obtain wider counters, by providing each lower-order counter with a carry output that indicates when it is about to roll over. There are two ways to obtain the carry output:

combinational carry output

- *Combinational.* The carry equation indicates that the counter is enabled and currently in its last state before rollover. For a 5-bit binary up counter, we have

```
COUT = CNTEN * Q4 * Q3 * Q2 * Q1 * Q0;
```

  Since **CNTEN** is included, this approach allows carries to be rippled through cascaded counters by connecting each **COUT** to the next **CNTEN**.

registered carry output

- *Registered.* The carry equation indicates that the counter is about to enter its last state before rollover. For a 5-bit binary up counter with load and clear inputs, we have

```
COUT := /CLR * /LD * CNTEN
            * Q4 * Q3 * Q2 * Q1 * /Q0
      + /CLR * /LD * /CNTEN
            * Q4 * Q3 * Q2 * Q1 * Q0
      + /CLR * LD
            * D4 * D3 * D2 * D1 * D0;
```

  This approach has the advantage of producing **COUT** with less delay than the combinational approach. A disadvantage is that external gates may be required to ripple the carries (unless the higher-order counters have multiple enable inputs).

* 7.3.7 X-Series Design Example

Our next PLD design example is based on a variation of the timing waveforms of Figure 7–37 in which each phase signal is asserted for only one clock tick instead of two, as shown in Figure 7–41. Although this is just a small modification to the waveforms, it requires a much different PLD design. In the original design, we effectively used an output-coded state assignment, so that the next state could always be determined from the current outputs and an auxiliary variable T. This is not possible with the new waveforms, since there are six states (corresponding to counts 0, 2, ..., 10 in the figure) that all produce the same output.

A simple approach to generating the new waveforms recognizes that they repeat every 12 clock ticks, and uses a modulo-12 counter and additional logic to assert the proper /Pi output as a function of the counter state indicated in the last line of the figure. Thus, four registered outputs are required for a modulo-12 counter, and six more outputs are needed for /P1–/P6.

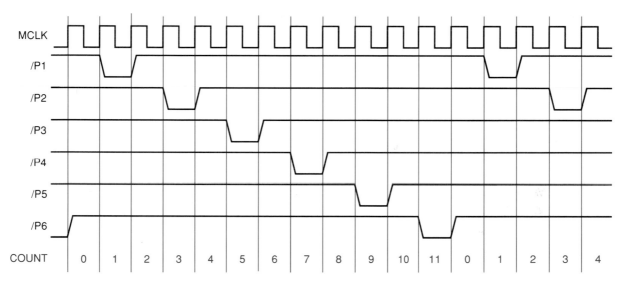

Figure 7–41 Modified timing waveforms for a digital system.

The modulo-12 counter can be designed according to the rules of the preceding subsection using an X-series PLD. The PAL20X4 is an obvious choice, since it has four registered outputs for the counter and also has six combinational outputs for /P1–/P6; Figure 7–42 shows the required connections. The ABEL program in Table 7–25 produces the required behavior, and includes /RESET, RUN, and /RESTART inputs similar to the ones in our original six-phase timing generator of Table 7–23.

Since /P1–/P6 are decoded from counter states, they may contain glitches at certain counter transitions, as discussed in Section 6.3.4. Alternatively, we can

Figure 7–42
PLD connections for modified
six-phase timing generator.

Table 7–25 Program for modified six-phase timing generator.

```
module Pulse_Generator_6
title 'Six-phase Master Timing Generator with 1-Tick Pulses'
PULSGEN6 device 'P20X4';
@ALTERNATE

" Input pins
MCLK, /OE                      pin 1, 13;
/RESET, RUN, /RESTART          pin 2, 3, 4;

" Output pins
/P1, /P2, /P3, /P4, /P5, /P6   pin 23, 22, 21, 16, 15, 14;
/Q3, /Q2, /Q1, /Q0             pin 20, 19, 18, 17;

" Sets
QSTATE = [Q3,Q2,Q1,Q0];

equations

Q0 := (/RESET * /RESTART * Q0) :+:          " Hold
      (/RESET * /RESTART * RUN);             " Toggle

Q1 := (/RESET * /RESTART * Q1) :+:          " Hold
      (/RESET * /RESTART * RUN * Q0);        " Toggle

Q2 := (/RESET * /RESTART * Q2 +             " Hold
       /RESET * /RESTART * RUN * (QSTATE==11)) :+:  " Cancel
      (/RESET * /RESTART * RUN * Q1 * Q0);   " Toggle

Q3 := (/RESET * /RESTART * Q3) :+:          " Hold
      (/RESET * /RESTART * RUN * Q2 * Q1 * Q0 +   " Toggle
       /RESET * /RESTART * RUN * (QSTATE==11));    " Extra

P1 = /RESET * (QSTATE == 1);

P2 = /RESET * (QSTATE == 3);

P3 = /RESET * (QSTATE == 5);

P4 = /RESET * (QSTATE == 7);

P5 = /RESET * (QSTATE == 9);

P6 = /RESET * (QSTATE == 11);

end Pulse_Generator_6
```

use the *registered* outputs of a PAL20X10 to produce /P1–/P6 without glitches; in this case, we must decode the *predecessor* of the state in which each signal is asserted. In the program of Table 7–25, we need only to change the `device` declaration and the /P1–/P6 equations:

```
PULSGENR device 'P20X10';

...
P1  := /RESET *  (QSTATE == 0);
P2  := /RESET *  (QSTATE == 2);
P3  := /RESET *  (QSTATE == 4);
P4  := /RESET *  (QSTATE == 6);
P5  := /RESET *  (QSTATE == 8);
P6  := /RESET *  (QSTATE == 10);
```

7.4 STATE-MACHINE DESIGN WITH SEQUENTIAL PLDS

As we showed in the preceding section, many different sequential circuits can be built with registered PLDs. However, probably the most important PLD applications are state machines, for two reasons:

(1) An SSI/MSI state-machine design often requires a large number of chips for the next-state and output equations. A PLD saves a lot of PCB area in a board-level design.

(2) An experienced designer can usually get the data paths right the first time in a system design. However, state machines that control the data paths almost always have bugs. Unlike an SSI/MSI design, a PLD-based design can often be salvaged just by changing the PLD equations, without modifying the PCB at all.

In this section we'll describe two approaches to designing PLD-based state machines that are natural extensions of the transition-list and ASM-chart approaches that we introduced in Sections 6.6.2 and 6.6.3.

7.4.1 Timing and Packaging of PLD-Based State Machines

Before looking at PLD-based design approaches, we'll examine the timing behavior and packaging considerations for state machines that are built from PLDs. Figure 7–43 shows how a generic PLD with both combinational and registered outputs might be used in a state-machine application. The timing parameters t_{CLK} and t_{PD} were explained in Section 7.3.2; t_{CLK} is the flip-flop clock-to-output delay and t_{PD} is the delay through the AND-OR array.

State variables must be assigned to registered outputs, of course, and are   *Mealy-type outputs*

Figure 7–43
Structure and timing of a PLD used as a state machine.

stable at time t_{CLK} after the rising edge of CLOCK. Mealy-type outputs must be assigned to combinational outputs, and are stable at time t_{PD} after any input change that affects them. Mealy-type outputs may also change as a result of a state change, in which case they are stable at time $t_{CLK} + t_{PD}$ after the rising edge of CLOCK.

Moore-type outputs   Moore-type outputs may be obtained in either of two ways in a PLD-based state machine. Since a Moore-type output is, by definition, a combinational logic function of the current state, the obvious way is to use a combinational output that is programmed with the output function, for example,

```
z = /Q1 * Q2 + /Q2 * Q3;
```

Such an output is stable at time $t_{CLK} + t_{PD}$ after the rising edge of CLOCK.

A Moore-type output can also be produced on a registered output, since it is supposed to be stable for the entire clock period. However, the registered output must be programmed differently. For example, if the required output function is $Z = Q1' \cdot Q2 + Q2' \cdot Q3$, we can't just substitute the clocked assignment operator in the equation and write

```
Z := /Q1 * Q2 + /Q2 * Q3;
```

Since the output is registered, output values would appear one clock tick later than in the previous case, where we used a combinational output. Instead, we must write a "lookahead" output equation that predicts what the output should be in the next state, based on the current state and inputs. If the next-state equations in the current example are

lookahead output equation

```
Q1 := expr1;
Q2 := expr2;
Q3 := expr3;
```

then the required output equation is

```
Z := /(expr1) * (expr2) + /(expr2) * (expr3);
```

An obvious disadvantage of this approach is that the output equation may "blow up," requiring more product terms than are available. However, it has two advantages: first, the registered outputs are stable sooner, only t_{CLK} after the rising edge of the clock; second, they are glitch free. Combinational outputs have an additional delay of t_{PD}, and they may have glitches because of combinational-logic static hazards during single-variable state changes and function hazards during multiple-variable state changes.

In a PLD-based design, it is also possible to use a registered output to generate a signal that is neither a state variable nor a Moore-type output. A good example of such an output is any conventional Mealy or Moore output signal that is fed back into a register to delay it by one or more clock ticks.

PLD-based state-machine designs are often limited by the number of input and output pins available in a single PLD. According to the model of Figure 7–43, one PLD output is required for each state variable and for each Mealy- or Moore-type output. For example, the T-bird tail-lights machine of Table 6–8 on page 501 requires three registered outputs for the state variables and six combinational outputs for the lamp outputs, too many for most of the PAL16xx and PAL20xx devices that we've described.

On the other hand, an output-coded state assignment usually requires a smaller total number of PLD outputs. Using an output-coded state assignment, the T-bird tail-lights machine can be built in a single PAL16R6, as we'll show in the next subsection.

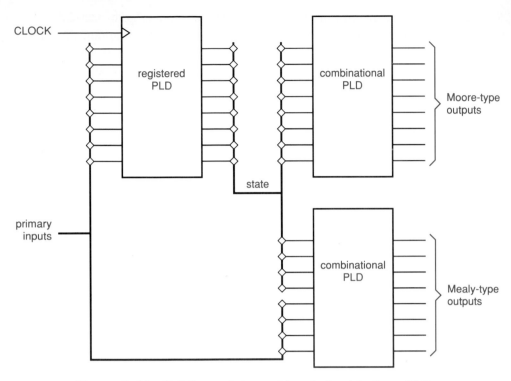

Figure 7–44 Splitting a state-machine design into three PLDs.

Like any state variable or Moore-type output, an output-coded state variable is stable at time t_{CLK} after the clock edge. Thus, an output-coded state assignment improves speed as well as packaging efficiency. In T-bird tail lights, turning on the emergency flashers 25 ns sooner doesn't matter very often, but in a high-performance digital system, an extra 25 ns of delay could make the difference between a maximum system clock rate of 20 MHz and a maximum of only 13.3 MHz.

If a design *must* be split into two or more PLDs, the best split in terms of both design simplicity and packaging efficiency is often the one shown in Figure 7–44. A single registered PLD is used to create the required next-state behavior, and combinational PLDs are used to create both Mealy- and Moore-type outputs from the state variables and inputs.

7.4.2 Synthesizing PLD-Based State Machines Using Transition Lists

In Section 6.6.2, we showed how to design state diagrams and convert them into transition lists from which a circuit could be synthesized. Recall that a transition list is a list of current states and transition expressions defining the next state of a

Table 7–26 Sum-of-products transition list for T-bird tail-lights machine.

S	Q2	Q1	Q0	LEFT	RIGHT	HAZ	S*	Q2*	Q1*	Q0*
IDLE	0	0	0	0	0	0	IDLE	0	0	0
IDLE	0	0	0	1	0	0	L1	0	0	1
IDLE	0	0	0	x	x	1	LR3	1	0	0
IDLE	0	0	0	1	1	x	LR3	1	0	0
IDLE	0	0	0	0	1	0	R1	1	0	1
L1	0	0	1	x	x	0	L2	0	1	1
L1	0	0	1	x	x	1	LR3	1	0	0
L2	0	1	1	x	x	0	L3	0	1	0
L2	0	1	1	x	x	1	LR3	1	0	0
L3	0	1	0	x	x	x	IDLE	0	0	0
R1	1	0	1	x	x	0	R2	1	1	1
R1	1	0	1	x	x	1	LR3	1	0	0
R2	1	1	1	x	x	0	R3	1	1	0
R2	1	1	1	x	x	1	LR3	1	0	0
R3	1	1	0	x	x	x	IDLE	0	0	0
LR3	1	0	0	x	x	x	IDLE	0	0	0

machine for every current state. When we drew state diagrams in Section 6.6.2, we allowed transition expressions to be written in any form. However, when we design the state diagram for a PLD-based state machine, it is convenient to write transition expressions in sum of products form. This makes it possible to write a transition list whose structure corresponds more closely to that of a PLD's AND OR array.

For example, consider the T-bird tail-lights machine, whose transition list was given in Table 6–8 on page 501. If each transition expression is written as a sum of products, we can rewrite the transition list as shown in Table 7–26. Here, the transition-expression column has been replaced with three columns, corresponding to the machine's three input variables. Each row of the table now represents a single product term, with 1s, 0s, and x's used to indicate how each variable appears in the product term. Of all the transition expressions in the original table, only one expanded into two product terms, the third and fourth rows of the new table.

We can write ABEL equations for a PLD realization of the state machine directly from the transition list. Recall that the transition/excitation equation for each next-state variable may be written as a sum for transition-list rows for which the variable is 1, or 0 at our option. Each row uses one product term in the PLD.

Table 7–27 shows the ABEL equations corresponding to the transition list of Table 7–26. The Q0* and Q1* equations can be written as a sum of four

Table 7–27 Program for T-bird tail-lights machine.

```
module T_Bird_Tail_Lights_Machine
title 'State Machine for T-Bird Tail Lights
Lee Iacocca, Ford Motor Company'
TBIRDSM device 'P16R4';
@ALTERNATE

" Input pins
CLOCK, /OE                              pin 1, 11;
LEFT, RIGHT, HAZ                        pin 2, 3, 4;

" Output pins
/Q0, /Q1, Q2                            pin 17, 16, 15;

equations

Q0   := /Q2 * /Q1 * /Q0 *  LEFT * /RIGHT * /HAZ
      + /Q2 * /Q1 * /Q0 * /LEFT *  RIGHT * /HAZ
      + /Q2 * /Q1 *  Q0                 * /HAZ
      +  Q2 * /Q1 *  Q0                 * /HAZ;

Q1   := /Q2 * /Q1 *  Q0                 * /HAZ
      + /Q2 *  Q1 *  Q0                 * /HAZ
      +  Q2 * /Q1 *  Q0                 * /HAZ
      +  Q2 *  Q1 *  Q0                 * /HAZ;

/Q2  := /Q2 * /Q1 * /Q0 * /LEFT * /RIGHT * /HAZ
      + /Q2 * /Q1 * /Q0 *  LEFT * /RIGHT * /HAZ
      + /Q2 * /Q1 *  Q0                 * /HAZ
      + /Q2 *  Q1 *  Q0                 * /HAZ
      + /Q2 *  Q1 * /Q0
      +  Q2 *  Q1 * /Q0
      +  Q2 * /Q1 * /Q0;

end T_Bird_Tail_Lights_Machine
```

rows each. Since the Q2* column has more 1s than there are product terms in a standard registered PLD, we sum 0s instead.

The ABEL compiler can minimize all the equations in Table 7–27; Q0 reduces to three product terms, Q1 to one, and /Q2 to four. For that matter, it wasn't really necessary to write the transition expressions in sum-of-products form in the first place, since the compiler can transform and minimize arbitrary expressions. The sum-of-products method is needed only if we don't have a minimizing PLD compiler, or if for some other reason we want to get a feel for how many product terms we're generating as we go along in a design.

Table 7–28
Structure of a "state diagram" in ABEL.

```
state_diagram state-variables
state state-value 1 :  equation ;
state state-value 2 :  equation ;
    ...
state state-value n :  equation ;
```

7.4.3 Designing with State-Machine Description Language

The method of the previous subsection is just a mechanical adaptation of the transition-list synthesis procedure that we described in Section 6.7. In it, we simply use the PLD compiler's facilities for "turning the crank" to combine and reduce logic expressions. However, most PLD programming languages have a notation for defining (and documenting!) state machines directly, without ever drawing a state diagram or writing a transition list. Such a notation is called a *state-machine description language.*

In ABEL, the keyword `state_diagram` indicates the beginning of a state-machine definition. The statements that define a state machine are actually more like the text equivalent of an ASM chart than like a state diagram. Table 7–28 shows the textual structure of an ABEL "state diagram." Here *state-variables* is an ABEL set that lists the state variables of the machine. If there are n variables in the set, then the machine has 2^n possible states corresponding to the 2^n different assignments of constant values to variables in the set. States are usually given symbolic names in an ABEL program; this makes it easy to try different assignments simply by changing the constant definitions.

An equation for each state variable is developed according to the information in the "state diagram." The keyword `state` indicates that the next states and current outputs for a particular current state are about to be defined; a *state-value* is a constant that defines state-variable values for the current state. An *equation* uses `IF-THEN-ELSE` and `GOTO` statements to define possible next states as a function of the state-machine inputs.

Table 7–29 is an ABEL "state diagram" for the T-bird tail-lights machine. There is a close correspondence between this program and the ASM chart in Figure 6–70. The program produces exactly the same reduced equations as the explicit transition equations in Table 7–27.

The program in Table 7–29 handles only the state variables of the tail-lights machine. The output logic requires six combinational outputs, but only four are available in the PAL16R4. A larger PLD, such as the PAL22V10, could provide enough combinational and registered outputs for a single-PLD design.

An alternative is to use an output-coded state assignment, which is possible because the output values of the tail-lights machine are different in each state. This requires only six registered outputs and no combinational outputs of

state-machine description language

`state_diagram`

state-variables

`state`
state-value

equation
`IF-THEN-ELSE`
`GOTO`

Table 7–29   Symbolic program for T-bird tail-lights machine.

```
module T_Bird_Tail_Lights_Machine
title 'State Machine for T-Bird Tail Lights, Symbolic Version'
TBIRDSD device 'P16R4';
@ALTERNATE

" Input pins
CLOCK, /OE                              pin 1, 11;
LEFT, RIGHT, HAZ                        pin 2, 3, 4;

" Output pins
/Q0, /Q1, Q2                            pin 17, 16, 15;

" Definitions
QSTATE = [Q2,Q1,Q0];                   " State variables
IDLE   = [ 0, 0, 0];                   " States
L1     = [ 0, 0, 1];
L2     = [ 0, 1, 1];
L3     = [ 0, 1, 0];
R1     = [ 1, 0, 1];
R2     = [ 1, 1, 1];
R3     = [ 1, 1, 0];
LR3    = [ 1, 0, 0];

state_diagram QSTATE
state IDLE: IF ((HAZ + LEFT*RIGHT) == 1) THEN LR3
            ELSE IF (LEFT == 1) THEN L1
            ELSE IF (RIGHT == 1) THEN R1
            ELSE IDLE;
state L1:   IF (HAZ == 1) THEN LR3 ELSE L2;
state L2:   IF (HAZ == 1) THEN LR3 ELSE L3;
state L3:   GOTO IDLE;
state R1:   IF (HAZ == 1) THEN LR3 ELSE R2;
state R2:   IF (HAZ == 1) THEN LR3 ELSE R3;
state R3:   GOTO IDLE;
state LR3:  GOTO IDLE;

end T_Bird_Tail_Lights_Machine
```

a PAL16R6 (or PAL16R8), as shown in Figure 7–45. To use an output-coded state assignment, only the device, pin, and state definitions in the previous ABEL program must be changed, as shown in Table 7–30. The six resulting excitation equations use a total of 36 product terms.

Figure 7–45
A single-PLD design for T-bird tail-lights.

Table 7–30   T-bird tail-lights machine with output-coded state assignment.

```
module T_Bird_Tail_Lights_Machine
title 'Output-Coded State Machine for T-Bird Tail Lights'
TBIRDSDO device 'P16R6';
@ALTERNATE

" Input pins
CLOCK, /OE                      pin 1, 11;
LEFT, RIGHT, HAZ                pin 2, 3, 4;

" Output pins
L3Z, L2Z, L1Z, R1Z, R2Z, R3Z       pin 18, 17, 16, 15, 14, 13;

" Definitions
ZSTATE = [L3Z,L2Z,L1Z,R1Z,R2Z,R3Z];   " State variables
IDLE   = [  0,  0,  0,  0,  0,  0];  " States
L3     = [  1,  1,  1,  0,  0,  0];
L2     = [  0,  1,  1,  0,  0,  0];
L1     = [  0,  0,  1,  0,  0,  0];
R1     = [  0,  0,  0,  1,  0,  0];
R2     = [  0,  0,  0,  1,  1,  0];
R3     = [  0,  0,  0,  1,  1,  1];
LR3    = [  1,  1,  1,  1,  1,  1];

state_diagram ZSTATE
...
end T_Bird_Tail_Lights_Machine
```

Figure 7–46 Traffic sensors and signals at an intersection in Sunnyvale, California.

7.4.4 The World's Worst Traffic-Light Controller

Our next state-machine design example is also from the world of cars and traffic. Traffic-light controllers in California, especially in the city of Sunnyvale, are carefully designed to *maximize* the waiting time of cars at intersections. An infrequently used intersection (one that would have no more than a "yield" sign if it were in Chicago) has the sensors and signals shown in Figure 7–46. The state machine that controls the traffic signals uses a 1 Hz clock and a timer and has four inputs:

NSCAR   Asserted when a car on the north-south road is over either sensor on either side of the intersection.

EWCAR   Asserted when a car on the east-west road is over either sensor on either side of the intersection.

TMLONG   Asserted if more than five minutes has elapsed since the timer started; remains asserted until the timer is reset.

TMSHORT   Asserted if more than five seconds has elapsed since the timer started; remains asserted until the timer is reset.

Figure 7–47 Sunnyvale traffic-light controller using two PLDs.

The state machine has seven outputs:

NSRED, NSYELLOW, NSGREEN Control the north-south lights.

EWRED, EWYELLOW, EWGREEN Control the east-west lights.

TMRESET When asserted, resets the timer and holds TMSHORT and TMLONG
negated. The timer starts timing when TMRESET is negated.

A typical, municipally approved algorithm for controlling the traffic lights
is embedded in the ABEL program of Table 7–31. This algorithm produces two
frequently seen behaviors of "smart" traffic lights. At night, when traffic is light,
it holds a car stopped at the light for up to five minutes, unless a car approaches
on the cross street, in which case it stops the cross traffic and lets the waiting car
go. (The 'early warning' sensor is far enough back to change the lights before the
approaching car reaches the intersection.) During the day, when traffic is heavy
and there are always cars waiting in both directions, it cycles the lights every
five seconds, thus minimizing the utilization of the intersection, and maximizing
everyone's waiting time.

The equations for the TMRESET output are worth noting. This output is
asserted during the "double-red" states, NSDELAY and EWDELAY, to reset the
timer in preparation for the next green cycle. The desired output signal could
be generated on a combinational output pin by decoding these two states, but
we have chosen instead to generate it on a registered output pin by decoding the
predecessors of these two states.

Table 7–31 Sunnyvale traffic-lights program.

```
module Sunnyvale_Traffic_Machine
title 'State Machine for Sunnyvale, CA, Traffic Lights'
SVALETL device 'P16R4';
@ALTERNATE


" Input pins
CLOCK, /OE                        pin 1, 11;
NSCAR, EWCAR, TMSHORT, TMLONG     pin 2, 3, 8, 9;

" Output pins
/Q0, /Q1, /Q2, /TMRESET           pin 17, 16, 15, 14;

" Definitions
LSTATE  = [Q2,Q1,Q0];             " State variables
NSGO    = [ 0, 0, 0];             " States
NSWAIT  = [ 0, 0, 1];
NSWAIT2 = [ 0, 1, 1];
NSDELAY = [ 0, 1, 0];
EWGO    = [ 1, 1, 0];
EWWAIT  = [ 1, 1, 1];
EWWAIT2 = [ 1, 0, 1];
EWDELAY = [ 1, 0, 0];

state_diagram LSTATE
state NSGO:                       " North-south green
   IF (/TMSHORT) THEN NSGO        " Minimum green is 5 seconds.
   ELSE IF (TMLONG) THEN NSWAIT   " Maximum green is 5 minutes.
   ELSE IF (EWCAR * /NSCAR)       " If E-W car is waiting and no one
         THEN NSGO                "   is coming N-S, make E-W wait!
   ELSE IF (EWCAR * NSCAR)        " Cars coming in both directions?
         THEN NSWAIT              " Thrash!
   ELSE IF (/NSCAR)               " Nobody coming N-S and not timed out?
         THEN NSGO                " Keep N-S green.
   ELSE NSWAIT;                   " Else let E-W have it.

state NSWAIT:  GOTO NSWAIT2; " Yellow light is on for two ticks for safety.
state NSWAIT2: GOTO NSDELAY; " (Drivers go 70 mph to catch this turkey green!)
state NSDELAY: GOTO EWGO;    " Red in both directions for added safety.

state EWGO:  " East-west green; states defined analogous to N-S
   IF (/TMSHORT) THEN EWGO
   ELSE IF (TMLONG) THEN EWWAIT
   ELSE IF (NSCAR * /EWCAR) THEN EWGO
   ELSE IF (NSCAR * EWCAR) THEN EWWAIT
   ELSE IF (/EWCAR) THEN EWGO
   ELSE EWWAIT;
```

Table 7–31   (continued)

```
state EWWAIT:  GOTO EWWAIT2;
state EWWAIT2: GOTO EWDELAY;
state EWDELAY: GOTO NSGO;

equations

TMRESET := (LSTATE == NSWAIT2)       " Reset the timer when going into
        + (LSTATE == EWWAIT2);       "   state NSDELAY or state EWDELAY.

end Sunnyvale_Traffic_Machine
```

Table 7–32   Output logic for Sunnyvale traffic lights.

```
module Sunnyvale_Traffic_Light_Output
title 'Output logic for Sunnyvale, CA, Traffic Lights'
SVALETLO device 'P16L8';
@ALTERNATE

" Input pins
FLASHCLK, OVERRIDE, /Q0, /Q1, /Q2   pin 1, 2, 3, 5, 6;

" Output pins
NSRED, NSYELLOW, NSGREEN            pin 19, 18, 17;
EWRED, EWYELLOW, EWGREEN            pin 14, 13, 12;

" Definitions (same as in state machine SVALETL)
...

equations

NSRED = /OVERRIDE * (LSTATE != NSGO) * (LSTATE != NSWAIT) * (LSTATE != NSWAIT2)
      + OVERRIDE * FLASHCLK;
NSYELLOW = /OVERRIDE * ((LSTATE == NSWAIT) + (LSTATE == NSWAIT2))
NSGREEN  = /OVERRIDE * (LSTATE == NSGO);

EWRED = /OVERRIDE * (LSTATE != EWGO) * (LSTATE != EWWAIT) * (LSTATE != EWWAIT2)
      + OVERRIDE * FLASHCLK;
EWYELLOW = /OVERRIDE * ((LSTATE == EWWAIT) + (LSTATE == EWWAIT2))
EWGREEN  = /OVERRIDE * (LSTATE == EWGO);

end Sunnyvale_Traffic_Light_Output
```

Figure 7–48   Traffic-light state machine using output-coded state assignment a single PLD.

Table 7–33   Definitions for Sunnyvale traffic-lights machine with output-coded state assignment.

```
module Sunnyvale_Traffic_Machine
title 'State Machine for Sunnyvale, CA, Traffic Lights'
SVALETLB device 'P16R8';
@ALTERNATE

" Input pins
CLOCK, /OE                          pin 1, 11;
NSCAR, EWCAR, TMSHORT, TMLONG       pin 2, 3, 8, 9;

" Output pins
NSRED, NSYELLOW, NSGREEN            pin 19, 18, 17;
EWRED, EWYELLOW, EWGREEN            pin 16, 15, 14;
/TMRESET, XTRA                      pin 13, 12;

" Definitions
LSTATE   = [NSRED,NSYELLOW,NSGREEN,EWRED,EWYELLOW,EWGREEN,XTRA]; " State vars
NSGO     = [    0,       0,       1,    1,       0,       0,   0]; " States
NSWAIT   = [    0,       1,       0,    1,       0,       0,   0];
NSWAIT2  = [    0,       1,       0,    1,       0,       0,   1];
NSDELAY  = [    1,       0,       0,    1,       0,       0,   0];
EWGO     = [    1,       0,       0,    0,       0,       1,   0];
EWWAIT   = [    1,       0,       0,    0,       1,       0,   0];
EWWAIT2  = [    1,       0,       0,    0,       1,       0,   1];
EWDELAY  = [    1,       0,       0,    1,       0,       0,   1];
```

Table 7–34 Reduced equations for the output-coded traffic-light machine of Table 7–33.

```
/NSRED := /EWGREEN * EWRED * /EWYELLOW * /NSGREEN * NSRED * /NSYELLOW * XTRA
    + /EWGREEN * EWRED * /EWYELLOW * /NSGREEN * /NSRED * NSYELLOW * /XTRA
    + /EWGREEN * EWRED * /EWYELLOW * NSGREEN * /NSRED * /NSYELLOW * /XTRA;
/NSYELLOW := /EWGREEN * /EWRED * EWYELLOW * /NSGREEN * NSRED * /NSYELLOW
    + EWGREEN * /EWRED * /EWYELLOW * /NSGREEN * NSRED * /NSYELLOW * /XTRA
    + /EWGREEN * EWRED * /EWYELLOW * /NSGREEN * NSRED * /NSYELLOW
    + /EWGREEN * EWRED * /EWYELLOW * /NSGREEN * /NSRED * NSYELLOW * XTRA
    + /EWGREEN * EWRED * /EWYELLOW * /NSCAR * NSGREEN * /NSRED * /NSYELLOW
             * /TMLONG * /XTRA
    + /EWGREEN * EWRED * /EWYELLOW * NSGREEN * /NSRED * /NSYELLOW * /TMSHORT
             * /XTRA;
/NSGREEN := /EWGREEN * /EWRED * EWYELLOW * /NSGREEN * NSRED * /NSYELLOW
    + EWGREEN * /EWRED * /EWYELLOW * /NSGREEN * NSRED * /NSYELLOW * /XTRA
    + /EWGREEN * EWRED * /EWYELLOW * /NSGREEN * NSRED * /NSYELLOW * /XTRA
    + /EWGREEN * EWRED * /EWYELLOW * /NSGREEN * /NSRED * NSYELLOW
    + /EWGREEN * EWRED * /EWYELLOW * NSCAR * NSGREEN * /NSRED * /NSYELLOW
             * TMSHORT * /XTRA
    + /EWGREEN * EWRED * /EWYELLOW * NSGREEN * /NSRED * /NSYELLOW * TMLONG
             * TMSHORT * /XTRA;
/EWRED := /EWGREEN * /EWRED * EWYELLOW * /NSGREEN * NSRED * /NSYELLOW * /XTRA
    + EWGREEN * /EWRED * /EWYELLOW * /NSGREEN * NSRED * /NSYELLOW * /XTRA
    + /EWGREEN * EWRED * /EWYELLOW * /NSGREEN * NSRED * /NSYELLOW * /XTRA;
/EWYELLOW := /EWGREEN * /EWRED * EWYELLOW * /NSGREEN * NSRED * /NSYELLOW * XTRA
    + /EWCAR * EWGREEN * /EWRED * /EWYELLOW * /NSGREEN * NSRED * /NSYELLOW
             * /TMLONG * /XTRA
    + EWGREEN * /EWRED * /EWYELLOW * /NSGREEN * NSRED * /NSYELLOW * /TMSHORT
             * /XTRA
    + /EWGREEN * EWRED * /EWYELLOW * /NSGREEN * NSRED * /NSYELLOW
    + /EWGREEN * EWRED * /EWYELLOW * /NSGREEN * /NSRED * NSYELLOW
    + /EWGREEN * EWRED * /EWYELLOW * NSGREEN * /NSRED * /NSYELLOW * /XTRA;
/EWGREEN := /EWGREEN * EWRED * /EWYELLOW * /NSGREEN * NSRED * /NSYELLOW * XTRA
    + /EWGREEN * /EWRED * EWYELLOW * /NSGREEN * NSRED * /NSYELLOW
    + EWCAR * EWGREEN * /EWRED * /EWYELLOW * /NSGREEN * NSRED * /NSYELLOW
             * TMSHORT * /XTRA
    + EWGREEN * /EWRED * /EWYELLOW * /NSGREEN * NSRED * /NSYELLOW * TMLONG
             * TMSHORT * /XTRA
    + /EWGREEN * EWRED * /EWYELLOW * /NSGREEN * /NSRED * NSYELLOW
    + /EWGREEN * EWRED * /EWYELLOW * NSGREEN * /NSRED * /NSYELLOW * /XTRA;
/XTRA := EWGREEN * /EWRED * /EWYELLOW * /NSGREEN * NSRED * /NSYELLOW * /XTRA
    + /EWGREEN * EWRED * /EWYELLOW * /NSGREEN * NSRED * /NSYELLOW
    + /EWGREEN * EWRED * /EWYELLOW * /NSGREEN * /NSRED * NSYELLOW * XTRA
    + /EWGREEN * EWRED * /EWYELLOW * NSGREEN * /NSRED * /NSYELLOW * /XTRA;
TMRESET := /EWGREEN * /EWRED * EWYELLOW * /NSGREEN * NSRED * /NSYELLOW * XTRA
    + /EWGREEN * EWRED * /EWYELLOW * /NSGREEN * /NSRED * NSYELLOW * XTRA;
```

The ABEL program in Table 7–32 defines only the state variables and one registered Moore output for the traffic controller. Six more Moore outputs are needed for the lights, more than remain on the PAL16R4. Therefore, a separate combinational PLD is used for these outputs, yielding the complete design shown in Figure 7–47. An ABEL program for the output PLD is given in Table 7–33. We've taken this opportunity to add an OVERRIDE input to the controller. This input may be asserted by the police to disable the controller and put the signals into a flashing-red mode (at a rate determined by FLASHCLK), allowing them to manually clear up the traffic snarls created by this wonderful invention.

A traffic-light state machine including output logic can be built in a single PAL16R8, shown in Figure 7–48, if we choose an output-coded state assignment. Only the definitions in the original program of Table 7–32 must be changed, as shown in Table 7–34. This PLD does not include the OVERRIDE input and mode, which is left as an exercise (7.41).

Just to give you an appreciation for using a state-machine description language to define and design state machines, Table 7–34 shows the reduced equations that the ABEL compiler generates for the traffic-light machine with the output-coded state assignment of Table 7–33. Now aren't you glad you didn't have to do that by hand?

7.5 READ/WRITE MEMORY

The read/write memories described in this section aren't usually called "programmable logic devices." However, like PLDs, they *are* highly structured devices, and can be used along with other PLDs to perform "random" tasks in a structured way. For that reason, and lacking any more appropriate place in this book, we'll cover them here.

read/write memory (RWM)

random-access memory (RAM)

The name *read/write memory (RWM)* is given to memory arrays in which we can store and retrieve information at any time. Most of the RWMs used in digital systems nowadays are *random-access memories (RAMs)*, which means that the time it takes to read or write a bit of memory is independent of the bit's location in the RAM. From this point of view, ROMs are also random-access memories, but the name "RAM" is generally used only for read/write random-access memories.

static RAM (SRAM)

dynamic RAM (DRAM)

In a *static RAM (SRAM)*, once a word is written at a location, it remains stored as long as power is applied to the chip, unless the same location is written again. In a *dynamic RAM (DRAM)*, the data stored at each location must be periodically refreshed by reading it and then writing it back again, or else it disappears. We'll discuss both types in this section.

volatile memory

Most RAMs lose their memory when power is removed; they are a form of *volatile memory*. Some RAMs retain their memory even when power is removed;

they are called *nonvolatile memory*. Examples of nonvolatile RAMs are old-style   *nonvolatile memory*
magnetic core memories and modern CMOS static memories in an extra-large
package that includes a lithium battery with a 10-year lifetime.

7.5.1 Basic Static-RAM Structure

Like a ROM, a RAM has address and control inputs and data outputs, but it also
has data inputs. The inputs and outputs of a simple $2^n \times b$-bit static RAM are
shown in Figure 7–49. The control inputs are comparable to those of a ROM,
with the addition of a *write-enable (/WE) input*. When /WE is asserted, the data   *write-enable (/WE)*
inputs are written into the selected memory location.   *input*

    Each bit of memory (or *SRAM cell*) in a static RAM has the same functional   *SRAM cell*
behavior as the circuit in Figure 7–50. The storage device in each cell is a D
latch. When a cell's /SEL input is asserted, the stored data is placed on the cell's
output, which is connected to a bit line. When both /SEL and /WR are asserted,
the latch is open and a new data bit is stored.

    SRAM cells are combined in an array with additional control logic to form a
complete static RAM as shown in Figure 7–51 for an 8×4 SRAM. As in a simple
ROM, a decoder on the address lines selects a particular row of the SRAM to
be accessed at any time. An SRAM is typically accessed as follows:

Read An address is placed on the address inputs and /CS and /OE are asserted.
       The latch outputs in the selected row are delivered to DOUT.

Write An address is placed on the address inputs and a data word is placed on
       DIN; then /CS and /WE are asserted. The latches in the selected row open,
       and the input word is stored.

Figure 7–49
Basic structure of a $2^n \times b$ RAM.

Figure 7–50   Functional behavior of a static-RAM cell.

Although Figure 7–51 gives a somewhat simplified model of internal SRAM structure, it accurately portrays several important aspects of SRAM behavior:

- During read operations, the output data is a combinational function of the address inputs, just like in a ROM. No harm is done by changing the address lines while the output data bus is enabled. The access time for read operations is specified from the time that the last address input becomes stable.

- During write operations, the input data is stored in *latches*. This means that the data must meet certain setup and hold times with respect to the *trailing* edge of the latch enable signal. That is, input data need not be stable at the moment /WR is asserted internally; it must only be stable a certain time before /WR is negated.

- During write operations, the address inputs *must* be stable for a certain setup time before /WR is asserted internally and for a hold time after /WR is negated. Otherwise, data may be "sprayed" all over the array because of the glitches that may appear on the SEL lines when the address inputs of the decoder are changed.

- Internally, /WR is asserted only when both /CS and /WE are asserted. Therefore, a *write cycle* begins when both /CS and /WE are asserted, and ends when either is negated. Setup and hold times for address and data are specified with respect to these events.

write cycle

Figure 7–52 shows the timing parameters that are typically specified for read operations in a static RAM; they are described below:

t_{AA}
access time from address

t_{AA} *Access time from address.* Assuming that /OE and /CS are already asserted, or will be soon enough not to make a difference, this is how long it takes to get stable output data after a change in address. When designers talk about a "70-ns SRAM," they're usually referring to this number.

t_{ACS}
access time from chip select

t_{ACS} *Access time from chip select.* Assuming that the address and /OE are already stable, or will be soon enough not to make a difference, this is

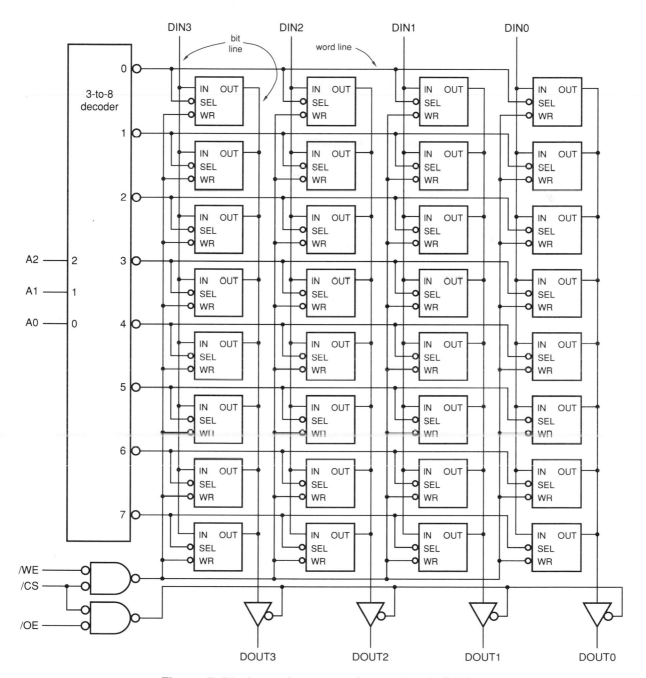

Figure 7–51 Internal structure of an 8×4 static RAM.

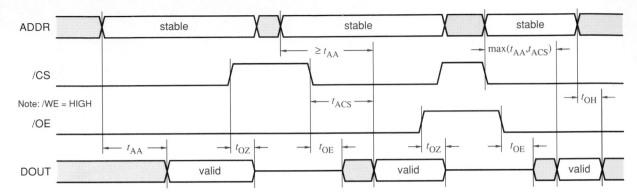

Figure 7–52 Timing parameters for read operations in a static RAM.

how long it takes to get stable output data after /CS is asserted. Most often, this parameter is identical to t_{AA}, but it is sometimes longer in SRAMs with a "power down" mode, and shorter in SRAMs without one.

t_{OE}
output-enable time

t_{OE} *Output-enable time.* This is how long it takes for the three-state output buffers to leave the high-impedance state when /OE and /CS are both asserted. This parameter is normally less than t_{ACS}, so it is possible for the RAM to start accessing data internally before /OE is asserted; this feature is used to achieve fast access times while avoiding "bus fighting" in many applications.

t_{OZ}
output-disable time

t_{OZ} *Output-disable time.* This is how long it takes for the three-state output buffers to enter the high-impedance state after /OE or /CS is negated.

t_{OH}
output-hold time

t_{OH} *Output-hold time.* This parameter specifies how long the output data remains valid after a change in the address inputs.

If you've been paying attention, you may have noticed that the timing diagram and timing parameters for SRAM read operations are identical to what we discussed for ROM read operations in Section 7.1.5. That's the way it is; when they're not being written, SRAMs can be used just like ROMs. The same is not true for DRAMs, as we'll see later.

Timing parameters for write operations are shown in Figure 7–53 and are described below:

t_{AS}
address setup time

t_{AS} *Address setup time before write.* All of the address inputs must be stable at this time before both /CS and /WE are asserted. Otherwise, the data stored at unpredictable locations may be corrupted.

t_{AH}
address hold time

t_{AH} *Address hold time after write.* Analogous to t_{AS}, all address inputs must be held stable until this time after /CS or /WE is negated.

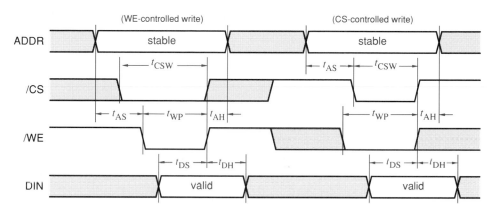

Figure 7–53   Timing parameters for write operations in a static RAM.

t_{CSW} *Chip-select setup before end of write.* /CS must be asserted at least this long before the end of the write cycle in order to select a cell.

t_{WP} *Write-pulse width.* /WE must be asserted at least this long to reliably latch data into the selected cell.

t_{DS} *Data setup time before end of write.* All of the data inputs must be stable at this time before the write cycle ends. Otherwise, the data may not be latched.

t_{DH} *Data hold time after end of write.* Analogous to t_{DS}, all address inputs must be held stable until this time after the write cycle ends.

Manufacturers of SRAMs specify two types of write cycles, *WE controlled* and *CS controlled*, as shown in the figure. The only difference between these cycles is whether /WE or /CS is the last to be asserted and the first to be negated when enabling the SRAM's internal write operation.

The write-timing requirements of SRAMs could be relaxed somewhat if, instead of using latches, the cells contained edge-triggered D flip-flops with a common clock input and enable inputs tied controlled by SEL and WR. However, this just isn't done, because it would at least double the chip area of each cell, since a D flip-flop is built from two latches. Thus, the logic designer is left to reconcile the SRAM's latch-type timing with the edge-triggered register and state-machine timing used elsewhere in a system.

A large SRAM does not contain a physical array whose dimensions equal the logical dimensions of the memory. As in a ROM, the SRAM cells are laid out in an almost square array, and an entire row is read internally during read operations. For example, the layout of a 32K×8 SRAM chip might be very similar to that of a 32K×8 ROM shown in Figure 7–9. During read operations, column multiplexers pass the required data bits to the output data bus as specified by a subset of the address bits (A5–A0 in the ROM example). For write operations, the write-enable circuitry is designed so that only one column in each subarray is enabled, as determined by the same subset of the address bits.

7.5.2 Standard Static RAMs

Static RAMs are available in many sizes and speeds. In 1989, the largest commonly available SRAMs used CMOS technology, were organized as 128K×8 bits (1 Mbit), and had access times as fast as 70 ns. The fastest TTL/CMOS compatible SRAMs also used CMOS technology and contained only 256×4 bits (1 Kbit), but had access times of 7 ns!

6164   Shown in Figure 7–54, standard 8K×8 and 32K×8 SRAMs have pinouts that are patterned after the like-size EPROMs in Figure 7–11. The *6164* 8K×8 SRAM is similar to a 2764 EPROM, except that it has a write-enable input and two chip-select inputs instead of one; /CS1 and CS2 both must be asserted

5256   to enable the chip. The *5256* 32K×8 SRAM is similar to a 27256 EPROM, with the addition of a write-enable input. Both chips are available from many manufacturers, sometimes with different part numbers, and have access times in the range of 50–200 ns.

   The SRAMs in Figure 7–54 have bidirectional data buses, that is, they use the same data pins for both reading and writing. This necessitates a slight change in their internal control logic, shown in Figure 7–55. The output buffer

Figure 7–54
Logic symbols for standard
SRAMs in 28-pin dual in-line
packages.

Figure 7–55 Output-buffer control in an SRAM with a bidirectional data bus.

is automatically disabled whenever /WE is asserted, even if /OE is asserted. However, the timing parameters and requirements for read and write operations are practically identical to what we described in the previous subsection.

A common use of SRAMs is to store data in small microprocessor systems, often in "embedded control" applications (telephones, toasters, electric bumpers, etc.). General-purpose computers more often use DRAMs, discussed in the next subsection, because their density is greater and their cost per bit lower. However, very fast SRAMs are often used in the "cache" memories of high-performance computers to store frequently used instructions and data.

7.5.3 Dynamic RAM

The basic memory cell in an SRAM, a D latch, requires four gates in a discrete design, and four to six transistors in a custom-designed SRAM LSI chip. In order to build RAMs with higher density (more bits per chip), chip designers invented memory cells that use as little as one transistor per bit.

It is not possible to build a bistable element with just one transistor. Instead, the memory cells in a *dynamic RAM (DRAM)* store information on a tiny capacitor accessed through an MOS transistor. Figure 7–56 shows the storage cell for one bit of a DRAM, which is accessed by setting the word line to a HIGH voltage. To store a 1, a HIGH voltage is placed on the bit line, which charges the capacitor

dynamic RAM (DRAM)

Figure 7–56
Storage cell for one bit in a DRAM.

Figure 7–57
Voltage stored in a DRAM cell after writing and refresh operations.

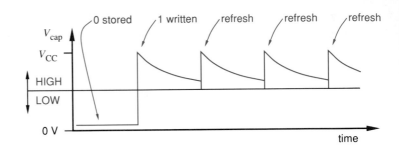

precharge

sense amplifier

through the "on" transistor. To store a 0, LOW voltage on the bit line discharges the capacitor. To read a cell, the bit line is first *precharged* to a voltage halfway between HIGH and LOW, and then the word line is set HIGH. Depending on whether the capacitor voltage is HIGH or LOW, the precharged bit line is pulled slightly higher or slightly lower. A *sense amplifier* detects this change and recovers a 1 or 0 accordingly. Note that reading a cell destroys the original voltage stored on the capacitor, so that the recovered data must be written back into the cell after reading.

The capacitor in a DRAM cell has a very small capacitance, but the MOS transistor that accesses it has a very high impedance. Therefore, it takes a relatively long time (many milliseconds) for a HIGH voltage to discharge to the point that it looks more like a LOW voltage. In the meantime, the capacitor stores one bit of information.

Naturally, using a computer would be no fun if you had to re-boot every few milliseconds because its memory contents disappeared. Therefore, memory systems built with DRAMs use *refresh cycles* to update every memory cell every so often, usually once every four milliseconds. This involves sequentially reading the somewhat degraded contents of each cell into a D latch, and writing back a nice solid LOW or HIGH value from the latch. Figure 7–57 illustrates the electrical state of a cell after a write and a sequence of refresh operations.

refresh cycle

The first DRAMs, introduced in the early 1970s, contained only 1024 bits, but modern DRAMs are available containing four *megabits* or more. If you had to refresh every cell, one at a time, in four milliseconds, you'd have a real problem—that works out to just 1 ns per cell, and includes no time for useful read and write operations. Fortunately, as we'll show next, DRAMs are organized using two-dimensional arrays, and a single operation refreshes an entire row of the array. Typical DRAM arrays have 256 rows, requiring 256 refresh operations every four milliseconds, or one about every 15.6 μs. A refresh operation typically takes less than 300 ns, so the DRAM is available for useful read and write operations over 98% of the time.

Figure 7–58 is a block diagram of the internal structure of a 64K×1 DRAM. The logical array has 64K×1 bits, but the physical array is square, containing

Figure 7–58
Internal structure of a 64K×1
DRAM.

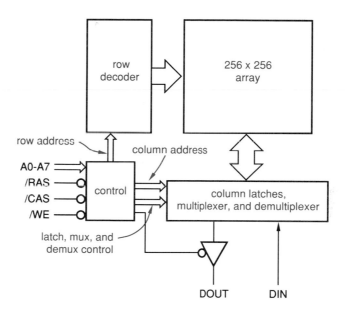

256×256 bits. Although the memory has 64K locations, the chip has only eight *multiplexed address inputs*. A complete 16-bit address is presented to the chip in two steps controlled by two signals—/RAS *(row address strobe)* and /CAS *(column address strobe)*. Multiplexing the address inputs saves pins, important for compact design of memory systems, and also fits quite naturally with the two-step DRAM access methods that we'll describe shortly.

Timing waveforms for three representative DRAM cycles are shown in Figure 7–59. The first is called a *RAS-only refresh cycle*, and is used to refresh a row of memory without actually reading or writing any data at the external pins of the DRAM chip. The cycle begins when a row address is applied to the multiplexed address inputs (8 bits in the case of a 64K×1 DRAM) and /RAS is asserted. The DRAM stores the row address in an internal *row-address register* on the falling edge of /RAS and reads the selected row of the memory array into an on-chip *row latch*.[11] When /RAS is negated, the contents of the row are written back from the row latch. To refresh an entire 64K×1 DRAM, the logic designer must ensure that 256 such cycles, using all 256 possible row addresses,

*multiplexed address
    inputs*
/RAS *(row address
    strobe)*
/CAS *(column
    address strobe)*
*RAS-only refresh
    cycle*

row-address register

row latch

---

[11] You may notice that the valid intervals of row and column addresses in the figure are skewed to the right with respect to the falling edges of /RAS and /CAS, which load the addresses into on-chip edge-triggered registers. DRAMs typically have a short setup-time requirement for the address inputs (often 0 ns), but a relatively large, positive hold-time requirement (10–20 ns). This can cause some problems in the design of an otherwise fully synchronous system that uses a single clock and flip-flops with zero hold times, and is often solved by distasteful methods involving tapped delay lines or using the opposite edge of the clock. This is one of the reasons that many designers consider DRAM system design a tricky, black art.

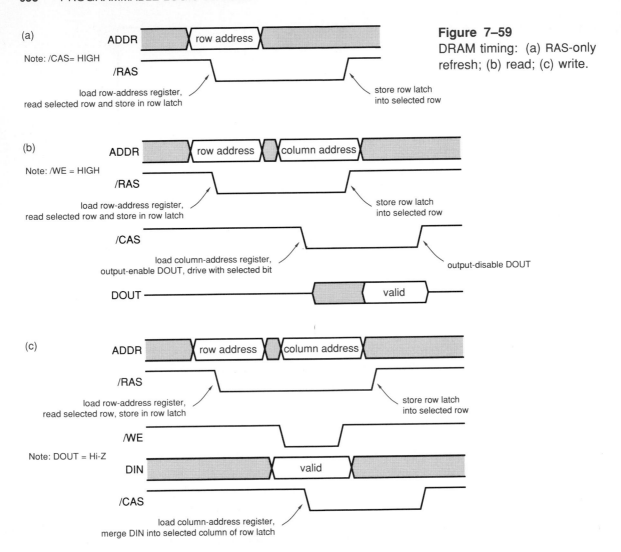

Figure 7–59
DRAM timing: (a) RAS-only refresh; (b) read; (c) write.

are executed every four milliseconds. Typically, an external 8-bit counter is used to generate the row addresses, and a timer is used to initiate a refresh cycle once every 15.6 μs.

read cycle

column-address register

A *read cycle*, shown in Figure 7–59(b), begins like a refresh cycle, where a selected row is read into the row latch. Next, a column address is applied to the multiplexed address inputs, and is stored in an on-chip *column-address register* on the falling edge of /CAS. The column address is used to select one bit of the just-read row, which is made available on the DRAM's DOUT pin. A three-state pin, DOUT is output-enabled as long as /CAS is asserted. In the meantime, the entire row is written back into the array as soon as /RAS is negated.

A *write cycle*, shown in (c), also begins like a refresh or read cycle. However, */WE (write enable)* must be asserted before /CAS is asserted to select a write cycle. One thing this accomplishes is to disable DOUT for the rest of the cycle, even though /CAS will be asserted subsequently. Once the selected row is read into the row latch, /WE also forces the input bit on DIN to be merged into the row latch, in the bit position selected by the column address. When the row is subsequently written back into the array on the rising edge of /RAS, it contains a new value in the selected column.

write cycle
/WE (write enable)

Several other types of cycles, not shown in the figure, are available in typical DRAMs:

- A *CAS-before-RAS refresh cycle* performs a refresh operation without requiring a row address to be supplied by an external counter. Instead, the DRAM chip maintains an internal row-address counter. If /CAS is asserted before /RAS is asserted, the DRAM refreshes the row selected by the counter and then increments the counter. This simplifies DRAM system design, eliminating the external refresh counter and reducing from three (row, column, refresh) to two the number of multiplexed sources that must drive the DRAM address inputs.

  CAS-before-RAS refresh cycle

- A *read-modify-write cycle* begins like a read cycle, and allows data to be read on DOUT when /CAS is asserted. However, /WE may be asserted subsequently to write new data back at the same location.

  read modify-write cycle

- A *page-mode read cycle* allows up to an entire row ("page") of data to be read without repeating the entire RAS-CAS cycle. Since the entire row is already stored in the row latch, this cycle simply requires /CAS to be pulsed LOW multiple times while /RAS remains asserted continuously. A new column address is provided and a new bit becomes available on DOUT on each falling edge of /CAS. This cycle provides much faster memory access during a sequence of reads to sequential or otherwise nearby addresses; such sequences often occur in computers when they're fetching instructions or filling caches.

  page mode read cycle

- Similarly, a *page-mode write cycle* allows multiple bits of a row to be written with a single /RAS and multiple /CAS cycles.

  page-mode write cycle

- A *static-column-mode read cycle* is similar to a page-mode read cycle, except that the column address is not stored in an on-chip register; instead it is applied directly to the column multiplexer. Therefore, another bit in the same row may be read at any time simply by changing the column address, *without* pulsing /CAS. This simplifies the external DRAM control logic.

  static-column-mode read cycle

Figure 7–60 Logic symbols for standard DRAMs in dual in-line packages.

static-column-mode write cycle
- Similarly, a *static-column-mode write cycle* allows the column address to be changed at any time when writing multiple bits in the same row. However, it is necessary to negate either /CAS or /WE when the address is changing, to avoid "spraying" data all over the selected row because of glitches in the column-address decoder.

nibble-mode read cycle
- A *nibble-mode read cycle* is similar to a page-mode read cycle, except that a new column address is not supplied on /CAS pulses after the first. Instead, the DRAM chip generates a sequence of addresses from the address supplied at the initial RAS-CAS cycle. The sequence repeats after four /CAS pulses, hence the name.

nibble-mode write cycle
- Similarly, a *nibble-mode write cycle* allows a sequence of four addresses to be written.

Note that page-mode, static-column-mode, and nibble-mode cycles are mutually exclusive. That is, a given DRAM chip (with a specific part number) supports at most one of these modes.

4164
4256
4464
41000
44256

Figure 7–60 shows the logic symbols for commonly used DRAMs in dual in-line packages. Note that DRAMs are available in 4-bit-wide as well as in 1-bit-wide configurations. The "size" of a DRAM is the total number of bits it stores; thus, a 44256 256K×4 device is still called a "1-Mbit DRAM." The ×4 configurations are useful in applications that do not require a huge total amount of memory, but can benefit from reduced chip count using high-density DRAMs. Thus, in the 1990s we can expect to see low-cost PC clones with 2 Mbytes of memory built with four 1M×4 (4-Mbit) DRAM chips.

(handwritten) (4) x 4 Mbit = 16 Mbits

(handwritten) 2 x 8 = 16 Mbits

REFERENCES

Programmable logic devices have been around for a while, but it was only in the mid-to-late 1980s that their cost, reliability, and ease of use finally caused an explosion in PLD-based designs. The best resources for learning about PLD-based design still seem to be the PLD manufacturers. For example, AMD publishes a comprehensive, two-volume *PAL Device Handbook* and *PAL Device Data Book* (Sunnyvale, CA 94088) [ouch, there's that town again!]. The first volume contains over 700 pages of introductory material, design examples, and article reprints. The second volume describes AMD's PALASM II PLD programming language and contains data sheets for all of AMD's PLDs.

A much more detailed discussion of the internal operation of structured LSI and VLSI devices, including ROMs, RAMs, and PLDs, can be found in electronics texts such as *Microelectronics* by J. Millman and A. Grabel (McGraw-Hill, 1987, 2nd ed.) and *VLSI Engineering* by Thomas E. Dillinger (Prentice-Hall, 1988).

EXERCISES

7.1 Our discussion of ROM sneak paths in connection with Figure 7–6 claimed that with A2–A0 = 101, bit lines /D2 and /D0 are pulled LOW through the direct connections. That's not really correct, unless the 74LS138 is replaced with a decoder with open-collector outputs. Explain.

7.2 Describe the logic function of seven variables that is performed by the 128×1 ROM in Figure 7–8. Starting with the ROM pattern, one way to describe the logic function is to write the corresponding truth table and canonical sum. However, since the canonical sum has 64 7-variable product terms, you might want to look for a simple but precise word description of the function.

7.3 Show how to double the number of different attenuation amounts that can be selected in the digital attenuator of Figure 7–17 without increasing the size of the ROM.

7.4 Write the Pascal functions `UlawTOlinear` and `LinearTOulaw` that are required in Table 7–6. In `LinearTOulaw`, you should select the μ-law byte whose theoretical value is closest to the linear input. (*Hint:* The most efficient approach builds, at program initialization, a 256-entry table that is used by both functions.) Both functions should report out-of-range inputs.

7.5 Modify the Pascal program in Table 7–6 to perform clipping in cases where the designer has specified a particular attenuation amount to be a gain, and multiplying an input value by the attenuation factor produces an out-of-range result.

7.6   Write a Pascal program to generate the contents of a 256×8 ROM that converts from 8-bit binary to 8-bit Gray code. (*Hint:* Your program should embody the second Gray-code construction method in Section 1.11.)

7.7   Write a Pascal program to generate the contents of a 256×8 ROM that converts from 8-bit Gray code to 8-bit binary. (*Hints:* You may use the results of Exercise 7.6. It doesn't matter if your program is slow.)

7.8   A certain communication system has been designed to transmit ASCII characters serially on a medium that requires an average signal level of zero, so a 5-out-of-10 code is used to code the data. Each 7-bit ASCII input character is transmitted as a 10-bit word with five 0s and five 1s. Write a Pascal program to generate the contents of a 128×10 ROM that converts ASCII characters into coded words.

7.9   The receiving end of the system in Exercise 7.8 must convert each 10-bit coded word back into a 7-bit ASCII character. Write a Pascal program to generate the contents of a 1K×8 ROM that converts coded words onto ASCII characters. The extra output bit should be an "error" flag that indicates when a noncode word has been received.

7.10 How many ROM bits would be required to build a 16-bit adder/subtracter with mode control, carry input, carry output, and two's-complement overflow output? Be more specific than billions and billions, and explain your answer.

7.11 Repeat Exercise 7.10 assuming that you may use *two* ROMs, so that the delay for a 16-bit addition or subtraction is twice the delay through one ROM. Assume that the two ROMs must be identical in both size and programming. Try to minimize the total number of ROM bits required, and sketch the resulting adder/subtracter circuit. Can the number of ROM bits be further reduced if the two ROMs are allowed to be different?

7.12 Show how, using additional SSI/MSI parts, a 2764 can be used as a 64K×1 ROM. What is the access time of the 64K×1 ROM?

7.13 Show how, using additional SSI/MSI parts, a 2764 can be used as a 2K×32 ROM. You may assume the existence of a free-running clock signal whose period is slightly longer than the access time of the 2764. What is the access time of the 2K×32 ROM?

7.14 Show how to build the μ-law adder circuit of Figure 7–18 with a 32K×8 ROM and two XOR gates. Also, write a Pascal program to generate the ROM contents.

7.15 Determine the ROM size needed to build the fixed-point to floating-point encoder of Figure 4–112 on page 335. Draw a logic diagram using a commercially available ROM, and write a Pascal program to generate the ROM contents. Unlike the MSI circuit, your Pascal program should perform rounding; that is, for each fixed-point number it should generate the nearest possible floating-point number.

7.16 Write logic expressions for product terms P3–P6 in Figure 7–23.

7.17 Using a single PAL16L8, design a circuit similar to the 74LS139 dual 2-to-4 decoder, including an active-low output-enable and a polarity-control input for each section. Draw a logic diagram showing the assignment of signals to device pins, and write a complete ABEL program for the function. (Create a pin assignment that is similar to the 139's.)

7.18 Design a 3-input, 5-bit multiplexer in a single PAL20L10. The select inputs are S1 and S0, the data inputs are A1–A5, B1–B5, and C1–C5, and the data outputs are Y1–Y5. The Y outputs should equal the A, B, or C for select-input combinations 00, 01, and 10, respectively, and should be zero when the select inputs are 11. Draw a logic diagram showing the assignment of signals to device pins, and write a complete ABEL program for the function. Use sets and relations in the program. Does the PAL20L10 have enough product terms for the function? If not, what changes could you make to the design specification to make it fit?

7.19 Indicate how the ABEL program in Exercise 7.18 must be modified if the Y outputs are to be floating when the select inputs are 11.

7.20 Rewrite the ABEL program in Table 7–14 using sets and relations to handle the X and Y inputs.

7.21 Write an equation for a /XGEY output for the MODECOMP PLD of Table 7–14 that is asserted when $X > Y$. Does your equation fit in a PAL16L8?

7.22 How many product terms are required for each output in Table 7–15 if each is assigned an active level opposite the one shown?

7.23 Modify the logic diagram and the ABEL program for the four-way, 2-bit transceiver of Figure 7–33 so that the data bits appear on output pins 19–12 in the order A1, A2, ..., D1, D2. (*Hint:* It takes more than a change in the declarations.)

7.24 Design an SSI/MSI circuit equivalent to the modified multiplexer of Figure 7–36. Your circuit should have the same input and output signals and active levels as the PLD. Compare the chip count of your circuit with that of the PLD design.

7.25 Draw an IEEE-standard symbol for the modified multiplexer of Figure 7–36.

7.26 Design a multiplexer PLD, similar to the ones in Figure 7–36, with four control inputs: LDA, LDB, CLR, and COM. The register should be loaded with the A bus if LDA is asserted and with B if LDB is asserted, and held otherwise. However, if CLR is asserted, then the register is cleared. Finally, if COM is asserted, the next state of the register is the complement of what it would have been otherwise. It is also desirable for the register to be loaded with A⊕B if LDA and LDB happen to be asserted simultaneously. However, you are limited to using a single PAL16R4 in this design. Draw a logic diagram showing the signal connections to the PAL16R4, and an ABEL program that produces the required behavior. Also write a truth table that shows the operation performed by your device for all 16 combinations of the control inputs.

7.27   Determine the number of product terms required for each output of the RING8 PLD in Table 7–22. Does it fit in a PAL16R8?

7.28   The timing-generator PLD of Table 7–23 can produce a 1-tick-wide pulse on the /P1–/P6 outputs if /RESET or /RESTART is asserted in the middle of a phase. This may be undesirable, depending on what the outputs are controlling. Modify the equations so that the output pulses are always at least two ticks wide.

7.29   Suppose the timing generator of Table 7–23 is used to control a dynamic memory system, such that all six phases must be completed to read or write the memory. If the timing generator is reset during a write operation without completing phase 6, the memory contents may be corrupted. Modify the equations in Table 7–23 to avoid this problem.

7.30   Modify the timing-generator equations in Table 7–23 so that phase 3 lasts for three ticks of MCLK and phase 4 lasts for four.

7.31   Write an ABEL program to generate the timing waveforms of Exercise 6.35 using a single PAL16R8.

7.32   How many fusible links are contained in a PAL20X8 PLD?

7.33   Write a "generic" equation that handles bit i of a synchronous binary up counter with synchronous count-enable, load, clear, and preset inputs in an X-series PLD. Your equation should assume active-high signals at the PLD output pins.

7.34   Write the equations for building, in a PAL20X8, an 8-bit synchronous binary down counter with count-enable, load, and clear inputs, and an active-low /RCO output that is asserted whenever the counter is in state 00000000. Your equation should assume active-low signals at the PLD output pins.

7.35   Design a binary up/down counter for the elevator controller in a 20-story building, using a single PAL16R6 or PAL20X4. The counter should have enable and up/down control inputs. It should stick at state 1 when counting down, stick at state 20 when counting up, and skip state 13 in either mode. Draw a logic diagram and write ABEL equations for your design.

7.36   Modify the ABEL program in Table 7–25 to use a PAL20X4 with combinational outputs for /P1–/P6. Does this approach have any advantages compared to the original PAL20X10-based design?

7.37   Design a circuit equivalent to the modulo-193 counter of Figure 6–27 using a single PAL20X8.

7.38   Redesign the T-bird tail-lights machine to include parking-light and brake-light functions. When the BRAKE input is asserted, all of the lights should go on immediately, and stay on until BRAKE is negated, independent of any other function. When the PARK input is asserted, each lamp is turned on at 50% brightness at all times when it would otherwise be off. This is achieved by driving the lamp with a 100 Hz signal DIMCLK with a 50% duty cycle. Draw a logic diagram for the circuit using one or two PLDs, write an ABEL program for each PLD, and write a short description of how your system works.

7.39 Write an ABEL program for the guessing-game machine described on page 510 using a single PAL16R6. Use an output-coded state assignment and ABEL's "state-diagram" notation.

7.40 Design the five-minute timer required traffic-light controller of Figure 7–47 using two 74LS163 counters. (You may reduce the "long" timeout to 255 seconds if you wish.) Use spare PLD outputs and modify the ABEL programs as required to design the entire traffic-light controller without using any additional SSI or MSI parts. Draw a logic diagram for your circuit and write the required ABEL programs.

7.41 Add an OVERRIDE input to the traffic-lights state machine of Figure 7–48, still using just a single PAL16R8. When OVERRIDE is asserted, the red lights should flash on and off, changing once per second. Write a complete ABEL program for your machine.

7.42 Write an ABEL program for the state machine described in Exercise 5.19. Use ABEL's "state-diagram" notation.

7.43 In the MODECOMP PLD of Figure 7–30, we assigned outputs to device pins so as to maximize the possible uses of the unused pins (compare with the situation if we used pins 13–14 for the outputs). Experience dictates this approach, since after a PCB is designed and built, we can often fix bugs or add last-minute features without too much pain if there are spare PLD resources available on the board. Find five other instances in this chapter where we assigned outputs to maximize flexibility in this way, and list the possible uses of each spare pin.

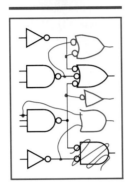

8

ADDITIONAL REAL-WORLD TOPICS

This chapter contains a collection of "real-world" topics that are certain to interest you at some point if you are (or become) a working digital designer. We barely scratch the surface of each of these topic areas, so be sure to consult the References in areas of particular interest.

8.1 COMPUTER-AIDED ENGINEERING TOOLS

"If it wasn't hard, they wouldn't call it hardware."

Many digital designers with twenty years of experience consider this statement to be indisputable. Yet more and more, digital design is being carried out using *software*, and is getting easier as a result.

computer-aided design (CAD)
computer-aided engineering (CAE)
CAD
CAE

The terms *computer-aided design (CAD)* and *computer-aided engineering (CAE)* are used to refer to software tools that aid the development of circuits, systems, and many other things. *CAD* is the more general term and applies to tools both inside and outside the electronics area, including architectural and mechanical design tools, for example. Within electronics, "CAD" often refers to *physical-design* tools, such as IC- and PCB-layout programs. *CAE* is used more often to refer to *conceptual-design* tools, such as schematic editors, circuit simulators, and PLD compilers. However, a lot of people in electronics (including the author) tend to use the two terms interchangeably. In this section, we'll introduce some CAD/CAE tools that are often used by digital designers.

8.1.1   Schematic Capture

Schematics used to be drawn by hand, but most are now prepared using *schematic editors*, CAE programs that run on engineering workstations.[1] The process of creating a schematic on a computer is often called *schematic capture*. (Huh? How did it get loose?) This term is used because the schematic editor captures more information than just a drawing. Most information in the schematic has a "type" associated with it, so that selected information can be retrieved automatically as required later in the design process.

schematic editor

schematic capture

In order to simplify both entry and retrieval of information in a schematic, a schematic editor typically provides *fields* for each information type, either hidden or drawn automatically nearby the associated schematic element. Some typical fields and their uses are described below.

fields

- *Component type.* By specifying a component type (e.g., resistor, capacitor, 74LS00, 74LS374), the designer can call out a pre-drawn symbol from a *component library*. Some components have user-selectable alternate symbols (e.g., DeMorgan equivalents for gates). The component type is used both for manufacturing documentation and for simulation.

  component library

- *Component value.* Most analog components have a value that must be specified by this field (e.g., 2.7 kΩ). Additional distinguishing information may appear in a "rating" or "tolerance" field (e.g., 1/4-W, 1%).

- *Approved-part number.* Using all of the distinguishing information from the previous fields, the CAE software may automatically select a part number from the company's "approved parts list" (e.g., 126-100117-0272 for a 2.7-kΩ, 1/4-W, 1% metal-film resistor), or a warning if no such part is available. (No, in large companies you can't use just any part in the catalog; all parts must be approved through a *parts-qualification* process.)

  parts qualification

- *Reference designator.* This alphanumeric label distinguishes among multiple instances of similar parts, and is used just about everywhere.

- *Component location.* Using a coordinate system, this field may indicate the physical position of the component on an assembled PCB, simplifying the task of finding it during debugging. Of course, the final component location generally is not known during the initial circuit design. However,

---

[1]Note that an engineering workstation nowadays can be anything from a $500 PC clone to a $100,000 parallel processor. The "canonical" engineering workstation in terms of cost (what an average company will spend per engineer) and performance (among the best) is a $15,000 Sun workstation, whatever you get for that price this year.

back annotation   CAE software with *back-annotation* capability can insert this information into the schematic when the PCB layout is completed.

- *Pin numbers*. These specify the assignment of the signal pins on each component, and are usually pre-specified in the symbol that is called up from the component library. However, for SSI and other parts that have multiple identical sections in a single package, pin numbers (and even reference designators) might not be specified until the PCB layout is completed; this is another case where back annotation is handy.

- *Wire type*. Usually the only wire types are "ordinary" and "bus," where a bus is just a set of ordinary wires that are grouped for drawing purposes. However, other wire types, such as extra-wide PCB traces for analog signals that carry high currents, may be indicated in some systems; such information is passed to the PCB-layout program.

- *Signal name*. The user may (and should) associate a name with each signal. The name is especially useful in simulation and debugging, but it does not affect the physical PCB layout.

- *Connection flag*. This drawing element indicates the connection of one signal line to another one on the same or a different page. Software can ensure that all outgoing connections are matched by incoming ones.

Once you've "captured" a circuit's schematic in a CAE system, there are lots of things you can do with it besides printing it. At least two documents can be generated for manufacturing purposes:

parts list   - *Parts list*. This is a list of components and their reference designators.

net   - *Net list*. A *net* is a set of pins that are all connected to the same electrical
net list   node or signal. A *net list* specifies all such connections that are required by the schematic, typically using an alphabetically sorted list of signal names. For each signal name, it lists the device pins (identified by reference designator and pin number) where that signal appears. Alternatively, the net list may be sorted by reference designator and pin number, and show
pin list   the signal name connected to each pin (sometimes this is called a *pin list*).

The parts list and net list are the main inputs to the PCB-layout process. The designer may also use the parts list to estimate the cost and reliability of the circuit, and the manufacturing department obviously can use it to order the parts.

Other interesting things you can do once you have a schematic are to analyze and simulate the circuit, as discussed in the next subsection.

8.1.2 Circuit Analysis and Simulation

The component library of a sophisticated CAE system contains more than just a symbol for each component. For ICs, the library may contain a *component model* describing the device's logical and electrical operation. Component models, along with the interconnection information in the schematic, can be used to find design errors.

component model

As a minimum, the model for an IC indicates whether each pin is an input or output. With just this information, a *design-rule checker* can detect some of the more common "dumb errors" in a design, such as shorted outputs and floating inputs. With the addition of input loading and output driving characteristics for each pin, the program can also determine whether any output's fanout capability is exceeded.

design-rule checker

The next step is *timing verification*. Even without a detailed model of an IC's logical behavior, the component library can provide a worst-case delay value for each input-to-output path, and setup and hold times for clocked devices. Using this information, a *timing verifier* can find the worst-case delay paths in the overall circuit, so the designer can determine whether the timing margins are adequate.

timing verification

timing verifier

Finally, the library may contain a detailed model of each component's logical behavior, in which case *simulation* may be used to predict the overall circuit's behavior for any given sequence of inputs. The designer provides an input sequence, and a *simulator* program determines how the circuit would respond to that sequence. The simulator output is usually displayed graphically, in the form of timing waveforms that the designer could see on an oscilloscope or logic analyzer if the same inputs were applied to a real circuit. In such an environment, it is possible to debug an entire circuit without "breadboarding" it, and to assemble a PCB that works on the very first try.

simulation

simulator

Simulators use different component models depending on the type of component. Analog simulators like SPICE may use mathematical models with just one parameter for a resistor, and with anywhere from a few to dozens of parameters for a bipolar transistor. Based on these models and the circuit's structure, the simulator develops and solves equations to determine the circuit's behavior.

Digital simulators normally use one of the following models for each logic element:

- *Behavioral model.* In such a model, the software "understands" the basic function of the logic element. For example, an AND gate is understood to produce a 1 output with delay t_{pLH} after all inputs are 1, and a 0 output with delay t_{pHL} after any input is 0.

  behavioral model

- *Structural model.* Larger elements may be modeled as a collection of smaller elements that have behavioral models. For example, a 74LS138

  structural model

decoder may be modeled as a set of gates connected according to the internal logic diagram of Figure 4–29 on page 253. Alternatively, the '138 could have a behavioral model consisting of a truth table and timing information.

software model

Both of the foregoing models are called *software models*, because the chip's operation is simulated entirely by software. Some chips, including microprocessors and other LSI parts, require huge software models. In most cases, the chip manufacturers don't provide such models (they're considered to be proprietary information), and the models would take months or years for a user to develop. To get around this problem, innovative CAE companies have developed a third type of model:

physical model
hardware model

- *Physical model*, sometimes called a *hardware model*. A real, working copy of the chip is connected to the computer that's running the simulation. The simulator applies inputs to the real chip and observes its outputs, which are then used as inputs to the other models in the simulation.

event queue

Regardless of the model type(s), the simulator's operation is driven by an *event queue*, a time-ordered list of signal transitions. For example, when the simulator determines that a 0-to-1 transition occurs at time t at the input of an inverter with a 5-ns delay, it inserts a 1-to-0 transition at the inverter's output at time $t+5$. Eventually, after processing all of the other events that occur between times t and $t+5$, the simulator propagates the change on the inverter output to the inputs that it affects.

A digital designer's input to a logic simulator is a schematic and a sequence of input vectors to be applied to the circuit in simulated time. The first output that the designer usually sees is "XXXXX," meaning that the simulator can't figure out what the output will be. This usually occurs because the circuit contains sequential circuits (flip-flops and latches) that are not initialized. To obtain meaningful simulator output, the designer must ensure that all circuits are explicitly reset to a known value, even if reset is not required for correct operation in the real system. For example, the self-synchronizing ring counters of Figure 6–47 on page 480 cannot be simulated, because the simulator has no way of knowing their starting state.

uninitialized
sequential circuits

Once the simulated circuit has been initialized, the designer's ability to probe its operation is limited only by the simulator's speed. Simulation of complex circuits can be very slow, compared to the speed of the real thing. Depending on the level of detail being simulated, the simulation time can be 10^3 to 10^8 times longer than the interval being simulated. A strictly functional simulation, in which all components are assumed to have zero delay, is fastest. A much more realistic simulation, one that calculates and considers a worst-case range of delays for all signal paths, is the slowest.

simulation
inefficiency

8.1.3 Hardware Description Languages

Most logic design is performed graphically, using block diagrams and schematics, but many different textual languages, called *hardware description languages*, have been developed for describing logic circuits and systems. PLD programming languages like ABEL are examples of hardware description languages for a predefined, limited collection of circuit structures. General state-machine description languages with features similar to ABEL's "state diagrams" can be used to describe arbitrary clocked, synchronous state machines. Other languages can describe an even wider range of circuit structures.

hardware description language

  Register-transfer languages have been used for decades to describe the operation of synchronous systems, including the simple structure of Figure 6–82 on page 525. Such a language combines the control-flow notation of a state-machine description language with a means for defining and operating on multibit registers. Register-transfer languages have been especially useful in computer design, where individual machine-language instructions are defined as a sequence of more primitive steps involving loading, storing, combining, and testing registers.

register-transfer language

  In the mid-1980s, the U.S. Department of Defense and the IEEE sponsored the development of a highly capable hardware-description language called VHDL,[2] that includes the following features:

VHDL

- Designs may be decomposed hierarchically.

- Each element of a design has both a well-defined interface (for connecting it to other elements) and a precise behavioral specification (for simulating it).

- Behavioral specifications can use either an algorithm or an actual hardware structure to define an element's operation. An element may initially be described by an algorithm, to enable design verification of the higher-level elements that use it; the algorithm can be replaced with a hardware structure later.

- Concurrency, timing, and clocking can all be modeled. VHDL handles asynchronous as well as synchronous circuit structures.

- Circuit operation of a design can be simulated.

- Advanced VHDL-based compilers can synthesize hardware structures directly from algorithms, without designer intervention. For example, it's possible to synthesize an entire VLSI chip from a VHDL description.

---

[2]VHDL stands for "VHSIC Hardware Description Language." VHSIC, in turn, stands for "Very High Speed Integrated Circuit," a DOD program to encourage the development of high-performance IC technology (using Very Healthy Sums of Instant Cash?).

Since VHDL is a government standard, you'd naturally expect to see its use grow. More important, VHDL circuit descriptions are attractive because they can provide the input to CAD tools that automatically synthesize the corresponding chip or board-level circuit and fulfill documentation, manufacturing, and testing requirements.

8.2   DESIGN FOR TESTABILITY

When you buy a new piece of equipment, you expect it to work properly "right out of the box." However, even with modern automated equipment, it's impossible for an equipment manufacturer to guarantee that every unit produced will be perfect. Some units don't work because they contain individual components that are faulty, or because they were assembled incorrectly or sloppily, or because they were damaged in handling. Most manufacturers prefer to discover these problems in the factory, rather than get a bad reputation with their customers. *testing* This is the purpose of *testing*.

go/no-go test   The most basic test, a *go/no-go test* yields just one bit of information—is the system 100% functional or not? If the answer is "yes," the system can be shipped. If not, the next action depends on the size of the system. The appropriate response for a digital watch would be to throw it out and try again. If many watches are turning up faulty, it's more efficient to repair what must be a problem in the assembly process than to try to salvage individual units.

diagnostic test   On the other hand, you wouldn't want to toss out a $10,000 computer that fails to boot. Instead, a more detailed *diagnostic test* may be run to locate the particular subsystem that is failing. Depending on cost, the failed subsystem may be either repaired or replaced. A typical digital subsystem is a PCB whose assembled cost ranges from $50 to $1000. The repair/replace decision is an economic trade-off between the cost of the assembled PCB and the estimated time to locate and repair the failure.

design for testability   This is where *design for testability (DFT)* comes in. "DFT" is a general
(DFT)   term applied to design methods that lead to more thorough and less costly testing. Many benefits ensue from designing a system or subsystem so that failures are easy to detect and locate:

- The outcome of a go/no-go test is more believable. If fewer systems with hidden faults are shipped, fewer customers get upset, which yields obvious economic as well as psychological benefits.

- Diagnostic tests run faster and produce more accurate results. This reduces the cost of salvaging a subsystem that fails the go/no-go test, making it possible to manufacture more systems at lower cost.

- Both go/no-go and diagnostic tests require less test-engineering time to develop.

- Although the savings in test-engineering time may be offset by added design-engineering effort to include DFT, any increase in overall product development cost can be expected to be outweighed by decreased manufacturing cost.

8.2.1 Testing

Digital circuits are tested by applying *test vectors* (input combinations) and observing the results. In the worst case, an n-input combinational circuit requires 2^n test vectors. However, if we know something about the circuit's physical realization and make some assumptions about the type of failures that can occur, the number of vectors required to test the circuit fully can be greatly reduced. The most common assumption is that failures are *single stuck-at faults*, that is, that they can be modeled as a single input or output signal stuck at logic 0 or logic 1. Under this assumption, an 8-input NAND gate, which might otherwise require 256 test vectors, can be fully tested with just nine—11111111, 01111111, 10111111, ..., 11111110.

test vector

single stuck-at fault

Under the single-fault assumption, it's easy to come up with test vectors for individual logic elements. However, the problem in practice is *applying* the test vectors to logic elements that are buried deep in a circuit, and *seeing* the results. For example, suppose that a circuit has a dozen combinational and sequential logic elements between its primary inputs and the inputs of an 8-input NAND gate that we want to test. It's not at all obvious what primary-input vector, or sequence of primary-input vectors, must be applied to generate the test vector 11111111 at the NAND-gate inputs. Furthermore, it's not obvious what else might be required to propagate the NAND gate's output to a primary output of the circuit.

Sophisticated *test generation programs* attempt to deal with this complexity and create a *complete test set* for a circuit, that is, a sequence of test patterns that fully tests each logic element in the circuit. However, the amount of computation required can be huge, and it's quite often just not possible to generate a complete test set.

test generation program
complete test set

DFT methods attempt to simplify test-pattern generation by enhancing the "controllability" and "observability" of logic elements in a circuit. In a circuit with good *controllability*, it's easy to produce any desired values on the internal signals of the circuit by applying an appropriate test vector to the primary inputs. Similarly, good *observability* means that any internal signal can be easily propagated to a primary output by the application of an appropriate primary-input combination. The most common method of improving controllability and

controllability

observability

test points

observability is to add *test points*, additional primary inputs and outputs that are used during testing.

8.2.2 Bed-of-Nails and In-Circuit Testing

test fixture
nail
bed of nails
automatic tester
in-circuit testing

overdrive

In a digital circuit that is fabricated on a single PCB, the "ultimate" in observability is obtained by using every pin of every IC as a test point. This is achieved by building a special *test fixture* that matches the underside (solder side) of the PCB and contains a spring-loaded pin (*nail*) at each IC-pin position. The PCB is placed on this *bed of nails*, and the nails are connected to an *automatic tester* that can monitor each pin as required by a test program.

Going one step further, *in-circuit testing* also achieves the "ultimate" in controllability. This method monitors the signals on the bed of nails for observability, but also connects each nail to a very low impedance driving circuit in the tester. In this way, the tester can override (or *overdrive*) whatever circuit on the PCB normally drives each signal, and directly generate any desired test vector on the internal signals of the PCB. Overdriving an opposing gate output causes excessive current flow in both the tester and the opposing gate, but the tester gets away with it for short periods (milliseconds).

To test an 8-input NAND gate (e.g., 74LS30), an in-circuit tester needs only to provide the nine test vectors mentioned previously, and can ignore whatever values the rest of the circuit is trying to drive onto the eight input pins. And the NAND-gate output can be observed directly on the output pin, of course. With in-circuit testing, each logic element can be tested in isolation from the others.

Although in-circuit testing greatly enhances the controllability and observability of a PCB-based circuit, logic designers must still follow a few DFT guidelines for the approach to be effective. Some of these are listed below.

- *Initialization.* It must be possible to initialize all sequential circuit elements to a known state.

Since the preset and clear input pins of registers and flip-flops are available to an in-circuit tester, you would think that this is no problem. However, Figure 8–1(a) shows a classic case of a circuit (a Gray-code counter) that cannot be initialized, since the flip-flops go to an unpredictable state when /PR and /CLR are asserted simultaneously. The correct way to handle the preset and clear inputs is shown in (b).

- *Clock generation.* The tester must be able to provide its own clock signal *without* overdriving the on-board clock signals.

Testers usually must override an on-board clock, for several reasons: the speed at which they can apply test vectors is limited; they must allow extra time for

Figure 8–1
Flip-flops with pull-up resistors
for unused inputs: (a) untestable;
(b) testable.

overdriven signals to settle; and sometimes they must stop the clock. However, overdriving the clock is a no-no, since an overdriven signal may "ring" and make several transitions between LOW and HIGH before finally settling to the level that the tester wants. On a clock signal, such transitions can create unwanted state changes.

Figure 8–2 shows a recommended clock driver circuit. To inject its own clock, the tester pulls CLKEN LOW and inserts its clock on /TESTCLK. Since the tester is not overdriving any gate outputs, the resulting CLOCK signal is clean. In general, any normally glitch-free signal that is used as a clock input or other asynchronous input must not be overdriven by the tester, and would have to be treated in a way similar to Figure 8–2. This is another reason that synchronous design with a single clock is so desirable.

Figure 8–2
Clock driver circuit that allows
a tester to cleanly override the
system clock.

- *Grounded inputs.* In general, ground should not be used as a constant-0 logic source.

The in-circuit tester can overdrive *most* signals, but it can't overdrive ground. Therefore, signals that require a constant-0 input during normal operation should be tied to ground through a resistor, which allows the tester to set them to 1 as required during testing. A good example of this situation occurs when a PLD is used to create a set of clock signals, as in Figure 7–38 on page 616. Pin 11 of the PAL16R8 should be connected to ground through a resistor, so that the tester can float the PLD outputs and provide its own clock signals on /P1–/P6. Although the tester could theoretically overdrive /P1–/P6 directly, it shouldn't if they are used as clocks.

- *Bus drivers.* In general, it should be possible to disable the drivers for wide buses so that the tester can drive the bus without having to overdrive all the signals on the bus.

That is, it should be possible to output-disable all of the three-state drivers on a bus, so that the tester drives a "floating" bus. This reduces electrical stress both on the tester and on multibit drivers (e.g., 74LS244) that might otherwise be overheated and damaged by having all of their outputs overdriven simultaneously.

8.2.3   Scan Methods

In-circuit testing works fine, up to a point. It doesn't do much good for custom VLSI chips and ASICs, because the internal signals simply aren't accessible. Even in board-level circuits, high-density packaging technologies such as surface mounting greatly increase the difficulty of providing a test point for every signal on a PCB. As a result, an increasing number of designs are using "scan methods" to provide controllability and observability.

scan method
scan-path method

A *scan method* attempts to control and observe the internal signals of a circuit using only a small number of test points. A *scan-path method* considers any digital circuit to be a collection of flip-flops or other storage elements interconnected by combinational logic, and is concerned with controlling and observing the state of the storage elements. It does this by providing two operating

scan mode

modes: a normal mode, and a *scan mode* in which all of the storage elements are reorganized into a giant shift register. In scan mode, the state of the circuit's *n* storage elements can be read out by *n* shifts (*observability*), and a new state can be loaded at the same time (*controllability*).

two-port D flip-flop
scan enable (SE)
scan data (SD)

Figure 8–3 shows a circuit designed using a scan-path method. Each storage element in this circuit is a *two-port D flip-flop* that can be loaded from one of two sources. The *scan enable (SE)* selects the source—normal data (D) or *scan data (SD)*. The SD inputs are daisy-chained to create the scan path shown in

Figure 8–3   Circuit containing a scan path, shown in color.

color. By asscrting ENSCAN for 11 clock ticks, a tester can read out the current
state of the flip-flops and load a new state. The test engineer is left with the
job of deriving test sets for the individual combinational logic blocks, which can
be fully controlled and observed using the scan path and the primary inputs and
outputs.

Scan-path design is used most often in custom VLSI and ASIC design,
because of the impossibility of providing a large number conventional test points.
The two-port flip-flops used in scan-path design add chip area. For example, in
LSI Logic Corp.'s LCA10000 series of CMOS gate arrays, an FD1 D flip-flop
macrocell uses 7 "gate cells," while an FD1S two-port D flip-flop macrocell uses
9 gate cells, almost a 30% increase in silicon area. However, the overall increase
in chip area is much less, since flip-flops are only a fraction of the chip, and
large "regular" memory structures (e.g., RAM) may be tested by other means.
In any case, the improvement in testability may actually *reduce* the cost of the
packaged chip when the cost of testing is considered.

8.3 ESTIMATING DIGITAL SYSTEM RELIABILITY

Qualitatively, the *reliability* of a digital system is the likelihood that it works   *reliability*
correctly when you need it. Marketing and sales people like to say that the
systems they sell have "high reliability," meaning that the systems are "pretty

Figure 8–4

Typical reliability function for a system.

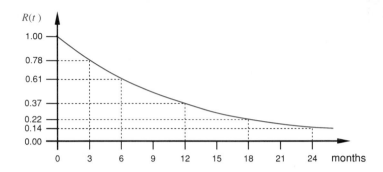

likely to keep working." However, savvy customers ask questions that require more concrete answers, such as, "If I buy 100 of these systems, how many will fail in a year?" To provide these answers, digital design engineers are often responsible for calculating the reliability of the systems they design, and in any case they should be aware of the factors that affect reliability.

Quantitatively, reliability is expressed as a mathematical function of time:

$R(t)$ = Probability that the system still works correctly at time t

Reliability is a real number between 0 and 1; that is, at any time $0 \le R(t) \le 1$. We assume that $R(t)$ is a monotonically decreasing function; that is, failures are permanent and we do not consider the effects of repair. Figure 8–4 is a typical reliablility function.

The foregoing definition of reliability assumes that you know the mathematical definition of probability. If you don't, reliability and the equivalent probability are easiest to define in terms of an experiment. Suppose that we were to build and operate N identical copies of the system in question. Let $W_N(t)$ denote the number of them that would still be working at time t. Then,

$$R(t) \quad = \quad \lim_{n \to \infty} W_N(t)/N$$

That is, if we build *lots* of systems, $R(t)$ is the fraction of them that are still working at time t. When we talk about the reliability of a single system, we are simply using our experience with a large population to estimate our chances with a single unit.

It would be very expensive to compute $R(t)$ if the only way to do it was to perform an experiment—to build and monitor N copies of the system. Worse, for any t, we wouldn't know the value of $R(t)$ until time t had elapsed in real time. Thus, to answer the customer's question posed earlier, we'd have to build a bunch of systems and wait a year; by then, our potential customer would have purchased something else.

Instead, we can estimate the reliability of a system by combining reliability information for its individual components, using a simple mathematical model.

The reliability of mature components (e.g., 74LS TTL chips) may be known and published based on actual experimental evidence, while the reliability of new components (e.g., an 80786 microprocessor) may be estimated or extrapolated from experience with something similar. In any case, a component's reliability is typically described by a single number, the "failure rate" described next.

8.3.1 Failure Rates

The *failure rate* is the number of failures that occur per unit time in a component or system. In mathematical formulas, failure rate is usually denoted by the Greek letter λ. Since failures occur infrequently in electronic equipment, the failure rate is measured or estimated using many identical copies of a component or system. For example, if we operate 10,000 microprocessor chips for 1000 hours, and eight of them fail, we would say that the failure rate is

failure rate

λ

$$\lambda = \frac{8 \text{ failures}}{10^4 \text{ chips} \cdot 10^3 \text{ hours}} = (8 \cdot 10^{-7} \text{ failures}/\text{hour})/\text{chip}$$

That is, the failure rate of a single chip is $8 \cdot 10^{-7}$ failures/hour.

The actual process of estimating the reliability of a batch of chips is not nearly as simple as we've just portrayed it; for more information, see the References. However, we can use individual component failure rates, derived by whatever means, in a straightforward mathematical model to predict overall system reliability, as we'll show later in this section.

Since the failure rates for typical electronic components are so small, there are several scaled units that are commonly used for expressing them: percent failures per 10^3 hours, failures per 10^6 hours, and failures per 10^9 hours. The last unit is called a *FIT*:

FIT

$$1 \text{ FIT} = 1 \text{ failure}/10^9 \text{ hours}$$

In our earlier example, we would say that $\lambda_{\text{microprocessor}} = 800$ FITs.

The failure rate of a typical electronic component is a function of time. As shown in Figure 8–5, a typical component has a high failure rate during its early life, during which most manufacturing defects make themselves visible; failures during this period are called *infant mortality*. Infant mortality is why manufacturers of high-quality equipment perform *burn-in*—operating the equipment for 24 to 168 hours before shipping it to customers. With burn-in, most infant mortality occurs at the factory rather than at the customer's premises. Even without a thorough burn-in period, the seemingly stingy 90-day warranty offered by most electronic equipment manufacturers does in fact cover most of the failures that occur in the first few *years* of the equipment's operation (but if it fails on the 91st day, that's tough!). This is quite different from the situation with an automobile

infant mortality
burn-in

Figure 8–5

The "bathtub curve" for electronic-component failure rates.

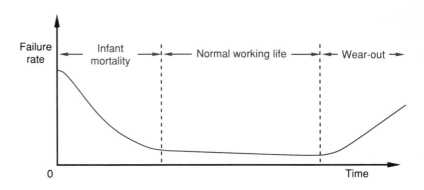

wear-out

or other piece of mechanical equipment, where "wear and tear" increases the failure rate as a function of time.

Once an electronic component has successfully passed the burn-in phase, its failure rate can be expected to be pretty much constant. Depending on the component, there may be *wear-out* mechanisms that occur late in the component's life, increasing its failure rate. In the old days, vacuum tubes often wore out after a few thousand hours because their filaments deteriorated from thermal stress. Nowadays, most electronic equipment reaches obsolescence before its solid-state components start to experience wear-out failures. For example, even though it's been over ten years since the widespread use of EPROMs began, we haven't seen a rash of equipment failures caused by their stored bits leaking away. (Do you know anyone with a 10-year-old PC or VCR?[3])

Thus, in practice, the infant-mortality and wear-out phases of electronic-component lifetime are ignored, and reliability is calculated on the assumption that failure rate is constant during the normal working life of electronic equipment. This assumption, which says that a failure is equally likely at any time in a component's working life, allows us to use a simplified mathematical model to predict system reliability, as we'll show later in this section.

There *are* some other factors that can affect component failure rates, including temperature, humidity, shock, vibration, and power cycling. The most significant of these for ICs is temperature. Many IC failure mechanisms involve chemical reactions between the chip and a contaminant, and these are accelerated by higher temperatures. Likewise, electrically overstressing a transistor, which heats it up too much and eventually destroys it, is worse if the device temperature is high to begin with. Both theoretical and empirical evidence support the following widely used rule of thumb:

- The failure rate of an IC approximately doubles for every 10°C rise in temperature.

---

[3]"No, but my 10-year-old has a PC and a VCR!"—Yuppie parent

This rule is true to a greater or lesser degree for most other electronic parts.

Note that the temperature of interest in the foregoing rule is the *internal* temperature of the IC, not the ambient temperature of the surrounding air. A power-hogging component in a system without forced-air cooling may have an internal temperature as much as 40–50°C higher than ambient. A well-placed fan may reduce the temperature rise to 10–20°C, reducing the component's failure rate by perhaps a factor of 10.

8.3.2 Reliability and MTBF

For components with a constant failure rate λ, it can be shown that reliability is a simple exponential function of time:

$$R(t) = e^{-\lambda t}$$

The reliability curve in Figure 8–4 is such a function, using the value $\lambda = 1$ failure / year.

Another measure of the reliability of a component or system is the *mean time between failures (MTBF)*, the average time that it takes for a component to fail. For components with a constant failure rate λ, it can be shown that MTBF is simply the reciprocal of λ:

mean time between failures (MTBF)

$$MTBF = 1/\lambda$$

Actually, the acronym *MTTF* (mean time *to* failure) should be used to talk about IC failures. Strictly speaking, "MTBF" should only be used when speaking about repairable equipment; ICs are never repaired and given the opportunity to fail again. However, most people ignore this minor semantic nit.

mean time to failure (MTTF)

8.3.3 System Reliability

Suppose that we build a system with m components, each with a different failure rate, $\lambda_1, \lambda_2, \ldots, \lambda_m$. Let us assume that for the system to operate properly, *all* of its components must operate properly. Basic probability theory says that the system reliability is then given by the formula

$$
\begin{aligned}
R_{sys}(t) &= R_1(t) \cdot R_2(t) \cdot \; \cdots \; \cdot R_m(t) \\
&= e^{-\lambda_1 t} \cdot e^{-\lambda_2 t} \cdot \; \cdots \; \cdot e^{-\lambda_m t} \\
&= e^{-\lambda_1 t - \lambda_2 t - \; \cdots \; - \lambda_m t} \\
&= e^{-\lambda_{sys} t}
\end{aligned}
$$

where

$$\lambda_{sys} = \lambda_1 + \lambda_2 + \cdots + \lambda_m$$

Table 8–1
Typical component failure rates
at 55°C.

Component	Failure rate (FITs)
SSI IC	90
MSI IC	160
LSI IC	250
VLSI microprocessor	500
Resistor	10
Decoupling capacitor	15
Connector (per pin)	10
Printed-circuit board	1000

Thus, system reliability is also an exponential function, using a composite failure rate λ_{sys} that is the sum of the individual component failure rates.

The constant-failure-rate assumption makes it very easy to determine the reliability of a system—simply add the failure rates of the individual components to get the system failure rate. Individual component failure rates can be obtained from manufacturers, reliability handbooks, or company standards. For example, a portion of one company's standard is listed in Table 8–1. Since failure rates are just estimates, this company has simplified the designer's job by considering only broad categories of components, rather than listing "exact" failure rates for each component. Other companies use more detailed lists, and some CAE systems maintain a database of component failure rates, so that a circuit's composite failure rate can be calculated automatically from its parts list.

Suppose that we were to build a single-board system with a VLSI microprocessor, 16 memory and other LSI ICs, 2 SSI ICs, 4 MSI ICs, 10 resistors, 24 decoupling capacitors, and a 50-pin connector. Using the numbers in Table 8–1, we can calculate the composite failure rate,

$$\lambda_{sys} = 1000 + 16 \cdot 250 + 2 \cdot 90 + 4 \cdot 160 + 10 \cdot 10 + 24 \cdot 15 + 50 \cdot 10 + 500 \text{ FITs}$$
$$= 7280 \text{ failures} / 10^9 \text{ hours}$$

The MTBF of our single-board system is $1/\lambda$, or about 15.6 years.

8.4 TRANSMISSION LINES, REFLECTIONS, AND TERMINATION

Nothing happens instantly, especially where digital circuits are concerned. In particular, consider the fact that the "speed-of-light" propagation delay of electrical signals in *wire* is on the order of 1.5–2 ns per foot (the exact delay depends on characteristics of the wire). When wire delays are on the same order as the transition times of the signals that they carry, we must treat wires not as zero-delay, perfect conductors, but as the "transmission lines" that they really are.

Transmission-line behavior includes signal changes that would not be predicted by a "DC" analysis of circuit operation. The most significant changes occur for an interval of approximately $2T$ after an output changes state, where T is the delay from the output to the far end of the wire that it drives.

Transmission-line behavior usually doesn't affect the logical operation of devices whose signal transition times and propagation delays are much longer than $2T$. Thus, transmission-line behavior typically is considered only for LS-TTL connections over a couple feet in length, for AS-TTL and Advanced CMOS connections over a foot, and for ECL connections over 3–6 inches. Naturally, you can sometimes get away with longer connections depending on the application, but there are also situations where transmission-line theory must be applied to even shorter connections. Even VLSI chips are beginning to achieve high enough speeds that transmission-line behavior must sometimes be considered in the internal chip design.

8.4.1 Basic Transmission-Line Theory

Two conductors in parallel constitute the simplest *transmission line*. Consider a pair of conductors with infinite length, as shown in Figure 8–6(a). If we instantaneously place a voltage source across the pair, a certain current flows to create a voltage wave that travels along the pair. The ratio of voltage to current, V_{out}/I_{out} depends on the physical characteristics of the conductors, and is called the *characteristic impedance Z_0* of the conductor pair.

transmission line

characteristic impedance Z_0

The magnitude of V_{out} is determined by viewing the series combination of R_{src} and Z_0 as a voltage divider, so that

$$V_{out} = V_{src} \cdot \frac{Z_0}{R_{src} + Z_0}$$

Of course, we don't have any infinitely long conductors. However, suppose that we have a 5-foot-long pair of conductors, and place a resistor equal to Z_0 across the far end. If we instantaneously place a voltage source across the pair as shown in (b), the same current flows forever as in the infinite-length case.

The situation is different for a transmission line of finite length that is not terminated in its characteristic impedance. An extreme case, in which the far end is short-circuited, is shown in (c). For simplicity, we assume that $R_{src} = Z_0$. Initially, all the driver sees is the characteristic impedance of the line, and a voltage wave with amplitude $V_{src}/2$ happily propagates down the line. However, when the wave hits the far end at time T, it sees the short-circuit. In order to satisfy Kirchoff's laws, a voltage wave of the *opposite* polarity propagates back down the line, canceling the original wave. The far end has *reflected* the original wave, and the driver sees the short-circuit at time $2T$.

reflection

Figure 8–6   A simple transmission line: (a) with infinite length; (b) with finite length, terminated with characteristic impedance; (c) short-circuited; (d) open-circuited.

Another extreme case, in which the far end is open-circuited, is shown in Figure 8–6(d). Everything starts out as before. However, when the initial voltage wave hits the far end, the current has nowhere to go, and so a voltage wave of the *same* polarity propagates back up the line, adding to the original voltage. When the reflected wave reaches the driver at time $2T$, the voltage everywhere is V_{src}, and nothing more happens.

In the general case, the amplitude of the wave reflected at the end of a transmission line is determined by the *reflection coefficient*, ρ (rho). The value of ρ depends on Z_0 and Z_{term}, the *termination impedance* at the end of the line:

reflection coefficient, ρ

termination impedance

$$\rho = \frac{Z_{term} - Z_0}{Z_{term} + Z_0}$$

When a voltage wave with amplitude V_{wave} hits the end of a transmission line, a wave with amplitude $\rho \cdot V_{wave}$ is reflected. Note that three simple cases match our previous discussion:

$Z_{term} = Z_0$ When a transmission line is terminated in its characteristic impedance, the reflection coefficient is 0.

$Z_{term} = 0$ The reflection coefficient of a short-circuited line is −1, producing a reflection of equal magnitude and opposite polarity.

$Z_{term} = \infty$ The reflection coefficient of an open-circuited line is +1, producing a reflection of equal magnitude and the same polarity.

If the source impedance (R_{src}) does not equal Z_0, then reflections occur at the near end of the line as well as at the far end. Each end of the line has its own value of ρ. The *principle of superposition* applies, so that voltage at any point on the line and instant in time is the sum of that point's initial condition and all waves that have passed it so far. The end of a transmission line is said to be *matched* if it is terminated with its characteristic impedance. Figure 8–7 shows the behavior of a transmission line that is not matched at either end. (Note that the driver's ideal voltage source V_{src} is considered to have a resistance of 0 Ω for this analysis.) Reflections occur at both ends of the line, with smaller and smaller waves reflecting back and forth. The voltage everywhere on the line asymptotically approaches $0.9V_{src}$, the value that would be predicted by a "DC" analysis of the circuit using Ohm's law.

principle of superposition

transmission-line matching

8.4.2 Logic-Signal Interconnections as Transmission Lines

So, how does transmission-line theory affect logic signals? Let us consider the case of a TTL driver producing a LOW-to-HIGH transition on a transmission line consisting of a signal line and ground, as illustrated in Figure 8–8. An LS-TTL

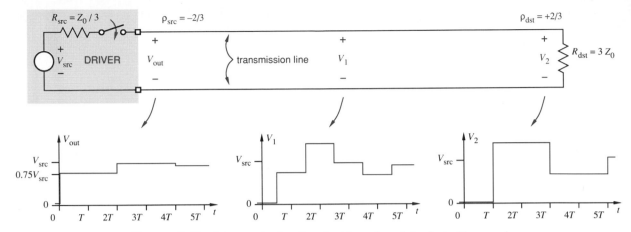

Figure 8–7   A transmission line that is not matched at either end.

output in the HIGH state is roughly equivalent to a 4 V source in series with a 120-Ω resistor. The characteristic impedance of a typical PCB trace with respect to ground is on the order of 100–150 Ω; let's assume 120 Ω for convenience, so the reflection coefficient on the driving end of the line is 0. An LS-TTL input has an impedance of about 20 kΩ, so the reflection coefficient on the receiving end is very nearly +1.

When the TTL driver switches from LOW to HIGH, the 4 V source in the driver sees the 120-Ω resistance of the driver in series with the 120-Ω Z_0 of the line, so a 2-volt wave propagates down the line. After time T, this wave reaches the receiving gate U2 on the far end and is reflected. After time $2T$, the reflected

Figure 8–8   Reflections on a TTL signal line changing from LOW to HIGH.

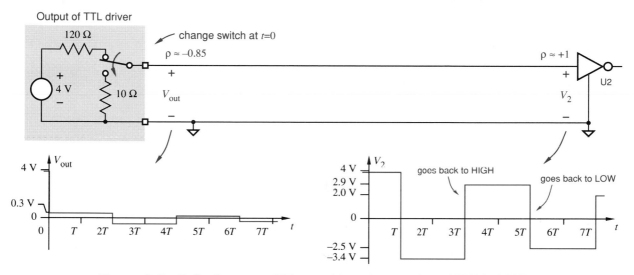

Figure 8–9   Reflections on a TTL signal line changing from HIGH to LOW.

wave reaches the sending end and is absorbed without a reflection because $\rho = 0$ at that end.

Everything works fine as far as receiving gate U2 is concerned—it sees an instantaneous transition from 0 V to 4 V at time T after the driver switches. However, consider the waveform seen by another receiving gate U1 positioned halfway between the driver and U2. As shown in the figure, U1 sees an input of only 2 V for an interval T. A receiving gate positioned closer to the driver would see this input voltage even longer. This is a problem, because 2 V is right on the borderline for a valid HIGH signal. If the driver resistance were a little higher, or if the line impedance were a little lower, or if a little noise occurred, the voltage could be even lower, which could cause the receiver to oscillate or produce a nonlogic-voltage output.

Any logic-signal connection, no matter how long or how short, is a transmission line. However, the foregoing analysis is quite idealized. In practice, transmission-line behavior causes no problem if T is somewhat shorter than the transition times of logic signals or the propagation delays of logic gates. In such a case, the reflections typically settle out before the receivers on the line have a chance to notice them.

A somewhat nastier case occurs when a TTL output switches from HIGH to LOW, as illustrated in Figure 8–9. In the LOW state, an LS-TTL output looks roughly like a 10-Ω resistor to ground, so the reflection coefficient on the sending end is about −0.85. When the output is first switched LOW it sees a voltage divider consisting of the 10-Ω resistor in series with the line impedance of 120 Ω. Since the line was originally at 4 V, the new voltage at the output of the divider must be

$$V_{out} = \frac{10 \ \Omega}{120 \ \Omega + 10 \ \Omega} \cdot 4 \ V = 0.3 \ V$$

Therefore, a voltage wave $(0.3 - 4.0) = -3.7$ V is propagated down the line. When this wave hits the receiving end at time T, it produces a reflection of equal sign and magnitude (since $\rho = 1$). Thus, the voltage at the receiving end is now $V_2 = 4.0 - 3.7 - 3.7 = -3.4$ V, a negative voltage! This is called *undershoot*.

undershoot

When the -3.7 V reflection gets back to the sending end at time $2T$, yet another reflection occurs. This time, since ρ is negative, the reflected wave has a positive amplitude, $-0.85 \cdot -3.7 = +3.15$ V. The output voltage is now modified by the sum of the incoming reflection and the new outgoing reflection: $V_{out} = 0.3 - 3.7 + 3.15 = -0.25$ V. This is no problem. However, when the $+3.1$ V reflection reaches the receiving end at time $3T$, the ensuing positive reflection yields $V_2 = -3.4 + 3.15 + 3.15 = 2.9$ V, back to positive again!

As shown in the figure, the reflections continue, with the voltage at both the sending and receiving ends asymptotically approaching 0 V, the value one would predict with a "DC" analysis of the circuit. This oscillating pattern is called *ringing*.

ringing

The large magnitude of the ringing on the receiving end may be a problem, since V_2 does not settle to a range that is less than the LOW-state input voltage (0.8 V) until time $17T$. Thus, the effective propagation delay of the circuit has been increased by many times the wire delay. Worse, if the signal is a clock, extra edges will be detected on the ringing transitions, falsely clocking the flip-flops at the receiving end.

Once again, the transmission-line behavior that we've described causes no problems if T is much shorter than the transition times of logic signals or the propagation delays of logic gates. In addition, TTL input structures include *clamping diodes*, the normally reverse-biased diodes from each input to ground that we showed in Figure 2–29 on page 84. Since these diodes change the receiver's input impedance from very high to very low for negative voltages, they limit the negative excursion at time T to about 1 V. This reduces the reflection back to the sending end, which in turn reduces the excursion at time $3T$ to less than 1 V.

clamping diodes

REFERENCES

CAE tools are just tools, and you won't find any scholarly references on how to use them. However, *IEEE Design & Test of Computers (D&T)* and *High-Performance Systems* magazines are excellent ongoing sources of information about them. The best source of information for a particular CAE tool is, naturally, the supplier's documentation. When a particular tool is used *very* widely, an independent tutorial guide may appear; a good example is *SPICE: A Guide to Circuit Simulation and Analysis Using Pspice®* by Paul W. Tuinenga (Prentice-Hall, 1988).

Register-transfer languages are discussed in most computer design books, including *Computer Architecture and Organization* by John P. Hayes (McGraw-Hill, 1988, 2nd ed.). *IEEE Computer* magazine had a special issue on hardware description languages in February 1985, and *D&T* had a special issue on VHDL in April, 1986. The emergence of VHDL has led to the publication of several descriptive books, such as *Chip-Level Modeling with VHDL* by James R. Armstrong (Prentice-Hall, 1989) and *The VHDL Handbook* by David R. Coelho (Kluwer Academic Publishers, 1989).

The authoritative treatment of design for testability in the context of logic design is in Edward J. McCluskey's *Logic Design Principles* (Prentice-Hall, 1986). A thorough treatment of the subject from the point of view of manufacturing and test engineers is *Digital Test Engineering* by J. Max Cortner (Wiley, 1987). Once again, *D&T* is an excellent ongoing information source. In the August 1989 special issue on built-in self-test methods, an article by Kenneth P. Parker, "The Impact of Boundary Scan on Board Test," gives a very practical perspective on the growing importance of scan methods in board-level design.

The required mathematical background and general methods for analyzing system reliability can be found in a handy little book by J. C. Cluley, *Electronic Equipment Reliability* (Halsted Press/Wiley, 1974). To see how ICs actually fail, and how their failure rates can be predicted in less than a lifetime (yours or an IC's), read *Failure Mechanisms in Semiconductor Devices* by E. A. Amerasekera and D. S. Campbell (Wiley, 1987).

An excellent introduction to transmission-line theory and practice from the point of view of a digital designer can be found in Thomas R. Blakeslee's *Digital Design with Standard MSI and LSI* (Wiley, 1979, 2nd ed.).

A

ELECTRICAL CIRCUITS REVIEW

by Bruce M. Fleischer

This appendix is recommended reading if you want a review of electronics' basic ideas. Why, you may ask, does a digital circuit designer need to know this analog "stuff"? You need it to understand the abilities and shortcomings of components available in the real world, which must be used in building circuits that will work in the real world.

A practical problem that motivates this appendix is the calculation of the rise and fall times of a TTL circuit driving another TTL circuit (see Section 2.9). That problem is really one of analog, not digital circuits. Therefore, much of this appendix consists of building up enough ideas from analog electronics to make a simple model for rise and fall times.

A.1 FUNDAMENTALS

A.1.1 Charge

electron
neutron
proton

In the Bohr theory of the atom (named after Niels Bohr, 1885–1962), *electrons* orbit a nucleus containing *neutrons* and *protons*. Attraction between the opposite charges of electrons and protons keeps atoms together. Particles with the same charge repel each other. You may wonder what attraction holds the nucleus together, with all those protons. The answer is that the physicists have it all under control, and be glad this isn't an appendix on particle physics.

The nearly equal numbers of electrons and protons in most objects, such as a piece of fur, cancel out, so electrons in a neighboring object, such as an amber rod, usually don't feel any overall attraction for the fur. In some atoms, however, the electrons are not as tightly held as in others, and there are situations in which some electrons come loose.

For example, if we rub the amber rod on the fur, some of the electrons from the fur end up on the amber. The fur is missing some electrons, while the amber has extras. You can see the attraction of these charges for each other; the hairs in the fur are pulled towards the rod. If the two are brought close enough, some electrons in the rod will jump back to the fur, making a spark. The ancient Greeks didn't know about the Bohr model of the atom, but they were familiar with fur and amber (which is just petrified tree sap); in fact, our word "electron" comes from the Greek for amber.

Electric charge is measured in *coulombs*. An individual electron or proton *coulomb* has much much less than one coulomb of charge, -1.6×10^{-19} coulomb on an electron, $+1.6 \times 10^{-19}$ on a proton. The decision to call the electrons negative is simply a convention started by Benjamin Franklin (1706–1790). Nature says only that an electron's charge is the opposite of a proton's; there is nothing inherently negative about electrons, they could just as easily be called positive and protons negative.

A.1.2 Voltage

The attraction of opposite charges means that *energy* is required to pull them *energy* apart, and that energy can be recovered when they come together again. In between, we say that the energy is kept as *potential energy*. When a system has *potential energy* potential energy, it has the potential to do work. *Work* here just means a more *work* visible form of energy, not necessarily something we want done.

The most familiar form of potential energy is gravitational potential energy. Because of the gravitational attraction between the earth and the objects on it, lifting objects gives them potential energy. When an object is dropped, that potential energy is rapidly converted into kinetic energy, which is even more rapidly converted into other forms of energy when the object hits something (e.g., sound, heat, kinetic energy of broken pieces). The more mass something has and the higher we lift it, the more potential energy it has. Fans of David Letterman (1947–) may remember just how much potential for mayhem a bowling ball has when lifted up six stories.

In electricity, the quantity analogous to height is *voltage*, usually indicated *voltage, V* by the symbol *V*. Voltage is measured in *volts* (symbol: V), named after Alessan- *volt (V)* dro Volta (1745–1827). (Notational example: $V_{object} = 2.5$ V.) In digital circuits, we sometimes measure voltage in thousandths of a volt, or *millivolts (mV)*. In- *millivolt (mV)* creasing the voltage of one coulomb of charge by one volt gives it one *joule (J)* *joule (J)* of electrical potential energy, named after James P. Joule (1818–1889).

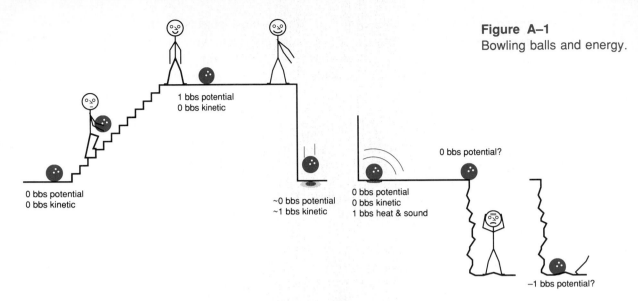

Figure A–1
Bowling balls and energy.

Because electrons have negative charge, we have to take electrons from a higher to a lower voltage to increase their potential energy. This is what happens in a battery: electrons go into the positive terminal and come out of the negative terminal with more potential energy, which is converted to other forms of energy as the electron travels "up" through the circuit. There is a special unit for the amount of energy one electron gets when its potential is reduced by one volt, an *electron-volt* (symbol: eV), which equals 1.6×10^{-19} J.

electron-volt (eV)

There are several lessons to remember about voltage and potential energy. One is that nearly all everyday systems have tremendous reservoirs of potential energy of various kinds, and any give and take of potential represents a minuscule fraction of a system's total potential energy.

bbs

As a simple example, shown in Figure A–1, get a fresh bowling ball and carry it up one story, giving it one bowling-ball-story (*bbs*) of gravitational potential energy. Then drop it back to ground level to let loose that 1 bbs of energy. Does the bowling ball now have no potential energy? Certainly not—if you stand in a ditch under it, it clearly has lots of potential!

Stepping out of the way, let the ball drop into the ditch. Does it have −1 bbs of potential energy now? Clearly, some sort of convention needs to be set up here. We could try to find an absolute scale, say with 0 corresponding to ball at the center of the earth, but that would cause many more problems than it would solve.

There are two workable conventions. The first is to consider potential energy as a purely relative term, and always speak of potential differences: the ball on the second floor has 1 bbs more potential than the one on the ground, and the one in the ditch is at −2 bbs relative to the second floor.

The second, shortcut convention uses the fact that many things sit at ground level. Thus, when we talk about the ball's potential at a given spot, we can imply the comparison, "relative to ground level." In this convention, "zero potential" just means that something is at ground level, not that it has absolutely no potential energy.

In electrical terms, we can use the full form to describe the voltage differ-ence, or voltage drop, between two points, or we can use the short form and talk about the voltage at a point. The short form always implies an unspoken com-parison with a common reference point. In most circuits that common reference point is called *ground*, and may actually be connected to the ground through a water pipe, but it is not at zero volts in any absolute sense. (Strange problems, and sometimes fireworks, can result from connecting circuits whose "grounds" have a voltage difference between them.)

ground

A.1.3 Current

Electrical *current* has the same relation to charge as a river's current does to water; it is the rate at which charge crosses a line cutting through part of the circuit. Current is usually indicated by the symbol I and is measured in *amperes*, or *amps* (symbol: A), named after A. M. Ampère (1775–1836). One ampere equals one coulomb per second (s). In digital circuits, we seldom have currents as high as one amp flowing into a single device. More often, we deal with currents of *milliamperes (mA)* and *microamperes (µA)*.

current, I

ampere (A)
amp
second (s)

milliampere (mA)
microampere (µA)

Like the flow of water, current has a size and direction. Usually, we decide on a direction beforehand, and then call current flowing that way positive and the other way negative. The one tricky thing about electrical current is that in electronic circuits, the protons stay put and the (negatively charged) electrons move around. If 6.25×10^{18} electrons flow through a wire from left to right, we can call that current −1 A flowing to the right or 1 A flowing to the left.

As shown in Figure A–2, the convention is that a certain current flows in the direction in which positive charge would move to make that current, which is the opposite of the direction in which electrons actually flow. To explicitly indicate the use of this convention, current is sometimes called *positive current*. If this seems confusing at first, blame it on Ben Franklin. We're all too committed to this system to flip things around, so you might as well get used to it.

positive current

Figure A–2
Conventions for current flow.

1 A   −1 A

1 coulomb/sec 6.25×10^{18} −1 coulomb/sec
electrons/sec

Figure A–3
A graph.

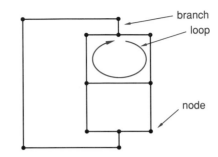

A.1.4   Circuits

The schematic diagrams that we use to represent circuits on paper are similar to figures studied in graph theory, a topic in abstract mathematics. We can use some ideas from graph theory to make formal rules about how circuits work. Fortunately, the abstract ideas we need can be directly related to physical circuits; there isn't any heavy math here.

graph
node
branch
loop

An example of a *graph* is shown in Figure A–3. It is made of points, called *nodes*, connected by lines, called *branches*. The only other abstract concept we need is a *loop*, which is just a path starting at some node, passing through a sequence of nodes and branches, and returning to the starting node. A direction is associated with each branch in the loop when we specify the loop's direction.

The graph in Figure A–3 looks like a circuit diagram, albeit a boring one. We could use it to represent a circuit, though, with some rules for transforming schematics into graphs. It's simpler, however, just to refer to schematics as if they were graphs. When we do, there are two concepts we need to consider.

The first concept is the difference between nodes and connection dots. A node is any "point" in a circuit, as shown in Figure A–4. Part (a) indicates that there is a node between the resistors, although we wouldn't put a connection dot there. The points where we do put connection dots, as in (b), generally correspond to nodes. The circled part of the schematic in (c) would usually be considered to be a single node, unless the connecting wiring is so nonideal that it must be treated as additional components.

The second concept is how to treat components, as shown in Figure A–5. Two-lead components, such as resistors, are like single branches with a node at

Figure A–4
Nodes and branches.

Figure A–5
Components.

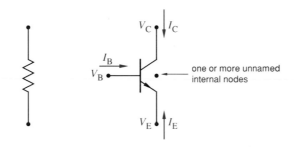

each end. Transistors and other larger components have at least one node per lead. Depending on how much detail we're interested in, we might imagine that there are one or more internal nodes also; that way we can give names to the voltages at and currents in each of the components' leads.

If a circuit diagram includes a node for ground, then the "single-ended" voltage at a node means the voltage difference from there to ground. The branches in the circuit diagram indicate physical paths through which current can flow, such as wires. Because current can't jump out of the circuit, the current flowing in a branch is the same anywhere along the branch. In other words, there is a definite value of current in each branch of the circuit. There is also a voltage drop (a "differential" voltage) along each branch, the voltage difference between the nodes it connects. The one term you should avoid is the current at a node; always refer to the current through some branch of a circuit.

Once we've picked a direction for a branch, its current (or voltage drop) is a positive or negative number depending on whether the current flows (or voltage decreases) in the chosen direction. Figure A–6 shows part of a circuit diagram with several named voltages and currents. The + and − signs and the arrows show which polarity of voltage or current is called positive.

Using the terms defined above, we can write down two basic laws that apply to all circuits. They are called *Kirchhoff's current law (KCL)* and *Kirchhoff's voltage law (KVL)*, after Gustav Robert Kirchhoff (1824–1887).

KVL The sum of the voltage drops around any loop is zero.

KCL The sum of the currents into any node of a circuit is zero.

Kirchhoff's current law (KCL)
Kirchhoff's voltage law (KVL)

All these laws really say is that the business of assigning voltages to nodes, and currents to branches, is consistent.

KVL is like saying: if you start at any point in a building and follow any path (up and down stairs, elevators, and escalators) that comes back to the same point, the net number of stories you went up or down is zero.

One way of restating KCL is to say that nodes don't store charge; any current that comes into a node through one connection has to go out through another. Implicit in KCL is a rule about components: any current going into a

Figure A–6
A circuit diagram with explicit voltages and currents.

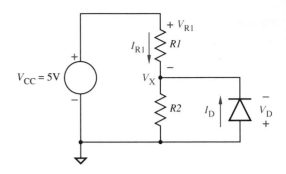

component through one lead must come out elsewhere. A resistor, for example, has two leads. Current can go through a resistor, but can't accumulate there.[1]

Kirchhoff's laws may seem obvious. In a way, that's their point: to formalize the basic ideas we take as axioms. They're not proof that our picture makes sense, just a formal statement of the rules we use. In other words, they're a way of avoiding lame answers like "that's how electricity works." When we calculate series and parallel resistances in the next section and you ask "why can you make that step?" we can point to Kirchhoff's laws.

What makes Kirchhoff's laws true? That's how electricity works. This may not seem like much of an improvement, but at least we know exactly what we need to have faith in, which is about the most you can hope for in science.

A.2   RESISTORS AND EQUIVALENT CIRCUITS

A.2.1   Ohm's Law and Resistors

Take a chunk of any material and hook up two wires to it, to make a two-lead component. Call the voltage across the component V_{chunk}, and the positive current flowing through the component I_{chunk}. For most materials, the current in a component made in this way would be proportional to the voltage. The ratio

resistance, R
ohm
Ohm's law

V_{chunk}/I_{chunk} is R, the component's *resistance*, and is measured in *ohms* (symbol: Ω), named after Georg Simon Ohm (1787–1854).

The equation $V = I \cdot R$ is called *Ohm's law*. Either current or voltage can be thought of as the driving term. A resistance of one ohm means that one volt is required to get one amp to flow, or that a current of one amp produces a voltage drop of one volt. Because such a component can be described by

---

[1]Those of you who are ahead of the game may protest this statement, because we can always force some charge onto any finite-sized component. The response is that to do so changes the voltage on that component, and the proper model includes a capacitance to ground from at least one interior node. When current through this capacitor is included, the terminal currents of the component sum to zero.

Figure A–7
Water-pipe analogy for resistance.

(a)   (b)   flow

+   pressure   −

giving its resistance, it is called a *resistor*. Most of the resistors used in real   *resistor*
circuits have resistances of many ohms, so larger units are used for convenience
(1 kΩ = 1000 Ω, 1 MΩ = 1000 kΩ). In circuit diagrams, a resistor is usually
drawn together with a name and its resistance, often with the "Ω" omitted from
the resistance (e.g., 470, 10k) since no other units would be appropriate.

   If any number of resistors are connected in a network with two external
terminals, the resulting object is also a resistor. That is, the total current flowing
through the network from one terminal to the other is proportional to the voltage
between the two terminals. Careful application of Kirchhoff's laws and some
linear algebra can be used to calculate the resistance of any such network, but
most of the circuits you'll work with don't need this level of math. Usually, a
couple of simple rules about series and parallel combinations, given below, are
enough. These rules can be derived from Kirchhoff's laws, but will be easier to
apply and will seem almost obvious given the right intuitive picture of current
and resistance.

   The bowling-ball analogy, used earlier to introduce the concept of voltage,
suggests that current flowing in a resistor might be something like a bowling-
ball waterfall. This is not the right intuitive picture; in fact, it is best to avoid
picturing individual electrons at all when thinking about macroscopic currents.

   Instead, water flowing in pipes provides a better analogy for electric current.
Pressure represents voltage, current is current, and resistance is resistance (see
Figure A–7). Fat pipes are the wires, and capillary tubes or pipes packed with
porous material are resistors. The flow through such a resistor is proportional
to the pressure imposed across it. To make the resistance greater, finer packing
can be used, or the tube can be made narrower or longer. Coarser packing or a
shorter or fatter piece of tubing reduces the resistance. The discussion to follow
is given in electrical terms, but you can translate it to this "water in pipes"
vocabulary if you want.

   The simplest resistor combination is two identical resistors in *series*, as   *series resistance*
shown in Figure A–8(a). The current is the same in the two resistors, so the
total voltage drop is just double the drop across one. Therefore, the resistance
(V/I ratio) of the combination is twice that of a single resistor, or $2R$. In the
more general case in (b), the resistance of any number of resistors in series is
the sum of the individual resistances:

Figure A–8
Series resistance: (a) identical
resistors; (b) different resistors.

$$R_{series} = R_1 + R_2 + \cdots + R_n$$

If one of the resistors has a much larger value than the others, it will dominate the series resistance (a plugged pipe has about the same resistance regardless of whether the unplugged section is short or long).

Take another look at the $R + R$ series combination. If one end is grounded and the other is connected to a known voltage, as in Figure A–9(a), what's the voltage at the node between the resistors? The answer is obvious: $V_{out} = V_{CC}/2$. This circuit is an example of a *voltage divider*. A voltage divider takes two voltages (one of which can be ground, but doesn't have to be) and produces a new voltage that is between them, dividing the original voltage difference into two smaller voltages.

voltage divider

Given the two input voltages, a divider's output voltage is set by the ratio of the resistors. It is halfway between the inputs if the resistors are equal, and is closer to one input if the resistor on that side is smaller. The divider in Figure A–9(b), for example, has an output voltage of 0.05 V. This result is easy to calculate, assuming that the current is the same through both resistors. If any current enters or leaves the divider at V_{out}, this assumption does not hold, and the output voltage is harder to see intuitively. We'll return to this point later.

parallel resistance

When resistors are arranged to offer *parallel* paths for the current, the combination's resistance is less than that of the individual resistors. Once again, the simplest case uses two identical resistors, as in Figure A–10(a). Both resistors see the same voltage drop, but now the total current is the sum of the two currents. Therefore, the net resistance is $R/2$.

The general case in Figure A–10(b) is easiest to work out if we think in terms of $1/R = I/V$, called the *conductance* or *admittance*. Because the total current through parallel resistors is just the sum of the individual currents (all

conductance
admittance

Figure A–9
Voltage dividers: (a) identical
resistors; (b) different resistors.

Figure A–10   Parallel resistance: (a) identical resistors; (b) different resistors.

resulting from the one applied voltage), the conductance of resistors in parallel can be written as the sum of individual conductances:

$$\frac{1}{R_{\text{parallel}}} = \frac{1}{R_1} + \frac{1}{R_2} + \cdots + \frac{1}{R_n}$$

The symbol "∥" is frequently used as shorthand for "in parallel":   ∥

$$R_{\text{parallel}} = R_1 \| R_2 \| \cdots \| R_n$$

When there are only two resistors, the formula is often written in this form:

$$R_1 \| R_2 = \frac{R_1 \cdot R_2}{R_1 + R_2}$$

If one resistor has a much smaller resistance (larger conductance) than the others, it dominates the parallel combination. Resistors in parallel can be viewed as a current divider, although they're seldom used that way.

Sometimes we find a network of resistors that is not a simple series or parallel combination, but that can be solved by successive applications of the series and parallel formulas above (see Exercise A.3). Some networks are just too complicated to analyze with these rules. As noted above, however, the resistance can always be found through more explicit use of Kirchhoff's laws (see Exercise A.4).

A.2.2   Power

Batteries and power supplies give energy to the electrons passing through. In most circuits, the electrons' energy is converted to heat as the electrons flow through the circuit back to the supply. Heat is produced any time that current flows through a voltage drop, so the electrons come out having less potential energy than when they went in.

The *power dissipation, P*, is the rate at which electrical energy is converted   *power dissipation, P*
to heat energy, and is given by the voltage (energy/charge) times the current (charge/s). The unit for power is the *watt* (symbol: W), named after James E.   *watt*
Watt (1736–1819); 1 W = 1 V-A = 1 joule/s .

Most of the heat dissipated by TTL circuits comes from the circuits' internal resistors. Because the current through a resistor is proportional to the voltage across it, power goes as the square of the voltage or current. One line of algebra gets you from Ohm's law to $P = I^2 \cdot R = V^2 / R$.

A.2.3 Thévenin Equivalent

ideal voltage source

The symbol for an *ideal voltage source* is shown in Figure A–11. This hypothetical device creates a voltage difference between its terminals that is independent of the current through those terminals. Like irresistible forces and immovable objects, ideal sources are only abstractions, but are useful in describing the behavior of real sources, such as batteries. The voltage difference between a battery's terminals is nearly independent of the current taken from it, but if you pull enough current out of a battery, its voltage will drop noticeably.

One way to describe this behavior of a real battery is to say that it acts like, or is modeled by, the circuit in Figure A–12, an ideal voltage source in series with a resistor. If the circuit doesn't have to supply any current, there is no drop across the resistor, and the source's voltage appears across the terminals

open-circuit voltage

(this is called the battery's *open-circuit voltage*, V_{oc}). If the battery's terminals are shorted together, the source's (unchanged) voltage is dropped entirely in the

short-circuit current

resistor, and the current through the short (the battery's *short-circuit current*) is $V_{oc} / R_{out} = (1.5\ \text{V}) / (1\ \Omega) = 1.5\ \text{A}$.

More completely, we can write a general formula relating the model's current and output voltage:

$$V_{out} = V_{oc} - I_{out} \cdot R_{out}$$

This linear model has two parameters, V_{oc} and R_{out}, that can be found from two measurements of V_{out} and I_{out}, such as the open-circuit voltage (V_{out} given $I_{out} = 0$) and the short-circuit current (I_{out} at which $V_{out} = 0$).[2]

[2]Real batteries or voltage sources are even less ideal than our model allows—the voltage is not necessarily a linear function of the current. One way to get around this is to use a linear model, but state that it only applies over a limited range. That is, we could say that the battery acts like the model in the figure, with $V_{oc} = 1.5$ V and $R = 1\ \Omega$, for output currents between 0 and 100 mA. The measurements from which the parameters are found (or extracted) must be made within the stated range. Outside that range, the battery's behavior would have to be described by other parameter values, or perhaps even by a different model, such as one including flames or oozing chemicals.

Figure A–11
An ideal voltage source.

Figure A–12
Model for a real voltage source—
a battery.

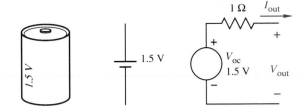

Thévenin's theorem (named for Charles Leon Thévenin, 1857–1926) proves
that any network of resistors and ideal voltage sources with only two terminals
is equivalent to a single source in series with a resistor (see Figure A–13). The
smaller circuit, called the *Thévenin equivalent*, is equivalent in the sense that   *Thévenin equivalent*
it has the same *I*-*V* characteristics (the full circuit may dissipate more power
than its Thévenin equivalent). As shown in the figure, the voltage and resistance
used in the equivalent circuit are called V_{Th} and R_{Th}, the *Thévenin voltage* and   *Thévenin voltage*
Thévenin resistance. These parameters can be found from two measurements of   *Thévenin resistance*
the circuit's voltage and current, as described in the preceding paragraph.

To illustrate Thévenin equivalent circuits, let's take another look at the
voltage divider in Figure A–9(a). With no load (nothing else connected to V_{out}),
its output voltage is obviously 2.5 V. What happens when we try to use that
voltage, though? Figure A–14(a) shows a similar divider driving a load resistance
(R_{load}) of 2 kΩ. It's not impossible to find V_{out} now, thinking of all the resistors
as one network, but it's easier to replace the source-plus-divider with its Thévenin
equivalent, as shown in Figure A–14(b). Looking at the resulting divider, we
see that V_{out} is 2.0 V. You might try putting in different load resistances and
recalculating V_{out}. Which approach is easier when you do this?

Thévenin equivalents make calculations easier once we have them, but you
may wonder where the foregoing values of V_{Th} and R_{Th} came from. A network's

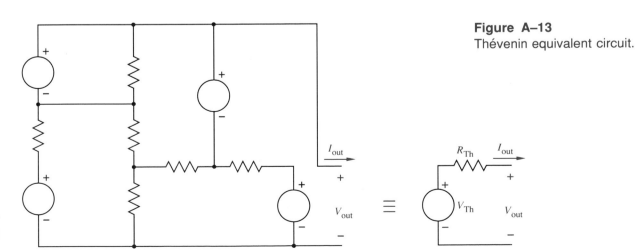

Figure A–13
Thévenin equivalent circuit.

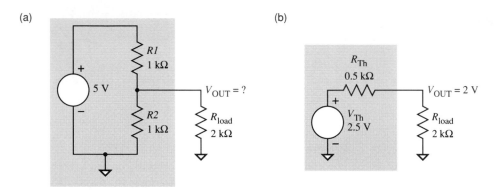

Figure A–14   Using the Thévenin equivalent: (a) a voltage divider under load;
(b) Thévenin equivalent.

Thévenin voltage is just the network's open-circuit voltage, and is often pretty easy to figure out. One way to find the Thévenin resistance, which equals R_{out}, is to calculate the network's short-circuit current and then solve for R_{out} from that and the open-circuit voltage. This approach is a bit messy, but if you work through it, you'll get

$$R_{Th} \;=\; \frac{R_1 \cdot R_2}{R_1 + R_2}$$

At least you only have to do that once, and the answer is easy to remember because it equals $R_1 \| R_2$. Actually, that equality is no coincidence. It's a consequence of a theorem that says: If you replace all voltage sources in a network with shorts (i.e., set them to 0 V sources) and open all current sources, the resistance of the resulting network equals the Thévenin resistance of the original network.

A.3   CAPACITORS

capacitor

A *capacitor* is a component that stores charge. The simplest capacitor is made from a pair of parallel metal plates, as shown in Figure A–15 (it's called a

parallel-plate capacitor

parallel-plate capacitor).   Let's say that we take some electrons out of one

Figure A–15
A parallel-plate capacitor.

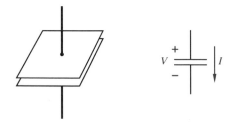

plate, leaving it positively charged, and put them on the other plate, making it negatively charged. It's not hard to see that there's now a voltage difference between the two plates.

Try grabbing an electron on the positive plate and moving it over to the negative plate. It is attracted to the positive charge behind it and repelled by the electrons in front, so you've got to push to move it. That push gives it potential energy, which means that its voltage is decreasing. The more charge there is on the plates, the more push is needed, and therefore the bigger the voltage difference between the plates. In a resistor, current and voltage are proportional, but in a capacitor, charge and voltage are proportional. The proportionality is usually expressed as

$$Q \;=\; C \cdot V$$

where Q represents the charge on the first plate (in coulombs), V is the voltage between the plates (positive if the first plate is at the higher voltage), and C is the *capacitance* of the capacitor, which is measured in *farads* (symbol: F), named after Michael Faraday (1791–1867).

capacitance, C
farad

A capacitance of one farad means that putting $+1$ and -1 coulomb on the plates makes the voltage drop 1 V. You can also think of it the other way around: at a voltage of 1 V, a 1-farad capacitor holds 1 coulomb. Most capacitors are much smaller than 1 farad, so they're usually measured in *picofarads* (1 pF = 10^{-12} F) or *microfarads* (1 μF = 10^{-6} F).

picofarad, pF
microfarad, μF

The capacitance of a parallel-plate capacitor is directly related to the size and spacing of the plates. With the amount of charge fixed, pulling the plates farther apart takes work, and therefore increases the voltage drop. Said the other way, the farther apart the plates are, the less charge is accumulated per volt. Capacitance decreases for farther plates and increases for closer plates. Also, the more area each plate has, the more charge can be stored, so a bigger area also means more capacitance.

One way to get a large capacitance in a small space is to use large sheets of very thin foil for the plates with a thin spacer to hold them apart, and roll everything up like a jellyroll. Larger capacitors, 1 μF and above, are often made this way. Smaller capacitors look more like the parallel-plate model, although their plates are usually separated by a solid material, such as ceramic or mica, rather than air. This material, called the *dielectric*, supports the metal plates and helps the capacitor hold more charge.

dielectric

Can other components, such as resistors or transistors, build up charge? If we stick electrons on a component, they'll make it harder to stick more on—and the component's voltage will drop. But remember, that's the voltage relative to ground. We can describe this effect by imagining that there are little capacitors between every part of every circuit and ground. The size of each capacitor tells us how much charge gets stored on that part of the circuit per volt.

As mentioned previously, changing the voltage across a capacitor requires changing the charge stored on each plate of the capacitor, which requires current through the capacitor. Each electron that goes in one lead of a capacitor can't come out the other lead because of the gap between the plates, but it does "push" one electron from the other plate out through the other lead. In a resistor, a fixed current creates a fixed voltage drop, but in a capacitor, a fixed current creates a steadily increasing voltage drop as charge accumulates on the plates. It's easy to put this relationship is quantitative terms. First, since the charge on one plate has to come through the attached lead, we can write

$$I = dQ/dt$$

We can take the derivative of charge equation $Q = C \cdot V$ and combine it with the preceding equation to derive the following equations:

$$dQ/dt = C \cdot dV/dt$$
$$I = C \cdot dV/dt$$
$$dV/dt = I/C$$

Thus, we can't instantly change the voltage across a capacitor, because that would require an infinite current. Capacitors are often explicitly placed at nodes (such as the power supply) where we want to hold a steady voltage. On the other hand, unavoidable stray circuit capacitances can prevent voltages from changing as fast as we might like them to, as we'll see in the next section.

A.4 RC CIRCUITS

Changing the voltage on a capacitor requires current. If that current must flow through a resistor, it requires a voltage across the resistor. If that voltage decreases as the capacitor charges, the current and the rate of charging will decrease exponentially with time.

For example, consider the circuit in Figure A–16. The circuit starts with switch $S1$ closed and switch $S2$ open, so the capacitor is charged to $V_{CC} = 5$ V and

Figure A–16
RC discharge.

Figure A–17
RC charge.

no current can flow through the resistor. Now, suppose that we disconnect the voltage source by opening *S1* and then start our stopwatches (or oscilloscopes) as we close *S2*. With *S2* closed, notice that $V_R = V_{cap}$ and $I_R = -I_{cap}$ (if you agree, then you understand Kirchhoff's laws). We can write another two equations relating these variables, from the rules for resistors and capacitors:

$$I_R = V_R / R$$
$$dV_{cap}/dt = I_{cap}/C$$

Substituting into the last equation we get

$$dV_{cap}/dt = I_{cap}/C$$
$$= -I_R/C$$
$$= -V_{cap}/RC$$

The solution to this equation is

$$V_{cap}(t) = V_{cap}(0) \cdot e^{-t/RC}$$
$$= V_{CC} \cdot e^{-t/RC}$$

In words, the capacitor's voltage decreases exponentially, with a time constant equal to *RC*. In order for the solution to make sense, the units for an *RC* product must be units of time, and this is indeed the case (1 Ω × 1 F = 1 s, 1 MΩ × 1 μF = 1 s, 1 kΩ × 1 pF = 1 ns).

Now open *S2*, and close *S1* again, as in Figure A–17, so V_{cap} goes back to 5 V. Does V_{cap} change instantaneously? Not in a real circuit, it doesn't. To make our figure a more accurate model, we need to include some resistance in series with the switch. This resistor, labeled R_2 in Figure A–17, can represent the resistance of the switch itself, or of a transistor serving as the switch, or perhaps of the voltage source. In any case, using the voltages and currents indicated in the figure, and ignoring *R* since no current can flow through it, we can now write

$$dV_{cap}/dt = (V_{CC} - V_{cap})/R_2 C$$

Figure A–18
Charging circuit: (a) redrawn;
(b) Thévenin equivalent.

(a)   (b)

If the capacitor starts with no charge at $t = 0$, the solution to this equation is

$$V_{cap}(t) = V_{CC} \cdot (1 - e^{-t/R_2 C})$$

The capacitor's voltage approaches 5 V with an exponential time constant equal to $R_2 C$. We can also write down the solution for an arbitrary initial voltage:

$$V_{cap}(t) = V_{CC} + (V_{cap}(0) - V_{CC}) \cdot e^{-t/R_2 C}$$

This solution reduces to the previous one if $V_{cap}(0) = 0$. Actually, the simple *RC* discharge is also a special case of this solution, with $V_{CC} = 0$.

The natural next step is to close both switches at the same time. It is possible, using Kirchhoff's laws, to write equations relating all the voltages and currents, and get a new differential equation for V_{cap}. There is, however, a much easier and much more intuitive way to figure out what happens, using the Thévenin equivalent of the divider. First, redraw the schematic so that the capacitor is separate, as in Figure A–18(a). Now the rest of the circuit can be replaced by its Thévenin equivalent, as in (b). The equivalent circuit has the same form as the charging circuit that we just analyzed, so the capacitor's voltage is

$$V_{cap}(t) = V_{Th} + (V_{cap}(0) - V_{Th}) \cdot e^{-t/R_{Th} C}$$

As discussed above, a divider's Thévenin voltage is just its open-circuit voltage, and its Thévenin resistance is the parallel combination of the two resistors.

More complicated problems can often be solved with this approach. Just move the capacitor off to one side, and draw a box around everything else. Since "everything else" is a two-terminal circuit, a Thévenin equivalent can be found, and the equation above can be used.

E X E R C I S E S

A.1   Over the lifetime of a D-size battery (2000 mA-hr capacity), how many electrons flow out of the negative terminal? Be more specific than "billions and billions."

A.2   There is one electronic device that is something like a bowling-ball waterfall in the "electrons as bowling balls" analogy. That device is a cathode ray tube (CRT), used in televisions and computer displays. The electron gun in a CRT gives electrons a whole lot of potential energy with a high-voltage power supply, then converts all of that potential to kinetic energy as the electrons are fired at the face of the tube. The molecules of *phosphor*, a chemical coated on the inside of the tube, receive   *phosphor* the electrons' kinetic energy and radiate it away as visible light.

Consider a computer display that accelerates the electrons through a voltage drop of 18,500 volts, has a beam current of 85 μA, and scans the beam across a display with 640×864 pixels, 69 times each second. An electron's mass is 9.1×10^{-31} kg. Ignoring relativistic effects, how fast do the electrons go? What fraction of the speed of light is this? About how many electrons are used to light up one pixel during one scan?

A.3   Calculate the resistance of the network in Figure XA.3.

Figure XA.3

A.4   What is the resistance of the resistor cube in Figure XA.4? You could put a 1-volt source across it, name the currents and voltage drops in every branch, use KCL to write equations for every node, use KVL to write equations for every loop, and solve them all to find the total current through the network. (*Hint:* There's also a sneaky but easier way.)

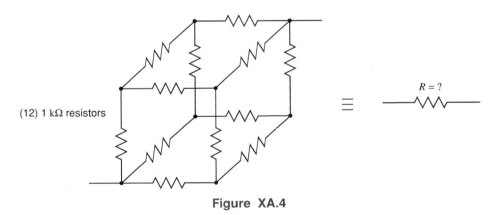

(12) 1 kΩ resistors

Figure XA.4

A.5   The circuit in Figure XA.5 is a good current source, in that its output current is nearly independent of its output voltage. When its output voltage is 0 V, it supplies 1 mA of current, and at 3 V, it supplies 0.97 mA. Draw the Thévenin equivalent of this circuit (called a *current mirror*). (You don't have to understand transistor   *current mirror* circuits to do this—you already have all the information you need.) Where does high-voltage source come from? (You *do* need to understand transistor circuits for this part.)

Figure XA.5

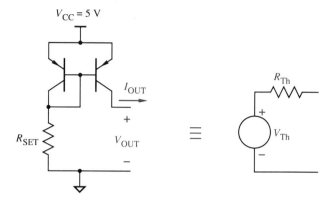

A.6   Thévenin equivalents are most natural for networks that we think of as almost ideal voltage sources. As Exercise A.5 showed, the Thévenin equivalent of a circuit that is close to an ideal current source can seem pretty strange.

    The symbol for an ideal current source (current independent of voltage) is shown in Figure XA.6(a). Real current sources, of course, do not meet this ideal, although we can often model them with a *Norton equivalent* (named after E. L. Norton, 1898–), an ideal current source (I_{Nort}) in parallel with a resistance (R_{Nort}), as in (b). The Norton resistance makes the current vary somewhat with voltage. Mathematically, Thévenin and Norton equivalents are equally applicable; either can replace any network of resistors, voltage sources, and current sources.

    Write I_{Nor} and R_{Nor} in terms of V_{Th} and R_{Th}. Draw the Norton equivalent of the current source in Exercise A.5.

Norton equivalent

Figure XA.6   (a)   (b)

A.7   Show that the units of an *RC* product are time.

A.8   In the circuit of Figure A–18, assume that $R_{Th} = 1 \ k\Omega$ and $C = 1\mu F$, and that V_{Th} is instantaneously changed from 0 V to 5 V. How long does it take for V_{cap} to reach 2.0 V?

A.9   In the circuit of Figure A–18, assume that $R_{Th} = 50 \ k\Omega$ and $C = 1\mu F$, and that V_{Th} is instantaneously changed from 5 V to 0 V. How long does it take for V_{cap} to reach 0.8 V?

INDEX

Boldface type is used to indicate page numbers on which the corresponding index terms are defined or used in an important way. The index terms usually appear in the outside margin on such pages.

TTL Sequential MSI Devices

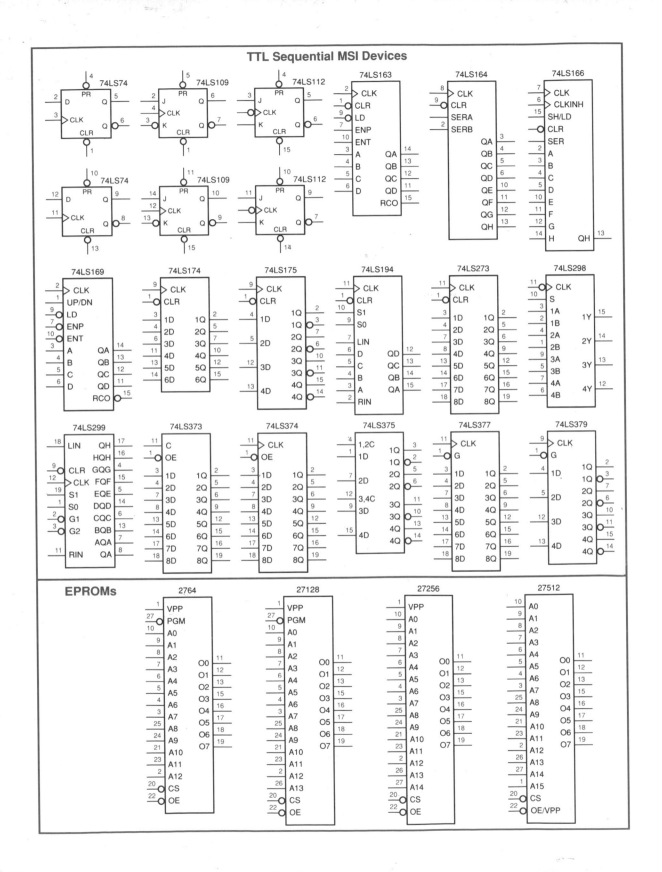

EPROMs